Lecture Notes in Computer Science 8181

Commenced Publication in 1973
Founding and Former Series Editors:
Gerhard Goos, Juris Hartmanis, and Jan van Leeuwen

Xuemin Lin   Yannis Manolopoulos
Divesh Srivastava   Guangyan Huang (Eds.)

Web Information Systems Engineering – WISE 2013

14th International Conference
Nanjing, China, October 13-15, 2013
Proceedings, Part II

 Springer

Volume Editors

Xuemin Lin
The University of New South Wales, Sydney, NSW, Australia
E-mail: lxue@cse.unsw.edu.au

Yannis Manolopoulos
Aristotle University of Thessaloniki, Greece
E-mail: manolopo@csd.auth.gr

Divesh Srivastava
AT&T Labs-Research, Florham Park, NJ, USA
E-mail: divesh@research.att.com

Guangyan Huang
Victoria University, Melbourne, VIC, Australia
E-mail: guangyan.huang@vu.edu.au

ISSN 0302-9743   e-ISSN 1611-3349
ISBN 978-3-642-41153-3   e-ISBN 978-3-642-41154-0
DOI 10.1007/978-3-642-41154-0
Springer Heidelberg New York Dordrecht London

Library of Congress Control Number: 2013948675

CR Subject Classification (1998): H.3, H.4, H.5, H.2.8, J.1, D.2, I.2, C.2

LNCS Sublibrary: SL 3 – Information Systems and Application, incl. Internet/Web and HCI

Typesetting: Camera-ready by author, data conversion by Scientific Publishing Services, Chennai, India

Printed on acid-free paper

Springer is part of Springer Science+Business Media (www.springer.com)

Preface

Welcome to the proceedings of the 14th International Conference on Web Information Systems Engineering (WISE 2013), held in Nanjing, China in October 2013. The series of WISE conferences aims to provide an international forum for researchers, professionals, and industrial practitioners to share their knowledge in the rapidly growing area of web technologies, methodologies, and applications. The first WISE event took place in Hong Kong, China (2000). Then the trip continued to Kyoto, Japan (2001); Singapore (2002); Rome, Italy (2003); Brisbane, Australia (2004); New York, USA (2005); Wuhan, China (2006); Nancy, France (2007); Auckland, New Zealand (2008); Poznan, Poland (2009); Hong Kong, China (2010); Sydney, Australia (2011); and Paphos, Cyprus (2012). This year, for a sixth time, WISE was held in Asia, in Nanjing to be precise.

WISE 2013 hosted four well-known keynote and invited speakers: Wen Gao of Peking University, who gave a talk on "Towards Web-Based Video Processing"; Divy Agrawal of the University of California at Santa Barbara, who gave a lecture on "Data-Driven Methodologies for Understanding, Managing, and Analyzing Online Social Networks"; Chengqi Zhang of the University of Technology Sydney, who gave a talk on "Big Data Related Research Issues and Progresses"; and Marek Rusinkiewicz of the Florida International University, who talked on "Security of Cyber-physical Systems (a Case Study)".

A total of 198 research papers were submitted to the conference for consideration, and 48 submissions were selected as full papers (with an acceptance rate of 24% approximately), plus 29 as short papers. The program also featured 10 demonstration papers and 5 challenge papers. The research papers cover the areas of web mining; web recommendation; web services; data engineering and databases; semi-structured data and modeling; web data integration and hidden web; social web; information extraction and multilingual management; networks, graphs, and web-based business processes; event processing, web monitoring and management; and innovative techniques and creations.

We wish to take this opportunity to thank the honorary general chair, Jian Lv; the industry program co-chairs, Min Wang and Lei Chen; the demo co-chairs, Hong Gao, Yoshiharu Ishikawa and Rui Zhang; the challenge program co-chairs, Weining Qian and Yabo (Arber) Xu; the panel co-chairs, Guoren Wang and Junzhou Lou; the tutorial co-chairs, Wojciech Cellary and Jeffrey Xu Yu; the workshop co-Chairs, Zhisheng Huang and Chengfei Liu; the publication chair, Guangyan Huang; the local arrangement chair, Jie Cao; the financial chair, Jing He; the publicity co-chairs, Haolan Zhang, Athena Vakali, and Wenjie Zhang; the WISE society representative, Xiaofang Zhou, and finally the webmaster, Zhiang Wu.

In addition, special thanks are due to the members of the International Program Committee and the external reviewers for a rigorous and robust reviewing process. All the papers were reviewed by at least three academic referees. In total, 727 reviews were uploaded by the members of the International Program Committee and the external reviewers.

Finally, we are also grateful to the UCAS (University of Chinese Academy of Sciences)-VU (Victoria University) Joint Lab for Social Computing and E-Health Research, to the Jiangsu Provincial Key Laboratory of E-Business at Nanjing University of Finance and Economics, and to the Nanjing Science and Technology Commission for their sponsorship of WISE 2013. We expect that the ideas that have emerged here will result in the development of further innovations for the benefit of science and society.

October 2013

Ricardo Baeza-Yates
Yanchun Zhang
Jie Cao
Xuemin Lin
Yannis Manolopoulos
Divesh Srivastava

Organization

Honorary General Chair

Jian Lv — Nanjing University, China

General Co-chairs

Ricardo Baeza-Yates — Yahoo!Research Lab, Spain
Yanchun Zhang — Victoria University, Australia
Jie Cao — Nanjing University of Finance and Economics,
  China

PC Co-chairs

Xuemin Lin — University of New South Wales, Australia
Yannis Manolopoulos — Aristotle University of Thessaloniki, Greece
Divesh Srivastava — AT&T Labs-Research, USA

Industry Program Chairs

Min Wang — Google Research, USA
Lei Chei — HKUST, Hong Kong, China

Demo Co-chairs

Yoshiharu Ishikawa — Nagoya University, Japan
Rui Zhang — University of Melbourne, Australia
Hong Gao — Harbin Institute of Technology, China

WISE Challenge Program Chairs

Weining Qian — East China Normal University, China
Yabo(Arber) Xu — Sun Yat-sen University, China

Panel Co-chairs

Guoren Wang — Northeastern University, China
Junzhou Lou — Southeastern University, China

Tutorial Co-chairs

Jeffrey Xu Yu   Chinese University of Hong Kong, China
Wojciech Cellary   Poznan University of Economics, Poland

Workshop Co-chairs

Zhisheng Huang   Vrije University of Amsterdam,
                                         The Netherlands
Chengfei Liu   Swinburne University of Technology,
                                         Australia

Publication Chair

Guangyan Huang   Victoria University, Australia

Local Arrangements Chair

Jie Cao   Nanjing University of Finance and
                                         Economics, China

Financial Chair

Jing He   Victoria University, Australia

Publicity Co-chairs

Haolan Zhang   NIT, Zhejiang University, China
Athena Vakali   Aristotle University of Thessaloniki, Greece
Wenjie Zhang   University of New South Wales, Australia

WISE Society Representative

Xiaofang Zhou   The University of Queensland, Australia

Webmaster

Zhiang Wu   Nanjing University of Finance and Economics,
                                         China

Program Committee

Alex Delis	University of Athens, Greece
Alexandros Ntoulas	Zynga, USA
Alfredo Cuzzocrea	University of Calabria, Italy
Anastasios Kementsietsidis	IBM T.J. Watson Research Center, USA
Anne Ngu	Texas State University-San Marcos, USA
Armin Haller	CSIRO, Australia
Athena Vakali	Aristotle University of Thessaloniki, Greece
Athman Bouguettaya	RMIT, Australia
Azer Bestavros	Boston University, USA
Bin Cui	Peking University, China
Birgitta König-Ries	Friedrich Schiller University of Jena, Germany
Brahim Medjahed	University of Michigan-Dearborn, USA
Bruno Martins	Instituto Superior Técnico (IST), Portugal
Chao Peng	East China Normal University, China
Chengfei Liu	Swinburne University of Technology, Australia
Chengkai Li	University of Texas at Arlington, USA
Claus Pahl	Dublin City University, Ireland
Costin Badica	University of Craiova, Romania
Dana Zhang	Google, USA
Dimitrios Katsaros	University of Thessaly, Greece
Dimitris Plexousakis	Universitat Oberta de Catalunya, Spain
Dimka Karastoyanova	University of Stuttgart, Germany
Elisa Bertino	Purdue University, USA
Evaggelia Pitoura	University of Ioannina, Greece
Fei Chen	HP Labs, USA
Fei Chiang	McMaster University, Canada
Ge Yu	Northeastern University, China
Georgia Koutrika	HP Labs, USA
Gill Dobbie	University of Auckland, New Zealand
Gong Zhang	Oracle Corporation, USA
Guangyan Huang	Victoria University, Australia
Guoliang Li	Tsinghua University, China
Hady Lauw	Singapore Management University, Singapore
Hao Yan	LinkedIn Corp, USA
Helen Karatza	Aristotle University of Thessaloniki, Greece
Hong Cheng	The Chinese University of Hong Kong, China
Hui Wang	Stevens Institute, USA
Ibrahim Kamel	University of Sharjah, UAE
Ilaria Bartolini	DEIS - University of Bologna, Italy
Ingmar Weber	Qatar Computing Research Institute, Qatar
Iraklis Varlamis	Harokopio University of Athens, Greece
Ismail Toroslu	Middle East Technical University, Turkey
Jian Yang	Macquarie University, Australia
Jiao Tao	Rensselaer Polytechnic Institute, USA

Jing He	Victoria University, Australia
Jinho Kim	Kangwon National University, South Korea
Joao Eduardo Ferreira	University of São Paulo, Brazil
John Shepherd	The University of New South Wales, Australia
Josiane Xavier Parreira	DERI - National University of Ireland, Ireland
Jun Shen	University of Wollongong, Australia
Kai-Uwe Sattler	TU Ilmenau, Germany
Karl Aberer	École Polytechnique Fédérale de Lausanne, Switzerland
Katja Hose	Aalborg University, Denmark
Kjetil Nørvåg	Norwegian University of Science and Technology, Norway
Kostas Stefanidis	Foundation for Research and Technology - Hellas, Greece
Kunal Verma	Accenture Technology Labs, USA
Ladjel Bellatreche	LIAS/ENSMA, France
Laure Berti	Institut de Recherche Pour le Développement, France
Lei Chen	Hong Kong University of Science and Technology, China
Luciano Baresi	DEI - Politecnico di Milano, Italy
Lyublena Antova	EMC Greenplum, USA
Marco Aiello	University of Groningen, The Netherlands
Mirjana Ivanovic	University of Novi Sad, Serbia
Miyuki Nakano	University of Tokyo, Japan
Mohamed Sharaf	University of Queensland, Australia
Monica Scannapieco	Istituto Nazionale di Statistica (ISTAT), Italy
Mourad Ouzzani	Qatar Computing Research Institute, Qatar
Muhammad Cheema	The University of New South Wales, Australia
Murali Mani	Flint, USA
Natwar Modani	IBM India Research Lab, India
Nick Bassiliades	Aristotle University of Thessaloniki, Greece
Oscar Corcho	Universidad Politécnica de Madrid, Spain
Panagiotis Karras	Rutgers University, USA
Panagiotis Symeonidis	Aristotle University of Thessaloniki, Greece
Panayiotis Andreou	University of Cyprus, Cyprus
Panayiotis Tsaparas	University of Ioannina, Greece
Pei Li	University of Zurich, Switzerland
Peiquan Jin	University of Science and Technology of China, China
Peter Scheuermann	Northwestern University, USA
Peter Triantafillou	University of Patras, Greece
Peter Yeh	Accenture Technology Labs, USA
Philippe Cudré-Mauroux	University of Fribourg, Switzerland
Pingpeng Yuan	Huazhong University of Science and Technology, China

Prasad Deshpande	IBM Research, India
Qi Yu	Rochester Institute of Technology, USA
Qing Liu	CSIRO, Australia
Quan Z. Sheng	The University of Adelaide, Australia
Ralf Schenkel	Saarland University, Germany
Rik Eshuis	Eindhoven University of Technology, The Netherlands
Rizos Sakellariou	University of Manchester, UK
Rui Zhang	University of Melbourne, Australia
Salima Benbernou	Université Paris Descartes, France
Samir Tata	Institut Télécom, France
Sangeetha Seshadri	IBM Almaden Research Center, USA
Schahram Dustdar	TU Wien, Austria
Sebastian Michel	Saarland University, Germany
Shawn Bowers	Gonzaga University, USA
Sherif Sakr	The University of New South Wales, Australia
Solmaz Kolahi	University of British Columbia, Canada
Soon Ae Chun	The City University of New York, USA
Sourav S. Bhowmick	Nanyang Technological University, Singapore
Stelios Paparizos	Microsoft Research, USA
Tingjian Ge	University of Massachusetts Lowell, USA
Torsten Suel	Yahoo! Research, USA
Verena Kantere	Cyprus University of Technology, Cyprus
Vladimir Zadorozhny	University of Pittsburgh, USA
Walter Binder	University of Lugano, Switzerland
Wei Wang	Fudan University, China
Weiyi Meng	Binghamton University, USA
Wilfred Ng	Hong Kong University of Science and Technology, China
Wojciech Cellary	Poznan University of Economics, Poland
Wolf-Tilo Balke	TU Braunschweig, Germany
X. Sean Wang	Fudan University, China
Xiangliang Zhang	King Abdullah University of Science and Technology, Saudi Arabia
Xiaofang Zhou	University of Queensland, Australia
Xuemin Lin	The University of New South Wales, Australia
Yanchun Zhang	Victoria University, Australia
Ying Zhang	The University of New South Wales, Australia
Yinian Qi	Purdue University, USA
Yoshiharu Ishikawa	Nagoya University, Japan
Yuanyuan Tian	IBM Research - Almaden, USA
Yuqing Wu	Indiana University, USA
Zhisheng Huang	Vrije University Amsterdam, The Netherlands
Zhiyuan Chen	University of Maryland Baltimore County, USA
Zhongqiang Chen	Yahoo Inc!, USA
Zi Huang	The University of Queensland, Australia

Industry Program Committee

Anastasios (Tasos) Kementsietsidis	IBM Watson, USA
Bin Liu	NEC Lab America, USA
Haixun Wang	Microsoft Research Asia, China
Honesty Yang	Intel, China
Howard Ho	IBM Almaden, USA
Jingren Zhou	Microsoft, USA
Jun Rao	LinkedIn, USA
Mingxuan Yuan	Huawei Noah's Ark Lab, China
Ping Luo	HP Labs, China
Wei Fan	Huawei Noah's Ark Lab, China

External Reviewers

Ahlers, Dirk
Akritidis, Leonidas
Alvanaki, Foteini
Aly, Ahmed
Araujo, Luciano
Basaras, Pavlos
Benouaret, Karim
Bento, Carolina
Bibi, Stamatia
Bourne, Scott
Bucur, Doina
Butt, Anila Sahar
Catasta, Michele
Choi, Jae-Yong
Christoforaki, Maria
Degeler, Viktoriya
Ebaid, Amr
Fehling, Christoph
Gaaloul, Walid
Garcia-Alvarado, Carlos
Gounaris, Anastasios
Grund, Martin
Gu, Tao
Gómez Sáez, Santiago
Haupt, Florian
Huang, Jin
Huang, Xin
Huynh, Tan Dat
Jain, Prateek
Jiang, Yu

Joshi, Salil
Kim, Nam-Soo
Koloniari, Georgia
Kontopoulos, Efstratios
Kritikos, Kyriakos
Lamba, Hemank
Li, Jianxin
Liakos, Panagiotis
Liang, Yongjiang
Liu, Xiaohui
Liu, Xin
Ma, Jiangang
Meditskos, Georgios
Midi, Daniele
Narayanam, Ramasuri
Neiat, Azadeh
Nie, Tiezheng
Nizamic, Faris
Nowak, Alexander
Oikawa, Marcio K.
Papadakis, Nikolaos
Podorozhny, Rodion
Podzimek, Andrej
Prokofyev, Roman
Qi, Jianzhong
Rahman, Rameez
Rezig, Elkindi
Shang, Haichuan
Shebaro, Bilal
Shugars, David

Spicuglia, Sebastiano
Stupar, Aleksandar
Tbahriti, Salah-Eddine
Tiakas, Eleftherios
Tonon, Alberto
Tsakalozos, Konstantinos
Tzouramanis, Theodoros
Vasirani, Matteo
Vassilakopoulos, Michael
Wang, Shenlu
Wang, Xiaoyang
Warriach, Ehsan Ullah
Weiß, Andreas
Xu, Lai
Xuan, Ming
Xue, Andy Yuan
Yang, Zhenglu
Yao, Lina
Ye, Zhen
Zeginis, Chris
Zhan, Liming
Zhang, Chengyuan
Zhao, Xiang
Zheng, Wei
Zheng, Yudi
Zhong, Youliang
Zhou, Rui
Zhou, Xiangmin
Zhou, Yang

Keynotes
(Abstracts)

Towards Web-Based Video Processing

Wen Gao

Peking University
wgao@pku.edu.cn

Image and video data is becoming the majority in big data era, a reasonable improvement of video coding efficiency may get a big cost saving in video transmission and/or storage, that is why so many researchers working on the new coding technologies and standards. For example, a team under IEEE standard association works on a new standard IEEE 1857 for internet/surveillance video coding, which targets to achieves about 50% bits saving than any of existing standards. However, the video coding story will be changed in the case of web application, because so many data we can reference comparing to the case of normal video coding, in that only a few frame of images can be referenced. In this talk, the recent developments of model-based video coding will be given, special on background picture model based surveillance coding and cloud-based image coding, and a on-going project that using web image and video to enhance the efficiency of video processing will be discussed also.

Data-Driven Methodologies for Understanding, Managing, and Analyzing Online Social Networks

Divy Agrawal

University of California at Santa Barbara
agrawal@cs.ucsb.edu

Online social networks provide unprecedented amounts of information about social interactions and therefore enable the study of various problems in the context of social networks on a scale and at a level of detail that has never been possible before. In this talk, we will consider ways of systematically exploring this vast space of online social network problems. Namely, we will consider three dimensions; understanding, managing and reporting on social networks and focus on example studies relating to these dimensions. To this end, we will consider three applications: modeling adoption behavior, limiting the spread of misinformation, and trend analysis in social networks. In modeling adoption behavior, we will challenge the common use of pure local models and revive research done in the context of diffusion of innovations and demonstrate the value of this technique in two different social networks. Next, we study the notion of competing campaigns in a social network and develop protocols whose goal is to limit the spread of misinformation by identifying a subset of individuals that need to be convinced to adopt the competing (or "good") campaign so as to minimize the number of people that adopt the "bad" campaign. And finally, relating to reporting on online social networks, we explore novel trend detection mechanisms. We propose two novel structural trend definitions that use friendship information to identify topics that are discussed among clustered and unconnected users respectively. Our analyses and experiments show that structural trends provide new insights into the way people share information online.

Table of Contents – Part II

Social Web

Information Extraction and Multilingual Management

Networks, Graphs and Web-Based Business Processes

Event Processing, Web Monitoring and Management

Innovative Techniques and Creations

Demo

Table of Contents – Part I

Web Mining

Web Recommendation

Web Services

Data Engineering and Database

Semi-Structured Data and Modeling

Web Data Integration and Hidden Web

Challenge

An Evaluation of Aggregation Techniques in Crowdsourcing

Nguyen Quoc Viet Hung, Nguyen Thanh Tam, Lam Ngoc Tran, and Karl Aberer

École Polytechnique Fédérale de Lausanne
{quocviethung.nguyen,tam.nguyenthanh,ngoc.lam,karl.aberer}@epfl.ch

Abstract. As the volumes of AI problems involving human knowledge are likely to soar, crowdsourcing has become essential in a wide range of world-wide-web applications. One of the biggest challenges of crowdsourcing is aggregating the answers collected from the crowd since the workers might have wide-ranging levels of expertise. In order to tackle this challenge, many aggregation techniques have been proposed. These techniques, however, have never been compared and analyzed under the same setting, rendering a 'right' choice for a particular application very difficult. Addressing this problem, this paper presents a benchmark that offers a comprehensive empirical study on the performance comparison of the aggregation techniques. Specifically, we integrated several state-of-the-art methods in a comparable manner, and measured various performance metrics with our benchmark, including *computation time, accuracy, robustness to spammers,* and *adaptivity to multi-labeling.* We then provide in-depth analysis of benchmarking results, obtained by simulating the crowdsourcing process with different types of workers. We believe that the findings from the benchmark will be able to serve as a practical guideline for crowdsourcing applications.

1 Introduction

In recent years, crowdsourcing becomes a promising methodology to overcome problems that require human knowledge such as image labeling, text annotation, and product recommendation [16]. Leveraging this methodology, a wide range of applications [5] (e.g. ESP game [1], reCaptcha [2], and SMART [3]) have been developed on top of more than 70 platforms[1] like Amazon Mechanical Turk and CloudCrowd. The rapid growth of such applications opens up a variety of technical challenges [18,9,8].

One of the most important technical challenges of crowdsourcing is answer aggregation [19], which aggregates a set of human answers into a single value. In our setting, we consider a broad class of problems in which there is an objective ground truth external to human judgment; i.e. each question has an exact answer but no one knows what it is. The goal of answer aggregation is to find this hidden ground truth from a set of answers given by the crowd workers. This goal is, however, difficult to achieve for two main reasons. First, the crowd workers have wide-ranging levels of expertise [23] , leading to high contradiction and uncertainty in the answer set. Second, the questions vary in different degrees of difficulty, resulting in an incorrect assessment of the true expertise

---

[1] http://www.crowdsourcing.org/

X. Lin et al. (Eds.): WISE 2013, Part II, LNCS 8181, pp. 1–15, 2013.

between truthful workers and malicious workers. To fully overcome this challenge, a rich body of research has proposed different techniques for the answer aggregation.

In general, the aggregation techniques are broadly classified into two categories according to their computing model:

- **Non-iterative:** uses heuristics to compute a single aggregated value of each question separately. One simple approach is Majority Decision (MD) [15], in which the answer with highest votes is selected as the final aggregated value. Other techniques are Honeypot (HP) [17] and ELICE [13].
- **Iterative:** performs a series of iterations, each consisting of two updating steps: (i) updates the aggregated value of each question based on the expertise of workers who answer that question, and (ii) adjusts the expertise of each worker based on the answers given by him. This incremental mechanism serves as the basis in EM [7], GLAD [25], SLME [21], and ITER [10].

While many aggregation techniques have been developed over the last decades, there has been no work on the evaluation of their performance altogether. The main reason is the lack of a common setting (i.e. no common dataset and no common metrics of success). As a result, understanding the performance implications of these techniques is challenging, since each of them has distinct characteristics. One, for example, may achieve very high accuracy over certain types of workers, while another is sensitive to spammers. Moreover, aggregation techniques have never been compared systematically, and each work often reported its superior performance generally using a limited variety of datasets or evaluation methodologies. Therefore, there is a need of common settings to test, research, and assess the advantage and disadvantage of these techniques.

The primary goal of this paper is to evaluate aggregation techniques within a common framework. To this end, we present a benchmark that offers an overview of comprehensive performance comparison among the aggregation techniques, describes in-depth analysis on the performance behavior of each method, and provides guidance on the selection of appropriate aggregation schemes. Moreover, potential users (e.g. researchers and developers) can utilize our benchmarking framework to assess their own techniques as well as reuse its components to reduce the development complexity. Specifically, the salient features of the benchmark are highlighted as follows:

- We developed or integrated, in a fair manner for comparisons, the most representative state-of-the-art techniques in each category of answer aggregation approaches, including[2] MD, HP, ELICE, EM, GLAD, SLME, and ITER.
- We designed a generic, extensible benchmarking framework to assist in the evaluation of different aggregation techniques, so that subsequent studies are able to easily compare their proposals with the state-of-the-art techniques.
- We simulated different types of crowd workers and questions. In addition, our benchmark allows users to customize the distribution of these workers. By this way, the users can predict the accuracy of worker answers and save their money before really posting the questions to the crowd.
- We offer extensive as well as intensive performance analyses. We believe that the analyses can serve as a practical guideline for how to select a well-suited aggregation technique on particular application scenarios.

---

[2] Full names of all abbreviations are given in section 2.

The remainder of the paper is organized as follows. Section 2 reviews state-of-the-art aggregation techniques. We then describe the methodology used in the benchmark in Section 3. Section 4 offers in-depth discussions on the benchmark results. Section 5 finally summarizes and concludes this study, where we provide important suggestions for the applications that consider employing an aggregation technique.

2 Answer Aggregation Techniques

In the domain of crowdsourcing, a large body of work has studied the problem of aggregating worker answers, which is formulated as follows. There are n objects $\{o_1, \ldots, o_n\}$, where each object can be assigned by k workers $[w_1, \ldots, w_k]$ into one of m possible labels $L = \{l_1, l_2, \ldots l_m\}$. The aggregation techniques take as input the set of all worker answers that is represented by an *answer matrix*:

$$M = \begin{pmatrix} a_{11} & \cdots & a_{1k} \\ \vdots & \ddots & \vdots \\ a_{n1} & \cdots & a_{nk} \end{pmatrix} \tag{1}$$

where $a_{ij} \in L$ is the answer of worker w_j for object o_i. The output of aggregation techniques is a set of aggregated values $\{\gamma_{o_1}, \gamma_{o_2}, \ldots \gamma_{o_n}\}$, where $\gamma_{o_i} \in L$ is the unique label assigned for object o_i. In order to compute aggregated values, we first derive the probability of possible aggregations $P(X_{o_i} = l_z)$, where X_{o_i} is a random variable of the aggregated value γ_{o_i} and its domain value is L. Each technique applies different models to estimate these probabilities. For simplicity sake, we denote γ_{o_i} and X_{o_i} as γ_i and X_i, respectively. After obtaining all probabilities, the aggregated value is computed by [3]:

$$\gamma_i = \arg \max_{l_z \in L} P(X_i = l_z) \tag{2}$$

In the following, we offer the details of aggregation techniques commonly used in the literature. We organize them into two categories: (i) *non-iterative aggregation*, including MD, HP, and ELICE; and (ii) *iterative aggregation*, including EM, GLAD, SLME, and ITER. Table 1 summarizes the important notations used in this paper.

2.1 Non-iterative Aggregation

The literature suggests various non-iterative techniques, including Majority Decision (MD)[15], Honeypot (HP)[17], and ELICE[13]. They differ in the preprocessing step as well as the probability computation. In particular, MD does not require preprocessing. HP filters the answers of spammers in advance, whereas ELICE considers both worker expertise and question difficulty. This section presents the details for these three techniques, which cover the characteristics of other non-iterative methods.

Majority Decision. Majority Decision (MD) is a straightforward method that aggregates each object independently. Given an object o_i, among k received answers for o_i, we count the number of answers for each possible label l_z. The probability $P(X_i = l_z)$

---

[3] Note that $\sum_{l_z \in L} P(X_i = l_z) = 1$.

Table 1. Summary of important notations

Symbol	Description
$M_{n \times k}$	answer matrix of n objects and k workers
o_i, w_j, l_z	an object, a worker, a possible label
a_{ij}	answer of worker w_j for object o_i
γ_{o_i} or γ_i	aggregated value of object o_i
$P(X_i = l_z)$	the probability of object o_i that its aggregated value γ_i is l_z
Ω	a set of trapping questions used to test worker expertise

Table 2. Characteristics of aggregation techniques

algo	trapping set	aggregation model	worker expertise	question difficulty	computing model
MD	no	non-iterative	no	no	online
HP	yes	non-iterative	no	no	online
ELICE	yes	non-iterative	yes	yes	offline
EM	no	iterative	yes	no	offline
SLME	no	iterative	yes	no	offline
GLAD	no	iterative	yes	yes	offline
ITER	no	iterative	yes	yes	offline

of a label l_z is the percentage of its count over k; i.e. $P(X_i = l_z) = \frac{1}{k} \sum_{j=1}^{k} \mathbb{1}_{a_{ij}=l_z}$. However, MD does not take into account the fact that workers might have different levels of expertise and it is especially problematic if most of them are spammers.

Honeypot. In principle, Honeypot (HP) operates as MD, except that untrustworthy workers are filtered in a preprocessing step. In this step, HP merges a set of trapping questions Ω (whose true answer is already known) into original questions randomly. Workers who fail to answer a specified number of trapping questions are neglected as spammers and removed. Then, the probability of a possible label assigned for each object o_i is computed by MD among remaining workers. However, this approach has some disadvantages: Ω is not always available or is often constructed subjectively; i.e truthful workers might be misidentified as spammers if trapping questions are too difficult.

Expert Label Injected Crowd Estimation. Expert Label Injected Crowd Estimation (ELICE) is an extension of HP. Similarly, ELICE also uses trapping questions Ω, but to estimate the expertise level of each worker by measuring the ratio of his answers which are identical to true answers of Ω. Then, it estimates the difficulty level of each question by the expected number of workers who correctly answer a specified number of the trapping questions. Finally, it computes the object probability $P(X_i = l_z)$ by *logistic regression* [6] that is widely applied in machine learning. In brief, ELICE considers not only the worker expertise ($\alpha \in [-1, 1]$) but also the question difficulty ($\beta \in [0, 1]$). The benefit is that each answer is weighted by the worker expertise and the question difficulty; and thus, the object probability $P(X_i = l_z)$ is well-adjusted. However, ELICE also has the same disadvantages about the trapping set Ω like HP as previously described.

2.2 Iterative Aggregation

Iterative aggregation is the approach that consists of a sequence of computational rounds. In each round, object probabilities–probability about possible labels of each object–are updated incrementally and this computation is repeated until convergence. This approach also differs from non-iterative one in the fact that the trapping set Ω is not required. The widely used techniques in this category includes EM, SLME, GLAD, and ITER. Each of them has different ways to initialize and update object probabilities. While EM and

SLME only concern about worker expertise, GLAD and ITER consider both worker expertise and question difficulty. The details are explained as follows.

Expectation Maximization. The Expectation Maximization (EM) technique [7] iteratively computes object probabilities in two steps: *expectation* (E) and *maximization* (M). In the (E) step, object probabilities are estimated by weighting the answers of workers according to the current estimates of their expertise. In the (M) step, EM re-estimates the expertise of workers based on the current probability of each object. This iteration is repeated until all object probabilities are unchanged. Briefly, EM is an iterative algorithm that aggregates many objects at the same time. Since it takes a lot of steps to reach convergence, running time is a critical issue.

Supervised Learning from Multiple Experts. In principle, Supervised Learning from Multiple Experts (SLME) [21] also operates as EM, but characterizes the worker expertise by *sensitivity* and *specificity*—two well-known measures from statistics—instead of the confusion matrix. Sensitivity is the ratio of positive answers which are correctly assigned, while specificity is the ratio of negative answers which are correctly assigned. One disadvantage of SLME is that it is incompatible with multiple labels since the sensitivity and specificity are defined only for binary labeling (aggregated value $\gamma \in \{0, 1\}$).

Generative Model of Labels, Abilities, and Difficulties. Generative Model of Labels, Abilities, and Difficulties (GLAD) [25] is an extension of EM. This technique takes into account not only the worker expertise but also the question difficulty of each object. It tries to capture two special cases. The first case is when a question is answered by many workers, the worker with high expertise have a higher probability of answering correctly. Another case is when a worker answers many questions, the question with high difficulty has a lower probability of being answered correctly. In general, GLAD as well as EM-based approaches are sensitive to arbitrary initializations. Particularly, GLAD's performance depends on the initial value of worker expertise α and question difficulty β. In fact, there is no theoretical analysis for the performance guarantees and it is necessary to have a benchmark for evaluating different techniques in the same setting.

Iterative Learning. Iterative Learning (ITER) is an iterative technique based on standard belief propagation [10]. It also estimates the question difficulty and the worker expertise, but slightly different in details. While others treat the reliability of all answers of one worker as a single value (i.e. worker expertise), ITER computes the reliability of each answer separately. And the difficulty level of each question is also computed individually for each worker. As a result, the expertise of each worker is estimated as the sum of the reliability of his answers weighted by the difficulty of associated questions. One advantage of ITER is that it does not depend on the initialization of model parameters (answer reliability, question difficulty). Moreover, while other techniques often assume workers must answer all questions, ITER can divide questions into different subsets and the outputs of these subsets are propagated in the end.

2.3 Summary

To sum up, we already implemented seven aggregation techniques—MD, HP, ELICE, EM, SLME, GLAD, ITER—which aggregate worker answers by computing the probability of possible labels. Each technique exhibits various aggregation characteristics. In fact, often these characteristics are not exclusive; a technique might have multiple ones. Table 2 features each implemented technique with following key characteristics.

- **Trapping set:** the set of trapping questions, whose answers are known before-hand. It is mainly used to filter spammers and initialize the expertise of other workers.
- **Aggregation model:** computation model of answer aggregation. It provides the basic categorization of aggregation techniques and the indication of their complexity.
- **Worker expertise:** the ability to capture the behavior of a worker; i.e. the accuracy and reliability of his answers. This ability is important since human workers often have wide-ranging levels of knowledge.
- **Question difficulty:** the ability to measure the difficulty degree of questions. This ability is a supplement of worker expertise: answering an easy question incorrectly is worse than answering a difficult question incorrectly.
- **Computing model:** the ability to perform (online or offline) in response to the new arrival of worker answers. An online technique can process answer-by-answer in a serial fashion, whereas offline ones have to re-compute the whole aggregation.

One interesting point to note is that all of the above techniques support aggregation on questions with binary choices (i.e. yes/no questions). For the questions with multiple choices, only three algorithms—MD, HP, and EM—are applicable. Another worth-noting point is that estimating worker expertise can serve as a quality indicator in practical scenarios such as payment mechanism and worker profiling.

3 Benchmark Setup

This section describes the setup used in our benchmark. We first present the details for our benchmarking framework as well as the simulation of crowdsourcing process. We then offer an insight of the implementation of aggregation techniques followed by descriptions of the measures used to assess their performance.

3.1 Framework

A primary goal of this study is to provide a flexible and powerful tool to support the comparison and facilitate the benchmarking analysis of aggregation techniques. To this end, we have developed a framework that employs original performance studies of each technique. Figure 1 illustrates the simplified architecture of the framework. It is built upon a component-based architecture having three layers. (1) The data access layer abstracts the underlying data objects, and loads the data to the upper layer. (2) The application layer interacts with users to receive configurable parameters and visualize outputs from the computing layer. (3) The computing layer consists of two modules: (i) *aggregation module* and (ii) *simulation module*. On one hand, the aggregation module is responsible for invoking plugged algorithms (algorithm component) upon inputs from

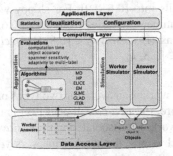

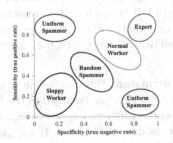

Fig. 1. Benchmarking framework   **Fig. 2.** Characterization of worker types

data access layer and delivering summarized information (evaluation component) to the application layer. On the other hand, the simulation module simulates the crowdsourcing process in which the workers (worker simulator) label a set of objects by answering various questions (answer simulator). This simulation will be described in Section 3.2.

We believe that subsequent studies are able to easily compare their algorithms with the state-of-the-art techniques by using our framework. It is flexible and extensible, since a new technique as well as a new measurement can be easily plugged in. Moreover, users are also supported to use their crowd simulators or real datasets. The framework is described in details in [11] and available for download from our website[4].

3.2 Crowd Simulation

The simulation module helps benchmark users simulate the crowdsourcing process in the literature. It is implemented with two components: (i) worker simulator—simulates different types of workers—and (ii) answer simulator—generates numbers of objects (questions) and their true labels (answers). Both of them demonstrate an online process where each worker is assigned to answer a set of questions. Details are provided below.

Worker Simulator. While some applications relied on expert workers only [14,20], many previous studies [12,24] characterized various types of crowd workers with different expertise levels. Based on the classification in [24], we simulate 5 worker types as depicted in Figure 2. (1) *Experts:* who have deep knowledge about specific domains and answer questions with very high reliability. (2) *Normal workers:* who have general knowledge to give correct answers, but with few occasional mistakes. (3) *Sloppy workers:* who have very little knowledge and thus often give wrong answers, but unintentionally. (4) *Uniform spammers:* who intentionally give the same answer for all their own questions. (5) *Random spammers:* who carelessly give the random answer for any question. To model these types of workers, we use two parameters: *sensitivity*—the proportion of actual positives that are correctly identified—and *specificity*—the proportion of negatives that are correctly identified. Following the statistical result in [12], we set randomly the sensitivity and specificity of each type of workers as follows. For experts,

---

[4] https://code.google.com/p/benchmarkcrowd/

the range is $[0.9, 1]$. For normal workers, it falls into $[0.6, 0.9]$. For sloppy workers, the range $[0.1, 0.4]$ is selected. For random spammers, it varies from 0.4 to 0.6. Especially for uniform spammers, there are two regions: (i) $sensitivity \in [0.8, 1]$, $specificity \in [0, 0.2]$ and (ii) $sensitivity \in [0, 0.2]$, $specificity \in [0.8, 1]$.

Answer Simulator. This component generates worker answers for two types of questions. (1) *Binary-choice (yes/no):* in the literature, the *two-coin* model [22] is used to generate worker answers for each object. Each worker is associated with *sensitivity* and *specificity*, as described above. If the true label is *yes*, the worker answers *yes* with the probability *sensitivity*. If the true label is *no*, the worker answers *no* with the probability *specificity*. (2) *Multiple-choice:* since the two-coin model is only compatible with binary-choice questions, we adapt to multi-choice questions by using a reliability degree $r \in [0, 1]$ for each worker. Given a question with k choices, the probability of the worker answer being the same as and being different from the true label is r and $(1 - r)/k$, respectively. Note that the reliability degree is a special case of sensitivity and specificity; i.e. if $sensitivity = specificity$ then $sensitivity = specificity = r$. It is important to note that real objects can also be used instead of simulated ones. Users can plug in their own datasets under different formats. For benchmarking purposes, we also provide well-known datasets of the data integration domain in our website [4].

3.3 Evaluation Measures

We characterize the aggregation methods compared in the benchmark using four measures: *computation time, accuracy, robustness to spammers*, and *adaptivity to multi-labeling*. We describe the details for each of the measures in the sequel.

Computation Time. A simple metric for evaluating aggregation techniques is computation time. Various applications (e.g. CrowdSearch [26]) often have constraints on computing speed, or limitations in using server resources. As a result, the computation time becomes an important aspect, when we characterize an aggregation method. In our benchmark, all techniques are evaluated on the same standard. Specifically, we randomly generate the answer matrix M $(n \times k)$, while varying its size: $n = 10, 50, 100$ and $k = 10, 50, 100$. For each setting, we measure the average computation time—from when M is processed until aggregated values are computed—over 100 runs.

Accuracy. Obviously, the most important aspect of an aggregation technique is its accuracy. It is straightforward how to measure that—accuracy is defined as the percentage of input objects that are correctly labeled:

$$accuracy = \frac{\#correctly\ labeled\ objects}{\#total\ objects} \tag{3}$$

The higher accuracy, the higher power of aggregation method. In experiments, we measure accuracy of each method while varying the number of answers per question and the number of questions per worker. In that, we find which algorithm requires least answers and which algorithm requires least workers to achieve the accuracy requirements.

Robustness to Spammers. In reality, spammers always exist in online community, especially crowdsourcing. Many experiments [24,4] in the literature showed that the proportion of spammers could be up to 40%. As a result, it is important for crowdsourcing applications to know how each aggregation technique performs when the worker answers are not trustworthy. In the benchmark, we studied the robustness to spammers by recording the accuracy, while varying the ratio of spammers. To this end, we artificially included spammers to the worker population, while applying different appearance ratio of spammers $p_{spam} = 5\%, 10\%, \ldots, 40\%$.

Adaptivity to Multi-labeling. In the literature, many applications are designed for multiple-choice questions. Therefore, it is important to know the adaptivity of aggregation techniques to this setting; i.e. which one is compatible and which one is not. Moreover, we would like to examine if there are significantly differences of their performance characteristics between the binary and the multiple setting. In the benchmark, we study the adaptivity to multi-labeling in terms of three aspects—computation time, accuracy, and robustness to spammers—while varying the number of possible labels equals to 2 or 4. Studying more than 4 labels is out of interest since these kinds of questions might be overwhelming to human workers.

4 Experimental Evaluation

We proceed to report results of applying the benchmark to the seven aggregation techniques presented in Section 2. The main goal of the experiments is not only to compare the aggregation performances, but also to analyze the effects of worker characteristics on the performance behavior. In order to compare them in a fair manner, we provide the key insights under a wide range of settings to verify their performance. All the experiments ran on an Intel Core i7 processor 2.8 GHz system with 4 GB of main memory.

4.1 Computation Time

This experiment helps to choose the right techniques for a particular input size under time constraints. It takes server resources to process worker answers. In some real-time applications like CrowdSearch [26], final aggregations need to be returned within minutes or even seconds. As a result, quickly aggregating the worker answers is a key factor. Table 3 shows the computation time of each technique, averaged over 100 runs, when varying the input size from 10×10 to 100×100 (#questions $\times$ #workers).

MD, HP, and ELICE are clear winners on this concern. They by far outperform the others (their computation time is less than one minute with the size 100×100 of M). This result is straightforward to understand—these techniques are one-time computation and do not execute any expensive routines. In contrast, EM and ITER exhibits high computing time (with 100×100 input size, more than 15 min). While, SLME and GLAD exhibit satisfactory performance. In fact, we had expected slower performance from SLME and GLAD before having the results, since they take relatively sophisticated computation to update the worker expertise and the question difficulty in their iterations. However, the updating formulas in the EM steps of SLME and GLAD are

Table 3. Average computation time (s) over 100 runs (the lower, the better)

Size of M [*]	MD	HP	ELICE	EM	SLME	GLAD	ITER
10 × 10	1	1	1	11	1	12	1
10 × 50	1	1	2	51	3	59	15
10 × 100	1	1	2	153	5	108	45
50 × 10	1	1	2	33	3	45	19
50 × 50	1	2	3	234	12	141	102
50 × 100	1	2	6	928	27	238	355
100 × 10	1	1	3	52	7	91	53
100 × 50	1	2	9	529	24	272	336
100 × 100	1	2	15	1591	46	473	915

[*] $n \times k$: n questions and k workers

less complex than EM and ITER. Briefly, this experiment suggests that MD, HP, and ELICE are fast enough for applications that prefer low response time, while the others should not probably be used for large inputs. Moreover, recall that iterative techniques (EM, SLME, GLAD, ITER) must re-compute the whole input when a new answer is received. We recommend using them for off-line analyses, when the answer set is fixed.

4.2 Accuracy

In order to reflect the accuracy of aggregation, in which the intuition behind this metric was explained in Section 3.3, our benchmark studies two dimensions of interest: *number of answers per question (#apq)* and *number of questions per worker (#qpw)*. On one hand, #apq is the number of answers received for each question (i.e. #columns of input matrix M). When we have more answers from workers, the accuracy of aggregation increases since these answers will justify each other. This factor is important to study the trade-off between the cost (of paying workers) and the accuracy (of aggregated values). On the other hand, #qpw is the number of questions assigned for each worker (i.e. #rows of M). It should not be too large to ask a human worker or too small to assess his expertise. Some aggregation techniques (e.g. EM) consider the quality of answers of each worker to justify aggregated values. This factor is crucial for this purpose. Our benchmark will help potential users opt for appropriate values of these two factors.

The Number of Answers Per Question (#apq). The experiment was conducted with #apw varying from 10 to 30. The worker types follow the distribution as previously described in Section 3.2. Figure 3 illustrates the results obtained by computing the average over 100 runs. In general, the accuracy of all techniques increases with the increase of #apw. However, each algorithm behaves with the changes of #apw very differently.

Overall, the iterative techniques perform significantly better when the #apw is higher. This is because the same questions are answered by multiple workers (overlapping between workers). As a result, the answers of each worker can be justified by the answers of others through iterations. Among iterative techniques, EM is the best performer in this experiment. This is because EM captures the worker expertise by a confusion matrix, whereas the other iterative algorithms use a single parameter α. Subsequently, the characteristics of workers are more specific. Moreover, we can see that EM's accuracy

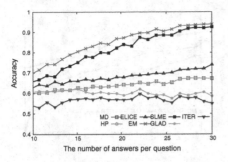

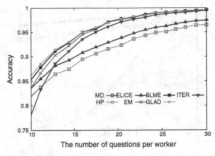

Fig. 3. Accuracy: effects of #apq (the higher, the better)

Fig. 4. Accuracy: effects of #qpw (the higher, the better)

is at least 25% higher than others in the end. In brief, we suggest using EM for high-accurate results, in case the computation time is not concerned.

The Number of Questions Per Worker (#qpw). In this experiment, we vary the number of questions per worker—hereby denoted as $#qpw$—from 10 to 30. The same worker population is used. In general, all techniques achieve higher accuracy when the $#qpw$ increases. But there is no significant difference between them. When the $#qpw > 20$, the accuracy of all techniques is more than 90%. Figure 4 depicts the result.

At starting points ($#qpw$ 10), ITER, HP, and MD are the worst techniques. For MD, this is because the majority effect: not enough trustworthy answers to dominate un-trustworthy ones. For HP, this effect is more severe since truthful workers have too few correct answers to pass trapping questions. For ITER, it is due to the lacks of initial information. However, as the $#qpw$ increases, the difference among all techniques is reduced (less than 0.05 with 30 $#qpw$). In addition, each of them has a "convergent point": continue increasing $#qpw$ above this point will not improve the accuracy significantly (e.g. EM achieves 95% accuracy at $#qpw \approx 18$, doubling $#qpw$ only increases the accuracy up to 5% more). Another interesting observation is that iterative techniques are slightly better non-iterative ones: the difference is only 5% when the $#qpw$ reaches to 30. This can be explained by the fact that in iterative techniques, worker answers are more refined by multiple of computational rounds.

4.3 Robustness to Spammers

In this experiment, we will increase the ratio of spammers to study its effects on accuracy. First, we remove sloppy workers from the crowd due to their lacks of knowledge, which generates many wrong answers in the input. By this way, we can see a clear effect of spammers. The spammer ratio is varied from 5% to 40%. Based on previous results, we fix the number of answers per object to 20 because this gives a high starting point of accuracy. Figure 5 and 6 illustrate the effects of uniform spammers and random spammers on accuracy, respectively. In general, the accuracy of all techniques decreases when the spammer ratio is higher. But their behaviors are significantly different.

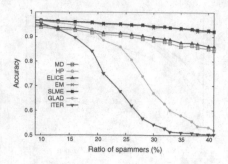

Fig. 5. Effects of uniform spammers (the higher, the better)

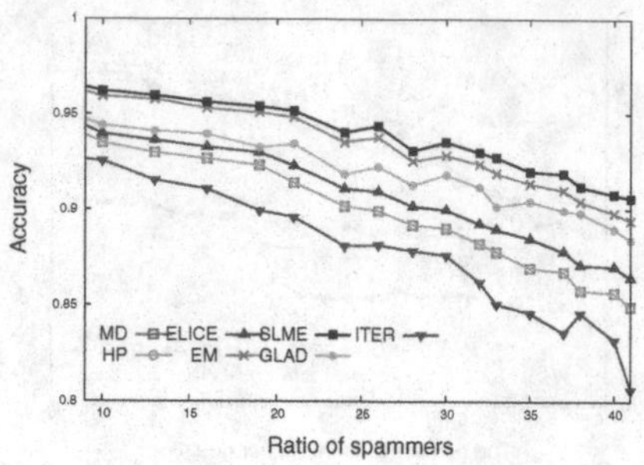

Fig. 6. Effects of random spammers (the higher, the better)

Uniform Spammers. The effects of uniform spammers are presented in Figure 5. An interesting observation is that ITER and GLAD are the worst in this setting. At the starting point (5% spammers), their accuracies are already lower than the others'— about 0.6 and 0.75 respectively. After the spammer ratio rises to 25%, ITER and GLAD drops rapidly to nearly 0.5. By the time more than 25% spammers, their behaviors are like random (accuracy converges to 0.5). This observation could be explained by their underlying models. First, ITER's algorithm depends on the entropy of a worker's answers. Since uniform spammers always give identical answers, the uncertainty of the input is high and it ends up with a poor accuracy. Second, GLAD is not able to identify and prioritize truthful workers (i.e. all workers are initially weighted as equal), resulting in a random accuracy in the end. In brief, ITER and GLAD are very sensitive to the spammers. Another key finding is that among the five remaining techniques, we can observe two distinct groups. The first group consists of EM and SLME, which are the better than the second group including MP, HP, and ELICE. However, the difference between them is not significant (less than 0.1 with 40% uniform spammers).

Random Spammers. The effects of random spammers are depicted in Figure 6. Similar to random spammers case, most of the techniques are robust to random spammers— their accuracy decreases up to 10% at the end. However, their accuracies are better in this case. This is reasonable because the answers of spammers are different from each other, which cannot dominate the answers of other workers. Another noticeable observation is that ITER and GLAD perform better than before. For ITER, although starting with high accuracy, its accuracy reduces more than 15% at the end. This could be explained by the same way. Since the answers are random but different, the entropy value is lower, resulting in a higher accuracy. For GLAD, it uses the answers of other workers to justify the spammers', rendering the similar performance like EM and SLME.

4.4 Adaptivity to Multi-labeling

In this experiment, we study the adaptivity to multi-labeling of aggregation techniques. Only three techniques—MD, HP, and EM—are retained while the others fail to adapt

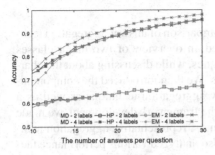
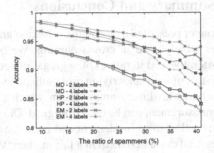

Fig. 7. Multi-label effect on accuracy   **Fig. 8.** Multi-label effect on spammer robustness

this setting. SLME models worker expertise by *sensitivity* and *specificity*, which are applicable for binary question only. Regarding ITER and ELICE, their original papers indicate that they are only applied for binary questions. Besides, they use the sign (positive or negative) of aggregated value to classify object. Regarding GLAD, we checked the source code and confirmed that it was implemented for only binary-choice questions. Similar to previous experiments, we proceed to report the performance characteristics of applicable techniques (MP, HP, and EM) in three aspects below.

Computation Time. In general, MD and HP are not affected by #labels—their computation time keeps unchanged when #labels increases. This is because the complexity of majority rule only depends on the number of answers per object (the label with highest number of answers wins). For EM, its completion time increases a little bit since it uses a confusion matrix to capture worker expertise. The size of this matrix is $n \times n$, where n is the number of labels. However, as n only up to 4, this shows no significant difference. Therefore, the results of computation time are omitted due to page limits.

Accuracy. Using the same worker distribution of Section 4.2, we measure the accuracy against different numbers of answers per question. The result for different numbers of questions per worker is omitted due to similar findings. Figure 7 illustrates the result. In general, with more labels, the accuracy is better. EM is still the winner in both cases—2 labels and 4 labels. However, the superiority of EM in comparison with MD and HP is reduced when the number of labels increase. For example, the difference is between 0.12 and 0.28 with 2 labels, whereas this difference is less than 0.02 with 4 labels. This is, in fact, reasonable. The incorrect answers are now distributed among several choices, which is unlikely to dominate the majority of the true answers.

Sensitive to Spammers. Like previous experiments in Section 4.3, we increase the spammer ratio to study the accuracy reduction. Results are presented in Figure 8. In both 2-label and 4-label settings, EM is more robust to spammers than MD and HP. Surprisingly, MD and HP become better with 4 labels. This can be explained by the majority property: answers given by spammers no longer dominate those of other workers. Since there are more choices, it is unlikely that spammers give the same answer together.

5   Summary and Conclusions

This paper presented a thorough evaluation and comparison of answer aggregation techniques widely used in crowdsourcing. We offered an overview of two major classes (non-iterative and iterative) of aggregation techniques, while discussing about the characteristics of their underlying probabilistic models. We then introduced the component-based benchmarking framework, in which a new aggregation technique as well as a new measurement can be easily plugged. During the framework development, we made the best effort to re-implement and integrate the most representative aggregation techniques, and evaluated them in a fair manner. We also analyzed various performance factors for each technique, including *computation time, accuracy, robustness to spammers,* and *adaptivity to multi-labeling.* The crowdsourcing process is simulated by letting five different types of workers answer binary or multiple-choice questions.

We here summarize our principal findings as a set of recommendations for how to select a well-suited aggregation technique on particular application scenarios:

- Overall, EM and SLME achieve highest accuracy and work robustly against spammers. In particular, they outperform the others when #answers per question is high. Regarding #questions per worker, there are two runner-ups (GLAD and ITER).
- If the crowd contains many spammers ($\geq 30\%$), we suggest using SLME or EM. Interestingly, the performance of non-iterative techniques (MD, HP, ELICE) is not significantly lower than SLME and EM. If accuracy is not highly required, they are best-suited for applications that require fast computation. In contrast, we strongly suggest not using GLAD and ITER since they are most sensitive to spammers.
- Only MD, HP, and EM can adapt to multi-labeling. For binary labeling, EM is the winner. In case of 4 labels, MD and HP are also appropriate choices since the difference between them and EM is not distinguishable.
- For applications that require fast computation, MD and HP are the winners. Oppositely, we strongly suggest not using iterative techniques. Not only is their computation time much higher than the non-iterative techniques, but also they require to re-compute the whole answer set upon the new arrival of worker answers.

category	winner	2nd best	worst
computation time	**MD**	HP	EM
accuracy	**EM**	SLME	HP
robustness to spammers	**SLME**	EM	ITER
adaptivity to multi-labeling [*]	**EM**	MD	HP

[*] other techniques (ELICE, SLME, GLAD, ITER) only work
with binary labeling due to their implementation limitation

As a concluding remark, we recommend potential applications to use our benchmarking framework as a tool to find out the best-suited aggregation technique accordingly, since there is no absolute winner that outperforms the others in every case. As the source codes as well as datasets used in the benchmark are publicly available, we expect that the experimental results presented in this paper will be refined and improved by the research community, in particular when more data become available, more experiments are performed, and more techniques are integrated into the framework in the future.

References

1. von Ahn, L., et al.: Labeling images with a computer game. In: CHI (2004)
2. von Ahn, L., et al.: recaptcha: Human-based character recognition via web security measures. Science (2008)
3. Hung, N.Q.V., Tam, N.T., Miklós, Z., Aberer, K.: On leveraging crowdsourcing techniques for schema matching networks. In: Meng, W., Feng, L., Bressan, S., Winiwarter, W., Song, W. (eds.) DASFAA 2013, Part II. LNCS, vol. DASFAA, pp. 139–154. Springer, Heidelberg (2013)
4. Difallah, D.E., et al.: Mechanical cheat: Spamming schemes and adversarial techniques on crowdsourcing platforms. In: CrowdSearch (2012)
5. Doan, A., et al.: Crowdsourcing systems on the world-wide web. CACM (2011)
6. Hosmer, D.W., et al.: Applied logistic regression. Wiley-Interscience Publication (2000)
7. Ipeirotis, P.G., et al.: Quality management on amazon mechanical turk. In: HCOMP (2010)
8. Kamar, E., et al.: Combining human and machine intelligence in large-scale crowdsourcing. In: AAMAS (2012)
9. Kamar, E., et al.: Incentives for truthful reporting in crowdsourcing. In: AAMAS (2012)
10. Karger, D., et al.: Iterative learning for reliable crowdsourcing systems. In: NIPS (2011)
11. Nguyen, Q.V.H., et al.: Batc - a benchmark for aggregation techniques in crowdsourcing. In: SIGIR (2013)
12. Kazai, G., et al.: Worker types and personality traits in crowdsourcing relevance labels. In: CIKM (2011)
13. Khattak, F., et al.: Quality Control of Crowd Labeling through Expert Evaluation. In: NIPS (2011)
14. Nguyen, Q.V.H., et al.: Collaborative schema matching reconciliation. In: CoopIS (2013)
15. Kuncheva, L., et al.: Limits on the majority vote accuracy in classifier fusion. Pattern Anal. Appl. (2003)
16. Law, E., et al.: Human Computation. Morgan & Claypool Publishers (2011)
17. Lee, K., et al.: The social honeypot project: protecting online communities from spammers. In: WWW (2010)
18. Mason, W., et al.: Conducting behavioral research on amazon mechanical turk. BRM (2012)
19. Quinn, A.J., et al.: Human computation: a survey and taxonomy of a growing field. In: CHI (2011)
20. Nguyen, Q.V.H., et al.: Minimizing Human Effort in Reconciling Match Networks. In: ER (2013)
21. Raykar, V., et al.: Supervised learning from multiple experts: Whom to trust when everyone lies a bit. In: ICML (2009)
22. Raykar, V.C., et al.: Learning from crowds. Mach. Learn. Res. (2010)
23. Ross, J., et al.: Who are the crowdworkers?: shifting demographics in mechanical turk. In: CHI (2010)
24. Vuurens, J., et al.: How much spam can you take? an analysis of crowdsourcing results to increase accuracy. In: CIR (2011)
25. Whitehill, J., et al.: Whose vote should count more: Optimal integration of labels from labelers of unknown expertise. In: NIPS (2009)
26. Yan, T., et al.: CrowdSearch: exploiting crowds for accurate real-time image search on mobile phones. In: MobiSys (2010)

Propagated Opinion Retrieval in Twitter

Zhunchen Luo, Jintao Tang, and Ting Wang

College of Computer, National University of Defense Technology,
410073 Changsha, Hunan, China
{zhunchenluo,tangjintao,tingwang}@nudt.edu.cn

Abstract. Twitter has become an important source for people to collect opinions to make decisions. However the amount and the variety of opinions constitute the major challenge to using them effectively. Here we consider the problem of finding propagated opinions – tweets that express an opinion about some topics, but will be retweeted. Within a learning-to-rank framework, we explore a wide of spectrum features, such as retweetability, opinionatedness and textual quality of a tweet. The experimental results show the effectiveness of our features for this task. Moreover the best ranking model with all features can outperform a BM25 baseline and state-of-the-art for Twitter opinion retrieval approach. Finally, we show that our approach equals human performance on this task.

Keywords: Opinion Retrieval, Twitter, Retweet, Propagation Analysis.

1 Introduction

Twitter is the most popular micorblogging service which attracts over 500 million registered users[1] and generates over 340 million tweets daily[2]. Within Twitter, people like to share their information or opinions about personalities, politicians, products, companies, events, etc. Indeed Twitter has became an enormous repository which can not only help other people to make decisions, but also help business and government to collect valuable feedback.

However, the sheer volume of available opinions as well as the large variations present a big impediment to the effective use of the opinions in Twitter. First, the users can experience information overload due to the high volume of opinions in Twitter. Second, the importance of opinions might not be equal and the users dealing with a large number of opinions are likely to miss some important tweets. See the following tweets which are both opinions related to the topic "Obama":

(a) *"RT@KG_NYK: The fact that Obama "lost" the debate b/c he didnt call Romney's lies out well enough is pretty harrowing commentary on surf ".*
(b) *"MyNameisGurley AND I HATE OBAMA.*

---

[1] http://techcrunch.com/2012/07/30/analyst-twitter-passed-500m-users-in-june-2012-140m-of-them-in-us-jakarta-biggest-tweeting-city/
[2] http://blog.twitter.com/2012/03/twitter-turns-six.html

X. Lin et al. (Eds.): WISE 2013, Part II, LNCS 8181, pp. 16–28, 2013.

Users may consider tweet (a) is more important than tweet (b), since tweet (a) introduces the *First Presidential Debate* event which is related to *Obama* and gives an opinion on *Obama*'s performance. Whereas tweet (b) shows a general opinion uninteresting to most users. Moreover tweet (a) is a retweet of *KG_NYK*'s opinion by its author, which shows the agreement of the author to the original one.

Estimating the importance of a tweet is very subjective. In Twitter, however, information can deemed important by the community propagates through *retweets* [4]. This is based on human behavioral patterns for propagating microblog posts, and follows from a simple assumption: users of microblogs will propagate a post when they consider it to be important and thus worthy of being shared with other users. In this paper, we present a study of finding propagated opinions in Twitter. **Relevant tweets** should satisfy three criteria: (1) be relevant to the query; (2) contain opinions or comments about the query, irrespective of being positive or negative and (3) will be retweeted.

Previous work of predicting whether a tweet will be propagated is largely about identifying the topics of interest, and it is conceivable that unigram representation of full-length document can reasonably capture that information [4,16]. In our case, most tweets are already of interest to that user topically, which ones the user ends up retweeting may depend on several non-topical aspects of the text: whether the tweet is convincing, whether the tweet is well written, etc. Previous work has shown that such analysis can be more difficult than topic-based analysis [15], and we have the additional challenge that tweets are typically much shorter. However, the difficulty in analyzing the textural information in tweets can be alleviated by additional contextual information such as the tweets' specific information and the authors' information which potentially can improve this task.

In this paper, we use a standard machine learning approach to learn a ranking function for tweets that uses a wide spectrum of features which can recover propagated opinions in Twitter. These features include the retweetability, opinionatedness and textural quality of a tweet. The retweetability feature is the confidence score of a tweet in general being retweeted. Additionally, we proposed an approach which using social and structural information to estimating the opinionateness score of a tweet. We integrated these two features into our ranking model for this new task. Finally, we develop some features which refer to the textual quality of a tweet, including the length, the linguistic properties and the fluency of the text for a tweet. The experimental results show that the three feature sets are effective for finding propagated opinions in Twitter. Our approach integrating all feature sets performs significantly better than two baselines, one is based on the BM25 score (BM25) and the other is a state-of-the-art Twitter opinion retrieval (TOR) [13]. Moreover, a comparison of our best ranking model with human performance shows our approach does well as humans on this task.

The contributions of this paper can be summarized as follows:

1) We define a new ranking task aiming at finding opinionated tweets that will be propagated in the future.
2) We develop a set of features derived from the field of Twitter for this task and the effectiveness factors are evaluated over real-world Twitter dataset.
3) The results show the performance of our best ranking model is significantly better than the TOR baseline [13] and a BM25 baseline.
4) Furthermore, our approach for identifying the propagated opinion in Twitter can achieve human subjects' ability as well.

2   Related Work

We review related work on three main areas: message propagation and opinion mining in Twitter, review quality evaluation.

2.1   Message Propagation in Twitter

In Twitter, message deemed important by the community propagates through retweets. There is much work which is related to predicting whether a tweet in general will be retweeted. Petrovic et al. [16] used a machine learning approach based on the passive-aggressive algorithm to predict whether a tweet would be retweeted in the future. They found the content of the tweet, listed number, followers number and whether the author was verified were more effective features for this task. Hong et al. [4] proposed a method to predict the volume of retweets for a tweet. Luo [12] considered the task of finding who will retweet a message posted on Twitter. They found that followers who retweeted or mentioned the author's tweets frequently before and have common interests are more likely to be retweeters. Liu [8] investigated information propagation in Twitter from the geographical view on the global scale. They discovered that the retweet texts are more effective than common tweet texts for real-time event detection. Stieglitz [17] examined whether sentiment occurring in politically relevant tweets had an effect on their retweetability. They found a positive relationship between the quantity of words indicating affective dimensions, including positive and negative emotions associated with certain political parties or politicians, in a tweet and its retweet rate. Their work investigated whether the sentiment in a tweet could affect retweetability, but our study examines which factors affect the retweetability of opinions in Twitter.

2.2   Opinion Mining in Twitter

Twitter has attracted hundreds of millions of users who post opinions on this platform and it is also a hot research domain for academic. For example, Jansen et al. [5] investigated tweets as a form of electronic word-of-mouth for sharing consumer opinions concerning brands; O'Connor et al. [14] proposed explicitly link measurement of textural sentiment in Twitter for public opinion polls; Bollen *et al.* [1] used Twitter mood to predict the stock market, etc. However, most of these work concentrates on analyzing opinions expressed in tweets for a given topic, none on how to obtain opinions towards some persons, products or events. Luo *et al.* [13] firstly studied finding opinionated tweets for a given topic. They integrated social information and opinionatedness information into a learning to rank model. The experimental result showed that opinion retrieval performance was improved when links, mentions, author information such as the number of statues or followers and the opinionatedness of the tweet were taken into account. We take their approach as one of our baselines for comparison.

2.3 Review Quality Evaluation

Ranking reviews (opinions) according to the quality is an important problem for many online sites such as Amazon.com and Ebay.com. However, most of websites use manual votes of the *helpfulness*, such as 'thumbs up' and 'thumbs down', to assess the quality. Kim *et al.* [7] and Zhang *and* Varadarajan [19] measured the helpfulness automatically and solved it with regression model. They adopted feature sets such as lexical and syntactically oriented. The results showed that the shallow syntactic features, e.g., the counts of proper nous, modal verbs, and adjectives were correlated with the quality. Liu *et al.* [10] studied the quality of movie reviews and found, besides textural information, reviews' expertise and the timeliness of the reviews were related to the review quality. All of these work deals with reviews in websites, Twitter, however, is a novel domain with varied short text and its rich social environment should be considered when estimating the quality.

3 Data

To investigate the factors that affect the propagation of opinions in Twitter, we use Luo *et al.* [13]'s opinion retrieval dataset[3]. It contains 50 queries and 5000 judged tweets. For each query, there are average of 16.62 opinionated tweets which are related to a given topic (query). This dataset was collected through the Twitter streaming API in November 2011. The purpose of our study is finding the opinionated tweets which will be propagated in the future. Hence, we crawled these tweets again using Twitter statuses API[4] in April 2012. Based on the principle about the relevant tweet introduced in Section 1, we take the opinionated tweets which have been retweeted within sixth months as relevant tweets and the other tweets as irrelevant tweets. We consider the state of these tweets is stable and they are not likely to be retweeted any more[5]. The task of

---

[3] https://sourceforge.net/projects/ortwitter/

[4] https://dev.twitter.com/docs/api/1/get/statuses/show/%3Aid

[5] When we crawled these tweets again, we found some of tweets have been deleted. We consider that if an opinionated tweet is deleted, it is not a propagated tweet any more. Therefore, we take the deleted tweets as irrelevant tweets.

this study is to show how to find these relevant tweets. The average number of relevant tweets per query is 3.4. It shows that there are only a small part of opinions which have been retweeted in Twitter and most of opinions are not be propagated. Interestingly, the percentage of opinions which have been retweeted is 20.5%, which is larger than the percentage of general tweets that have been retweeted (the value is 16.6%) in this new dataset. It shows opinions are more likely to be propagated than general tweets.

4   Overview of Our Approach

To generate a good function which ranks the tweets according our principle for finding propagated opinions in Twitter, we investigate the features concerning retweetability, opinionatedness and textural quality of a tweet. We develop a bag of features into a learning-to-rank scenario which demonstrated excellent power for ranking problem [9].

4.1   Learning to Rank Framework

Learning to rank is a data driven approach which effectively incorporates a bag of features in a model for ranking task. First, a set of queries and related tweets were used as training data. Every tweet is labeled whether it is a relevant tweet or not. A bag of features related to the relevance of a tweet is extracted to form a feature vector. Then a learning to rank algorithm is used to train a ranking model. For a new query, their related tweets, which extract the same features to form feature vectors, can be ranked by the rank function based on this model. The ranking performance of the model using a particular of feature sets in testing data can reflect the effect of these features for finding propagated opinions in Twitter.

4.2   Features for Tweets Ranking

For propagated opinion retrieval in Twitter, we consider a retweetability feature, opinionatedness feature and textural quality features for tweets ranking.

1) *Retweetability feature* refers to whether a tweet in general will be retweeted.
2) *Opinionatedness feature* refers to estimating the opinionatedness score of a tweet.
3) *Textural quality features* refer to textural information of a tweet.

In the next section, we will describe these features in details.

5   Features

5.1   Retweetability Feature

In Twitter, retweeting is an important way for information diffusion and there is a lot of work about predicting if a tweet will be retweeted [4,16]. Therefore, we develop a feature which can predict whether a tweet in general will be retweeted. We set this feature based on Petrovic *et al.* [16]. We used a machine learning approach based on the passive-aggressive algorithm to predict the retweetability score of a tweet. A set of features was developed for this prediction. It contains:

Content: the actual words in a tweet. It captures the topic of a tweet and some tweets refer to the specific topic are more likely to be retweeted. For example, people might pay more attention to the tweets related to "iran nuclear" than the tweets about "systems biology".

Followers: the number of followers about the author of a tweet. This indicates the popularity of the user. The tweets associated with the popular authors are more likely to be retweeted.

Listed: the number of times the author of a tweet has been listed. It also indicates the popularity of the user.

Verified: whether the author of a tweet is verified. It is used by Twitter mostly to confirm the authenticity of celebrity. 91% of tweets written by verified users are retweeted, compared with 6% for tweets where the author is not verified [16].

For retweeting, the time is a critical factor. For example, people may pay more attention about the tweets related to the "American Music Awards" in November 2011 than in April 2012. Therefore, we train the prediction model on the stream of tweets crawled from the Twitter streaming API[6] throughout November 2011. We gathered a total of 30 million tweets and used them as training data. In this training data, we take the tweets which were retweeted by *retweet* button as positive samples and the other tweets as negative samples. We test the performance of our model for retweet prediction in 100,000 samples. The accuracy is 95.99%. To our **retweetability** feature, we use the margin value calculated by the passive-aggressive algorithm as the confidence of a tweet in general being retweeted.

5.2   Opinionatedness Feature

Obviously estimating the opinionatedness score of a tweet is essential for propagated opinion retrieval in Twitter. We adopt the lexicon-based approach, since it is simple and non-dependence on machine learning techniques. However, a lexicon such as *MPQA Subjectivity Lexicon*[7] which is widely used might not be effective in Twitter, since the textual content of a tweet is often very short, and lacks reliable grammatical style and quality. Therefore, we propose an approach which can automatically construct opinionated lexical from sets of tweets matching specific patterns indicative of opinionated message.

In Twitter, when people retweet another user's tweet and give a comment before this tweet, this tweet is likely to be a subjective tweet. For example, the tweet *"I thought we were isolated and no one would want to invest here! RT @BBCNews: Honda announces 500 new jobs in Swindon bbc. in/vT12YY"* is a subjective tweet. Here, we call this tweet *Pseudo Subjective Tweet (PST)*. Many

---

[6] http://stream.twitter.com/

[7] http://www.cs.pitt.edu/mpqa/

tweets posted by news agencies are likely to be objective tweets and these tweets usually contain links. For example, a tweet "*#NorthKorea:#KimJongil died after suffering massive heart attack on train on Saturday, official news agency reports bbc.in/vzPGY5*" is an objective tweet. We define a tweet satisfies two criteria: (1) it contains links and (2) the user of this tweet posted many tweets before and has many followers as *Pseudo Objective Tweet (POT)*.

According to the definition introduced above, it is easy for us to design patterns and collect a large number of PSTs and POTs from Twitter. Using a PSTs set and a POTs set, we can automatically construct opinionated lexica. We use the chi-square value to estimate the opinion score of a term, which measures how dependent a term is with respect to the PSTs set and the POTs set. For the **opinionatedness feature**, we estimate the opinionatedness score of a tweet by summing all the terms with a chi-square value no less than m. The estimated formula as follows:

$$Opinion_{avg}(d) = \sum_{t \in d, \chi^2(t) \geq m} p(t|d) \cdot Opinion(t)$$

where $p(t|d) = c(t,d)/|d|$ is the relative frequency of a term t in tweet d. $c(t,d)$ is the frequency of term t in tweet d. $|d|$ is the number of terms in tweet d.

$$Opinion(t) = sgn(\frac{O_{11}}{O_{1*}} - \frac{O_{21}}{O_{2*}}) \cdot \chi^2(t)$$

where $sgn(*)$ is sign function. $\chi^2(t)$ calculates chi-square value of a term.

$$\chi^2(t) = \frac{(O_{11}O_{22} - O_{12}O_{21})^2 \cdot O}{O_{1*} \cdot O_{2*} \cdot O_{*1} \cdot O_{*2}}$$

O_{ij} in Table 1 is counted as the number of tweets having term t in the PSTs set or POTs set respectively. For example O_{12} is the number of tweets not having term t in the PSTs set.

Table 1. Table for pearson's chi-square. $O_{1*} = O_{11} + O_{12}$; $O_{2*} = O_{21} + O_{22}$; $O_{*1} = O_{11} + O_{21}$; $O_{*2} = O_{12} + O_{22}$; $O = O_{11} + O_{12} + O_{21} + O_{22}$.

	t	$\neg t$	Row total
PSTs set	O_{11}	O_{12}	O_{1*}
POTs set	O_{21}	O_{22}	O_{2*}
Column total	O_{*1}	O_{*2}	O

5.3   Textural Quality Features

Twitter is a social network that contains various content such as personal updates, babbles, conversations, etc. They are less carefully edited than other formal text (e.g., news reports) and therefore contain more misspellings and typographical errors. We develop some features which refer to the textural quality of a tweet affecting the propagation in Twitter.

Length: The total number of tokens in a tweet. Kim *et al.* [7] found the **length** feature is effective for estimating high quality reviews. Intuitively, a long tweet is apt to contain more information than a short one. We use this feature to indicate information richness for a tweet.

PosTag: Luo *et al.* [13] found the personal content is more likely to be the opinionated tweets. These tweets usually contain personal pronoun (e.g., "i", "u" and "my") and emotions (e.g., ":)", ":(" and ":d"). However there is a lot of garbage which has less open-class words (i.e., nouns, verbs, adjectives and adverbs) in these tweets. E.g., the tweet *"@fayemckeever Jennifer Aniston :)"* is not a high quality opinion. Therefore we develop some features aiming to capture the linguistic properties of a tweet which include the percentage of tokens that are open-class, the percentage of tokens that are nouns, the percentage of tokens that are verbs and the percentage of tokens that are adjectives or adverbs. We use the *Twitter Part-of-Speech Tagging*[8] to tag the tweets [3].

Fluency: The fluency of a text can capture the readability of a tweet and we use language model to tackle the fluency of text. We take the probability of a tweet t under a particular language model (LM) as the fluency score $F(t)$. It is determined by:

$$F(t) = \frac{1}{m} P(w^m) = \frac{1}{m} \prod_i^m P(w_i | w_{i-N+1}, w_{i-N+2}, ..., w_{i-1})$$

where a tweet t can be expressed as a sequence of words $w^m = (w_1, w_2, ..., w_m)$. To deal with length bias, we normalize the probability by the number of tokens. We work with the N-gram based language model (N = 4) using 30 million tweets from November 2011.

6 Experiment

6.1 Human Experiments

Before estimating the performance of our approach for finding propagated opinion in Twitter, we first conduct an experiment judging whether propagated opinion can be detected by human subjects. We presented two human subjects with 100 pairs of tweets produced from our dataset, and asked them to judge which tweets were propagated opinions based on the principle introduced in Section 1. Every pair of tweets are associated to the same topic (query) and exactly one of tweets is a relevance tweet and the other is irrelevant (see the definition of relevant tweets in Section 1). The order of the two tweets in each pair was chosen randomly to avoid bias. We evaluate the performance as accuracy: the number of pairs where the human can judge which tweet is the propagated opinion correctly. In our experiment, both human subjects beat the random baseline (which is a 50% accuracy): the first subject is 75% and the other is 69%. It shows that humans are capable of judging which tweets are propagated opinions from those which are not.

---

[8] http://www.ark.cs.cmu.edu/TweetNLP/

6.2   Experimental Settings and Baselines

We investigate the effect of features introduced above for propagated opinion retrieval in Twitter. For learning to rank, SVM light [6] which implements the ranking algorithm is used. We use a linear kernel for training and report results for the best setting of parameters. In order to avoid overfitting the data we perform 10 fold cross-validation in our new dataset. Thus for each fold we have 45 queries with the related tweets in the training set and 5 queries with the related tweets in the testing set. We use Mean Average Precision (MAP) as the evaluation metric.

To automatically generate PSTs and POTs, we design some simple patterns: For PSTs generation, we choose the tweets uses the convention "RT @username", with text before the first occurrence of this convention. Additionally we find that the length of the preceding text should be no less than 10 characters. For POTs generation, we choose the tweets which contain a link, the author for each tweet has no less than 1,000 followers and has posted at least 10,000 tweets. In our one-month tweets dataset, 4.64% tweets are high quality PSTs and 1.35% tweets are POTs. We use 4500 PSTs and POTs[9] as opinion corpus. In our corpus-derived approach, we use the Porter English stemmer and stop words to preprocess the text of tweets. Using these tweet datasets we can calculate the value of opinionatedness score for a new tweet. To achieve the best performance of tweets ranking, we set the threshold of m is 5.02 corresponding to the significance level of 0.025 for each term in the opinion corpus. This setting is the same as [13,18].

We choose two approaches as our baselines for comparison. One is using the Okapi BM25 score of each tweet as a feature for modeling. This approach has been widely used as a baseline of Twitter retrieval [2,11,13]. We call this baseline **BM25**. The other baseline we used is based on Luo *et al.* [13]. This method integrates some social features and an opinionatedness feature for Twitter opinion retrieval. We call this baseline **TOR**. The detail of the features in **TOR** baseline are shown in Table 2.

Table 2. TOR Baseline Features

TOR Features	Description
BM25	The Okapi BM25 score
Mention	A binary feature whether a tweet contains "@username"
URL	A binary feature whether a tweet contains a link
Statuses	The number of tweets (statuses) the author has ever written
Followers	The number of followers the author has
Opinionatedness	The opinionatedness score of a tweet

---

[9] We test that 4500 PSTs and POTs as corpus for estimating opinionatedness feature can achieve high performance for propagated opinion ranking in Twitter and there is no significant improvement adding more PSTs and POTs.

6.3   Result

We investigate whether the features introduced in Section 5 are effective for propagated opinion retrieval in Twitter. We integrate each feature with the two baselines features into our tweets ranking systems respectively. Table 3 and Table 4 show the performance of each ranking model.

We can see that using **Retweetability, PosTag** and **Fluency** features significantly improve the results when integrated with the **TOR**. It suggests the retweetability, the linguistic properties and the readability of a tweet can indeed help finding propagated opinions in Twitter. We can also see that the performance **TOR** is significantly better than **BM25** (p<0.01). It is not surprising that the opinionatedness information and some social information of tweets are essential for this task. Although the performance results of the **BM25** ranking model integrated with **Retweetability, PosTag** and **Fluency** respectively are higher than **BM25**, they are not significant. The reason may be that just using these features alone are not enough for improving tweets ranking. Interestingly, we find the **Length** feature can help finding propagated opinion integrated with **TOR**, but the result is decreased combined with the **BM25**. It shows the length information is not very effective for finding propagated opinions in Twitter as other review websites [7]. This is because each tweet has to follow the 140-characters limitation, therefore the diversity of length between propagated opinions and the other tweets is not obvious. We integrate the **Textual Quality** features (combine **Length, PosTag** and **Fluency** together) into the two baselines and find the performance is improved more. All these show that our **Retweetability, Opinionatedness** and **Textual Quality** are all effective for finding propagated opinions in Twitter.

Table 3. BM25 is a baseline. A significantly improvement with $\triangle$ and $\blacktriangle$ (for p < 0.05 and p < 0.01 respectively). BM25+All combines BM25, Retweetability, Opinionatedness and Textural Quality features together.

	MAP
BM25	0.0997
BM25+Retweetability	0.1077
BM25+Opinionatedness	0.1146
BM25+Length	0.0881
BM25+PosTag	0.1157
BM25+Fluency	0.1046
BM25+Textural Quality	0.1277
BM25+All	0.1317

At last we add all the features based on **TOR** baseline into a ranking model (**TOR+Retweetability+Textural Quality**). Table 4 shows its best result achieved the MAP value 0.1992. The best result improves MAP by 30.97% over the **TOR** method and 99.80% over the **BM25** method. All these show our **Best** ranking model can not only find the opinionated tweets to a given topic, but these tweets are also more likely to be propagated in the future. For example, the query *American Music Awards* yields three tweets in our data:

(a) *Watch Olnine Free— The 39th Annual American Music Awards (TV 2011): The 39th Annual American Music Awards (TV 20... http: // t. co/ SxrjVVmx.*
(b) *We're so excited for the American Music Awards this weekend.*
(c) *That awkward moment when the American Music Awards is really the American Minaj Awards.*

In our experiment, the **BM25** method ranks tweet (a) higher than tweet (b) and tweet (c), but this tweet is an objective message without opinions. **TOR** ranks tweet (b) higher than the other tweets, since it contains the author's opinion about the *American Music Awards*, however it was not propagated. Our **Best** ranking model ranks tweet (c) higher and this funny opinion had been propagated 143 times within six months.

Table 4. TOR is a baseline. A significantly improvement with $\triangle$ and $\blacktriangle$ (for $p < 0.05$ and $p < 0.01$ respectively).

	MAP
TOR	0.1521
TOR+Retweetability	0.1806$^\blacktriangle$
TOR+Length	0.1580
TOR+PosTag	0.1917$^\blacktriangle$
TOR+Fluency	0.1875$^\triangle$
TOR+Textural Quality	0.1930$^\blacktriangle$
TOR+Retweetability+Textural Quality (Best)	0.1992$^\blacktriangle$

6.4 Opinion Propagation Prediction vs General Message Propagation Prediction

There are much work which predicts whether a tweet in general will be retweeted [4,16]. We are interested in the relationship of propagation predictions between opinions and general message in Twitter. We investigate whether only using the **Retweetability** feature is enough to find the propagated opinionated tweets. Table 5 gives the result that the performance of **Retweetability** ranking model is worse than **Best** ranking model significantly. It shows the task of predicting whether an opinion will be propagated is different to the related task of predicting whether a tweet in general will be propagated. Therefore, to the task in this study, we should consider more information such as the opinionatedness and textual quality of tweets.

6.5 Comparison with Humans

Finally, using the **Best** ranking model for finding propagated opinions in Twitter, we turn back to see human experiment in Section 6.1. We use our ranking

Table 5. Retweetability is a baseline. A significant improvement with $^\triangle$ and $^\blacktriangle$ (for p < 0.05 and p < 0.01 respectively).

	MAP
Retweetability	0.0936
TOR+Retweetability+Textural Quality (Best)	0.1992$^\blacktriangle$

model to judge which tweets presented are more likely to be propagated opinions. This model achieved an accuracy of 71%, which is slightly lower than human subjects (average 72%), but not significantly different from either subject at p=0.05. This result shows that for the task of finding propagated opinion in Twitter our approach is able to do as well as humans.

7   Conclusion

In this paper we study the task aiming at finding propagated opinions in Twitter. A set of features, including the retweetability, opinionatedness and textural quality of a tweet, are developed and integrated into learning to rank model for solving this task. The experimental results show these features are effective for finding propagated opinions in Twitter. Moreover, our best ranking model integrating all features is significantly better than the start-of-the-art TOR baseline and a BM25 baseline. Finally, we are encouraged by the performance of our ranking model, which can achieve the human subjects' ability as well, in identifying the propagated opinions in Twitter.

Acknowledgements. We would like to thank Zheng Wang for tagging the data. This research is supported by the National Natural Science Foundation of China (Grant No. 61170156 and 61202337).

References

1. Bollen, J., Mao, H., Zeng, X.-J.: Twitter mood predicts the stock market. J. Comput. Science 2(1), 1–8 (2011)
2. Duan, Y., Jiang, L., Qin, T., Zhou, M., Shum, H.Y.: An empirical study on learning to rank of tweets. In: Proceedings of the 23rd International Conference on Computational Linguistics, COLING 2010, pp. 295–303. Association for Computational Linguistics, Stroudsburg (2010)
3. Gimpel, K., Schneider, N., O'Connor, B., Das, D., Mills, D., Eisenstein, J., Heilman, M., Yogatama, D., Flanigan, J., Smith, N.A.: Part-of-speech tagging for twitter: annotation, features, and experiments. In: Proceedings of the 49th Annual Meeting of the Association for Computational Linguistics: Human Language Technologies: Short Papers, HLT 2011, vol. 2, pp. 42–47. Association for Computational Linguistics, Stroudsburg (2011)
4. Hong, L., Dan, O., Davison, B.D.: Predicting popular messages in twitter. In: Proceedings of the 20th International Conference Companion on World Wide Web, WWW 2011, pp. 57–58. ACM, New York (2011)

5. Jansen, B.J., Zhang, M., Sobel, K., Chowdury, A.: Twitter power: Tweets as electronic word of mouth. J. Am. Soc. Inf. Sci. Technol. 60(11), 2169–2188 (2009)
6. Joachims, T.: Making large scale svm learning practical (1999)
7. Kim, S.M., Pantel, P., Chklovski, T., Pennacchiotti, M.: Automatically assessing review helpfulness. In: Proceedings of the 2006 Conference on Empirical Methods in Natural Language Processing, pp. 423–430. Association for Computational Linguistics (2006)
8. Liu, P., Tang, J., Wang, T.: Information current in twitter: which brings hot events to the world. In: Proceedings of the 22nd International Conference on World Wide Web Companion, pp. 111–112. International World Wide Web Conferences Steering Committee (2013)
9. Liu, T.Y.: Learning to rank for information retrieval. Found. Trends Inf. Retr. 3(3), 225–331 (2009)
10. Liu, Y., Huang, X., An, A., Yu, X.: Modeling and predicting the helpfulness of online reviews. In: ICDM, pp. 443–452 (2008)
11. Luo, Z., Osborne, M., Petrovic, S., Wang, T.: Improving twitter retrieval by exploiting structural information. In: AAAI 2012: Proceedings of the Twenty-Sixth AAAI (2012)
12. Luo, Z., Osborne, M., Tang, J., Wang, T.: Who will retweet me? finding retweeters in twitter. In: Proceedings of the 36th International ACM SIGIR Conference on Research and Development in Information Retrieval. ACM (2013)
13. Luo, Z., Osborne, M., Wang, T.: Opinion retrieval in twitter. In: ICWSM (2012)
14. O'Connor, B., Balasubramanyan, R., Routledge, B.R., Smith, N.A.: From tweets to polls: Linking text sentiment to public opinion time series. In: ICWSM (2010)
15. Pang, B., Lee, L.: Opinion mining and sentiment analysis. Found. Trends Inf. Retr. 2(1-2), 1–135 (2008)
16. Petrovic, S., Osborne, M., Lavrenko, V.: Rt to win! predicting message propagation in twitter. In: ICWSM (2011)
17. Stieglitz, S., Dang-Xuan, L.: Political communication and influence through microblogging-an empirical analysis of sentiment in twitter messages and retweet behavior. In: HICSS, pp. 3500–3509 (2012)
18. Zhang, W., Yu, C., Meng, W.: Opinion retrieval from blogs. In: Proceedings of the Sixteenth ACM Conference on Conference on Information and Knowledge Management, CIKM 2007, pp. 831–840. ACM, New York (2007)
19. Zhang, Z., Varadarajan, B.: Utility scoring of product reviews. In: Proceedings of the 15th ACM International Conference on Information and Knowledge Management, pp. 51–57. ACM (2006)

Diversifying Tag Selection Result for Tag Clouds by Enhancing both Coverage and Dissimilarity

Meiling Wang[1,2], Xiang Zhou[1], Qiuming Tao[1], Wei Wu[1,2], and Chen Zhao[1]

[1] Institute of Software, Chinese Academy of Sciences, Beijing, China
[2] Graduate University, Chinese Academy of Sciences, Beijing, China
{meiling,zhouxiang,wuwei}@nfs.iscas.ac.cn
{qiuming,zhaochen}@iscas.ac.cn

Abstract. Tag cloud has been a popular facility used by social sites for online resource summarization and navigation. Tag selection, which aims to select a limited number of representative tags from a large set of tags, is the core task for creating tag clouds. Diversity of tag selection result is an important factor that affects user satisfaction. Information coverage and item dissimilarity are two major perspectives for exploring the concept of diversity, while existing tag selection approaches usually consider diversification from single perspective. In this paper, we propose a new approach for diversifying tag selection result, which takes into account both information coverage and tag dissimilarity. We design two sub-objective functions about information coverage and tag dissimilarity, respectively, and construct an objective function as a convex combination of the two sub-objective ones. We also give out a greedy algorithm that can well approximate the objective function. We conduct experiments on 17 datasets extracted from the website of CiteULike to compare our approach with existing ones. The experiment results show that our approach can achieve promising performance of diversification.

Keywords: Tag Cloud, Tag Selection, Result Diversification, Coverage, Dissimilarity, Submodularity, Greedy Algorithm.

1 Introduction

Tagging has become a common feature of most social sites. In these sites, users are allowed to annotate information resources (e.g., texts, pictures, and videos) with free-form tags, and meanwhile they could explore resource space through tags contributed by all users. Tag cloud is one major facility to help users access resources through tags and has been applied by many social sites, such as Flickr, CiteULike, and Delicious. A tag cloud is a visual depiction of a limited set of tags, which summarizes a group of resources. To create a tag cloud, one key step is to select representative tags from all the tags associated to resources [1][2][3]. In this paper we concentrate on tag selection problem for tag clouds.

Tag clouds convey information about resources mainly through their selected tags, and thus to make tag clouds more informative, it is necessary to diversify

X. Lin et al. (Eds.): WISE 2013, Part II, LNCS 8181, pp. 29–42, 2013.
© Springer-Verlag Berlin Heidelberg 2013

tag selection result. Result diversification has been much investigated in multiple domains, and in the related literature it is common to introduce diversity by constructing result sets that contain dissimilar items, as in [4][5][6][7], or by constructing result sets that cover different information categories, as in [8][9]. Two perspectives above could be adapted to diversify tag selection result. Actually tag selection approaches considering tag dissimilarity have been studied in [1][2]. Besides, since one resource could constitute one information nugget, tag selection approaches that aim to cover more information nuggets have been proposed in [2][3]. However, it is more interesting to us whether it is feasible to combine both of the two perspectives for diversifying tag selection result because when considering diversification from single perspective, the achieved tag selection result may be unsatisfactory on the other perspective.

In this paper, we treat tag selection result diversification as an optimization problem of maximizing a certain objective function subject to cardinality constraints. The objective function is defined as the convex combination of two sub-objective functions: one quantifies information coverage of any set of tags associated to resources, and the other is tightly related to tag dissimilarity. It is interesting that our objective function satisfies properties of monotonicity and submodularity (to be introduced in Sect. 3), and as a result we can use a greedy algorithm to approximately solve the optimization problem, and the tag set obtained is guaranteed to be not far from the best possible solutions.

To evaluate our approach, we conduct an experimental study based on 17 datasets that are extracted from the website of CiteULike and compare our approach with existing ones in terms of evaluation metrics about information coverage and tag dissimilarity. The experiment results demonstrate that our approach performs well in both information coverage and tag dissimilarity.

The rest of the paper is organized as follows: in Sect. 2, related work is introduced; in Sect. 3, some preliminary notations are introduced; in Sect. 4, our approach for diversifying tag selection result is presented; in Sect. 5, results of our experiments are analyzed; and finally in Sect. 6, conclusion is drawn.

2   Related Work

Result diversification has been extensively studied in different domains, such as web search [8][10][11][12], databases [13][14], text summarization [15][9][16], and recommendation [7][17]. General issues of result diversification are investigated in [4][5][18][6].

There are many tag selection approaches for tag clouds currently, and they consider result diversification more or less:

- The earliest tag selection approach (hereinafter referred to as POP) selects tags according to their popularity, which are the numbers of resources they annotate. The result of POP often contains many similar tags and covers few aspects of resources.
- Hassan-Montero and Herrero-Solana [1] propose a tag selection approach (hereinafter referred to as USE) that aims to decrease resource overlap among

different tags. USE selects tags based on the concept of tag usefulness, which emphasizes the "discrimination value" of tags for those resources that are annotated with few different tags.

- Venetis et al. [3] propose a tag selection approach (hereinafter referred to as COV) based on maximum coverage problem in combinatorial optimization. COV selects one tag each time based on the numbers of resources that are annotated with the candidate tags and previously uncovered.
- Skoutas and Alrifai [2] propose two tag selection approaches that consider result diversification explicitly. One approach (hereinafter referred to as POP+DIS) computes utility score of a tag as the convex combination of its popularity and its minimum distance from currently selected tags. And the other one (hereinafter referred to as NOV) computes utility score of a tag as its "novelty" to currently selected tags. Moreover, NOV obtains the same selection result as the approach COV when the emphasis on novelty is maximized. In this paper, we regard NOV and COV as same category of approaches, which emphasize on improvement of information coverage.

3   Preliminaries

3.1   Social Tagging System

A social tagging system [19][20] is typically modeled as a tripartite hypergraph $G = (V, E)$ with $V = U \dot{\cup} T \dot{\cup} R$ and $E \subseteq U \times T \times R$, where U, T, and R denote user set, tag set, and resource set, respectively, and an edge $(u, t, r) \in E$ represents that user u annotates resource r with tag t. Further, for each $t \in T$ and each $r \in R$, let

- $R(t) = \{r \in R | \exists u(u \in U \wedge (u, t, r) \in E)\}$,
- $T(r) = \{t \in T | \exists u(u \in U \wedge (u, t, r) \in E)\}$,
- and $U(r, t) = \{u \in U | (u, t, r) \in E\}$.

3.2   Submodularity

Let X be a finite set. A set function $\mathcal{F} : 2^X \to \mathbb{R}$ is a *submodular* function if

$$\mathcal{F}(A \cup \{x\}) - \mathcal{F}(A) \geq \mathcal{F}(B \cup \{x\}) - \mathcal{F}(B),$$

for every $A, B \subseteq X$ with $A \subseteq B$ and every $x \in X \backslash B$. $\mathcal{F}(A \cup \{x\}) - \mathcal{F}(A)$ is called *marginal returns* of x given A. Submodular functions have diminishing marginal returns.

A set function $\mathcal{G} : 2^X \to \mathbb{R}$ is *monotone* if

$$\mathcal{G}(A) \leq \mathcal{G}(B) \text{ for all } A, B \subseteq X \text{ with } A \subseteq B.$$

It is not hard to prove that monotone submodular functions have the following two properties:

Lemma 1. *Let X be a finite set and $\mathcal{F} : 2^X \to \mathbb{R}$ be a monotone submodular function and $\phi : \mathbb{R} \to \mathbb{R}$ be nondecreasing concave, then the function $\mathcal{G} : 2^X \to \mathbb{R}$, defined as $\mathcal{G}(A) = \phi(\mathcal{F}(A))$, is a monotone submodular function.*

Lemma 2. *Let X be a finite set and $\alpha_1, \alpha_2, \ldots, \alpha_n$ be nonnegative numbers and $\mathcal{F}_1, \mathcal{F}_2, \ldots, \mathcal{F}_n : 2^X \to \mathbb{R}$ be monotone submodular functions, then the function $\mathcal{G} : 2^X \to \mathbb{R}$, defined as $\mathcal{G}(A) = \sum_{i=1}^{n} \alpha_i \cdot \mathcal{F}_i(A)$, is also a monotone submodular function.*

4   Our Approach

Let $G = (U \dot{\cup} T \dot{\cup} R, E)$ be a social tagging system, and let $\mathcal{D} : 2^T \to \mathbb{R}$ be an objective function measuring "diversity" of subsets of T. Given a positive integer k, the goal of *tag selection result diversification problem* is to find a subset $S \subseteq T$ that:

$$\text{maximizes } \mathcal{D}(S) \quad \text{subject to} \quad |S| = k. \tag{1}$$

4.1   Objective Function

Our objective function incorporates both information coverage and tag dissimilarity. Since information coverage and tag dissimilarity may compete with each other, our objective function will be in the following form:

$$\mathcal{D}(S) = \lambda \cdot \mathcal{H}(S) + (1 - \lambda) \cdot \mathcal{L}(S), \tag{2}$$

where $0 \le \lambda \le 1$ and $\mathcal{H} : 2^T \to \mathbb{R}$ is a sub-objective function about information coverage and $\mathcal{L} : 2^T \to \mathbb{R}$ is a sub-objective function about tag dissimilarity. Next are the concrete construction of the two sub-objective functions.

Coverage Function. In social tagging systems, each resource could be seen as an information nugget, and thus it is natural to quantify information coverage of tags in terms of resources they annotate. Further, since different tags may annotate some same resources, it is intuitive that as a tag set increases, information gained from the tag set can be increased, but at a decreasing rate. This indicates that coverage function should be monotone and submodular. Here, we define $\mathcal{H}$ as follows:

$$\mathcal{H}(S) = \sqrt{\left| \bigcup_{t_i \in S} R(t_i) \right|} \quad \text{for each } S \subseteq T, \tag{3}$$

where square root function is used to avoid too large value of $\mathcal{H}$ when compared with $\mathcal{L}$.

$\mathcal{H}$ has a nice property:

Proposition 1. *$\mathcal{H}$ is a monotone submodular function.*

Proof. $\mathcal{G}(S) = \left| \bigcup_{t_i \in S} R(t_i) \right|$ is monotone and submodular according to definition. Besides, $\sqrt{x}$ is nondecreasing concave according to definition. Thus $\mathcal{H}(S) = \sqrt{\mathcal{G}(S)}$ is a monotone submodular function according to Lemma 1.   □

Remark: In practice, the square root function in the definition of $\mathcal{H}$ can be replaced with some other nondecreasing concave functions (e.g., *log* function) to obtain similar effect.

Dissimilarity Function. Since the role of tags is to annotate resources, it is natural to identify a tag with its set of annotated resources. Thus dissimilarity of each pair of tags depends on the overlap of resources they annotate, and further the design of dissimilarity functions should aim to reduce resource overlap among different tags.

There exist several concepts closely related to reduction of resource overlap among different tags: the concept of distance is adopted in the approach POP+DIS [2] and the concept of tag usefulness is introduced in the approach USE [1]. Considering that tag usefulness expresses more comprehensive information about tags, we draw on this concept to design our dissimilarity function.

In [1], the usefulness of a tag t is defined as

$$\sum_{r_j \in R(t)} d(t, r_j), \tag{4}$$

where $d(t, r_j) = \dfrac{log(|U(r_j, t)|)}{|T(r_j)|^2}$:

- $log(|U(r_j, t)|)$ measures the *representation value* of t for r_j, and the *log* function is used to diminish the effect of high frequency;
- $|T(r_j)|^2$ measures the potential *discrimination value* of t for r_j, and the square function is used to increase the effect of low frequency.

The objective function of USE could be regarded as

$$\mathcal{M}(S) = \sum_{t_i \subset S} \sum_{r_j \in R(t_i)} d(t_i, r_j) = \sum_{r_j \in R} \sum_{t_i \subset S \cap T(r_j)} d(t_i, r_j) \quad \text{for each } S \subseteq T. \tag{5}$$

Since according to formula (4) and the definition of $d(t, r_j)$, tags could get high usefulness value from the resources that are annotated with few tags, and through selecting these tags of high usefulness, USE could decrease resource overlap of the result set.

Making use of quantity $d(t, r_j)$, we define our dissimilarity function $\mathcal{L}$ as follows:

$$\mathcal{L}(S) = \sum_{r_j \in R} \sqrt{\sum_{t_i \in S \cap T(r_j)} d(t_i, r_j)^2} \quad \text{for each } S \subseteq T. \tag{6}$$

The difference between function $\mathcal{L}$ and function $\mathcal{M}$ lies in the square root function and the square of $d(t_i, r_j)$:

- The square root function is significant for $\mathcal{L}$ to achieve the effect that the usefulness returns of a candidate tag (See the usefulness returns of a tag in Sect. 4.3) depends on currently selected tags.
- The square of $d(t_i, r_j)$ is used to make $\mathcal{L}$ comparable with $\mathcal{M}$ since $\mathcal{L}$ is the summation of some square roots.

$\mathcal{L}$ also satisfies the property of monotone submodularity:

Proposition 2. $\mathcal{L}$ *is a monotone submodular function.*

Proof. $\mathcal{G}_{r_j}(S) = \sum\limits_{t_i \in S \cap T(r_j)} d(t_i, r_j)^2$ is monotone submodular according to the definition, $\sqrt{\mathcal{G}_{r_j}(S)}$ is monotone submodular according to Lemma 1, and $\mathcal{L}(S) = \sum\limits_{r_j \in R} \sqrt{\mathcal{G}_{r_j}(S)}$ is a monotone submodular function according to Lemma 2.   □

4.2   Greedy Algorithm and Approximation Ratio

The problem (1) (defined at the beginning of Sect. 4) is NP-hard in general. Here we approximately solve it with a greedy algorithm, i.e., Algorithm 1. The algorithm starts with empty set and selects the members of result set through k iterations. In each iteration, the algorithm firstly selects a tag that has maximal marginal returns given the set of currently selected tags, and then constructs a new set of selected tags for the next iteration.

Algorithm 1.

Input: T, positive integer k
Output: $S \subseteq T$, $|S| = k$
1: $S = \emptyset$
2: **while** $|S| < k$ **do**
3:     $t^* = \underset{t \in T \setminus S}{\mathrm{argmax}}(\mathcal{D}(S \cup \{t\}) - \mathcal{D}(S))$
4:     $S = S \cup \{t^*\}$
5: **end while**
6: **return** S

Algorithm 1 has strong theoretical guarantees as follows:

Theorem 1 (Nemhauser, Wolsey, and Fisher(1978)[21]). *If $\mathcal{D}$ is monotone submodular and $\mathcal{D}(\emptyset) = 0$, then Algorithm 1 is guaranteed to find a set $\hat{S}$ with $|\hat{S}| = k$, such that $\mathcal{D}(\hat{S}) \geq (1 - 1/e)\mathcal{D}(S^*)$, where $S^* \in \underset{|S|=k}{\mathrm{argmax}} \mathcal{D}(S)$.*

The objective function of our approach is defined as $\mathcal{D}(S) = \lambda \cdot \mathcal{H}(S) + (1 - \lambda) \cdot \mathcal{L}(S)$, where $0 \leq \lambda \leq 1$, and thus $\mathcal{D}$ is monotone and submodular according to Proposition 1, Proposition 2, and Lemma 2. Moreover, $\mathcal{D}(\emptyset) = 0$ according to definition. Therefore, our approach obtains the theoretical approximation guarantees stated in Theorem 1, i.e., the tag selection result generated with Algorithm 1 achieves an objective value that is at least a factor $(1 - 1/e)(\approx 0.63)$ of the optimal score.

4.3   Analysis of λ-Cases

The performance of our approach completely depends on the two combined parts in our objective function and Algorithm 1.

When $\lambda = 1$, which means our objective function contains only the coverage function, our approach achieves the same performance as the approach COV because the objective function of COV is different from $\mathcal{H}$ only in the square root function and the greedy algorithm used in COV is an instance of Algorithm 1.

When $\lambda = 0$, which means our objective function contains only the dissimilarity function, our approach could be more effective than the approach USE in reducing resource overlap among different tags. In essence, two approaches adopt the same tag selection procedure as Algorithm 1, but with $\mathcal{D}$ replaced with $\mathcal{M}$ and $\mathcal{L}$, respectively. They select the same tag firstly, but in the following iterations, two approaches select tags based on different *usefulness returns* of tags. In the approach USE, the *usefulness returns of a tag t given S*, is equal to

$$\mathcal{M}(S \cup \{t\}) - \mathcal{M}(S) = \sum_{r_j \in R(t)} d(t, r_j),$$

and in our approach, the *usefulness returns of a tag t given S*, is equal to

$$
\begin{aligned}
&\mathcal{L}(S \cup \{t\}) - \mathcal{L}(S) \\
&= \sum_{r_j \in R} \left(\sqrt{\sum_{t_i \in (S \cup \{t\}) \cap T(r_j)} d(t_i, r_j)^2} - \sqrt{\sum_{t_i \in S \cap T(r_j)} d(t_i, r_j)^2} \right) \\
&= \sum_{r_j \in R(t)} \left(\sqrt{\sum_{t_i \in S \cap T(r_j)} d(t_i, r_j)^2 + d(t, r_j)^2} - \sqrt{\sum_{t_i \in S \cap T(r_j)} d(t_i, r_j)^2} \right) \\
&= \sum_{r_j \in R(t)} \frac{d(t, r_j)^2}{\sqrt{\sum_{t_i \in S \cap T(r_j)} d(t_i, r_j)^2 + d(t, r_j)^2} + \sqrt{\sum_{t_i \in S \cap T(r_j)} d(t_i, r_j)^2}}.
\end{aligned}
$$

We can see that the usefulness returns of a candidate tag in our approach depends on currently selected tags while it is not so in the approach USE, which means our approach considers resource overlap between the candidate tag and tags currently selected during tag selection while USE cannot. Thus our approach could be more effective than USE in reducing the degree of resource overlap.

When $0 < \lambda < 1$, both information coverage and tag dissimilarity, which are given the weights of λ and $1 - \lambda$, respectively, determine the returns of a tag on $\mathcal{D}$ and the selection of each tag in Algorithm 1. For instance, when $\lambda = 0.5$, information coverage and tag dissimilarity are given the same weight and contribute equally in selection of tags. Generally speaking, when λ range between 0 and 1, the performance of our approach varies between the two cases previously discussed.

5   Experiments

We conduct an experimental evaluation of our approach using 17 datasets that are extracted from the tagging dataset of the website of CiteULike. We first describe how we extract the datasets, then present the evaluation metrics we use, and finally analyze the results of the experiments.

5.1   Data Preparation

The whole tagging dataset of CiteULike consists of 17,481,632 annotations and each piece of annotation contains a user id, a tag, a resource id, and some other information. In total, there are 118,027 distinct users, 761,674 distinct tags, and 3,817,796 distinct resources in the dataset.

We first extract the top 35 popular non-noisy tags (e.g., "human", "animals", and "computer"), and we treat each one of the extracted tags as a topic [22]. For each topic, we first extract all the resources that are tagged with it and then extract all the annotations associated with these resources. Each set of the extracted annotations is treated as a dataset about the corresponding topic, and hence we obtain 35 datasets. For each dataset, to make the tags more meaningful, we delete noisy tags and the tags that annotate less than 6 resources. Further, we remove datasets that are either too small or too large and finally get 17 datasets whose numbers of annotations, users, tags, and resources are on the same order of magnitude.

Table 1 describes some statistics about the 17 datasets. When we treat all the datasets as a whole, the distribution of tags across the resources related to them and the datasets they appear in can be described as Fig. 1(a) and Fig. 1(b), respectively. From Fig. 1, we can see that the majority of tags have a relatively low resource frequency and dataset frequency. We also observe that those tags with more abstract levels (e.g., "humans", "govt", and "adult") annotate the majority of resources and appear in almost all of the datasets.

Table 1. Dataset statistics

	Average	Maximum	Minimum	Total
Number of annotations	425,405	762,598	228,463	2,915,759
Number of users	3466	7163	1054	15,079
Number of tags	5880	8678	4277	19,205
Number of resources	17,588	34,125	11,580	128,976

5.2   Evaluation Metrics

Currently, there does not exist a single and unified metric for evaluating the diversity of tag selection result. In the related literature[1][2][3], several metrics have been proposed for evaluating tag selection result, such as coverage, overlap, cohesiveness, and relevance. Among all these metrics, coverage and overlap reflect

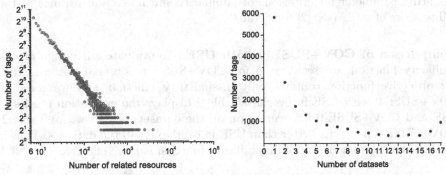

(a) Distribution of tags across resources   (b) Distribution of tags across datasets

Fig. 1. Distribution of tags across resources and datasets

information coverage and tag dissimilarity better. In this paper, we use them for evaluating the diversification of tag selection approaches.

Coverage of a tag selection result S measures the proportion of resources covered by S and is defined in [2] as

$$coverage(S) = \frac{\left| \bigcup_{t_i \in S} R(t_i) \right|}{|R|}. \tag{7}$$

According to our viewpoints about the information coverage of tag selection result, the higher $coverage(S)$ is, the more information nuggets S covers.

Overlap of S measures the degree of redundancy of S and is defined in [1] as

$$overlap(S) = \frac{\sum_{t_i, t_j \in S, t_i \neq t_j} J(R(t_i), R(t_j))}{|S| \cdot (|S| - 1)/2}, \tag{8}$$

where $J(R(t_i), R(t_j)) = \frac{|R(t_i) \cap R(t_j)|}{|R(t_i) \cup R(t_j)|}$, called *Jaccard similarity* of $R(t_i)$ and $R(t_j)$. According to our viewpoints about the tag dissimilarity of tag selection result, the lower $overlap(S)$ is, the more dissimilar the tags in S are.

5.3   Evaluation Results

We implement our approach described in Sect. 4 and refer to it as COV+SUSE. We also implement the existing approaches POP, USE, COV, and POP+DIS and compare COV+SUSE with them in both metrics of coverage and overlap.

In the implementation and experiments, we set k, the number of tags to be selected, as 30. For convenience, we refer to COV+SUSE as COV+SUSE-x when the weighting parameter λ is equal to x $(0 \leq x \leq 1)$. Similarly, we refer to POP+DIS as POP+DIS-x when λ is equal to x $(0 \leq x \leq 1)$, where λ is the weighting parameter for criterions of popularity and minimum distance in the utility score of a tag (see [2] for details).

Comparison of COV+SUSE-0 with USE. To evaluate our design of dissimilarity function, we set λ as 0 for COV+SUSE to observe the case that our objective function contains only dissimilarity function, and we compare COV+SUSE-0 with USE in overlap. Table 2 displays the evaluation results of USE and COV+SUSE-0 for overlap on all the datasets. As shown in Table 2, COV+SUSE-0 performs better than USE in overlap on all the datasets. Thus it can be seen that our design of dissimilarity function is more effective.

Table 2. Overlap evaluation of USE and COV+SUSE-0

Approach	Dataset ID								
	#1	#2	#3	#4	#5	#6	#7	#8	#9
USE	0.0190	0.0285	0.0261	0.0236	0.0277	0.0375	0.0248	0.0186	0.0506
COV+SUSE-0	0.0181	0.0218	0.0104	0.0213	0.0179	0.0132	0.0213	0.0183	0.0221

Approach	Dataset ID								Average
	#10	#11	#12	#13	#14	#15	#16	#17	
USE	0.0326	0.0271	0.0294	0.0272	0.0271	0.0457	0.0292	0.0229	0.0293
COV+SUSE-0	0.0232	0.0242	0.0250	0.0232	0.0243	0.0162	0.0231	0.0154	0.0199

Comparison of COV+SUSE-0.5 with POP, USE, COV and POP+DIS-0.5. We set λ as 0.5 for both POP+DIS and COV+SUSE. Table 3 and Table 4 display the evaluation results of POP, USE, COV, POP+DIS-0.5, and COV+SUSE-0.5 for coverage and overlap on all the datasets, respectively.

As shown in Table 3 and Table 4:

- POP performs the worst in overlap on all the datasets. Moreover, POP performs worse than COV and COV+SUSE-0.5 in coverage on all the datasets and performs worse than POP+DIS-0.5 in coverage on almost all of the datasets.
- USE outperforms POP in overlap on all the datasets and outperforms COV, POP+DIS-0.5, and COV+SUSE-0.5 in overlap on almost all of the datasets, but it achieves poor performance in coverage.
- COV achieves the best performance in coverage on all the datasets, but it performs worse than COV+SUSE-0.5 in overlap on all the datasets and performs worse than USE in overlap on almost all of the datasets.
- POP+DIS-0.5 performs worse than USE, COV and COV+SUSE-0.5 in overlap on almost all of the datasets. Meanwhile, POP+DIS-0.5 outperforms POP and USE in coverage on almost all of the datasets, but it achieves worse coverage than COV and COV+SUSE-0.5.
- COV+SUSE-0.5 is closer to COV than all of the other approaches in coverage on all the datasets. Moreover, COV+SUSE-0.5 outperforms POP and COV in overlap on all the datasets and outperforms POP+DIS-0.5 in overlap on almost all of the datasets.

Table 3. Coverage evaluation of five approaches

Dataset ID	Approach				
	POP	USE	COV	POP+DIS-0.5	COV+SUSE-0.5
#1	0.8430	0.6387	0.9140	0.8475	0.8976
#2	0.8716	0.6390	0.9687	0.9089	0.9571
#3	0.9681	0.9423	0.9909	0.9675	0.9832
#4	0.9323	0.7816	0.9807	0.9579	0.9621
#5	0.9165	0.5413	0.9697	0.9208	0.9562
#6	0.9651	0.6457	0.9846	0.9672	0.9754
#7	0.8763	0.4783	0.9619	0.9154	0.9443
#8	0.8525	0.7899	0.9610	0.8891	0.9400
#9	0.9468	0.7488	0.9768	0.9504	0.9757
#10	0.9720	0.9155	0.9881	0.9751	0.9787
#11	0.9375	0.7028	0.9765	0.9454	0.9601
#12	0.9444	0.3201	0.9853	0.9721	0.9764
#13	0.8345	0.5827	0.9324	0.8714	0.9005
#14	0.8091	0.8188	0.9007	0.8494	0.8846
#15	0.9745	0.9865	0.9980	0.9803	0.9947
#16	0.9845	0.8769	0.9955	0.9852	0.9897
#17	0.8372	0.4912	0.9212	0.8592	0.8957
Average	0.9098	0.7000	0.9651	0.9272	0.9513

Table 4. Overlap evaluation of five approaches

Dataset ID	Approach				
	POP	USE	COV	POP+DIS-0.5	COV+SUSE-0.5
#1	0.1054	0.0190	0.0397	0.0223	0.0299
#2	0.1460	0.0285	0.0347	0.0761	0.0299
#3	0.1547	0.0261	0.0462	0.1060	0.0357
#4	0.1346	0.0236	0.0668	0.0741	0.0375
#5	0.1046	0.0277	0.0381	0.0308	0.0285
#6	0.1499	0.0375	0.0658	0.0941	0.0421
#7	0.0948	0.0248	0.0267	0.0362	0.0266
#8	0.1288	0.0186	0.0376	0.0446	0.0265
#9	0.1060	0.0506	0.0570	0.0346	0.0506
#10	0.1457	0.0326	0.0783	0.0982	0.0430
#11	0.1415	0.0271	0.0570	0.0637	0.0410
#12	0.1380	0.0294	0.0503	0.0705	0.0388
#13	0.1268	0.0272	0.0372	0.0379	0.0265
#14	0.1222	0.0271	0.0499	0.0350	0.0332
#15	0.1443	0.0457	0.0308	0.0813	0.0243
#16	0.1587	0.0292	0.0763	0.1234	0.0472
#17	0.1071	0.0229	0.0427	0.0256	0.0292
Average	0.1299	0.0293	0.0491	0.0620	0.0347

With respect to coverage, COV performs the best, COV+SUSE-0.5 is close to COV, and POP and USE performs worse. With respect to overlap, USE and COV+SUSE-0.5 performs better, and POP+DIS-0.5 and POP performs worse. We summarize the results of the comparison among the five approaches as Table 5 and from the table, we can see that COV+SUSE-0.5 performs relatively better in both coverage and overlap.

Table 5. Summary of comparison among five approaches

Metrics	POP	USE	COV	POP+DIS-0.5	COV+SUSE-0.5
Coverage	bad	bad	good	middle	good
Overlap	bad	good	middle	bad	good

Comparison of COV+SUSE with POP+DIS as λ varies. The performance of COV+SUSE and POP+DIS is affected by the values of λ. We let λ range from 0 to 1 at intervals of 0.05. Fig. 2 displays the coverage and overlap of COV+SUSE and POP+DIS for increasing λ on the datasets #1 and #2.

As shown in Fig. 2(a), when λ increases from 0 to 1, the coverage of POP+DIS increases slowly at the beginning and starts to decrease when λ is close to 1; in contrast, the coverage of COV+SUSE increases quickly to a high level at the beginning and then gradually reach the maximal value. As shown in Fig. 2(b), when λ increases from 0 to 1, both the overlap of POP+DIS and the overlap of COV+SUSE increase from a minimal value to a maximal value overall, and the overlap of COV+SUSE stays at a lower level throughout. From Fig. 2(a) and Fig. 2(b) we can see that, compared with POP+DIS, COV+SUSE can simultaneously achieve better performance in coverage and overlap for most values of λ. We also can see that the performance of COV+SUSE is less sensitive to the value of λ, which is favorable to application of this approach in practice.

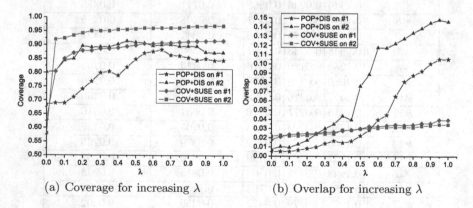

(a) Coverage for increasing λ   (b) Overlap for increasing λ

Fig. 2. Coverage and overlap of COV+SUSE and POP+DIS for increasing λ on the datasets #1 and #2

6 Conclusion

We focus on diversifying tag selection result by enhancing both information coverage and tag dissimilarity in this paper. We regard this diversification task as a bi-criterion optimization problem, in which we design sub-objective functions about information coverage and tag dissimilarity, respectively. We also approximately solve the optimization problem with a greedy algorithm, which generates results that are not far from the best possible solutions. We evaluate the diversification performance of our approach experimentally and the results show that our approach performs well in terms of both information coverage and tag dissimilarity.

In the future, we plan to evaluate our approach on more datasets, such as the datasets from the website of Delicious. We also plan to study more comprehensive metrics for evaluating the diversity of tag selection result.

Acknowledgement. This work is supported by the National Natural Science Foundation of China under Grant No.61100067, the National Science and Technology Major Project under Grant No.2012ZX01039-004, and the Special Funds for Strategic and Prospective Sci & Tech Program of Chinese Academy of Sciences under Grant No.XDA06010600.

References

1. Hassan-Montero, Y., Herrero-Solana, V.: Improving tag-clouds as visual information retrieval interfaces. In: Proceedings of International Conference on Multidisciplinary Information Sciences and Technologies 2006, pp. 25–28 (2006)
2. Skoutas, D., Alrifai, M.: Tag clouds revisited. In: CIKM 2011, pp. 221–230 (2011)
3. Venetis, P., Koutrika, G., Garcia-Molina, H.: On the selection of tags for tag clouds. In: WSDM 2011, pp. 835–844 (2011)
4. Borodin, A., Lee, H.C., Ye, Y.: Max-sum diversification, monotone submodular functions and dynamic updates. In: PODS 2012, pp. 155–166 (2012)
5. Carbonell, J., Goldstein, J.: The use of mmr, diversity-based reranking for reordering documents and producing summaries. In: SIGIR 1998, pp. 335–336 (1998)
6. Gollapudi, S., Sharma, A.: An axiomatic approach for result diversification. In: WWW 2009, pp. 381–390 (2009)
7. Yu, C., Lakshmanan, L., Amer-Yahia, S.: It takes variety to make a world: diversification in recommender systems. In: EDBT 2009, pp. 368–378 (2009)
8. Agrawal, R., Gollapudi, S., Halverson, A., Ieong, S.: Diversifying search results. In: WSDM 2009, pp. 5–14 (2009)
9. Liu, K., Terzi, E., Grandison, T.: Highlighting diverse concepts in documents. In: SDM 2009, 545–556 (2009)
10. Bansal, N., Jain, K., Kazeykina, A., Naor, J(S.): Approximation algorithms for diversified search ranking. In: Abramsky, S., Gavoille, C., Kirchner, C., Meyer auf der Heide, F., Spirakis, P.G. (eds.) ICALP 2010. LNCS, vol. 6199, pp. 273–284. Springer, Heidelberg (2010)
11. Clarke, C., Kolla, M., Cormack, G., Vechtomova, O., Ashkan, A., Büttcher, S., MacKinnon, I.: Novelty and diversity in information retrieval evaluation. In: SIGIR 2008, 659–666 (2008)

12. Zhai, C., Cohen, W., Lafferty, J.: Beyond independent relevance: methods and evaluation metrics for subtopic retrieval. In: SIGIR 2003, 10–17 (2003)
13. Demidova, E., Fankhauser, P., Zhou, X., Nejdl, W.: Divq: diversification for keyword search over structured databases. In: SIGIR 2010, 331–338 (2010)
14. Fraternali, P., Martinenghi, D., Tagliasacchi, M.: Top-k bounded diversification. In: SIGMOD 2012, 421–432 (2012)
15. Lin, H., Bilmes, J.: A class of submodular functions for document summarization. In: ACL-HLT 2011, 510–520 (2011)
16. Tsaparas, P., Ntoulas, A., Terzi, E.: Selecting a comprehensive set of reviews. In: KDD 2011, 168–176 (2011)
17. Hurley, N., Zhang, M.: Novelty and diversity in top-n recommendation–analysis and evaluation. TOIT 10(4), 14 (2011)
18. Drosou, M., Pitoura, E.: Search result diversification. SIGMOD Record 39(1), 41–47 (2010)
19. Halpin, H., Robu, V., Shepherd, H.: The complex dynamics of collaborative tagging. In: WWW 2007, pp. 211–220 (2007)
20. Mika, P.: Ontologies are us: A unified model of social networks and semantics. In: Gil, Y., Motta, E., Benjamins, V.R., Musen, M.A. (eds.) ISWC 2005. LNCS, vol. 3729, pp. 522–536. Springer, Heidelberg (2005)
21. Nemhauser, G., Wolsey, L., Fisher, M.: An analysis of approximations for maximizing submodular set functions. Mathematical Programming 14(1), 265–294 (1978)
22. Song, Y., Zhuang, Z., Li, H., Zhao, Q., Li, J., Lee, W., Giles, C.: Real-time automatic tag recommendation. In: SIGIR 2008, 515–522 (2008)

Community Detection
in Multi-relational Social Networks

Zhiang Wu[1], Wenpeng Yin[1], Jie Cao[1,*], Guandong Xu[2],
and Alfredo Cuzzocrea[3]

[1] Jiangsu Provincial Key Laboratory of E-Business,
Nanjing University of Finance and Economics, Nanjing, China
[2] Advanced Analytics Institute, University of Technology, Sydney, Australia
[3] Institute of High Performance Computing and Networking,
Italian National Research Council, Italy
{zawuster,mr.yinwenpeng}@gmail.com, Jie.Cao@njue.edu.cn,
Guandong.Xu@uts.edu.au, cuzzocrea@icar.cnr.it

Abstract. Multi-relational networks are ubiquitous in many fields such
as bibliography, twitter, and healthcare. There have been many studies in
the literature targeting at discovering communities from social networks.
However, most of them have focused on single-relational networks. A
hint of methods detected communities from multi-relational networks
by converting them to single-relational networks first. Nevertheless, they
commonly assumed different relations were independent from each other,
which is obviously unreal to real-life cases. In this paper, we attempt
to address this challenge by introducing a novel co-ranking framework,
named *MutuRank*. It makes full use of the mutual influence between
relations and actors to transform the multi-relational network to the
single-relational network. We then present GMM-NK (Gaussian Mixture
Model with Neighbor Knowledge) based on local consistency principle
to enhance the performance of spectral clustering process in discovering
overlapping communities. Experimental results on both synthetic and
real-world data demonstrate the effectiveness of the proposed method.

Keywords: Social Networks, Community Detection, Multi-relational
Network, MutuRank, Gaussian Mixture Model.

1 Introduction

Community detection has become a fundamental yet difficult task ever since the
network science came into vogue. What is the nature of network communities?
So far, there is no standard answer to this question [23]. In general, actors in
a same community tend to interact with each other more frequently than with
those outside the community. The communities are also called *groups*, *clusters*,
cohesive subgroups or *modules* in different research fields [21].

---

* Corresponding author.

X. Lin et al. (Eds.): WISE 2013, Part II, LNCS 8181, pp. 43–56, 2013.

Although a large body of research efforts have been devoted to community detection [4,8,21], most of the existing methods are designed for *single-relational* networks. This kind of network is composed of a set of nodes (i.e., objects) connected by a set of edges (i.e., links) which represent relationships of a single type. Whereas, in many real-world situations, objects are usually associated with each other in multiple aspects. For example, in Twitter, users could be followers/followees of others, could retweet tweets of others, could produce topic relevant tweets with others and etc. Considering further the relationships among scholars in DBLP, we could treat co-authorship, citation and venue as distinct relation types between scholars.

In some literatures [19,20], multi-relational networks (a.k.a. *heterogeneous* or *multi-mode* networks) often contain more than one typed entities. However, the meaningful communities are still defined on the same typed entities. So, in this paper, we limit our scope to the multi-relational network containing multiple typed relations but one typed entities. Especially to deserve to be mentioned, the multi-relational network considered in this paper is a special case of the multi-mode network when only one typed entities are considered.

To date, general methods handling a multi-relational network, such as [1,14,20] and etc., first try to convert it to a single-relational network, and then employ existing methods for community detection. However, such approaches usually suppose that multiple typed relations are independent, which is not the case in real situations. Taking a look at the scholar circle, two researchers may have relevant research interest, co-author several papers, co-operate several projects, and even publish papers in the same conferences. How can we ignore an intuition that scholars with similar academic backgrounds are more likely to publish papers in same venues? How can we neglect the tendency that persons who co-direct a project are very likely to co-author some literature?

Motivated by that, in this work, we propose a novel co-ranking framework, *MutuRank*, to determine the weights of various relation types and objects simultaneously. The essential part of this framework lies in that it makes full use of the mutual influence between those relations and actors, with the aim of deriving equilibrium/stationary probability distributions as evaluation scores for actors and relations, respectively. To be specific, the mutual-feedback here means that: (i) the importance of a relation depends on the probability distribution of actors and the importance of other relations, i.e., a relation, selected by high-weight actors with high probabilities, deserves high weight itself; (ii) the importance of an actor depends on the probability distribution of relations and its neighbors' importance, i.e., an actor, linked by high-weight actors with strong and high-weight relations, deserves high-weight. Here, *strong relation* implies the intense closeness of two objects under that relation, and *high-weight relation* indicates that the relation type itself is very important. The probability distributions derived from such mutual-feedback are able to convey the intrinsic status of relations and objects more accurately.

We then combine the probability distributions of relations linearly to produce a single-relational network just as some typical literature did. Although any of the existing methods can be used to partition the network into crisp communities, this paper focus on discovering overlapping communities by which most real networks are characterized [11,22]. For instance, a scientist might belong to an academic community as well as some personal life communities (e.g., school, hobby, family). In this paper, we propose a novel soft clustering method named *GMM-NK* (Gaussian Mixture Model with Neighbor Knowledge), which originates from this inspiration: the probability of an object belonging to a community could be derived not only via its own gaussian value, but also from the probabilities of its near neighbors belonging to the same community. This idea is also in accordance with *the local consistency principle* of machine learning in which very related objects should have similar category attributes.

Finally, we perform experiments on simulated synthetic data as well as real-world DBLP dataset. Experiments results on both datasets validate the good performance of our proposed community detection algorithm.

The remainder of this paper is organized as follows. In Section 2, we present the related work. In Section 3, we introduce *MutuRank* for identification of relation distribution. In Section 4, we show how to mine communities using soft clustering method. Experimental results will be given in Section 5. We finally conclude this paper in Section 6.

2 Related Work

Multi-relational social network has attracted much attention in the few years. A great many studies attempted to integrate latent [18] or explicit [14,3,1] heterogeneous relations to form a single-relational network. In almost all of these studies, different relations are considered to be independent from each other. However, from a case study [16] conducted on a heterogeneous network consisting of about 300,000 game players with six different relations including three positive and three negative ones, a conclusion was drawn that positive links were highly reciprocal while negative links were not. It is somewhat consistent with our motivation that various relations interact with each other. Our *Mutu-Rank* is greatly inspired by the *MultiRank* framework presented in [9]. *MultiRank* employed a PageRank-like [10,24] random walk model to co-rank objects and relations in a multi-relational network. Whereas, *MultiRank* assumed that the probability for a node to select a relation is not only independent of the node's importance, but also independent of the network structure. This assumption is not strictly rational, and is remedied by our *MutuRank* model.

Community detection has been extensively studied [4]. So far, most of the existing methods can be classified into two main categories, in terms of whether or not explicit optimization objectives are being used. The methods with explicit optimization objectives typically consider the global topology of a network, and aim to optimize a criterion defined over a network partition. Some methods along this line include the Kernighan-Lin algorithm [7], stochastic block models [5], modularity optimization [8], and traditional clustering techniques [15] such as K-means, multi-dimensional scaling (MDS), and spectral clustering. On the other hand, the methods without using explicit optimization objectives discover communities based on predefined assumptions or heuristic rules. For example, Clique Percolation Method (CPM) [11] is based on the concept of k-clique, and a k-clique community is then defined as the union of all "adjacent" k-cliques, which by definition share k-1 nodes. To sum up, most of the methods using explicit optimization objectives often lead to crisp partitions, while our GMM-NK employs the spectral clustering to carry out soft community detection.

It is worthy of mentioning some work having the same target with this paper, i.e., discovering communities from multi-relational networks. Cai et al. [1] proposed a regression-based algorithm to learn the optimal relation weights, and then utilized threshold cut as the optimization objective for community detection. Tang et al. [20] proposed several integration strategies including network integration, utility integration, feature integration and partition integration for transforming the multi-relational network to single-relational network. However, they failed to consider the mutual influence between relations and actors.

3   Identification of Relation Distribution

In this section, we aim to introduce our *MutuRank* algorithm to identify the relation distribution, and thus show how to transform the multi-relational network into a single-relational graph.

Let's begin by introducing some notation conventions. Let $\mathcal{N}$ denote the node set with n elements, and use i or j to represent the index of a random node, hence, $1 \leq i, j \leq n$. Let $\mathcal{R}$ denote the relation set with totally m relation types $\{k|1 \leq k \leq m\}$ where k is the relation index. Further, let R be the real field, and represent the affinity tensor as $\mathcal{S} = (s_{i,j,k})$ where $s_{i,j,k}$ ($s_{i,j,k} \in R$) denotes the relation strength between nodes i and j under the k-th relation type. In most cases, *relation strength* is also treated as a kind of *similarity*. Additionally, vectors $\mathbf{p} = (p_1, p_2, \cdots, p_n)$ and $\mathbf{q} = (q_1, q_2, \cdots, q_m)$ denote the probability distributions of nodes and relation types, respectively.

3.1   Co-ranking Nodes and Relations in Multi-relational Network

As mentioned earlier, the rationale of our *MutuRank* framework is that the weight of a node is affected not only by its neighbors' weights, but also by the strength of various link types which are also endowed with different weights. The more important neighbors and more strong links with high importance, the more important a node will be. Accordingly, given a node i, the probability for information transiting from i's neighbor j to i is as follows:

$$\mathrm{Prob}_1(i|j) = \frac{\sum_k q_k \cdot s_{i,j,k}}{\sum_l \sum_k q_k \cdot s_{l,j,k}}, \tag{1}$$

where q_k means the current weight of the k-th relation, and $\mathrm{Prob}_1(\cdot|\cdot)$ represents the probability of a node selecting another node, just as the the random walk in PageRank. Above equation demonstrates that the transition probability between two nodes does not keep unchanged, which is different with the traditional PageRank. It is easy to understand that if the probability distribution of relations is consistent with the arrangement of link strengthes between two nodes, this pair of nodes may have relatively higher transition probability.

Furthermore, based on the probability distribution of nodes, the probability of the node i selecting the k-th relation is adjusted as:

$$\text{Prob}_2(k|i) = \frac{q_k \cdot \sum_j s_{i,j,k}}{\sum_k q_k \cdot \sum_j s_{i,j,k}}, \tag{2}$$

where $\text{Prob}_2(\cdot|\cdot)$ represents the probability of a node selecting a relation type. It is worth mentioning that the *MultiRank* [9] has a similar objective with us. However, *MultiRank* relaxed the joint probability of a node and a relation to be two mutually independent probabilities, which leads to the probability of a node selecting a relation has nothing to do with the similarity structure of the whole network. Let's illustrate it by a simple example: we assume the k-th relation has gotten a very high weight in previous iterations, while node i has very low similarities with all of its neighbors. This situation might probably happen when almost all of node pairs, except node i and its neighbors, have higher mutual similarities under relation k. *MultiRank* still assumed the probabilities of all nodes selecting relation k in the next iteration to be equal, which is obviously unfair to the node i. So, we believe that node i should reduce its probability of selecting relation k based on its true similarities to its neighbors.

Let $\mathbf{p}^* = (p_1^*, p_2^*, \cdots, p_n^*)$ and $\mathbf{q}^* = (q_1^*, q_2^*, \cdots, q_m^*)$ be the prior distributions of nodes and relations, respectively. To conclude, we have following iterative equations for computing the ranking scores of nodes and relations simultaneously:

$$p_i^{t+1} = \sum_j p_j^t \cdot \text{Prob}_1^t(i|j) + \alpha \cdot p_i^* \quad (1 \leq i \leq n), \tag{3}$$

$$q_k^{t+1} = \sum_i p_i^t \cdot \text{Prob}_2^t(k|i) + \beta \cdot q_k^* \quad (1 \leq k \leq m), \tag{4}$$

where t is the times of iteration, and α, β are two parameters to balance the knowledge coming from network structure and the prior knowledge.

3.2   Theoretical Analysis

In this section, we prove the existence and uniqueness of stationary probability distributions $\mathbf{p}$ and $\mathbf{q}$ so that they can be used to co-rank the nodes and relation types effectively.

First, we show why our iterative algorithm in Eqs.(3) and (4) will converge. Let $\Omega_n = \{\mathbf{p} = (p_1, p_2, \cdots, p_n) \in R^n | p_i \geq 0, 1 \leq i \leq n, \sum_{i=1}^n p_i = 1\}$ and $\Omega_m = \{\mathbf{q} = (q_1, q_2, \cdots, q_m) \in R^m | q_k \geq 0, 1 \leq k \leq m, \sum_{k=1}^m q_k = 1\}$. We also set $\Omega = \{[\mathbf{p}, \mathbf{q}] \in R^{n+m} | \mathbf{p} \in \Omega_n, \mathbf{q} \in \Omega_m\}$. We notice that Ω_n, Ω_m and Ω are closed convex sets. We call $\mathbf{p}$ and $\mathbf{q}$ to be positive if all their entries are greater than 0. For convenience, we represent Eqs.(3) and (4) as following simpler form:

$$\mathbf{p}^{t+1} = f_1(\mathbf{p}^t, \mathbf{q}^t), \tag{5}$$
$$\mathbf{q}^{t+1} = f_2(\mathbf{p}^t, \mathbf{q}^t). \tag{6}$$

We then have the following theorem:

Theorem 1. *For any* $\mathbf{p}^t \in \Omega_n$ *and* $\mathbf{q}^t \in \Omega_m$, *then* $f_1(\mathbf{p}^t, \mathbf{q}^t) \in \Omega_n$ *and* $f_2(\mathbf{p}^t, \mathbf{q}^t) \in \Omega_m$

PROOF. With Eq. (3), it is easy to prove that if $\mathbf{p}^t$, $\mathbf{q}^t$ and $\mathbf{p}^*$ are probability distributions, $\mathbf{p}^{t+1}$ is also a probability distribution. Similarly, if $\mathbf{p}^t$, $\mathbf{q}^t$ and $\mathbf{q}^*$ are probability distributions, $\mathbf{q}^{t+1}$ is also a probability distribution according to Eq. (4).   □

On the basis of Theorem 1, we next attempt to show the existence of positive solution for MutuRank. But before that, it is necessary for us to know the connectivity among the objects and the relations within that multi-relational network. We first give a definition:

Definition 1 (Irreducibility). $\mathcal{S} = (s_{i,j,k})$ *is called irreducible if* $s_{i,j,k}$ *(n-by-n matrices) for fixed* k *($1 \le k \le m$) are irreducible.*

Here, $\mathcal{S}$'s irreducibility suggests that two objects can be connected via some relations. *Irreducibility* is a reasonable assumption that we will use in following discussion. It was also adopted by literature [10] in the PageRank matrix for calculating PageRank values. Further, such an assumption contributes to following conclusion:

Theorem 2. *If* $\mathcal{S} = (s_{i,j,k})$ *is irreducible, then there exist* $\bar{\mathbf{p}} \in \Omega_n$ *and* $\bar{\mathbf{q}} \in \Omega_m$ *s.t.,* $\bar{\mathbf{p}} = f_1(\bar{\mathbf{p}}, \bar{\mathbf{q}})$ *and* $\bar{\mathbf{q}} = f_2(\bar{\mathbf{p}}, \bar{\mathbf{q}})$, *and* $\bar{\mathbf{p}} > 0$, $\bar{\mathbf{q}} > 0$.

PROOF. This problem can be addressed as a fixed point problem. Suppose a mapping $F : \Omega \to \Omega$ as follows:

$$F([\mathbf{p}, \mathbf{q}]) = [f_1(\mathbf{p}, \mathbf{q}), f_2(\mathbf{p}, \mathbf{q})]. \tag{7}$$

It is clear that $F(\cdot)$ is well-defined (i.e., when $[\mathbf{p}, \mathbf{q}] \in \Omega$, $F([\mathbf{p}, \mathbf{q}]) \in \Omega$) and continuous. According to the Brouwer Fixed Point Theorem, there exists $[\bar{\mathbf{p}}, \bar{\mathbf{q}}] \in \Omega$ such that $F([\bar{\mathbf{p}}, \bar{\mathbf{q}}]) = [\bar{\mathbf{p}}, \bar{\mathbf{q}}]$, i.e., $f_1(\bar{\mathbf{p}}, \bar{\mathbf{q}}) = \bar{\mathbf{p}}$ and $f_2(\bar{\mathbf{p}}, \bar{\mathbf{q}}) = \bar{\mathbf{q}}$.   □

Now, we will have a deep discussion that both $\bar{\mathbf{p}}$ and $\bar{\mathbf{q}}$ are positive. Let us re-write the Eq. (4) as:

$$q_k^{t+1} = q_k^t \cdot \sum_i p_i^t \cdot \frac{\sum_j s_{i,j,k}}{\sum_k q_k^t \cdot \sum_j s_{i,j,k}} + \beta \cdot \mathbf{q}^*. \tag{8}$$

If $q_k^T = 0$ (i.e., after the T−th iteration, the importance of the k-th relation type becomes 0), then, $\sum_j s_{i,j,k} = 0$. Note that $\sum_j s_{i,j,k}$ means the sum of the similarities between node i and all its neighbors. Apparently, $\sum_j s_{i,j,k} = 0$ indicates that node i is an isolated node under the k-th relation, which is in contradiction with the irreducibility of $\mathcal{S}$. Similarly, given the Equation (3), if $p_i^T = 0$, we can easily find that the resulting situation also violates the irreducibility of $\mathcal{S}$. Due to the space limit, we do not provide any detailed analysis.

Finally, we give the conditions under which our algorithm will converge to a unique solution. Literature [6] has guaranteed the uniqueness of the fixed point in the Brouwer Fixed Point Theorem with following prerequisites: (i) for each point in the domain boundary of the mapping, it is not a fixed point; (ii) 1 is not an eigenvalue of the Jacobian matrix of the mapping. As for the first condition, we have shown, in Theorem 2, that all the fixed points of $F(\cdot)$ are positive when $\mathcal{S}$ is irreducible, i.e., they do not lie on the boundary $\partial\Omega$ of Ω. As regards the second condition, we have following conclusion:

Theorem 3. *If* 1 *is not the eigenvalue of the Jacobian matrix of mapping F for all* $[\mathbf{p}, \mathbf{q}] \in \Omega/\partial\Omega$, *the probability distributions in Theorem 2 are unique.*

Note that Theorem 3 presents only a condition for the solution's uniqueness. How to prove or guarantee 1 is not the eigenvalue of a function's Jacobian matrix remains an open problem. Nevertheless, it does not affect *MutuRank*'s convergence and real-world application.

Up to now, we have elaborated how to calculate the relation distribution via our *MutuRank* algorithm. Based on the results of this step, we keep nodes unchanged and merge those pairwise relations linearly to construct a single-relational graph. Next, we utilize spectral clustering to perform community detection on our graph data.

4   Community Detection in Single-Relational Network

In this part, we exploit the widely-used spectral clustering framework to conduct clustering in single-relational graph. The rationale of spectral clustering is as follows: first represents graph nodes with vectors by means of some matrix operations over the graph's affinity matrix, then calls certain basic clustering algorithm, such as K-means or GMM (Gaussian Mixture Model), to do clustering. GMM should be more suitable than K-means in clustering objects in social network because most entities are unlikely interested in only one community.

In many real-world social networks, actors tend to exert influence to their friends. For instance, in a scientific coauthorship network, if most of a researcher's friends have much interest in data mining, it is reasonable to infer the researcher himself/herself is very likely interested in data mining too. When it turns to applying GMM to cluster objects in social networks, we believe that the probability of an object belonging to a community is decided not only by the value of Gaussian function of the object itself, but also according to the probabilities of the object's neighbors belonging to that community. Indeed, some literatures [12,13] aimed to assign a node with the label that most of its neighbors have, so that a consensus on a label will finally form a community. Hence, we attempt to modify the traditional GMM algorithm to be a novel model named GMM-NK(Gaussian Mixture Model with Neighbor Knowledge).

Similar with GMM, GMM-NK is also a linear superposition of Gaussian components for the purpose of providing a richer class of density models than the single Gaussian. Clustering based on GMM-NK is probabilistic in nature and aims at maximizing the likelihood function with regard to the parameters (comprising the means and covariances of the components and the mixing coefficients).

Consider n data points $\mathcal{X} = \{x_1, x_2, \cdots, x_n\}$ in d-dimensional space, the probability density of x_i can be defined as follows:

$$p(x_i|\pi, \mu, \Sigma) = \sum_{z=1}^{c} \pi_z \cdot N^*(x_i; \mu_z, \Sigma_z), \tag{9}$$

where c is the component number, π_z is the prior probability of the z-th Gaussian component. $N^*(x_i; \mu_z, \Sigma_z)$ is defined as:

$$N^*(x_i; \mu_z, \Sigma_z) = \gamma \cdot N(x_i; \mu_z, \Sigma_z) + (1 - \gamma) \cdot \sum_{j:j \neq i} s_{i,j} \cdot N(x_j; \mu_z, \Sigma_z), \tag{10}$$

where $0 < \gamma \leq 1$ is a parameter for controlling the impacts from neighbors, and $s_{i,j}$ is the similarity between node x_i and x_j. $\gamma = 1$ means that we do not consider the influence of neighbors and this algorithm reduces to the original GMM. Let $N(x_i; \mu_z, \Sigma_z)$ denote the standard Gaussian function, i.e.,

$$N(x_i; \mu_z, \Sigma_z) = \frac{\exp\{-\frac{1}{2}(x_i - \mu_z)^T \Sigma_z^{-1}(x_i - \mu_z)\}}{((2\pi)^d |\Sigma_z|)^{\frac{1}{2}}}. \tag{11}$$

Similar with the basic GMM, the log of the likelihood function is then given by:

$$lnP(\mathcal{X}|\pi, \mu, \Sigma) = ln \prod_{i=1}^{n} P(x_i|\pi, \mu, \Sigma) = ln \prod_{i=1}^{n} \sum_{z=1}^{c} \pi_z \cdot N^*(x_i; \mu_z, \Sigma_z)$$
$$= \sum_{i=1}^{n} ln\{\sum_{z=1}^{c} \pi_z \cdot N^*(x_i; \mu_z, \Sigma_z)\}. \tag{12}$$

An elegant and powerful method for finding maximum likelihood solutions for models with latent variables is Expectation Maximization algorithm (EM). EM is an iterative algorithm in which each iteration contains an E-step and a M-step. In the E-step, we compute the probability of the z-th Gaussian component given the data point x_i using the current parameter values:

$$p(z|x_i, \pi, \mu, \Sigma) = \frac{\pi_z \cdot N^*(x_i; \mu_z, \Sigma_z)}{\sum_{j=1}^{c} \pi_j \cdot N^*(x_i; \mu_j, \Sigma_j)}. \tag{13}$$

In the M-step, we re-estimate the parameters using the current responsibilities, as follows:

$$\mu_z^{new} = \frac{1}{n_z} \sum_{i=1}^{n} x_i \cdot p(z|x_i, \pi, \mu, \Sigma), \tag{14}$$

$$\Sigma_z^{new} = \frac{1}{n_z} \sum_{i=1}^{n} p(z|x_i, \pi, \mu, \Sigma)(x_i - \mu_z^{new})(x_i - \mu_z^{new})^T, \tag{15}$$

$$\pi_z^{new} = \frac{n_z}{n}, \tag{16}$$

where $n_z = \sum_{i=1}^{n} p(z|x_i, \pi, \mu, \Sigma)$. The EM algorithm runs iteratively until the log likelihood reaches (approximate) convergence. In experiments, we simply run GMM-NK until the increment of log likelihood is less than 10^{-7} and pick the one with largest log likelihood as the best estimate of the underlying communities.

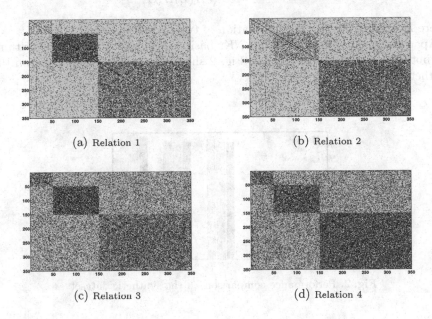

(a) Relation 1

(b) Relation 2

(c) Relation 3

(d) Relation 4

Fig. 1. The four relation networks on the synthetic dataset

5   Experimental Validation

In this section, we demonstrate the effectiveness of our *MutuRank* and *GMM-NK* algorithms for community detection in multi-relational social networks. For the sake of simplicity, $\alpha = \beta = 0.5$ in Eqs.(3) and (4), and $\gamma = 0.65$ in Eq.(10) are the default settings in these experiments.

5.1 Experiments on Synthetic Dataset

Generally, real-world corpus does not provide the ground truth information about the membership of component objects. So, we start with a synthetic dataset to illustrate some good properties of our framework. This dataset is generated by a synthetic data simulator coded in MATLAB [20]. The synthetic network contains 350 nodes which roughly form 3 communities containing 50, 100, 200 nodes, respectively. Furthermore, 4 different relations are constructed to look at the clustering structure in different angles, as shown in Fig. 1.

Since a priori community memberships (a.k.a. ground truth) is known, we then adopt commonly used *normalized mutual information* (NMI) [2] as the evaluation measure. Let L and G denote the label vector obtained by community detection methods and the ground truth. NMI is defined as:

$$NMI(L;G) = \frac{I(L;G)}{\sqrt{H(L)H(G)}},$$ (17)

where $I(L;G)$ is the mutual information of two variables.

Apart from our proposed GMM-NK, basic GMM and K-means algorithms are both implemented as baselines. Fig. 2 shows the overall comparison on the synthetic dataset.

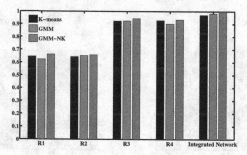

Fig. 2. Performance comparison on the synthetic dataset

Note that aggregated bars on R_1 to R_4 correspond to the results on 4 different relations, and "Integrated Network" corresponds to the results on the single relation network obtained by our *MutuRank*. As can be seen, NMI on the integrated network outperforms any of single relations, regardless of community detection algorithms. That implies that *MutuRank* can incorporate multi-relations to form a good quality single-relation network, in which community structures emerge more clearly than that on any original single-relation graphs. Furthermore, we observe that our GMM-NK indeed shows better performance than both GMM and K-means, and combining *MutuRank* with GMM-NK achieves perfect clustering, i.e., $NMI \approx 1$ on integrated network of GMM-NK.

5.2   Experiments on DBLP Dataset

We now proceed to provide experimental results on the real-world dataset, i.e., the DBLP dataset. We first discuss data collection and graph construction, and then report and analyze experimental findings.

Data Collection. According to the rankings of authoritative conferences in computer science[1], we crawled publication information of both "Top Tier" and "Second Tier" conferences of 13 different fields from the DBLP web site. Table 1 shows the legends of these 13 different fields. Note that all publication periods are from 2000 to 2010. In total, the extracted DBLP dataset contains 97 conferences, 185,490 researchers and 105,264 publications.

Table 1. The Legends of 13 Fields

Categories	Explanations	Categories	Explanations
DB	Databases	DP	Distributed and Parallel Computing
DM	Data Mining	GV	Graphics, Vision and HCI
AI	Artificial Intelligence	MM	Multimedia
NL	Natural Language Processing	NC	Networks, Communications and Performance
ED	Computer Education	SE	Security and Privacy
IR	Information Retrieval	OS	Operating Systems/Simulations
W3	Web and Information Systems		

Graph Construction. We treat researchers as nodes in network, and treat fields as distinct relation types. Then, for each relation k ($1 \leq k \leq 13$), we construct a *relation-graph* G_k where the link strength of two researchers i and j is calculated according to their publications in that field. More exactly, suppose the numbers of their publications in relation k are $p_{i,k}$ and $p_{j,k}$, respectively. Then their relevance can be determined by Eq. 18.

$$s_{i,j,k} = e^{-\left(\frac{2(p_{i,k}-p_{j,k})}{p_{i,k}+p_{j,k}}\right)^2}. \tag{18}$$

Baselines. In whole, as for relation distribution, we have three choices, i.e., (1) *Uniform*: setting to uniform distribution; or (2) *MultiRank*: using the MultiRank [9] algorithm; or (3) *MutuRank*: using our *MutuRank* algorithm. With respect to spectral clustering, it could be implemented on the basis of K-means, GMM, or our *GMM-NK*. Hence, we combine these choices to obtain 9 baselines for experimental comparison.

Experiment Results. First, we give the relation distribution derived from *MutuRank* algorithm, as shown in Fig. 3. Relation types "AI" and "DB" enjoy apparent dominance, while "ED" is the most unimportant one. It results from the tendency that researchers in "AI" and "DB" areas not only have huge

---

[1]   http://webdocs.cs.ualberta.ca/~zaiane/htmldocs/ConfRanking.html

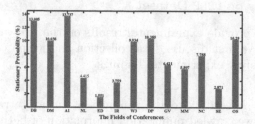

Fig. 3. The stationary relation distribution

Table 2. Performance Comparison on DBLP Dataset

NMI	K-means	GMM	GMM-NK
Uniform	0.618	0.622	0.687
MultiRank	0.757	0.762	0.776
MutuRank	0.824	0.871	**0.913**

amount, but also own rich publications. Contrarily, "ED" and "SE" only attract a few scholars to dedicate. Understandably, the more the high-weight edges in a dimension, the more important the relation type.

Second, we investigate the overall performance of afore-mentioned 9 baselines on DBLP dataset. Though K-means generates crisp communities, two soft clustering methods GMM and GMM-NK output the probabilities that every researcher belonging to all communities. It is straightforward to assign the community label to each researcher by choosing the highest probability. To facilitate the evaluation, we need to assign the ground-truth label to each researcher, and thus we select the field in which the researcher has the most publications as his/her ground-truth label. Table 2 demonstrates the comparison results in terms of NMI. Above statistics suggest the apparent regularity of performance trends. On the whole, once the clustering method is fixed, NMI increases with the algorithms deriving relation distribution, according to order "Uniform→ MultiRank→MutuRank". The similar phenomenon also happens to the order "K-means→GMM→GMM-NK", when the algorithm of computing relation distribution is fixed. The results shown in Table 2 are sufficient to validate the effectiveness of both our *MutuRank* in capturing relation distribution, and our GMM-NK on community detection.

Last but not the least, we look inside the membership probabilities calculated by GMM-NK. We sampled five famous scholars and showed their membership probabilities of 13 fields, as shown in Fig. 4. Note that the probability values below 0.001 are set to 0. These results well matched the "Research Interest" of these five scholars given by AMiner (http://arnetminer.org) [17]. For instance, Jiawei Han has always been very active on "DB" and "DM", and Chengxiang Zhai concentrates on the "IR" field.

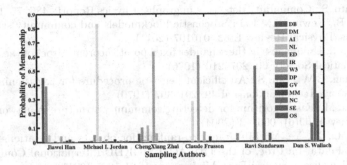

Fig. 4. The membership probabilities of sampling authors

6 Conclusions

In this paper, we addressed the issue of community detection in multi-relational network. Similar with some literature, we also attempted to first convert the original multi-relational network into a single-relational graph which has been studied more extensively. In this phase, the proposed *MutuRank* model enabled us to acquire relation distribution with the consideration of interdependency between relations and objects. Subsequently, we utilized spectral clustering on the basis of a novel GMM-NK algorithm to detect communities in the single-relational graph. Experimental results on both synthetic and real-world data demonstrate the effectiveness of the proposed method.

Acknowledgments. This research was partially supported by National Natural Science Foundation of China under Grants 71072172 and 61103229, National Key Technologies R&D Program of China under Grants 2013BAH16F01 and 2013BAH16F04, Industry Projects in Jiangsu S&T Pillar Program under Grant BE2012185, Key Project of Natural Science Research in Jiangsu Provincial Colleges and Universities under Grant 12KJA520001, and the Natural Science Foundation of Jiangsu Province of China under Grant BK2012863, and the Priority Academic Program Development of Jiangsu Higher Education Institutions.

References

1. Cai, D., Shao, Z., He, X., Yan, X., Han, J.: Community mining from multi-relational networks. In: Jorge, A.M., Torgo, L., Brazdil, P.B., Camacho, R., Gama, J. (eds.) PKDD 2005. LNCS (LNAI), vol. 3721, pp. 445–452. Springer, Heidelberg (2005)
2. Cao, J., Wu, Z., Wu, J., Xiong, H.: SAIL: Summation-based incremental learning for information-theoretic text clustering. IEEE Transactions on Cybernetics 43(2), 570–584 (2013)
3. Dai, B.T., Chua, F.C.T., Lim, E.P., Faloutsos, C.: Structural analysis in multi-relational social networks. In: Proc. of the International SIAM Conference on Data Mining (SDM 2012), 451–462 (2012)

4. Fortunato, S.: Community detection in graphs. Physics Reports 486, 75–174 (2010)
5. Karrer, B., Newman, M.E.J.: Stochastic blockmodels and community structure in networks. Physical Review E 83, 016107 (2011)
6. Kellogg, R.: Uniqueness in the schauder fixed point theorem. Proc. of the American Mathematical Society 60, 207–210 (1976)
7. Kernighan, B.W., Lin, S.: An efficient heuristic procedure for partitioning graphs. Bell Systems Technical Journal 49, 291–307 (1970)
8. Newman, M.: Fast algorithm for detecting community structure in networks. Physical Review E 69(6), 066113 (2004)
9. Ng, M., Li, X., Ye, Y.: Multirank: co-ranking for objects and relations in multi-relational data. In: Proc. of the 17th ACM SIGKDD International Conference on Knowledge Discovery and Data Mining, pp. 1217–1225 (2011)
10. Page, L., Brin, S., Motwani, R., Winograd, T.: The pagerank citation ranking: bringing order to the web (1999)
11. Palla, G., Derenyi, I., Farkas, I., Vicsek, T.: Uncovering the overlapping community structure of complex networks in nature and society. Nature 435(7043), 814–818 (2005)
12. Raghavan, U.N., Albert, R., Kumara, S.: Near linear time algorithm to detect community structures in large-scale networks. Physical Review E 76(3), 036106 (2007)
13. Rahimian, F., Payberah, A.H., Girdzijauskas, S., Jelasity, M., Haridi, S.: Ja-be-ja: A distributed algorithm for balanced graph partitioning. Technical Report, Swedish Institute of Computer Science (2013)
14. Rodriguez, M., Shinavier, J.: Exposing multi-relational networks to single-relational network analysis algorithms. Journal of Informetrics 4(1), 29–41 (2010)
15. Slater, P.B.: Established clustering procedures for network analysis. Technical Report, arXiv:0806.4168 (June 2008)
16. Szell, M., Lambiotte, R., Thurner, S.: Multirelational organization of large-scale social networks in an online world. Proc. of the National Academy of Sciences 107(31), 13636–13641 (2010)
17. Tang, J., Yao, L., Zhang, D., Zhang, J.: A combination approach to web user profiling. ACM Transactions on Knowledge Discovery from Data (TKDD) 5(1), 2 (2010)
18. Tang, L., Liu, H.: Relational learning via latent sociál dimensions. In: Proc. of the 15th ACM SIGKDD International Conference on Knowledge Discovery and Data Mining, pp. 817–826 (2009)
19. Tang, L., Liu, H., Zhang, J.: Identifying evolving groups in dynamic multimode networks. IEEE Transactions on Knowledge and Data Engineering 24(1), 72–85 (2012)
20. Tang, L., Wang, X., Liu, H.: Community detection via heterogeneous interaction analysis. Data Mining Knowledge Discovery 25(1), 1–33 (2012)
21. Tang, L., Liu, H.: Community detection and mining in social media. Synthesis Lectures on Data Mining and Knowledge Discovery 2(1), 1–137 (2010)
22. Wei, F., Qian, W., Wang, C., Zhou, A.: Detecting overlapping community structures in networks. World Wide Web Journal 12(2), 235–261 (2009)
23. Yang, B., Liu, J., Feng, J.: On the spectral characterization and scalable mining of network communities. IEEE Transactions on Knowledge and Data Engineering 24(2), 326–337 (2012)
24. Yu, W., Zhang, W., Lin, X., Zhang, Q., Le, J.: A space and time efficient algorithm for simrank computation. World Wide Web Journal 15(3), 327–353 (2012)

Community Detection in Social Media by Leveraging Interactions and Intensities

Maria Giatsoglou, Despoina Chatzakou, and Athena Vakali

Informatics Department, Aristotle University of Thessaloniki
{mgiatsog,deppych,avakali}@csd.auth.gr

Abstract. Communities' identification in topic-focused social media users interaction networks can offer improved understanding of different opinions and interest expressed on a topic. In this paper we present a community detection approach for user interaction networks which exploits both their structural properties and intensity patterns. The proposed approach builds on existing graph clustering methods that identify both communities of nodes, as well as outliers. The importance of incorporating interactions' intensity in the community detection algorithm is initially investigated by a benchmarking process on synthetic graphs. By applying the proposed approach on a topic-focused dataset of Twitter users' interactions, we reveal communities with different features which are further analyzed to reveal and summarize the given topic's impact on social media users.

Keywords: community detection, user weighted interaction networks.

1 Introduction

Events and topics emerging in the real world and social networks influence one another mutually due to social media users activities which have radically changed information dissemination and people's opinions communication. In social media frameworks, such as in microblogs, event-relevant information (in the form of news broadcasts or opinion snapshots) is directly propagated to the users' *followers* but it can be discovered by other users as well (by public posts). *User interaction networks* capture users' associations derived from their activities in social media such as: commenting on others' posts, replying to comments, referencing other users, etc. Variation in users' interaction frequency or intensity can be captured by the assignment of different strengths to the networks' associations, thus resulting in weighted networks. Earlier research has indicated that several types of user interaction networks (such as coauthorship networks [15], communication networks [12], friendship networks [21]) exhibit a structure of non random nature and organize around *communities*. Communities can be generally defined as *groups of users that are "closely-knit"*, in the sense that a group's interconnections are more dense compared to connections with the rest of the network.

Here we study user communities in social media users interaction networks as formed around a given theme/topic related to a real-world event. Our focus is on revealing the types of communities generated with respect to certain events by analyzing them in the dimensions of size, topic diversity and time span. This work builds on the idea that user interaction strengths are crucial in communities formation and that the detection and

X. Lin et al. (Eds.): WISE 2013, Part II, LNCS 8181, pp. 57–72, 2013.

qualitative characterization of communities can lead to a better understanding of the impact of real world events on society. To achieve this, both inherent structural measures and emergent features are needed. These structural properties and measures are here related to the well-known community detection algorithm SCAN [22], due to its scalability and its capability to detect hubs and outliers, adapted for weighted networks.

In summary this work's main contributions are as next:

- *adapts a weighted similarity measure* which encompasses both the structural properties and the weighted connectivity patterns existing in the locality of nodes. This measure extends *structural similarity* and brings closer nodes that not only share common neighbors, but are also connected to them with matching intensities.
- *deals with inherent limitations in local structural / density-based community detection algorithms* which characterize SCAN driven approaches. The proposed method reveals communities in weighted networks based on *weighted structure connected order of traversal* [4] and an approximate peak detection approach inspired by [18].
- *introduces a community meta-analysis approach* that highlights the usefulness of communities' detection on user interaction networks to reveal and summarize real world events' impact. Our Twitter case study validates the proposed methodology.

2   Related Work

Community detection has been applied to user interaction networks such as e.g. to the Enron e-mail exchange dataset [19] and coauthorhips networks [15], however few works have tackled community detection in social media users' interaction networks [9,12]. In social media's context, community detection has been mainly applied to friendship networks generated by the declared users' affiliations [8], resulting in easily-interpreted groups of users. Thus, interaction networks may be of more complex nature with their derived communities' interpretations being non-obvious since interactions within their members may indicate that they are both aware of each other in the web world, while also interested in common topics. Important aspects of community detection in social media are covered in [17] where the need to detect meaningful communities of nodes as well as identify *hub* and *outlier* nodes is highlighted. This requirement is addressed in SCAN [22], a community detection algorithm which builds on the density-based clustering algorithm DBSCAN [6]. While DBSCAN has been widely used for clustering spatial points based on their density distribution, SCAN operates on graphs based on a *structural similarity* measure. The main limitation of DBSCAN and SCAN is their sensiteivity to the selection of an initial similarity threshold parameter, whose fine-tuning requires repeated algorithm executions for several parameter values. An approach to alleviate this limitation in SCAN was given in [20] with the clustering quality *modularity* criterion [14] being used to find the optimal parameter's value.

Alternative efforts to address the problem of DBSCAN and SCAN parameter produced the so called *reachability plots* [1,4], which represent the algorithms' multiple clustering outcomes for every possible parameters' combination. A technique proposed in [18] operates on reachability plots produced by DBSCAN to automatically determine significant clusters. Up to now SCAN's applicability to weighted networks has only been addressed in [20] where a structural similarity measure for weighted networks is proposed, but no explicit experimental results are offered for such networks.

Thus, all previous efforts were tested on limited (unweighted) synthetic or *closed-world* networks. *Closed-world* networks are limited within the scope of a certain "community" (e.g. the Enron email network with internal company email exchanges), when on the contrary, social media users' interactions are of an open nature since they generate networks "connecting" people of different disciplines and wider scope. Topic- and event-specific networks can be derived from broader social media generated networks by keeping as edges only interactions relevant to the given topic/event. Users, in principle, interact with different intensities, the level of which can be inferred by the interactions' frequency, duration, etc. Although weighted networks are a natural representation of such interactions' intensities, many community detection approaches for real world datasets, operate on unweighted networks after preserving either all relationships, or only those whose intensity is above a cut-off threshold. Here, we examine whether SCAN approaches, which generally have desirable traits for application to user interaction networks, can successfully uncover the underlying community structure in real world networks, or they need to be adapted to leverage the interactions' intensity. SCAN and the proposed adaptation for weighted networks WSCAN (i.e. WeightedSCAN) are evaluated on a series of synthetic networks. The combination of both approaches' experimental results with the corresponding intrinsic network properties (the global *clustering* and *weighted clustering* coefficients [16]) leads to an empirical criterion for the selection of SCAN or WSCAN for the network at hand. WSCAN's limitation of parameter selection is also addressed by an automatic approach, AutoWSCAN, which detects communities from nodes' weighted structure connected order of traversal, inspired by [18], and is validated in synthetic and real-world event-centric networks.

3 Proposed Methodology

To generate users' interaction networks given an event-related topic T, we first aggregate for a given period of time ΔT user activity data from selected social media applications, then extract the observed interactions Int_t, and connect users who have interacted at least once. An edge connecting nodes u and v is weighted by $w_{u,v} = \sum_{t \in \Delta T} Int_t(u, w)$. To detect communities in user networks embedding interaction strengths, here we adapt existing SCAN-based algorithms, and propose the use of WSCAN and AutoWSCAN.

3.1 Getting from SCAN to WSCAN

SCAN [22] discovers cohesive network subclusters based on parameters μ and ϵ, which control the minimum community's size and the minimum *structural similarity* between two community's nodes, respectively. Generally, a larger μ value leads to fewer and larger communities, while a larger ϵ value to tighter communities and more outliers. Using structural similarity as a clustering criterion, nodes with several common neighbors are placed in the same (μ, ϵ)-core community. To adapt SCAN for weighted interaction networks we propose *weighted structure reachability* for (μ, ϵ)-cores' detection.

Definition 1. *Given a weighted undirected network (G, w), where $G = \{V, E\}$ and $w : E \rightarrow \mathbb{R}$, the **weighted structural similarity** $wSSim$ of two nodes, u and v, is defined as:*

$$wSSim(u,v) = \frac{\sum_{k\in\Gamma(u)\cap\Gamma(v)} w_{u,k} \cdot w_{v,k}}{\sqrt{\sum_{k\in\Gamma(u)} w_{u,k}^2} \sqrt{\sum_{k\in\Gamma(v)} w_{v,k}^2}} . \tag{1}$$

where $\Gamma(v)$ is the **neighborhood** of node v: $\Gamma(v) = \{k \in V | (v,k) \in E\} \cup \{v\}$, $w_{u,v} \in [0,1)|u \neq v$; $w_{u,v} = 1|u = v$.

Definition 2. *The ϵ-neighborhood of a given node u is the subset of its neighborhood containing only nodes that are at least ϵ-similar with u:*

$$N_\epsilon(u) = \{v \in \Gamma(u) | wSSim(u,v) \geq \epsilon\} . \tag{2}$$

Definition 3. *A vertex v is called a (μ, ϵ)-core if its ϵ-neighborhood contains at least μ vertices: $CORE_{\mu,\epsilon}(v) \Leftrightarrow |N_\epsilon(v)| \geq \mu$.*

Additional nodes are attached to (μ, ϵ)-cores based on *structural connectivity*. A node u is *structure-reachable* from a core node v if u can be reached from v through a chain of nodes each belonging to the ϵ-neighborhood of the previous one. Nodes u and v are *structure-connected* if they are *reachable* from the same core node k. A community is then defined as a set of structure-connected nodes that is maximal in terms of structure reachability. Nodes not assigned to any communities, are characterized as either outliers or hubs depending on whether they are linked to a single or multiple communities, respectively. For the calculation of wSSim it is important to ensure that all weights are < 1, since a weight of 1 is used as each node's self-similarity in the definition of wSSim. To achieve this, we scale all interactions' weights before community detection.

3.2 AutoWSCAN

Our experiments with WSCAN showed its high sensitivity to parameter ϵ. Finding an ϵ value that leads to a balanced community structure regarding outliers' number, coherence, and communities' separation is, though, tedious. A heuristic approach is proposed in [22] for selecting the ϵ value based on the "knee-point hypothesis" for the μ-nearest neighbor similarity plot. Thus, our application of this approach to real-world networks with both the "unweighted" and weighted structural similarity, did not reveal clear knee-points at such plots. We rather adopt the structure connected order of traversal which represents all *structure-connected* community sets detected in a network for all possible ϵ values [4]. To this end, nodes are re-ordered by structure-connected order of traversal based on *weighted core reachability* and *weighted reachability similarity*, defined next.

Definition 4. *Given a network (G, w), the **weighted core reachability** wCSim of node u is defined as:*

$$wCSim(u) = \begin{cases} wSSim(u, \mu NN(u)), if \, |\Gamma(u)| \geq \mu \\ UNDEFINED, else \end{cases}, \tag{3}$$

where $\mu NN(u)$ is the μ-nearest neighbor of node u.

Definition 5. *Given a network (G, w), the **weighted reachability similarity** wRSim of node v from node u is defined as:*

$$wRS\,im(v, u) = \begin{cases} max\,(wCS\,im(u), wS\,S\,im(u, v)), if\,|\Gamma(u)| \geq \mu \\ UNDEFINED, else \end{cases} \quad (4)$$

Weighted core reachability is calculated for each node, standing for the minimum ϵ value that would allow this node to become a core (Alg. 1). Then, each possible core node u ($|\Gamma(u)| \geq \mu$) is "visited", a process that involves finding the node's neighbors, calculating their weighted reachability similarity from the current core, and inserting them at a priority queue based on the $wRS\,im$ value (or reordering the queue if they have already been inserted). At each iteration, the node with the highest $wRS\,im$ value from any previously visited node is extracted from the queue to ensure that regions of higher weighted structural similarity are spanned before surrounding areas of lower similarity [4]. The node visiting order represents the weighted structure connected order of traversal. For a connected network, the algorithm will never return to its first loop, thus, since thematic social media users' interaction networks are often disconnected, this is probable. Our approach is to generate partial nodes' sequences based on structure-connected order of traversal for each disconnected component and detect communities in them.

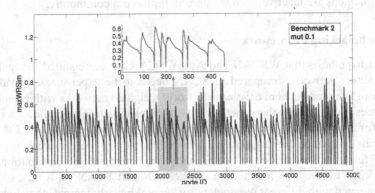

Fig. 1. Weighted reachability plot for Benchmark 2 with $\mu_w = 0.1$

The weighted structure-connected order of traversal can be depicted via a *reachability plot*, which illustrates, in the corresponding order, the maximum weighted reachability value of each node from its previously visited nodes (referred to as $maxWRS\,im$). Such a reachability plot is depicted in Fig. 1, where we can observe areas in which the $maxWRS\,im$ values steadily rise and then fall at a local minimum to rise again after a while. Such "hills" represent different communities, whereas areas of low $maxWRS\,im$ values are outliers. Such communities can be revealed by 'slicing' the plot horizontally at a selected global similarity threshold, and isolating the regions that lay above it.

Definition 6. *Given a sequence of nodes* $\{n_1, n_2..., n_{|V|}\}$ *ordered based on weighted structure-connected order of traversal, a community is defined with respect to* ϵ_{thres} *as a subsequence of nodes* $\{n_{a-1}, n_a, ..., n_b\}$ *where* $1 < a - 1 < b \leq |V|$, *iff* $\forall i \in [a, b], maxWRS\,im(n_i) \geq \epsilon_{thres}$ *and* $[a, b]$ *is maximal.*

Since in real world networks communities are usually of different cohesion and strength, a global ϵ_{thres} will fail to identify communities of different *similarity-range* scales. Thus, to detect communities at different (local) ϵ values, we apply AutoWSCAN, an algorithm inspired by [18]. AutoWSCAN (Alg. 2) detects communities as contiguous areas between two local minima, satisfying some desired properties that reflect the way a person would identify communities by observing a reachability plot. AutoWSCAN gets a weighted reachability plot and first identifies local minima points, ensuring that they have the lowest value in a subregion centered on them and spanning $2 \cdot \mu$ points. Then, it puts them in a priority queue by increasing value, and iteratively removes the first point from the queue and uses it to split the input sequence of nodes in two subregions. A split point is considered valid only if the generated subregions differ noticeably in their *maxWRS im* values compared to the split's value. Thus, we check that the maximum value in each region is "significantly" larger than the split point's *maxWRS im* (with use of a *minRatio* $\simeq 0.7$). AutoWSCAN is recursively called for each subregion whose size is larger than μ (active), and the same process is applied for the subregion based on the minima points within its span. If there are no more (valid) minima points or both subregions are inactive, then the current region is a community.

3.3   Benchmarking Framework

Our initial hypothesis that WSCAN and AutoWSCAN are more suited for real world user interaction networks compared to SCAN needs to be experimentally validated. Since, to our knowledge, there exist no real world weighted networks with ground truth communities, we utilize synthetic networks with planted partitioning of nodes in communities for the algorithm's evaluation. In specific, we use the well-known LFR benchmark graphs [10] since they support weights and possess some important real world networks' features (node degree and community size heterogeneous distributions). Our benchmarking involves the application of WSCAN and SCAN on a series of LFR graphs generated with different parameters for several linearly increasing values of the parameter ϵ, while maintaining the same value for parameter μ. The accuracy of each run is evaluated by the well known Normalized Mutual Information (NMI) score [5], which quantifies the closeness between the identified communities and the ground-truth communities in a scale of 0 to 1 (1 denotes identical assignment of nodes to communities). For each graph we record the best NMI score achieved and the corresponding ϵ value. To assess the performance of AutoWSCAN, we apply it on the same graphs, and also compare it with a modified implementation for unweighted graphs, AutoSCAN. The latter follows exactly the same process as AutoWCAN with the exception that is uses the classic (unweighted) measures of core reachability and reachability similarity.

SCAN-based approaches might characterize some nodes as outliers or hubs and not assign them to a community, as opposed to the LFR graphs which consider that each node belongs to at least one community. Since we are not aware of any weighted benchmark network with known community structure embedding also outliers and hubs, we adopt the LFR benchmark graphs and follow a workaround to extract NMI scores. Thus, upon the algorithms' execution, we assign i) outliers to the community with which they have at least one connection, and ii) hubs to the community towards which they are most strongly connected based on the (weighted) structural similarity or (weighted)

Algorithm 1. WeightedSCOT

Input: $G = (V, E, w), \mu$
Output: A sequence of nodes in structure-connected order of traversal.
  foreach *node* $v \in V$ **do**
    if *v not visited* **then**
      Cl=AutoS CAN (orderedList)
      Communities.append (cl)
      orderedList =null
    enqueueNeighbors (v)
    if *v is core* **then**
      while *visitQueue in not empty* **do**
        currNode = visitQueue.getNode ()
        visitNodes.add (curNode)
        enqueueNeighbors (curNode)
  Function enqueueNeighbors (v)
    v.visited = true
    orderedList.append (v)
    cs = computeWCoreS im (v)
    if *v is core* **then**
      foreach *vN in v.neighbors* **do**
        if *visitedNodes not contains vN* **then**
          ss = getWS tructuralS im (v, vN)
          newWRSim = min (cs, ss)
          if *vN.wReachSim is null* **then**
            vN.wReachSim = newWRSim
            visitQueue.update (vN, newWRS im)
          elif *newWRSim > vN.wReachSim* **then**
            vN.wReachSim = newWRSim
            visitQueue.setPriority (vN, newWRS im)
  EndFunction

---

reachability score for (W)SCAN and Auto(W)SCAN, respectively. This evaluation approach, although not optimal, reflects as accurately as possible how closely the given algorithm approximates ground-truth communities.

After obtaining the NMI scores for all approaches, we seek to reason their comparative performance by examining the benchmark graphs' structural properties. To this end, we employ two metrics: the global clustering coefficient and weighted clustering coefficient. The global clustering coefficient, CC, expresses the density of triplets of nodes in a network, where a *triplet* comprises three nodes connected by two (*open triplet*) or three edges (*closed triplet*). It is defined as 3 times the number of closed triplets (for each pair of the triangle's edges) over the total number of triplets at the network, and its value ranges from 1 for a fully connected network to 0 for random networks with sufficiently large size. A similar idea is followed by the global weighted clustering coefficient, wCC, in weighted networks [16]. By assigning a value to each triplet, wCC is defined as the sum of all closed triplets' values over the sum of all triplets' values. Four methods have been proposed for the calculation of a triplet's value: the arithmetic mean, geometric mean, maximum, and minimum of the weights of the corresponding two edges. Of all four approaches, we select to use the geometric mean since it is considered the most appropriate for alleviating sensitivity to extreme weights. The

---

Algorithm 2. AutoWSCAN

Input: partialWReachabilityPlot: *maxWRS im*
Output: Clusters
  find localMinima
  order localMinima from min to max
  pNode.setRange (*reachVal* (*start*), *reachVal* (*end*))
  return *findClusters* (*pNode, localMinima*)

  Function findClusters (*treeNode,localMinima*)
    if *localMinima is empty* **then**
      if *sizeOf* (*treeNode*) > μ **then**
        treeNode is a cluster
        return
    lMin = localMinima.pop
    [*leftNode, rightNode*] = *split* (*treeNode, lMin*)
    remove all points before/after lMin that have
    the same wReach value
    if *sizeOf* (*leftNode*) > μ **then** *leftNode: active*
    if *sizeOf* (*rightNode*) > μ **then** *rightNode: active*
    if *leftNode* & *rightNode inactive*
      then *treeNode is a cluster*
    foreach *activeNode* **do**
      find its maximum wReach$_{max}$ value
      if (*lMin/wReach$_{max}$*) > *minRatio* **then**
        ignore split point
        findClusters (*treeNode, localMinima*)
      actMinima = localMinima in activeNode's range
      findClusters(*activeNode, actMinima*)
  EndFunction

---

definition of *wCC* implies that for a random distribution of weights in the network, *wCC* equals to *CC*. Here, for each network we calculate the ratio of *wCC* to *CC* and observe the performance of the algorithms when this ratio is greater or lower than 1.

4   Experiments and Results

The proposed approaches, WSCAN and AutoWSCAN, are compared with their un-weighted counterparts, SCAN and AutoSCAN, in terms of their performance on the LFR benchmark framework. Our aim is to determine the validity of the proposed methods and their suitability for graphs that exhibit real world features. Since disregarding the variability of the intensity of interactions in real world networks is a common approach, here we try to identify how it affects performance and in which situations it can be safely followed. We also apply AutoWSCAN to a user interaction network from Twitter, focused on a real world event-related topic, and identify features of the detected communities that can be leveraged to gain insights regarding the event's impact.

4.1   Synthetic Networks

To evaluate the algorithms' performance, we apply them on four weighted LFR graphs, whose complexity is governed by the *topological mixing* (μ_t) and the network's *weighted mixing* (μ_w) parameters [10]. Since μ_t is the ratio of the number of a node's external neighbors to the node's total degree, its increasing values indicate mixed and difficult to separate communities. μ_w has a similar effect, since it is the ratio of the sum of the weights of the edges between a node and its neighbors in different communities to the sum of the all nodes' incident edges. Table 1 outlines the parameter combination for each generated benchmark graph. Benchmarks 1 and 3 refer to graphs with smaller communities (10-50 nodes per community) compared to Benchmarks 2 and 4 (20-100 nodes nodes per community). Also, graphs of Benchmarks 1 and 2 (with $\mu_t = 0.5$) have a more apparent community structure compared to Benchmarks 3 and 4 (with $\mu_t = 0.8$). Since we are interested in how weights affect the community detection results, we perform runs of SCAN, AutoSCAN, WSCAN and AutoWSCAN for varying values of μ_w. Fig. 2 depicts NMI scores for all runs on the four benchmark graphs (with $\mu = 4$).

Table 1. Synthetic Benchmark Graph Specification

	n	k	k_{max}	min_c	max_c	μ_t
Benchmark 1	5000	20	50	10	50	0.5
Benchmark 2	5000	20	50	20	100	0.5
Benchmark 3	5000	20	50	10	50	0.8
Benchmark 4	5000	20	50	20	100	0.8

As expected, the performance of (Auto)SCAN is invariable with respect to μ_t for all benchmarks, since its operation is not affected by changes at the edges' weights. The performance of (Auto)WSCAN is satisfactory for the NMI score, since its starts to decay at $\mu_w \approx 0.5$. Lower NMI values are expected for high μ_w values, since, then, the algorithms characterize more nodes as outliers/hubs and assign them to communities based on the workaround described in Sect. 3.3. For Benchmarks 1 and 2 the weighted algorithms perform better than (Auto)SCAN for $0.1 \leq \mu_t \leq 0.4$, while the corresponding set of graphs exhibit $wCC/CC > 1$ (as depicted in Fig. 3). For $\mu_t \geq 0.5$ unweighted graphs maintain a good performance for Benchmark 1, whereas they perform poorly for all graphs of Benchmark 2 (with bigger communities and $CC < 0.1$). On the contrary, larger community sizes do not significantly affect (Auto)WSCAN's performance, since NMI scores for Benchmarks 1 and 2, as well as for Benchmarks 3 and 4 are similar. NMI scores from Benchmarks 3 and 4 indicate that the weighted algorithms perform better for $\mu_t = 0.8$, rather than for $\mu_t = 0.5$ (Benchmarks 1 and 2). This may seem contradictory, however, as explained in [10], when $\mu_t < \mu_w$ inter-communities edges carry on average more weight, rather than when $\mu_t > \mu_w$. This is inconsistent with most community detection algorithms' hypothesis that intra-community nodes are connected with highly-weighted edges. For all graphs of Benchmarks 3 and 4 the unweighted algorithms fail to detect the community structure. An important observation is that for all these graphs $wCC/CC > 1$ (except for when $\mu_t = 0.8$, where $wCC/CC = 1$). Our results indicate that the decision of whether to apply (Auto)SCAN or (Auto)WSCAN on a given network could be based on the ratio of wCC to CC, selecting the first when it is < 1, or the second otherwise. In all cases, the automatic algorithms follow closely the best performance of their unweighted counterparts. This is a significant outcome considering the temporal cost induced by the search of the (ϵ) parameter space in (W)SCAN.

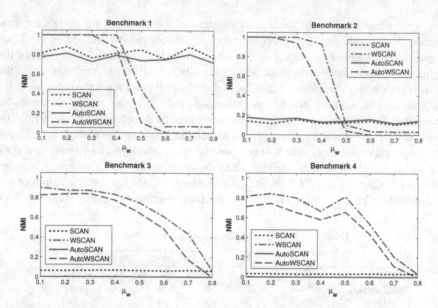

Fig. 2. NMI scores for the algorithms' benchmarks with varying value of μ_w

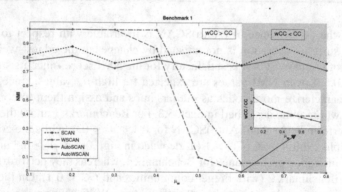

Fig. 3. NMI scores for Benchmark 1 combined with the evolution of the ratio of wCC to CC for increasing value of μ_w (depicted in the embedded plot)

In our experiments, while the selected value for SCAN is ~0.2 for all graphs, in WS-CAN it increases for rising value of μ_w with no common pattern over all graphs. The selected *epsilon* value for all runs where WSCAN performs satisfactorily ($NMI > 0.5$) ranges from 0.04 to 0.28, it thus appears difficult to estimate it in advance. AutoWSCAN emerges as a good alternative to WSCAN, since it is independent of this parameter and performs similarly to WSCAN under the parameter setting leading to the best results.

4.2 Real-World Networks

Our target case-study is to apply community detection in real-world user interaction networks and identify emergent community structure's features for the characterization of real world events based on their impact. For experimentation we have generated a network based on Twitter user interactions, (i.e. *mentions*, *replies*, *retweets*), extracted from data collected via the Twitter Streaming API with topic-related keywords. Our selected topic refers to the official Eurogroup meetings (of Eurozone's finance ministers), which have attracted major interest due to the recent financial crisis and the Eurogroup's role in important decision taking. Our EUROGROUP dataset (covering 8 meetings from 13/06/12 to 30/11/12) acts as an exemplary case study of a series of events held at different time instances, having the same participants with a common generic context (i.e. the Eurozone's monetary issues), but different focus (depending on the agenda). The dataset spans 227 days and comprises: 29529 tweets, 10305 interactions and 3015 different users. Regarding the interactions' type, retweets span more than 50% of the total interactions, thus they affect considerably the networks' shape (star-like forms). Statistical features such as tweet frequencies, depicted in Fig. 4(a), can be used to obtain some initial insights for an event's popularity in Twitter (e.g. more intense activity towards late November). Here, we are mostly interested in the users' clustering around such periods claiming that communities' emergent features reveal finer aspects of events.

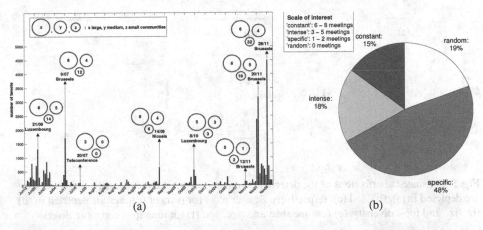

(a)   (b)

Fig. 4. EUROGROUP meetings, tweets, and communities: (a) depicts the daily number of tweets and is annotated by the meetings' dates and locations. The number of active communities per meeting is depicted above its corresponding day; (b) shows a distribution of the communities in a scale of users' interest based on their members' activity on the events' dates

Before we apply community detection, we normalize all weights and calculate wCC and CC for the user interaction network, resulting at a ratio of $wCC/CC = 1.22$. wCC is, thus, larger than CC implying that the intensities of user interactions are not random in this network, but play indeed an important role in communities' formation. Therefore, based on the observations of Section 4.1 we opt to apply AutoWSCAN for the detection of communities. AutoWSCAN reveals 67 communities which we further analyze on three feature axes: size, topic diversity, and time span. *Size* simply refers to the number of users that are members of a given community, and is indicative of the community's popularity. By analyzing the tweets corresponding to intra-community user interactions, we can infer more refined topics that interest each community's members. This analysis involves extracting the text of all inter-community tweets containing interactions between its members, and applying LDA [3] to detect topics within them. Since LDA requires specifying the number of topics to be detected, we empirically set this parameter to 100. Each document in LDA is a mixture of various topics with different probabilities. Here, due to the small length of tweets' text, a tweet is most likely to belong to a single topic, thus we assign it to the most probable (topic). Then, each community is characterized by the set of different topics expressed within its relevant tweets, and is associated with the feature of *topic diversity* which refers here to the size of its topic set. Finally, each community's *time span* is simply derived by taking the length of temporal duration covered by its corresponding tweets (at a day granularity).

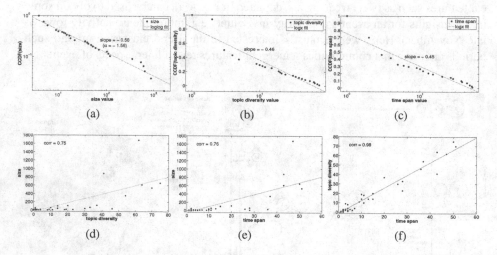

Fig. 5. Estimated distributions of the detected communities' size, topic diversity, and time span are depicted in (a), (b), and (c), respectively. Scatter plots for pairs of features are depicted in (d) for size and topic diversity, (e) for time span and size, and (f) for time span and topic diversity.

Figs. 5(a), (b), (c) depict an estimated distribution of communities' size, topic diversity, and time span, using their Complementary Cumulative Distribution Functions (CCDF), which represents for feature f_i the probability $P(f_i > x)$. Size's CCDF exhibits a slow, power law decay with exponent 0.56, and a p-value of 0.78, indicating good fit. Thus, it can be derived that communities' size also follows a power law distribution with an exponent of $\alpha = 1 + 0.56 = 1.56$ [13]. Similar results for community size have been documented in [2]. Careful observation of Fig. 5(a) reveals a knee at ~538, beyond which CCDF decays faster. This indicates that there are fewer communities of very large size in the dataset compared to these dictated by the power law distribution that fits communities of less that 538 members. Topic diversity and time span do not exhibit a power law distribution, but they have a logarithmic relationship, since their CCDFs can be both fitted best by a exponentially decaying line with slope ~0.45 in a lin-log plot. Fig. 5(b) reveals that 94% and 50% of the communities cover more that 2 and 3 topics, respectively, while the intra-community topics' number is best fitted by the exponential distribution after $f_{topic} = 3$. Fig. 5(c) shows that only 46% of the communities span more than 2 days, indicating that roughly half of the communities are short-lived.

Figs. 5(d), (e), and (f) depict the scatter plots generated for the features of size & topic diversity, size & time span, and time span & topic diversity, respectively. By plotting the least-squares line, we get a strong correlation of ~0.75 for both the size-topic diversity and size-time span feature pairs, as well as a very strong correlation of 0.98 for topic diversity and time span. These results indicate that larger communities cover, in general, more topics which was up to a point expected, but they are also active for a longer duration of time. This might be explained by the assumption that interest in small communities is focused on specific topics which correspond to a limited time period, whereas larger communities interact more frequently since they are interested in multiple relevant events. From both Figs. 5(d) and (e) we can observe that there is an outlier point at the largest community ($f_{size} = 1670$), which does not exhibit the expected magnitude in topic diversity and time span. This behavior could be attributed to the effect of retweets that may cause a significant increase in a community's size, but are focused on a single topic and are usually relevant to a single time-limited event.

To understand each Eurogroup meeting's impact, we associate them with the discovered communities and their features. We assume that each community expresses interest in an event, thus it is *active* on it, given that interactions between its members are observed on the current/previous/next day of the event. The number of active communities identified for each meeting are depicted in Fig. 4(a). To qualitatively characterize active communities, we further classify them as *small* (< 50 members), *medium* (50 ≤ members < 200), and *large* (≥ 200 members), and present their distribution for each event in the same figure. Since in total 6 large communities have been detected, we can observe that they are all active in 5 out of 8 events, which are also the events with the most tweets on the day they took place. This observation is inline with our previous analysis which indicated that larger communities generally cover more topics. Examination of the most popular events with respect to the tweets' number (20/11 and 26/11 in Brussels), reveals that although the latter one has attracted the most tweets, the earlier has more medium active communities. The meeting of 20/11 corresponds to the failure of European leaders to *reach an understanding of how to restructure Greece's aid package, thus delaying the next aid tranche*, whereas this of 26/11 to the *IMF's and eurozone's €40 billion debt-reduction agreement for Greece*[1]. Although apparently more buzz was generated on the day of the later event, it seems that the previous, a long critical meeting building up tension and failing to reach a result, has attracted the interest of more large and medium communities combined. The later event, on the contrary, has been of interest to more small communities, probably focused on its decision. By comparing the summer meetings of 21/6 and 9/7, we can observe that although the first has attracted less tweets than the second, it is related to more active communities. June meeting's target was *to discuss the latest developments in the eurozone, mainly in Greece, Spain, Portugal and Ireland*, whereas July's meeting aimed at *discussing*

---

[1] http://blogs.cfainstitute.org/investor/2011/11/21/european-sovereign-debt-crisis-overview-analysis-and-timeline-of-major-events/

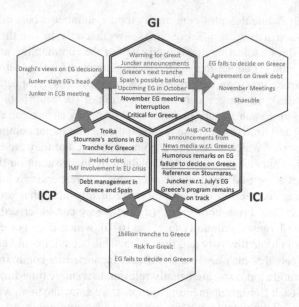

Fig. 6. Classification of the most significant topics based on interest intensity and diffusion. Horizontal lines separate different topics, while topics in red/blue correspond to the English/Greek language. Greek topics have been translated in English. (*Best viewed in color*)

EU/IMF's rescue programs for Spain, Greece and Cyprus[2]. More topics seem to be involved in the first event which may, up to a point, explain interest's dispersion in more communities. Some communities active on June's meeting might also be interested in a related topic: the announcement of the successful formation of a new government in Greece (which happened after a critical long election period associated with the question of Greece's continued eurozone membership), which took place one day before the event. Communities are also characterized in terms of their interest in the "Eurogroup" topic based on the number of meetings on which they are active. We assess interest expressed within a community in the following scale: *constant, intense, specific, random*, based on whether the community is active on 6-8, 3-5, 1-3, or 0 meetings, respectively. Most communities appear to have *specific* interest on few meetings, thus, a considerable percentage of them are indeed *focused* on the topic (with intense or constant interest). To identify the most popular topics within tweets, we resort to the following approach. We form 3 orderings of topics by ranking each topic based on: A) the number of tweets that express it over all communities, B) the number of communities that are related to at least one tweet that expresses it, C) the number of communities that are *strongly-related* to it. For ordering C, we assign each community to a single topic, i.e. the one expressed in most of its members' interactions, and then we rank each topic by the number of communities assigned to it. A set of 12 unique topics is generated by taking the top-5 from each ordering. We define 3 topic features: *General Intensity* (GI), *Inter-Community Popularity* (ICP), and *Inter-Community Intensity* (ICI), which characterize

---

[2] http://www.consilium.europa.eu/

topics that rank high (here, in the top-5) in ordering A, B, and C, respectively. In our set, there exist: 3 GI topics (which have the most intense user interest overall), 3 ICP topics (which reach out to the most communities), and 3 ICI topics (which play a major role in the most communities). There also exist 3 topics that combine two features, GI & ICP (attracting intense general interest while also being diffused in several communities), GI & ICI (attracting intense interest while also being major in several communities), and ICP & ICI (spanning several dedicated communities). Summaries of all 12 topics are depicted in Fig. 6, where the central hexagons correspond the GI, ICP, and ICI features, whereas the hexagons adjacent to two central ones represent the corresponding intersection. Topics are also divided based on their terms' language in English and Greek, since they are the ones represented in the set. It can be easily observed that all topics that combine two features (thus are more significant), are in English, indicating their significant impact on more users and communities. The most important illustrated topics along with their borderlines highlight users interest permutations.

5   Conclusions

In this work we propose a community detection approach for topic-focused interaction networks of social media users, which leverages both the structural properties of the network and the interactions' intensity. We investigate the role of weights in community detection approaches based on structural similarity and the possibility to combine them with automatic parameter selection. Our approach's correctness is validated on a series of synthetic networks. Moreover, its application on a real world network combined with a community meta-analysis process enables us to better understand the dual relationship between real world events / topics, and community formation.

Acknowledgements. This research was supported by Qualia, an on-line media monitoring and business intelligence company, situated in Athens, Greece, and the Research Committee of the Aristotle University of Thessaloniki, Greece.

References

1. Ankerst, M., et al.: Optics: ordering points to identify the clustering structure. In: SIGMOD, pp. 49–60 (1999)
2. Arenas, A., et al.: Community analysis in social networks. European Physics Journal B 38(2), 373–380 (2004)
3. Blei, D.M., et al.: Latent Dirichlet allocation. Journal of Machine Learning Research 3(4-5), 993–1022 (2003)
4. Bortner, D., Han, J.: Progressive clustering of networks using structure-connected order of traversal. In: ICDE, pp. 653–656 (2010)
5. Danon, L., et al.: Comparing community structure identification. J. Stat. Mech. P09008(0505245) (2005)
6. Ester, M., et al.: A density-based algorithm for discovering clusters in large spatial databases with noise. In: KDD, pp. 226–231 (1996)
7. Girvan, M., Newman, M.E.J.: Community structure in social and biological networks. Proc. of the National Academy of Sciences USA 99(12) (2002)

8. Jurgens, D., Lu, T.-C.: Friends, Enemies, and Lovers: Detecting Communities in Networks Where Relationships Matter. In: WebScience (2012)
9. Kamath, K.Y., Caverle, J.: Transient crowd discovery on the real-time social web. In: WSDM, pp. 585–594 (2011)
10. Lancichinetti, A., Fortunato, S.: Benchmarks for testing community detection algorithms on directed and weighted graphs with overlapping communities. Phys. Rev. E 80(016118) (2009)
11. Lancichinetti, A., et al.: Finding statistically significant communities in networks. PLoS ONE 6, e18961 (2011)
12. Morrison, D., et al.: Evolutionary clustering and analysis of user behaviour in online forums. In: ICWSM, 519–522 (2012)
13. Newman, M.E.J.: Power laws, Pareto distributions and Zipf's law. Contemporary Physics 46, 323–351 (2005)
14. Newman, M.E.J., Girvan, M.: Finding and evaluating community structure in networks. Phys. Rev. E 69 (2004)
15. Newman, M.E.J.: Finding community structure in networks using the eigenvectors of matrices. Phys. Rev. E 74(0602124v1) (2006)
16. Opsahl, T., Panzarasa, P.: Clustering in weighted networks. Social Networks 31(2), 155–163 (2009)
17. Papadopoulos, S., et al.: Community detection in social media. Data Mining & Knowledge Discovery 24, 515–554 (2012)
18. Sander, J., Qin, X., Lu, Z., Niu, N., Kovarsky, A.: Automatic extraction of clusters from hierarchical clustering representations. In: Whang, K.-Y., Jeon, J., Shim, K., Srivastava, J. (eds.) PAKDD 2003. LNCS (LNAI), vol. 2637, pp. 75–87. Springer, Heidelberg (2003)
19. Sun, J., et al.: Graphscope: parameter-free mining of large time-evolving graphs. In: KDD, pp. 687–696 (2007)
20. Sun, H., et al.: gskeletonclu: Density-based network clustering via structure-connected tree division or agglomeration. In: ICDM, pp. 481–490 (2010)
21. Ugander, J., et al.: The anatomy of the facebook social graph. CoRR abs/1111.4503 (2011)
22. Xu, X., et al.: Scan: a structural clustering algorithm for networks. In: KDD, pp. 824–833 (2007)

A Novel and Model Independent Approach for Efficient Influence Maximization in Social Networks

Hemank Lamba and Ramasuri Narayanam

IBM Research, India
{helamba1,ramasurn}@in.ibm.com

Abstract. Most of the recent online social media collects a huge volume of data not just about who is linked with whom *(aka link data)* but also, about who is interacting with whom *(aka interaction data)*. The presence of both *variety* and *volume* in these datasets pose new challenges while conducting social network analysis. In particular, we present a general framework to deal with both variety and volume in the data for a key social network analysis task - Influence Maximization. The well known influence maximization problem [15] (or viral marketing through social networks) deals with selecting a few influential initial seeds to maximize the awareness of product(s) over the social network. As it is computationally hard [15], a greedy approximation algorithm is designed to address the influence maximization problem. However, the major drawback of this greedy algorithm is that it runs extremely slow even on network datasets consisting of a few thousand nodes and edges [20,6]. Several efficient heuristics have been proposed in the literature [6] to alleviate this computational difficulty; however these heuristics are designed for specific influence propagation models such as linear threshold model and independent cascade model. This motivates the strong need to design an approach that not only works with any influence propagation model, but also efficiently solves the influence maximization problem. In this paper, we precisely address this problem by proposing a new framework which fuses both link and interaction data to come up with a backbone for a given social network, which can further be used for efficient influence maximization. We then conduct thorough experimentation with several real life social network datasets such as DBLP, Epinions, Digg, and Slashdot and show that the proposed approach is efficient as well as scalable.

Keywords: Social networks, influence maximization, viral marketing, sparsification.

1 Introduction

Most of the recent online social media collects a huge volume of data not just about who is linked with whom *(aka link data)* but also, about who is interacting with whom *(aka interaction data)*. The presence of both *variety* and *volume* in

X. Lin et al. (Eds.): WISE 2013, Part II, LNCS 8181, pp. 73–87, 2013.

these datasets pose new challenges while conducting social network analysis. In particular, we present a general framework to deal with both variety and volume in the data for a key social network analysis task, namely viral marketing. The phenomenon of viral marketing is to exploit the social interactions among individuals to promote awareness for new products. There are two well known operational models in the literature that capture the underlying dynamics of the information propagation in viral marketing. They are the linear threshold (LT) model [28,15] and the independent cascade (IC) model [13,15].

Domingos and Richardson [8] posed a fundamental algorithmic problem in the context of viral marketing with single product as follows. We are given the information about the extent individuals influence each other. We would like to market a new product that we hope will be adopted by a large fraction of individuals in the network. One of the key issues in viral marketing is to select a set of influential individuals (also called as *initial seeds*) in the social network and give them free samples of the product (or simply promotional offers on the product) to trigger cascade of influence over the network. The problem is, given a value for k, how should we choose a set of k influential individuals so that the cascade of influence over the network is maximized? Hereafter, we refer to this problem as *influence maximization problem* [15]. It is shown to be a NP-hard problem and this problem is well studied in the literature in the context of a single product [8,26,15,20,6], multiple independent products [7], and two competing companies [1,2]. For more details, we refer to Section 2 on the related work.

As the influence maximization problem is computationally hard, Kempe *et al.* [15] proposed a greedy approximation algorithm and it achieves an approximation guarantee of $\left(1 - \frac{1}{e}\right)$. However the sever drawback of this greedy algorithm is that it runs extremely slow even on network datasets consisting of a few thousand nodes and edges [20,6]. To circumvent the difficulties associated with the computational aspects of the greedy algorithm, several efficient heuristics have been proposed in the literature [6] to address the influence maximization problem. However, a major inadequacy of these efficient heuristics is that they are designed for specific influence propagation models such as linear threshold model and independent cascade model.

This motivates the strong need to design an approach that not only works with any influence propagation model, but also efficiently solves the influence maximization problem taking into account heterogeneous data sources that a social network provides. The primary contribution of this paper is to address this important research gap and our proposed approach complements the existing approaches in the literature for solving the influence maximization problem.

1.1 Our Approach

There are two key tasks associated with our proposed approach. We refer to the first task as *graph sparsification* and the second task as *influence maximization on sparse graphs*. In what follows, we now briefly describe each of the two tasks.

Graph Sparsification: In recent times, the proliferation of user activity on web-based social communities, micro-blogging sites, and other media such as Slashdot, Wikipedia, Facebook, Twitter, and Epinions [19,16] have offered a tremendous scope to mine not only the data about the link structure among the users, but also the rich information present in the content of interactions among users. The data about the link structure among users is popularly known as *link data* and the data about user interactions is known as *interaction data*[1]. A practical example for the link data is whether user x appears in the friends list of user y in online social communities such as Facebook and that of the interaction data is whether user x sends a message to user y (e.g., sending a wall-post in Facebook).

The major challenge here lies in dealing with the huge *volume* and a wide *variety* of datasets. While volume throws computational challenges, variety poses the challenge of data curation and information aggregation. The core idea of the sparsification task is to derive a backbone of the given social network by leveraging the link data and the interaction data (if it is available). At high level, our approach comprises of four steps: (i) For each of the link data and the interaction data, we induce ranking of the neighboring nodes for every node. Higher ranked neighbor means stronger association between these two nodes; (ii) Since different data sources convey different kinds of information at varying magnitude, we compute relative importance of these data sources after having derived the neighborhood rankings; (iii) Next, we aggregate different rankings of the neighborhood (along with their weights) induced by different data sources to arrive at a single aggregated ranking. (iv) Finally, we apply a sparsification trick to derive a sparse representation of the given graph by dropping edges that are less informative.

Influence Maximization using Sparse Graphs: For a given integer value for k, we determine the top-k initial seeds for the influence maximization problem by using the sparse graph obtained during the first task. To show the efficacy of the proposed approach, we then conduct thorough experimentation with several real life social network datasets such as Digg, Epinions, Slashdot, Amazon, and DBLP.

1.2   Novelty of the Paper

We believe that our contributions lead to the following novelty:

- Our approach is capable of handling volume as well as variety of the data while solving the influence maximization problem. Even in the absence of the interaction data, our approach also works with only the link data.
- The proposed approach is also very simple and extendible. At the core of our approach lies the notion of rank ordering the neighboring nodes for every node in a given graph. This rank ordering can be considered as a common

---

[1] Some authors call link data as *static data* and interaction data as *dynamic data* or *trace data*.

currency for capturing the information contained in the link data and the interaction data. For any given new type of data source, one just needs to figure out the way to compute ranking of neighborhood and rest of the framework remains unchanged.

- The proposed method to determine the sparse graph of the given social network is independent of the information propagation model. The proposed approach can be considered as a pre-processor step for the given dataset. Once the sparse graph is computed from the social network, one can just ignore the huge volume of the heterogenous data sources and just focus on this extracted structure as far as the influence maximization problem is concerned.

Note: We also use the phrase *data sources* to refer to both the link data and the interaction data in the rest of the paper.

2   Relevant Work

The topic of influence maximization over social networks has got a very significant attention from the research community and several algorithms are proposed in the literature. Here we only present the a few key results corresponding to the influence maximization problem in the literature and we divide them into several appropriate categories as follows.

Influence Maximization with Single Product: Domingos and Richardson [8,26] were the first to study influence maximization problem as an algorithmic problem. They modeled social networks as Markov random fields where the probability of an individual adopting a technology (or buying a product) is a function of both the intrinsic value of the technology (or the product) to the individual and the influence of neighbors. The computational aspects of the influence maximization problem are investigated by Kempe, Kleinberg, and Tardos [15]. The authors show that the optimization problem of selecting the most influential nodes is NP-hard and then proposed an approximation algorithm that achieves an approximation guarantee of $(1 - \frac{1}{e})$ where $e = \sum_{r=1}^{\inf} \frac{1}{r!}$. Followed by this, several efficient algorithms are proposed for the influence maximization problem [20,6,20,11,24].

Goyal *et al.* [12] proposed a new approach that leverages the traces of past action propagations while solving the influence maximization problem. Datta, Majumder, and Shrivastava [7] considered the influence maximization problem for *multiple independent products*. The work by [1,2,5,14,4] consider the algorithmic problem of how to introduce a new product into the market in the presence of a single or multiple competing products already in the market.

Our work in this paper is similar in spirit to that of Mathioudakis *et al.* [21]. The authors [21] proposed an efficient algorithm to determine the backbone of an influence network, given a social network and a log of past propagations. Our approach is very different from Mathioudakis *et al.* [21] in two fundamental aspects:

– Our work does not require any past propagations as in Mathioudakis *et al.* [21], and
– Our proposed approach is independent of the information propagation model, whereas the work [21] assumes the independent cascade (IC) model to be the underlying information propagation model.

3 Our Proposed Approach: Graph Sparsification

In this section, we describe our four step approach to discover the sparse graph of the social network. We begin with defining the notion of the sparse graph.

Definition 1 (Sparse Graph). *Let $G = (V, E)$ be a given directed / undirected social network. The sparse graph of a social network is a subgraph $G' = (V, E')$ so that we have $|E'| \subset |E|$ and the results of the influence maximization that we perform on G' is a good approximation of the result of the influence maximization problem when performed on the original graph G.*

Thus, in order to discover the sparse graph of a social network, we need to systematically delete a set of edges from the original given network. We identify this set of edges, i.e. $E \setminus E'$, in four steps - (1) Ranking the Neighborhood for each Data Source, (2) Computing Relative Importance of Rankings, (3) Rank Aggregation, (4) Graph Sparsification. In what follows, we describe each of these steps in detail.

3.1 Ranking the Neighborhood

For each data source D_s (link as well as interaction), we rank order the set of neighboring nodes, say $N(i)$, for each node $i \in V$. We denote this rank list by $R_i^s(\cdot)$ and the rank of a neighboring node j is given by $R_i^s(j)$ This rank list is a reflection of which edges are relatively more informative, as far as the corresponding data source is concerned. In other words, if j_1 and j_2 are two neighboring nodes for the node i and we have $R_i^s(j_1) < R_i^s(j_2)$ then it would mean that the edge (i, j_1) should be given priority over the edge (i, j_2) when it comes to remove an edge for the purpose of computing the sparse graph. Bear in mind that this recommendation is based on the data source D_s. For a different data source, say D_t, it could very well be possible that we have $R_i^t(j_1) > R_i^t(j_2)$.

Now the question is *"how to generate such a ranked list for a given data source D_s and a node $i \in V$"*. As far as link data is concerned, it comprises of graph structure and the edge weights. For link data, one can choose any structural property, for example *centrality measure*, as a criterion to rank order the neighborhood. For example, if one uses betweenness centrality as a criterion then the node $j \in N(i)$ having highest value of betweenness centrality would acquire rank one. In our experiments, we have used 11 different such measures to induce 11 different rankings of neighborhood for each node $i \in V$.

When it comes to a interaction data source, one can induce a ranking of the neighborhood either by simply counting the frequency of the neighboring node appearing in the interaction data or by applying a sophisticated *Learning to Rank* method [3]. However, the key idea behind any scheme is to assign higher ranking to a neighbor $j_1 \in N(i)$ as compared to the neighbor $j_2 \in N(i)$ if j_1 appears to be interacting relatively more with node i than what j_2 is interacting with i as per the given interaction data.

It is crucial to mention that the ranking criteria for link as well as interaction data are typically independent of the influence maximization problem.

3.2   Computing Relative Importance of Rankings

Suppose, in the previous step, we generated k different ranked lists $R_i^1(\cdot), R_i^2(\cdot), \ldots, R_i^k(\cdot)$ for any node $i \in V$. In this step, we assess the relative importance of these ranked lists so that we can avoid introducing bias in the sparse graph of the social network by having considered ranked lists of complementary nature multiple times. This steps is analogous to feature selection step in learning tasks.

Like feature selection problem, there can be multiple ways of assigning relative importance to the ranked list. However, we suggest one approach based on the idea of spectral clustering. For this, we define a distance $dist(R^s, R^t)$ between two different ranked lists $R^s(\cdot)$ and $R^t(\cdot)$ as follows.

$$dist(R^s, R^t) = \frac{1}{|V|} \sum_{i \in V} dist(R_i^s, R_i^t) \tag{1}$$

where $dist(R_i^s, R_i^t)$ is the distance between two different permutations of the nodes in the neighborhood $N(i)$ of the node i. There are several different ways for computing $dist(R_i^s, R_i^t)$ including *Spearman footrule* distance and *Kendall tau* distance [9]. We, in our experiments used Kendall Tau distance.

While there can be many ways to assign relative importance to the ranked lists from a given matrix of pairwise distances, we highlight two such approaches for the sake of illustration. In the first approach, we use techniques of spectral clustering or graph-cut to identify clusters of ranked lists where each cluster represents the set of similar ranked lists. We choose a centroid element as a representative element from each of these clusters and assign its relative importance as 1 and for others as 0. In the second approach, we construct a complete graph whose nodes represent ranked lists and whose edge weights represent the corresponding distance between the lists. For each node of this graph, we compute the closeness centrality. Closeness centrality contains information as to how close a particular node is to all the other nodes in the graph. Therefore, high closeness centrality will indicate that the node is quite close to other nodes and thus is not that unique. Therefore we assign $[closeness\ centrality]^{-1}$ as its relative importance score.

3.3   Rank Aggregation

In this step, we aggregate multiple ranked lists into one single ranked list in order to achieve an overall ordering of the neighborhood for every node. We also utilize the relative importance (aka weights) of the ranked list in the aggregation step. The rank aggregation is a well researched topic in Social Choice Theory and AI communities. There are large number of ways to perform rank aggregation. These methods differ interms of their underlying objectives. For our purpose, we prefer to use the well-known *Borda Rule* [9] because its simple, intuitive, and computationally easy to perform. Additionally, this rule can be augmented to take into account weights of the ranked lists as well.

As per Borda rule, if $R_i^1(\cdot), R_i^2(\cdot), \ldots, R_i^k(\cdot)$ are the k different rankings of the neighborhood for node $i \in V$, with the corresponding weights being $w_i^1, w_i^2, \ldots, w_i^k$, then the aggregated ranking $R_i(\cdot)$ is achieved by arranging them in a decreasing order of their *Borda score*. The Borda score of a neighboring node $j \in N(i)$ is given by

$$\alpha_i(j) = \sum_{t=1}^{k} w_i^t \, b\left(R_i^t(j)\right) \tag{2}$$

where $b\left(R_i^t(j)\right)$ is the Borda score function which assigns scores of $m-1, m-2, \ldots, 1, 0$ to the elements at the rank positions $1, 2, \ldots, m$, respectively, in any given ranked list of size m.

3.4   Graph Sparsification

In this step, we leverage aggregated ranked ordering of the neighborhood $N(i)$ for the purpose of deciding which of the edges incident on node i can be removed in order to compute the sparse graph. We essentially retain certain fraction of edges incident on every node $i \in V$. Note, the aggregated ranked list $R_i(\cdot)$ represents the importance of every edge (i.e. neighboring node) incident on node i. Therefore, it would be apt to retain those edges that appear in the top part of the aggregated ranked list. For this, we appeal to the method suggested by [27] where we retain $[deg(i)]^e$ number of top ranked edges (as per the list $R_i(\cdot)$) incident on node i. Here $deg(i)$ denotes the degree of the vertex i and $0 \le e \le 1$. One can also use more sophisticated sampling based techniques to select the edges.

4   Our Proposed Approach: Influence Maximization on Sparse Graphs

The second task of our proposed approach is to solve the influence maximization problem on the sparse graph. We measure the effectiveness of the initial seeds using the expected number of nodes in the social network that become active when we use these top k seeds as the initial active nodes [15]. In this paper, we refer to this as **reach**.

We work with both the LT model and the IC model. We consider two algorithms to determine the initial seeds for the influence maximization problem, namely the CELF algorithm [20] and the degree discount heuristic [6]. In particular, we deal with the following three configurations to address the influence maximization problem: (i) The CELF algorithm with LT model and we refer to this as *CELF - LT*, (ii) The CELF algorithm with IC model and we refer to this as *CELF - IC*, and (iii) Degree discount heuristic with IC model and we refer to this as *DDIC*. In each of these three configurations, for a given integer constant k, we first determine the top-k initial seeds using the sparse graph. Then we compute the reach of these top-k nodes on the original graph by running Monte-Carlo simulations. We finally compare this value with the reach of the top-k initial seeds determined using the original graph itself.

5   Experiments

In this section, we conduct thorough experimentation of the proposed approach to reveal its novelty. We first describe the datasets and then present the experimental setup. We next present the experimental results.

5.1   Datasets

We conduct experiments on 8 well known network datasets. These datasets are HepTh [22], Digg [17], FilmTip, Epinions [25], Amazon [29], DBLP [29], Netscience [23], and Slashdot [19]. Among these eight datasets, only Digg and FlimTip datasets have both link and interaction data. All the remaining five datasets have only link data. Table 1 provides a brief summary of these datasets and we now describe them as follows: (i) HepTh dataset is a co-authorship network of researchers in Physics. (ii) Digg dataset is a friendship network of users on Digg.com. (iii) Epinions dataset relates to *who-trust-whom* type relationship among users of a general purpose consumer review site Epinions.com. (iv) FilmTip dataset is a friendship network data. (v) Amazon dataset relates to the following feature of Amazon - *who bought this item also bought this*. In this kind of dataset, if a product i is frequently co-purchased with product j, the graph contains an undirected edge from i to j. (vi) DBLP dataset is a co-authorship network of researchers in computer science. In this dataset, two authors are connected if they publish at least one paper together. (vii) Netscience dataset is a co-authorship network of scientists working on network theory [23]. The vertices of the network represent authors of papers and edges join every pair of individuals whose names appear together as authors of a paper. (viii) Slashdot is a technology-related news website and since 2002 it allows for users to tag others as friends or foes.

5.2   Criteria for Ranking Neighbors

In our experiments, as far as the link data is concerned, we use following 12 criteria to rank the neighbors of each node in the network. We denote these 12 ranking criteria as *static criteria*. These criteria are applicable only for those

Table 1. Datasets used in the experiments. The datasets in italics are the ones that have the interaction data as well.

Dataset	Nodes	Edges	Type
Digg	*68,634*	*1,242,544*	*Directed*
FilmTip	*39,581*	*72,312*	*Undirected*
Amazon	334,863	925,872	Undirected
DBLP	317,080	1,049,866	Undirected
Epinions	75,879	508,837	Directed
HepTh	10,748	52,293	Directed
NetScience	1,589	2,742	Directed
Slashdot	82,168	948,464	Directed

datasets where links are directed (e.g. in the case for HepTh, Epinion, and Digg datasets). For an undirected dataset, different subsets of these criteria collapse to one single criterion and hence remaining number of different criteria becomes 6 (as described later). For a directed network, we adhere to the following convention when it comes to using these criteria: *if there is an edge from node a to node b, we say that node a is a follower of node b and node b is a friend of node a.* For undirected networks, if there is an edge between nodes a and b, it just means that nodes a and b are neighbors. For any node i, $d_{in}(i)$ and $d_{out}(i)$ represent its indegree and outdegree, respectively.

We start with defining first 5 criteria that are based on local graph structure for any node.

1. Average Indegree: It is the arithmetic mean of the number of incoming edges for each a and b, given by $[d_{in}(a) + d_{in}(b)]/2$.

2. Average Outdegree: It is the arithmetic mean of the number of outgoing edges for each a and b, given by $[d_{out}(a) + d_{out}(b)]/2$.

3. Common Followers: It is the number of common followers of a and b, given by $|d_{in}(a) \cap d_{in}(b)|$.

4. Common Friends: It is the number of common friends of a and b, given by $|d_{out}(a) \cap d_{out}(b)|$.

5. Common Neighbors: It is the number of common neighbors of a and b, given by $|(d_{out}(a) \cap d_{out}(b)) \cup (d_{in}(a) \cap d_{in}(b))|$.

We now introduce a few criteria based on network proximity measures [18]. These criteria measure the likelihood of a message originating from node a reaches node b regardless of the path it takes.

6. Common Neighbors Metric: Let us define $C = d_{out}(a) \cap d_{in}(b)$ and $C' = d_{in}(a) \cap d_{out}(b)$. Now, the common neighbors metric is the average number of nodes in C and C'. That is, $0.5|C| + 0.5|C'|$.

7. Jaccard Coefficient: $0.5 \frac{|d_{out}(a) \cap d_{in}(b)|}{|d_{out}(a) \cup d_{in}(b)|} + 0.5 \frac{|d_{out}(b) \cap d_{in}(a)|}{|d_{out}(b) \cup d_{in}(a)|}$

8. Adamic Adar Score: $0.5 \sum_{z \in C} [\log d(z)]^{-1} + 0.5 \sum_{z' \in C'} [\log d(z')]^{-1}$

We next consider a few more criteria where the proximity score is dependent not only on the neighborhood-overlap, but also on different types of interactions that can occur between network nodes [17].

9. Conservative Metric: $0.5 \sum_{z \in C} \frac{1}{d_{out}(a)d_{out}(z)} + 0.5 \sum_{z' \in C'} \frac{1}{d_{out}(b)d_{out}(z)}$

10. Conservative Attention Limited Metric:
$0.5 \sum_{z \in C} \frac{1}{d_{out}(a)d_{in}(z)d_{out}(z)d_{in}(b)} + 0.5 \sum_{z \in C} \frac{1}{d_{out}(b)d_{in}(z)d_{out}(z)d_{in}(a)}$.

11. Non-Conservative Metric: $0.5|C| + 0.5|C'|$.

12. Non-Conservative Attention Limited Metric: $0.5 \sum_{z \in C} \frac{1}{d_{in}(z)d_{in}(b)} + 0.5 \sum_{z \in C'} \frac{1}{d_{in}(z)d_{in}(a)}$.

For undirected networks, it is easy to verify that some of the above criteria would produce the same ranking of neighbors. Therefore, for undirected networks, these 12 criterion would boil down to following 6 criteria: (i) Average Degree, (ii) Common Neighbors Metric, (iii) Jaccard Coefficient, (iv) Adamic Adar Score, (v) Conservative Metric, (vi) Conservative Attention Limited Metric.

As far as the interaction data is concerned, we also consider a few more other criterion and we refer to them as dynamic criterion. Recall that only Digg and FilmTip datasets have interaction data. For Digg dataset, we use the method proposed in Goyal *et. al* [10]. We use the Jaccard coefficient metric $\frac{A_{a2b}}{A_{a|b}}$ where A_{a2b} are the actions (here votes on stories) performed by user a and seen by user b and $A_{a|b}$ are unique actions performed by either a or b. Similarly, for FilmTip dataset, we work with 3 dynamic criterion as follows: (i) It is the jaccard coefficient between the movies rated by user a and b. It is denoted by $\frac{M_a \cap M_b}{M_a \cup M_b}$. where M_a is the set of movies rated by user a and M_b is the set of movies rated by b; (ii) We divide the movie ratings into 3 categories, namely Bad (ones with score of 1 or 2), Neutral (those with score 3) and Good (ones with the score of 4 or 5). The metric is denoted by number of movies rated in the same category by both of the users; and (iii) The third criteria measures exact similarity of users. It is denoted by number of movies rated exactly the same by both of the users.

5.3   Experimental Setup

We conducted all our experiments on a PC with Linux OS, 3.05 GB of RAM, and Intel Core i7 CPU Q720 1.2 GHz. As we already mentioned earlier, we consider various algorithms and heuristics to determine the initial seeds and the various datasets to work with. Table 2 shows the list of algorithms that we ran on different datasets. For every instance of the IC model, we consider a constant probability $p = 0.01$ except for for NetScience dataset where we set $p = 0.25$.

5.4   Experimental Results

We categorize the experimental results into various subsections as follows. We first present the results obtained using the CELF algorithm and then present the results obtained using the degree discount heuristic.

CELF Algorithm: We ran CELF algorithm with both LT model and IC model on three datasets namely, FilmTip, HepTh, and Netscience. Figure 1 shows the

Table 2. Listing of dataset and algorithm (or heuristic) pair used to determine the initial seeds

Dataset	DDIC	CELF - IC	CELF - LT
Digg	✓	✗	✗
FilmTip	✓	✓	✓
Amazon	✓	✗	✗
DBLP	✓	✗	✗
Epinions	✓	✗	✗
HepTh	✓	✓	✓
NetScience	✓	✓	✓
Slashdot	✓	✗	✗

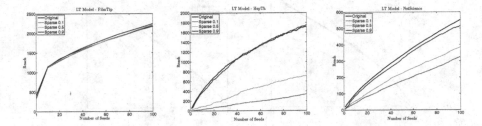

Fig. 1. Reach of the initial seeds obtained using CELF with LT Model on (i)FilmTip (ii)HepTh (iii) NetScience

reach obtained using CELF with LT model on FilmTip, HepTh and Netscience datasets. For the IC model, we set $p = 0.01$ for FilmTip and HepTh datasets and we set $p = 0.25$ for Netscience dataset. Similarly, Figure 5.4 shows the reach obtained using CELF with IC model on FilmTip, HepTh and Netscience datasets. The reach of the initial seeds using the CELF with both LT and IC models computed with the sparse graphs is almost similar to that of the original graph.

Results Based on DDIC: Since applying greedy algorithm on larger datasets takes long time, we used degree discount heuristic on larger datasets to measure the reach of the initial seeds computed using the sparse graphs. The results of degree discount heuristic with independent cascade probability 0.01 on DBLP, Amazon, Epinions and FilmTip is shown in Figure 3. We plot only reach of the seeds for the sparse files with $e = 0.1$, $e = 0.5$, and $e = 0.9$ and the rest are omitted in the interest of space constraints. Similar results were obtained even for other datasets. From the figure we can see that the seed quality obtained using sparse graphs is equivalent and in some cases even outperforms the original seeds. On Amazon and FilmTip, even by using just 34.23% 39.56% edges (e=0.1) respectively we obtain a solution that is matchable with the original.

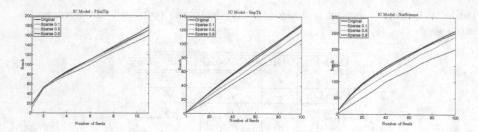

Fig. 2. Reach of the initial seeds obtained using CELF with IC Model on (i)FilmTip (ii)HepTh (iii) NetScience

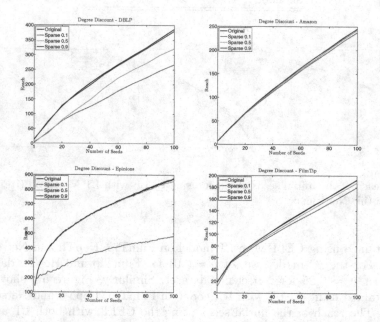

Fig. 3. Degree Discount Results: (i) DBLP (ii)Amazon (iii) Epinions (iv) FilmTip

Analysis of Running Time: The previous results show that the efficacy of the initial seeds computed using the sparse graphs in terms of *reach*. We now compare and contrast the running time taken by the CELF algorithm on the sparse graphs and the original graph respectively for each dataset to determine top-100 initial seeds. Figure 5.4 shows the running of the CELF algorithm with LT model on FilmTip, HepTh, and Netscience datasets to determine top-100 initial seeds. Similarly, Figure 5 shows the running of the CELF algorithm with IC model on FilmTip, HepTh, and Netscience datasets to determine top-100 initial seeds.

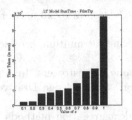
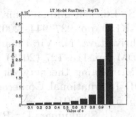
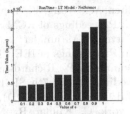

Fig. 4. LT Model Time Analysis: (i)FilmTip (ii)HepTh (iii) NetScience

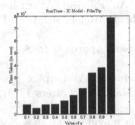
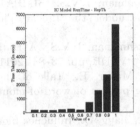
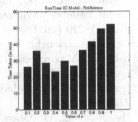

Fig. 5. IC Model Time Analysis: (i)FilmTip (ii)HepTh (iii) NetScience

From these results, it is easy to see that the time taken by sparse graphs to determine the top-100 initial seeds is much lesser than the time taken by the original graph. In other words, the proposed method to determine the sparse graphs essentially retains very rich information as far as the original graph is concerned while dropping a significant number of edges. These experimental results reveal that the sparse graphs obtained using the proposed approach not only produce the initial seeds with high quality, but also lead to significant speed-up in terms of computation time.

References

1. Bharathi, S., Kempe, D., Salek, M.: Competitive influence maximization in social networks. In: Deng, X., Graham, F.C. (eds.) WINE 2007. LNCS, vol. 4858, pp. 306–311. Springer, Heidelberg (2007)
2. Borodin, A., Filmus, Y., Oren, J.: Threshold models for competitive influence in social networks. In: Saberi, A. (ed.) WINE 2010. LNCS, vol. 6484, pp. 539–550. Springer, Heidelberg (2010)
3. Burges, C., Shaked, T., Renshaw, E., Lazier, A., Deeds, M., Hamilton, N., Hullender, G.: Learning to rank using gradient descent. In: Proceedings of the 22nd ICML, pp. 89–96 (2005)
4. Budak, C., Agrawal, D., Abbadi, A.E.: Limiting the spread of misinformation in social networks. In: Proceedings of WWW, pp. 665–674 (2011)
5. Carnes, T., Nagarajan, C., Wild, S., van Zuylen, A.: Maximizing in uence in a competitive social network: a follower's perspective. In: Proceedings of the 9th International Conference on Electronic Commerce (ICEC), pp. 351–360 (2007)

6. Chen, W., Wang, Y., Yang, S.: Efficient influence maximization in social networks. In: Proceedings of ACM SIGKDD, pp. 937–944 (2009)
7. Datta, S., Majumder, A., Shrivastava, N.: Viral marketing for multiple products. In: Proceedings of IEEE ICDM, pp. 118–127 (2010)
8. Domingos, P., Richardson, M.: Mining the network value of customers. In: Proceedings of the 7th SIGKDD International Conference on Knowledge Discovery and Data Mining (KDD), pp. 57–66 (2001)
9. Dwork, C., Kumar, R., Naor, M., Sivakumar, D.: Rank aggregation methods for the web. In: Proceedings of WWW, pp. 613–622 (2001)
10. Goyal, A., Bonchi, F.: Laks V. S. Lakshmanan. Learning influence probabilities in social networks. In: WSDM, pp. 241–250 (2010)
11. Goyal, A., Lu, W., Lakshmanan, L.V.S.: SIMPATH: An Efficient Algorithm for Influence Maximization under the Linear Threshold Model. In: ICDM, pp. 211–220 (2011)
12. Goyal, A., Bonchi, F., Lakshmanan, L.V.S.: A Data-Based Approach to Social Influence Maximization. In: PVLDB, pp. 73–84 (2011)
13. Goldenberg, J., Libai, B., Muller, E.: Talk of the network: A complex systems look at the underlying process of word-of-mouth. Marketing Letters 12(3), 211–223 (2001)
14. He, X., Song, G., Chen, W., Jiang, Q.: Influence blocking maximization in social networks under the competitive linear threshold model. In: Proceedings of the 12th SIAM International Conference on Data Mining, SDM (2012)
15. Kempe, D., Kleinberg, J., Tardos, E.: Maximizing the spread of influence through a social network. In: Proceedings of the 9th SIGKDD International Conference on Knowledge Discovery and Data Mining (KDD), pp. 137–146 (2003)
16. Kunegis, J., Lommatzsch, A., Bauckhage, C.: The slashdot zoo: Mining a social network with negative edges. In: Proc. of 18th WWW, pp. 740–750 (2009)
17. Lerman, K., Ghosh, R.: Information Contagion: An Empirical Study of Spread of News on Digg and Twitter Social Networks. In: Proceedings of 4th International Conference on Weblogs and Social Media, ICWSM (2010)
18. Lermann, K., Intagorn, S., Kang, J.H., Ghosh, R.: Using Proximity to Predict Activity in Social Networks. In: Proceedings of the 21st International World Wide Web Conference (poster) (December 2012)
19. Leskovec, J., Huttenlocher, D., Kleinberg, J.: Signed networks in social media. In: Proceedings of the 28th ACM SIGCHI Conference on Human Factors in Computing Systems (CHI), pp. 1361–1370 (2010)
20. Leskovec, J., Krause, A., Guestrin, C., Faloutsos, C., VanBriesen, J., Glance, N.: Cost-effective outbreak detection in networks. In: Proceedings of the 13th SIGKDD International Conference on Knowledge Discovery and Data Mining (KDD), pp. 420–429 (2007)
21. Mathioudakis, M., Bonchi, F., Castillo, C., Gionis, A., Ukkonen, A.: Sparsification of influence networks. In: Proceedings of the 17th SIGKDD International Conference on Knowledge Discovery and Data Mining, KDD (2011)
22. Newman, M.E.J.: The structure of scientific collaboration networks. Proceedings of the National Academy of Sciences (PNAS) 98, 404–409 (2001)
23. Newman, M.E.J.: Finding community structure in networks using the eigenvectors of matrices. Physical Review E 74(3), 036104 (2006)
24. Ramasuri, N., Narahari, Y.: A Shapley Value-Based Approach to Discover Influential Nodes in Social Networks. IEEE T. Automation Science and Engineering 8(1), 130–147 (2011)

25. Richardson, M., Agrawal, R., Domingos, P.: Trust management for the semantic web. In: Fensel, D., Sycara, K., Mylopoulos, J. (eds.) ISWC 2003. LNCS, vol. 2870, pp. 351–368. Springer, Heidelberg (2003)
26. Richardson, M., Domingos, P.: Mining knowledge-sharing sites for viral marketing. In: Proceedings of the 8th SIGKDD International Conference on Knowledge Discovery and Data Mining (KDD), pp. 61–70 (2002)
27. Satuluri, S., Parthasarathy, V., Ruan, Y.: Local graph sparsification for scalable clustering. In: Proc. of SIGMOD, pp. 721–732 (2011)
28. Schelling, T.: Micromotives and Macrobehavior. W.W. Norton and Company (1978)
29. Yang, J., Leskovec: Defining and evaluating network communities based on ground-truth. In: Proceedings of ICDM (2012)

iPLUG: Personalized List Recommendation in Twitter

Lijiang Chen[1], Yibing Zhao[2], Shimin Chen[3], Hui Fang[4], Chengkai Li[5],
and Min Wang[6]

[1] HP Labs, Beijing, China
lijiang.chen@hp.com
[2] John Hopkins University, Baltimore, MD, USA
zyb009988@gmail.com
[3] Chinese Academy of Sciences, Beijing, China
chensm@ict.ac.cn
[4] University of Delaware, Newark, DE, USA
hfang@udel.edu
[5] University of Texas at Arlington, Arlington, TX, USA
cli@cse.uta.edu
[6] Google Research, Mountain View, USA
minwang@google.com

Abstract. A Twitter user can easily be overwhelmed by flooding tweets from her followees, making it challenging for the user to find interesting and useful information in tweets. The feature of *Twitter Lists* allows users to organize their followees into multiple subsets for selectively digesting tweets. However, this feature has not received wide reception because users are reluctant to invest initial efforts in manually creating lists. To address the challenge of bootstrapping Twitter Lists, we envision a novel tool that automatically creates personalized Twitter Lists and recommends them to users. Compared with lists created by real Twitter users, the lists generated by our algorithms achieve 73.6% similarity.

1 Introduction

Twitter is an instant content sharing service, through which users share their opinions and status by posting *tweet*s, i.e., short messages with less than 140 characters. As one of the most popular social networking services, Twitter boasts over 140 million active users and more than 340 million tweets per day[1]. With Twitter, a user (*follower*) can follow other users' (*followees*) tweets. A follow operation establishes a subscription-dissemination channel to send messages from a followee to a follower.

Twitter Lists. Users can easily be overwhelmed by flooding tweets from their followees since both the number of followees they subscribe to and the number of tweets their followees post on a daily basis could be very large. Thus, it is critical to help users organize their followees in order to more effectively digest and access the information posted by their followees.

Twitter Lists is one useful feature in Twitter for coping with this information overload problem. A user can organize her followees into multiple lists, each containing a subset

---

[1] http://blog.twitter.com/2012/03/twitter-turns-six.html

X. Lin et al. (Eds.): WISE 2013, Part II, LNCS 8181, pp. 88–103, 2013.
© Springer-Verlag Berlin Heidelberg 2013

of followees. She can then opt to view the tweets from the followees in a particular list, thereby selectively digesting information from the list. A user can also subscribe to lists created by others without following the users in those lists.

Problem Definition. Albeit a potentially effective tool for social user organization and information selection in overloaded information space, Twitter Lists has not received wide reception from users [2]. One observation made from our empirical study is that over 80% of Twitter users have more than 100 followees, but only 35% of them have created at least one list and only 16% have more than 3 lists (cf. Observation 1). Moreover, among the users who have created lists, 49.6% of them included only less than 5% of their followees into their lists, and 76% of them included less then 10% of their followees in the lists (cf. Observation 2). This is not surprising, as other social media and social networks have encountered similar challenges in bootstrapping their services, including tagging, social bookmarking, and friend recommendation [5,1,4]. One particular reason for the low popularity of Twitter Lists is that users are reluctant to invest initial efforts in manually creating lists. Another reason is that users may have intrinsic difficulty in such a fuzzy grouping process.

To tackle the low popularity of Twitter Lists and help achieve its full utility, we propose to automatically recommend personalized lists to Twitter users. This study can also shed light on solving similar problems in other leading social network services, which have features similar to Twitter Lists (e.g., *friend lists* in Facebook, *circles* in Google Plus).

Various definitions of the list recommendation problem may be worth studying. As an initial step towards effective list recommendation, this paper focuses on a particular problem definition— ***Given a Twitter user, recommend to the user new lists consisting of only the user's current followees.*** We consider the problem of recommending new followees a separate issue. We also do not simply recommend lists subscribed by many users, as such globally popular lists do not necessarily match a particular user's personal interests and social relationships.

Our Solution: iPLUG. We propose a solution that combines two approaches— a structure-based method and a content-based method, which recommend lists based on how users are co-listed in existing lists and how similar the contents of users' tweets are, respectively. After obtaining the initial list recommendations by the structure-based method and the content-based method, we improve the accuracy of the recommendations by performing both inter-list optimization and intra-list optimization. For inter-list optimization, we diversify top k recommended lists, to achieve a degree of overlap among these lists similar to the overlap exhibited by existing lists. For intra-list optimization, we study ways to prune unimportant members from the recommended lists. We call the resulting solution ***integrated PLUG (iPLUG)***.

The effectiveness of the proposed techniques is demonstrated by extensive experimental results. We evaluate iPLUG as well as various schemes that consist of only subsets of the proposed techniques. In a nutshell, the similarity between lists recommended by the proposed methods and those created by real users is as high as 73.6% for the largest lists and 72.2% for the largest five lists.

---

[2] http://moreinmedia.com/2011/12/why-make-use-twitter-lists

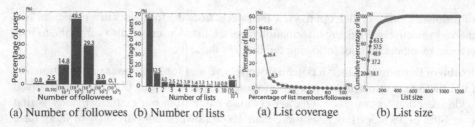

(a) Number of followees (b) Number of lists   (a) List coverage   (b) List size

Fig. 1. Statistics of Twitter lists   **Fig. 2.** Live coverage and list size

2   Characteristics of Twitter Lists Created by Real Users

To understand the characteristics of real lists created by Twitter users and to gain insights towards our technical solution for personalized list recommendation, we collected a large Twitter data set and conducted a comprehensive study on its various statistics. Our data set contains $810,769$ Twitter users, $53,922,948$ lists, and 324 million tweets. The tweets were posted between December 27th, 2007 and March 21st, 2011. Details on the collection of the data set shall be described in the experiment section. In the rest of this section, we introduce our observations from the data set and provide analysis of the observations.

Observation 1. A Twitter user often follows a large number of other users, but rarely creates lists.

Figure 1(a) and Figure 1(b) show the distribution of the number of followees of a user and the distribution of the number of lists created by a user, respectively. From Figure 1(a), we observe that over 80% of Twitter users have 100 or more followees and almost half of all users have between 100 and 1000 followees. However, Figure 1(b) shows that 65% of users do not create any list. Only 16% of users have more than 3 lists, and nearly 93% have less than 10 lists. The sharp contrast between the large number of followees per user and the low popularity of Twitter Lists motivates our study of mechanisms for automatic list recommendation.

Observation 2. For users who have created lists, their lists often cover only a small fraction of their followees.

For those users that have lists, we measured the percentages of their followees in their lists. The results are shown in Figure 2(a). We observe that the lists created by a user often cover only a small fraction of the followees of the user. More specifically, 49.6% of the lists contain only less than 5% of the corresponding list owners' followees and 76% of the lists cover less than 10% of their followees. We also observe that lists are typically small in size. Figure 2(b) shows the cumulative distribution of list sizes (by increment of every 10 lists). The largest list has around 1200 members, while the most typical list sizes are [1, 10) and [10, 20), accounting for 18.1% and 19.1% of all the lists, respectively. 63.5% of the lists have less than 50 members. The observed small coverage and small size of existing lists further motivate the need for an automatic list recommendation approach.

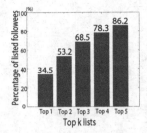

Fig. 3. The coverage of the largest 5 lists

User Set	Top 2	Top 3	Top 4	Top 5
Created > 1 lists	13.2%			
Created > 2 lists	13.7%	13.5%		
Created > 3 lists	13.1%	12.6%	10.9%	
Created > 4 lists	12.7%	11.8%	10.8%	10.0%

Fig. 4. Average redundancy in users' largest k lists

Observation 3. For users who have created multiple lists, a small number of their largest lists can cover most of their followees in their lists.

We further investigated the coverage of users' largest lists. We focused on the users who created at least 5 lists. For a user, we compute the coverage of her largest k lists by dividing the number of her followees in her largest k lists by all the followees in her lists. Figure 3 reports the average coverage across all these users while varying k from 1 to 5. As shown in Figure 3, the largest list of a user covers 34.5% of the followees in her lists on average, and the largest 5 lists cover an average of 86% of all followees in her lists. The statistics make it clear that a small number of lists cover most listed followees for a user. Assuming that creating longer lists takes more efforts, we speculate that users spend most efforts in creating largest 5 lists, which should be most important to them.

Observation 4. Although a user could put a followee into multiple lists, the number of overlapped followees among the lists is small.

We measured the redundancy in a user's largest k lists $l_1, \ldots, l_k$ as follows:

$$Redundancy = \frac{\sum_{j=1}^{k} |l_j| - |\bigcup_{j=1}^{k} l_j|}{|\bigcup_{j=1}^{k} l_j|} \qquad (1)$$

where $|l_j|$ is the size of list l_j, i.e., the number of members in l_j. The $\bigcup$ denotes the set union operation.

Note that we only considered the largest k lists because many lists in the long tail only contain a few members and thus do not have much impact on the measured redundancy. We investigated the redundancy for different k values, ranging from 2 to 5. For each k value, we calculated the redundancy in the largest k lists, averaged across all users that have at least k lists. The result is shown in Table 4. We see that overall the average redundancy is small (10% to 13%) and slightly decreases while k increases.

Observation 3 and Observation 4 provide important characteristics of the lists created by real users. Given these characteristics, we aim to develop methods that generate lists exhibiting similar characteristics.

3 Personalized Twitter Lists Recommendation

Given a Twitter user u, let FE_u and FR_u denote the set of *followees* and the set of *followers* of u, respectively. Hence, $u_1 \in FE_{u_2} \Leftrightarrow u_2 \in FR_{u_1}$. A *list* l created by

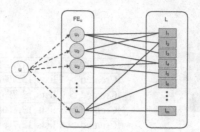

Fig. 5. The PLUG for a user u

user u is a subset of u's followees, i.e., $l \subseteq FE_u$. We use $L_u = \{l_1, l_2, \ldots, l_m\}$ to denote the set of lists created by u. Since a followee can be placed into multiple lists of u, it is possible that $l_i \cap l_j \neq \emptyset$. A followee of a user can also be followed by other users and included in their lists. The set of lists, to which a user u belongs, is denoted $L'_u = \{l \mid u \in l\}$.

Problem Statement: Given a user u and her followees FE_u, the problem of *list recommendation* is to generate k new lists $l_1, \ldots, l_k$ for u, where $l_i \subseteq FE_u$ and $l_i \notin L_u$ for $1 \leq i \leq k$.

Note that the problem focuses on recommending top-k lists. It is not necessary that every followee of a user u is assigned to a new or existing list.

3.1 Solution Overview

We propose two methods to form and rank candidate lists. The structure-based method exploits the personalized list-user graph (PLUG) of a user and recommends new lists containing her followees that are often included together in existing lists. The content-based method recommends lists based on content similarity of the tweets posted by list members. The two methods are different on two aspects. First, the structure-based method can only cover *listed followees*— a user u's followees that are included in at least one existing list created by any user, including u. The content-based method can also cover *non-listed followees*. Second, the content-based method is explicitly geared towards recommending interest-oriented lists, in which the list members share similar interests on topics (e.g., music and sports) shown by their tweets. The structure-based method can also recommend relation-oriented lists, in which the list members have common real-world relations with their follower u (e.g., family, friends, colleagues).

To improve the accuracy of the initial recommendations from these two methods, an inter-list optimization is applied to diversify recommended top-k lists, and an intra-list optimization is applied to prune unimportant members from the recommended lists.

We also combine the results from both structure- and content-based methods. The resulting method is termed *integrated-PLUG (iPLUG* for short). Suppose the top-k ranked lists for user u by structure-based and content-based methods are Rec^s and Rec^c, respectively. For each list $l \in Rec^s \cup Rec^c$, its ranking scores in Rec^s and Rec^c are $Score_l^s$ and $Score_l^c$, respectively. (Note that $Score_l^s = 0$ if $l \notin Rec^s$ and $Score_l^c = 0$ if $l \notin Rec^c$.) The iPLUG method computes the new ranking score of l by the following equation. The top-k lists with highest new scores are recommended to u.

$$Score_l = \eta Score_l^s + (1 - \eta)Score_l^c \tag{2}$$

In Equation (2), η is a regulatory factor (i.e. $0 \leq \eta \leq 1$) that controls the contributions of the two methods to the final result. Larger η values prefer more contributions from the structure-based method. In our implementation we tested an intuitive value for η, which is the percentage of u's listed followees, i.e., $\eta = |\{u_i | u_i \in FE_u \text{ and } L'_{u_i} \neq \emptyset\}|$ / $|FE_u|$. The rationale is that, since the structure-based method can only cover listed followees, the percentage is used to determine the extent of the method's effect.

3.2   Structure-Based Method

Personalized List-User Graph (PLUG). The structure-based method takes advantage of crowd intelligence of Twitter users to recommend new lists. We call two users u_1 and u_2 *co-listed* in list l, if both users are members of l, i.e., $u_1 \in l$ and $u_2 \in l$. If an existing list l co-lists u_1 and u_2, l's creator indicates that u_1 and u_2 are related from her perspective. Therefore, if u_1 and u_2 are frequently co-listed by existing lists, then many existing users agree that u_1 and u_2 are related. The more frequent that two users are co-listed, the stronger that they are considered related among Twitter users. We exploit this intuition in the structure-based method.

Specifically, to recommend lists to a user u, we build a *personalized list-user graph* (PLUG) for u. As illustrated in Figure 5, it is an undirected bipartite graph between the followees of u, depicted as circles, and existing lists created by all users (including u), depicted as rectangles. An edge between a followee node u_i and a list node l_j captures the membership relation $u_i \in l_j$. Two followees are co-listed in a list if they are both connected to the list node. The concept of PLUG is defined as follows.

Definition 1 (Personalized List-User Graph)
The personalized list-user graph of a user u, $PLUG_u = (FE_u, L, E)$, is a bipartite graph between two node sets FE_u and L, where FE_u is the set of u's followees and L is the set of lists containing u's followees, i.e., $L = \bigcup_{u_i \in FE_u} L'_{u_i}$. In the edge set E, each edge captures the membership of a followee u_i in a list l_j, i.e., $E = \{(u_i, l_j) \mid u_i \in FE_u, l_j \in L, u_i \in l_j\}$.

The algorithm exploits an iterative random walk model [3] to propagate scores on the bipartite graph. It ranks the followees FE_u and the lists L based on their connections in $PLUG_u$ according to the following equations.

$$s_{u_i}^{t+1} = \sum_{l_j \in L'_{u_i}} \frac{Weight(u_i, l_j)}{\sum_{u_k \in FE_u} Weight(u_k, l_j)} Score_{l_j}^t \tag{3}$$

$$s_{l_j}^{t+1} = \sum_{u_i \in l_j \cap FE_u} \frac{Weight(u_i, l_j)}{\sum_{l_t \in L} Weight(u_i, l_t)} Score_{u_i}^t \tag{4}$$

$$Score_{u_i}^{t+1} = \frac{s_{u_i}^{t+1}}{\sum_{u_k \in FE_u} s_{u_k}^{t+1}} \tag{5}$$

$$Score_{l_j}^{t+1} = \frac{s_{l_j}^{t+1}}{\sum_{l_k \in L} s_{l_k}^{t+1}} \tag{6}$$

In Equations (3) and (4), the score of a followee u_i is collected from all lists containing u_i, where the score of such a list l_j is distributed into its members. The score of u_i collected from l_j is according to the transition probability, which captures the importance of l_j to u_i. Recursively, the score of a list is collected from all its members, where the score of a member is distributed into its containing lists. The score of l_j collected from u_i is according to the transition probability. Then the scores of followees and lists are both normalized, by Equations (5) and (6), respectively. The initial score of a followee is 1 normalized to the total number of followees. The initial score of a list is the list size after removing unrelated members, $|l_j \cap FE_u|$, normalized to the aggregate size of the lists. With the initial scores of followees and lists, our method iteratively updates their scores using the above equations. After the scores converge, the algorithm selects and returns the top k lists with the highest scores as the recommended lists.

Tackling Sparsity of Co-listed Relation by List Clustering. For most users, we find that their PLUGs are sparse bipartite graphs, since many of the followees are placed in only a small number of existing lists because of the current low popularity of Twitter Lists. To tackle this sparsity problem of the co-listed relation, we improve the structured-based method by merging similar lists before the iterative computation of scores by Equation (3) and (4).

A natural similarity measure between two lists would have been based on the tweet contents of their members. However, such a similarity measure will only work for topic interest-oriented lists (e.g., food, music) rather than relation-oriented lists (e.g., colleagues in an organization), since the members of a relation-oriented list may not share the same interests and their tweets may not have similar contents. Such a similarity measure would increase the significance of interest-oriented lists, while relation-oriented lists will become relatively less significant. As a result, our algorithm would tend to recommend only interest-oriented lists, which is undesirable.

Instead, we merge lists according to the similarity of list names, thus making a balanced improvement for both kinds of lists. We pre-process list names by applying word stemming. Then we cluster the lists using the edit distance between stemmed list names as the distance function. Since we do not know the resulting number of clusters, we prefer hierarchical clustering to k-means. However, a regular hierarchical clustering algorithm is compute-intensive. Therefore, we instead use a greedy algorithm which aims to achieve the goal that the distance between any two final clusters is greater than or equal to a specified threshold T_{dist} ($T_{dist} = 0.4$ in our implementation). The algorithm randomly chooses a point (i.e., a list name) to form a single-point cluster, and then iteratively grows the cluster by adding other points whose average distances to the existing points in the cluster satisfy the threshold T_{dist}, until no more point can be included into the current cluster. The algorithm repeats this process to form multiple clusters. This greedy clustering algorithm computes the pair-wise distance between any two points at most once, and therefore has worst-case time complexity of $O(|L|^2)$.

Finally, we merge the multiple lists in each cluster into one list. The new PLUG of a user u is a bipartite graph between u's followees and the merged lists. Each merged list corresponds to a set of original lists. An edge exists between a followee u_i and a merged list l if u_i was in at least one of the constituent lists of l. In our data set, we start with $53, 922, 948$ original lists. The total number of resulting merged lists (clusters) for all users is $240,572$, a nearly 99.5% reduction from the original lists. The new PLUG of a user becomes more compact and the extended co-listed relation helps improve our recommendation algorithm.

A new PLUG may not distinguish the importance of different followees in the merged lists. It is possible that a followee was included in multiple original lists within a merged list. For example, consider a musician u_1 that is in many music-related lists created by her fans. Such lists with similar names may be clustered into the same cluster and merged into a single list. In contrast, another followee u_2 may be in only one of the original lists before merging. It is clear that these two followees u_1 and u_2 should bear different importance in the resulting PLUG.

To distinguish the importance of different followees in the merged lists, we enhance PLUG to edge-weighted PLUG. A weight on an edge between a followee u_i and a merged list l_j represents the number of l_j's constituent lists that contain u_i:

$$Weight(u_i, l_j) = |L_{l_j}^0 \cap L'_{u_i}| \tag{7}$$

where L_l^0 is the set of constituent lists of l, i.e., the original lists that were merged into l. As discussed in the previous section, the edge weights are used to compute the transition probabilities for the iterative computation.

3.3   Content-Based Method

A limitation of the structure-based method is that it cannot cover *non-listed followees*— a user's followees that are not included into any existing list. This limitation is amplified by the low popularity of Twitter Lists. To tackle this problem, we also propose a content-based method that recommends lists based on the semantic similarity between the followees' tweets. Hence, this method can cover a non-listed followee as long as the followee has posted tweets.

To recommend k lists to a user, this method applies the k-means clustering algorithm to cluster a user u's followees into k clusters by the semantic similarity between their tweets. A virtual document is generated for each followee u_i, by concatenating the tweets posted by u_i. TF-IDF weighting is applied to represent each virtual document as a vector. The similarity between two followees is the cosine similarity between the corresponding two vectors, following the standard vector space model [12]. The clusters, i.e., new lists of followees, are ranked by the similarity between their centroids and the vector representing u's virtual document.

Since the content-based method is based on similarity of tweets, it is explicitly geared towards recommending interest-oriented lists. It tends to include two followees into the same list if their tweets exhibit similar topics. The structure-based method can implicitly recommend both relation-oriented and interest-oriented lists, since it captures the otherwise complex reasons for two followees to be included into the same list. A relation-oriented list represents real-world relations between a user and her followees (e.g., family, friends, colleagues), where the followees in the same list may not share common interests.

3.4   Inter-list Optimization: Reducing List Redundancy through Diversified Ranking

Given the ranked lists by both structure- and content-based methods, we discovered that many top-ranked lists share a large fraction of common members. One reason is that popular users are likely to be placed into multiple popular lists which makes both popular users and lists ranked high by our methods. However, we have observed that real lists created by users do not overlap much (Observation 4). To reduce the overlap among recommended lists and improve their diversity, we employ a Maximum Marginal Relevance algorithm [2]. It re-ranks lists by trading off between their original scores and diversity.

$$Score_l^d = Arg \max_{l \in S} [\lambda Score_l - (1 - \lambda) \max_{l_i \in L} Sim(l, l_i)] \tag{8}$$

The new score of a list, $Score_l^d$, is calculated by Equation (8). Suppose L is the set of selected lists so far. We are to select the next list among a set of candidate lists S. $Score_l$ is the score of a list l, by Equation (2). $Sim(l, l_i)$ is the similarity between two lists. (Specifically, we use Jaccard similarity, i.e., $Sim(l, l_i) = \frac{|l \cap l_i|}{|l \cup l_i|}$.) λ is the parameter to balance the original list score and list diversity. A proper λ value is selected through experiments. According to Observation 4, the redundancy of the real lists created by users is about 10–13%. In choosing a proper value of λ, we set the list redundancy round 10%, because our goal is to make recommended lists exhibit similar statistics to that of the lists created by real users. Our diversification method chooses the list with the highest original score as the first recommended list, then iteratively applies Equation (8) to recommend the next list, until k lists are selected.

3.5   Intra-list Optimization: Pruning Unrelated Members from Merged Lists

Both structure-based and content-based methods may produce long lists, especially due to clustering and merging of original lists. The resulting merged lists may contain many members that have little relevance to the lists. They might be included just because they were co-listed with other members that are more relevant. For example, a followee u_f may be put into a list by another user u with whom she happened to play tennis once. The followee u_f in fact may not like tennis that much. However, since the list also contains many other followees that are true tennis fans, u_f may be included in a merged list corresponding to the tennis interest of u. In another example, user u_{1f} may be listed by a user u_1 in a list named "colleague", while user u_{2f} may be listed by a user u_2 in a list that happens to have the same name "colleague". Our list clustering algorithm will merge the two lists and u_{1f} and u_{2f} will be co-listed in the resulting merged list although they are not quite related.

Given a result list produced by the structure-based or content-based method, we compute a ranking score for each individual followee u_i in the list. With regard to the content-based method, the ranking score captures the relevance of an individual followee to the list. We compute the score as the cosine similarity between the vectors corresponding to the centroid of l and u_i. Recall that the vectors model the tweets posted by the users.

$$Score^c(u_i, l) = Cosine(Vector_{u_i}, Vector_l) \tag{9}$$

With regard to the structure-based method, the ranking score is the sum of co-occurrence counts for u_i and all other members in l, given by the following equation:

$$Score^s(u_i, l) = \sum_{u_j \in l \wedge u_i \neq u_j} |L'_{u_i} \cap L'_{u_j}| \qquad (10)$$

We sort the followees in a list l by their ranking scores according to the above definitions, and then find a threshold t to divide the sorted followees into two classes, members ($C_m = \{u_1, \ldots, u_t\}$) and non-members ($C_n = \{u_{t+1}, \ldots, u_{|l|}\}$). We then prune all non-members from the list. Instead of tuning a pruning threshold by experiments and using the same threshold invariably for all lists, we apply the Otsu Thresholding Method [10] to automatically select an optimum threshold. This method is widely used for threshold selection from gray-level histograms in computer vision and image processing. The idea is to exhaustively search for the threshold t that minimizes the weighted intra-class variance for the two classes, defined as follows:

$$\sigma_w^2(t) = \omega_{C_m}(t)\sigma_{C_m}^2(t) + \omega_{C_n}(t)\sigma_{C_n}^2(t) \qquad (11)$$

where weights $\omega_{C_m}(t)$ and $\omega_{C_n}(t)$ are the probabilities of the two classes separated by the threshold t, which are given by $\omega_{C_m}(t) = Pr(C_m) = \sum_{i=1}^{t} p_i$ and $\omega_{C_n}(t) = Pr(C_n) = \sum_{i=t+1}^{|l|} p_i$, where p_i is a normalized probability based on the followee's score $Score(u_i, l)$. $\sigma_{C_m}^2(t)$ and $\sigma_{C_n}^2(t)$ are the variances of scores in the member class and the non-member class, respectively. More details of this method can be found in [10].

4   Evaluation

4.1   Experimental Design

Twitter Data Set. We collected a Twitter data set through the Twitter API. Our implementation is based on an open source Java library– twitter4j[3].

We started with around 10 thousand randomly selected seed Twitter users. We call them level-1 users. We retrieved all the followees of the level-1 users. We call these followees level-2 users. We also retrieved all the followees of the level-2 users, which we call level-3 users. The union set of all level-1, level-2, and level-3 users is our Twitter user set. For each user in the set, we crawled the user's Twitter Lists, her followees, her followers, and her latest 3, 200 tweets[4]. The collected data set contains 810, 769 Twitter users, 53, 922, 948 Twitter lists, and 324 million tweets. The earliest tweet was created on December 27, 2007 and the latest one was dated March 21, 2011.

We studied the characteristics of Twitter lists on the full data set. To quantitatively evaluate the effectiveness of the proposed methods, we construct an evaluation data set based on the collected data. The evaluation set contains a focus group with 8, 614 users and together they have 550, 793 followees. Our task is to recommend personalized lists for each user in the focus group based on his or her followees. The real lists created by these users are used as ground truth for our evaluation. The number of lists created by these users is 20, 016. On average around 25% followees of a user in the focus group have been included in the lists.

---

[3] http://twitter4j.org/en/index.jsp
[4] Due to Twitter API constraint, we can only retrieve at most 3200 tweets of a given user.

Table 1. Overview of list recommendation methods to evaluate

| Algorithm | Basic Model | | List Merging (Sec 3.2) | Diversification (Sec 3.4) | Member Pruning (Sec 3.5) | | Integration Model (Sec 3.1) |
	Structure (Sec 3.2)	Content (Sec 3.3)			User Co-occurrence	Tweet Similarity	
Content		✓					
Content-Div		✓		✓			
Content-LMP		✓		✓		✓	
PLUG-Basic	✓						
PLUG-Merge	✓		✓				
PLUG-Div	✓		✓	✓			
PLUG-LMP	✓		✓	✓	✓		
iPLUG	✓	✓	✓	✓	✓	✓	✓

Methods To Be Evaluated. Several list recommendation methods are formed by combining the proposed techniques. Table 1 summarizes the techniques included in these methods which are compared in the evaluation.

Evaluation Methodology. When evaluating the above algorithms, we hide all the original lists created by users from the focus group and generate recommended lists for each of them. We then compare the recommended lists with the original ones. The reported performance is computed by taking the average of the performance for all the users in the focus group. Since we observed that the largest 5 lists of a user often cover 86% of the followees, we focus on comparing the top 5 recommended lists with the largest 5 lists created by users. In order to better understand the performance of our algorithms, we evaluate the results under two scenarios: (1) *Member Set (M-Set):* We recommend lists based on the followees who were list members in the original lists. (2) *Full Set (F-Set):* We recommend lists based on all the followees of a user.

We report the following measures in our evaluation:

- **Redundancy:** As defined in the previous section, it is a measure to quantify the overlaps among a set of lists. The statistics in Observation 4 show that the redundancy from the largest 2 to the largest 5 original lists are stably around 10%. Therefore, in our experiments, we choose a proper λ setting to keep the redundancy around 10%.

- **List Similarity:** We measure the similarity between two lists by Jaccard coefficient, i.e., $Sim(l_i, l_j) = \frac{|l_i \cap l_j|}{|l_i \cup l_j|}$. With this definition, the larger the similarity value is, the more similar the two lists are. To calculate the similarity between two sets of lists, we compute the average pairwise-similarity of all list pairs from the two sets.

4.2 Experiment Results

Top-1 List Recommendation Results. As shown in Observation 3, the largest list (top-1) covers more than $\frac{1}{3}$ followees of a user on average. Thus, it is the most representative list of a user. We compare the members of this list with the members of the top-1 recommended list, in terms of precision, recall, F1-measure and list similarity. The results on *M-Set* are shown in Table 2. We omit all results of Content-Div and PLUG-Div because in our experiments the diversification algorithm always selects the top-1 list as the first candidate to recommend. We also omit the results on *F-Set* because they exhibit similar trend.

Table 2. Experimental results of the top-1 lists

Table 3. List similarity results on Member Set

Algorithm	Precision	Recall	F1	Similarity
Content	0.321	0.276	0.297	0.295
Content-LMP	0.533	0.552	0.544	0.532
PLUG-Basic	0.051	0.065	0.095	0.048
PLUG-Merge	0.456	0.393	0.357	0.359
PLUG-LMP	0.694	**0.757**	0.724	0.727
iPLUG	**0.718**	0.742	**0.731**	**0.736**

Algorithm	top-2	top-3	top-4	top-5
Content	0.354	0.372	0.319	0.273
Content-Div	0.457	0.482	0.441	0.424
Content-LMP	0.556	0.581	0.612	0.574
PLUG-Basic	0.043	0.034	0.036	0.026
PLUG-Merge	0.456	0.393	0.357	0.359
PLUG-Div	0.563	0.592	0.601	0.552
PLUG-LMP	0.731	0.729	0.705	0.690
iPLUG	**0.735**	**0.746**	**0.733**	**0.722**

Table 4. List similarity results on Full Set

Table 5. Redundancy of recommended lists

Algorithm	top-2	top-3	top-4	top-5
Content	0.143	0.153	0.139	0.086
Content-Div	0.238	0.253	0.261	0.251
Content-LMP	0.321	0.357	0.365	0.343
PLUG-Basic	0.012	0.009	0.009	0.008
PLUG-Merge	0.216	0.203	0.197	0.199
PLUG-Div	0.275	0.291	0.286	0.285
PLUG-LMP	0.436	0.462	0.476	0.426
iPLUG	**0.452**	**0.508**	**0.542**	**0.524**

Algorithm	top-2	top-3	top-4	top-5
Content	73.5%	71.4%	69.2%	67.4%
Content-Div	18.6%	17.5%	16.6%	15.9%
Content-LMP	14.5%	13.1%	12.4%	11.3%
PLUG-Merge	72.4%	70.6%	68.2%	67.3%
PLUG-Div	16.2%	15.8%	14.6%	13.3%
PLUG-LMP	13.1%	12.4%	11.6%	9.9%
iPLUG	11.2%	10.3%	9.6%	9.3%

Table 2 shows that the iPLUG algorithm outperforms other algorithms in terms of precision, F1-measure, and list similarity. It achieves slightly worse recall compared to PLUG-LMP, but is dramatically better than other algorithms. Overall, iPLUG achieves 73.6% in list similarity and 73.1% in F1-measure. We also see that the performance of the PLUG-Basic model is poor due to the sparsity of the co-listed relation. List merging can effectively improve the performance of PLUG-Basic by about 5 to 10 times. Then list member pruning yields as much as 200% improvement in performance, indicating that irrelevant members often exist in the initial recommended lists.

Top-2 to top-5 Recommended Lists. The list similarity for top-2 to top-5 recommended lists for *M-Set* and *F-Set* are shown in Table 3 and Table 4. The two tables exhibit result trends similar to those in Table 2. Overall, the iPLUG algorithm achieves the best performance among all the methods. PLUG-LMP is quite close to iPLUG. There is only a 4% performance difference between iPLUG and PLUG-LMP on *M-Set*. This means that PLUG-LMP is good enough for most users in *M-Set*.

Comparing Table 3 with Table 4, we see that the performance of all algorithms deteriorates on *F-Set*. iPLUG is significantly better than all other methods, including the second best PLUG-LMP. The improvement of iPLUG compared with PLUG-LMP on *F-Set* is as much as 20%. It indicates that the content-based method has significant effect on iPLUG's performance when non-listed followees exist.

Diversification. To show the effectiveness of the technique of diversified list ranking, we compare the redundancy of the recommended lists of our major algorithms and show the results of top-2, 3, 4, 5 lists in Table 5. We find that the redundancy of PLUG-Div and Content-Div decreases by more than 70%, compared with PLUG-Merge and

Content, respectively. This demonstrates the effectiveness of our diversified list ranking technique.

From results shown in Table 3 and Table 4, we see that the list similarity of both PLUG-Div and Content-Div improves significantly compared to PLUG-Merge and Content. The performance gap increases as k increases. The reason is that the top-k recommended lists are highly redundant before diversification. For example, the redundancy is 67–73% in the lists recommended by PLUG-Merge and Content. With diversified ranking, we are able to keep the redundancy to a relatively low level (around 10%), which is near the level of the lists created by real users.

4.3   Parameter Sensitivity Analysis

Parameter Settings of λ. In our diversified list ranking algorithm, the parameter λ plays the role of balancing list ranking from basic models (structure-based and content-based) and list diversification. As discussed earlier, the redundancy of lists captures list diversification. Two lists with less redundant members are considered more diverse. Therefore, we vary λ from 0 to 1, and report the redundancy of the top-k ($k = 3, 4, 5$) recommended lists, as shown in Figure 6(a). We see that λ has significant impact on the redundancy of the results. According to the statistics discussed in Observation 4, the redundancy of original lists in the real data is around 10%. From the results of Figure 6(a), it is clear that when λ is around 0.3, the redundancy of our recommended lists is close to 10%.

To understand the impact of the settings of λ on list similarity, we compute list similarity between the largest 5 original lists and the top-5 recommended lists while varying λ values for both *M-Set* and *F-Set*. The results of both structure-based and content-based algorithms are shown in Figure 6(b). The results indicate that the PLUG-Div and Content-Div algorithms achieve the best performance on *M-Set* when $\lambda = 0.3$. On *F-Set* Content-Div achieves the best performance when $\lambda = 0.4$. That is to say, setting λ to 0.3 (or 0.4) results in recommending lists most similar to the ones created by real users. Therefore, we set $\lambda = 0.3$ in our experiments.

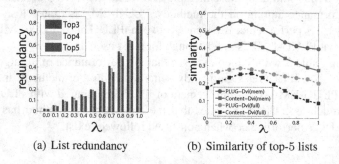

(a) List redundancy   (b) Similarity of top-5 lists

Fig. 6. Impact of λ values on list recommendation

Parameter Settings of η. We study the impact of different settings of η on the performance of the integration algorithm (iPLUG). η is the factor that controls the contributions of the structure-based model and the content-based model to iPLUG. A larger η indicates that iPLUG sees higher contribution from the structure-based model, and lower contribution from the content-based model. We vary η from 0 to 1 and report the corresponding list similarity of iPLUG. Figure 7 shows the list similarity results of top-1 and top-5 lists. When $\eta = 0$, iPLUG becomes the content-based model. On the other hand, when $\eta = 1$, iPLUG uses only the structure-based model.

As we see in Figure 7, the performance monotonically increases from $\eta = 0$ to the peak ($\eta = 0.7$ on *M-Set* and $\eta = 0.6$ on *F-Set*). Afterwards, it decreases and finally is equal to the performance of PLUG-LMP algorithm ($\eta = 1$). That is, η has significant impact on the performance of iPLUG. It also shows the trend that generally taking in more contribution of the structure-based model results in better performance for iPLUG. The best setting of η for both top-1 and top-5 cases on *M-Set* are 0.7 while on *F-Set* are 0.6, we therefore set $\eta = 0.7$ and 0.6 respectively in our experiments.

As mentioned before, an intuitive value for η is the proportional of the followees who are listed, i.e., $\eta = \frac{\tau}{|FE_u|}$, where τ is the number of the followees that appear in at least one list. This method prefers more contribution from the structure-based model if a user has a larger fraction of listed followees. We set η separately for each individual user using this method. The results are shown as the black dash line in Figure 7. We see that in all four figures, the performance of this setting is between the performance of the settings of $\eta = 0.3$ and $\eta = 0.4$. The reason is that in our data set only 24.52% of followees of the focus group users have been listed. That is on average the iPLUG model takes only 24.52% of contribution from the PLUG model, and the fine-grained η settings for each user improves the performance more than a fixed setting of $\eta = 0.3$. However, we observe that it is far from the best performance result in our data set. Hence, $\eta = \frac{\tau}{|FE_u|}$ is not an optimal setting for η.

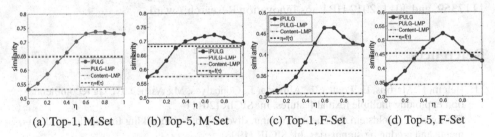

(a) Top-1, M-Set   (b) Top-5, M-Set   (c) Top-1, F-Set   (d) Top-5, F-Set

Fig. 7. Impact of η values on lists similarity

5   Related Work

Our goal is to help Twitter users automatically group their followees and put them into different lists. The problem is similar to a recent study on classifying users accordingly to their interests [11]. This study is limited to bifacial interests detection such as detecting whether a user is democrat or republican or whether a user likes Starbucks or

not in business affinity detection. Twitter List can be considered as a tag for each followee. Tag and Tag-based entity (e.g. photos, music, videos, web pages) classification
or clustering [9,6,15] have been used to improve the performance of web page indexing, music categorization, news filtering and content recommendation. Compared with
these studies, the difference of our method is to take advantage of both structural and
content information for list recommendation.

Social recommendation is becoming prevalent in online services [1,14,7]. Social
tags help users better understand, interact with and propagate variety of content. However, user-generated tags with uncontrolled vocabulary can sometimes be ambiguous,
obscure, inadequate and redundant. To tackle this problem, [13] proposed a personalization algorithm based on content-dependent variant of hierarchical tag clustering. We
observed the same drawback of list names as [13] discovered with user-generated tags.
Also, the topic selection algorithm in [8] helps to eliminate the redundancy between
topics and focus on the important ones. In our algorithm, instead of using the semantic
information from list name, we take advantage of the information of users and lists.

6   Conclusions and Future Work

In this paper, we propose the iPLUG algorithm for recommending lists for Twitter users,
which leverages both the structure of list-user graphs and the tweets of users. Experiments over a Twitter dataset show that iPLUG is capable of recommending lists similar
to the ones created by real users. There are many interesting directions for future work.
First, it is interesting to study how to help users automatically discover and add new
followees to existing lists. Second, it would be interesting to study how list recommendation changes user behavior on Twitter.

Acknowledgments. The work of Li is partially supported by NSF grants 1018865,
1117369, and 2011, 2012 HP Labs Innovation Research Award.

References

1. Belém, F., Martins, E., Pontes, T., Almeida, J., Gonçalves, M.: Associative tag recommendation exploiting multiple textual features. In: SIGIR (2011)
2. Carbonell, J., Goldstein, J.: The use of mmr, diversity-based reranking for reordering documents and producing summaries. In: SIGIR (1998)
3. Deng, H., Lyu, M.R., King, I.: A generalized co-hits algorithm and its application to bipartite
   graphs. In: SIGKDD (2009)
4. Guan, Z., Wang, C., Bu, J., Chen, C., Yang, K., Cai, D., He, X.: Document recommendation
   in social tagging services. In: WWW (2010)
5. Guy, I., Zwerdling, N., Ronen, I., Carmel, D., Uziel, E.: Social media recommendation based
   on people and tags. In: SIGIR (2010)
6. Khudyak, A., Kurland, O.: Cluster-based fusion of retrieved lists. In: SIGIR (2011)
7. Lu, C., Hu, X., Chen, X., Park, J.-R., He, T., Li, Z.: The topic-perspective model for social
   tagging systems. In: SIGKDD (2010)

8. Rowe, M., Wagner, C., Strohmaier, M., Alani, H.: Measuring the topical specificity of online communities. In: Cimiano, P., Corcho, O., Presutti, V., Hollink, L., Rudolph, S. (eds.) ESWC 2013. LNCS, vol. 7882, pp. 472–486. Springer, Heidelberg (2013)
9. Meeder, B., Karrer, B., Sayedi, A., Ravi, R., Borgs, C., Chayes, J.: We know who you followed last summer: inferring social link creation times in twitter. In: WWW (2011)
10. Otsu, N.: A threshold selection method from gray-level histograms. In: IEEE TSMC (1979)
11. Pennacchiotti, M., Popescu, A.-M.: Democrats, republicans and starbucks afficionados: user classification in twitter. In: SIGKDD (2011)
12. Salton, G., Wong, A., Yang, C.S.: A vector space model for automatic indexing. Commun. ACM (1975)
13. Shepitsen, A., Gemmell, J., Mobasher, B., Burke, R.: Personalized recommendation in social tagging systems using hierarchical clustering. In: RecSys (2008)
14. Venetis, P., Koutrika, G., Garcia-Molina, H.: On the selection of tags for tag clouds. In: WSDM (2011)
15. Yin, Z., Li, R., Mei, Q., Han, J.: Exploring social tagging graph for web object classification. In: SIGKDD (2009)

The Irreducible Spine(s) of Undirected Networks

John L. Pfaltz

Dept. of Computer Science, University of Virginia

Abstract. Using closure and neighborhood concepts, we show that within every undirected network, or graph, there is a unique irreducible subgraph which we call its "spine". The chordless cycles which comprise this irreducible core effectively characterize the connectivity structure of the network as a whole. In particular, it is shown that the center of the network, whether defined by distance or betweenness centrality, is effectively contained in this spine.

By counting the number of cycles of length $3 \leq k \leq max\_length$, we can also create a kind of signature that can be used to identify the network.

Performance is analyzed, and the concepts we develop are illustrated by means of a relatively small running sample network of 379 nodes, although they have been applied to networks of 4,764 and 5,242 nodes as well.

1 Introduction

It is hard to describe the structure of large networks. If the network has fewer than 100 nodes, then we can hope to draw it as a graph and visually comprehend it [6]. But, with more than 100 nodes this becomes increasingly difficult.

Simply counting the number n of nodes and number e of edges, or connections, provides essential basic information. Other combinatorial measures include counting the number of triangles, the number of edges incident to a node v, or degree $d(v)$, and the total number of nodes such that $d(v) = k$. There exist data representations that effectively keep these kinds of counts, even in rapidly changing dynamic networks [12].

More sophisticated methods involve treating the defining adjacency matrix as if it were a linear transformation and employing an eigen analysis [14]. All these techniques convey information about a network. In this paper, we present a rather different approach.

First in Section 2 we reduce the network to an irreducible core, or "spine", which is shown to be unique (upto isomorphism) for any network. Then in Section 3 we show that the irreducible spine is comprised exclusively of chordless cycles of length k, or k-cycles. In Section 3.1 we show that the "center" of the network, whether defined in terms of distance, or betweenness centrality [3], can always be found in this spine. In addition, the distribution of these k-cycles, $3 \leq k \leq max\_length$, can provide a "signature" for the network.

X. Lin et al. (Eds.): WISE 2013, Part II, LNCS 8181, pp. 104–117, 2013.

2   Irreducible Networks

For this paper we regard a network $\mathcal{N}$ as an undirected graph on a set N of n nodes with a set E of e edges, or connections. Many of these results can be applied to directed networks as well, but we will not explore these possibilities here. The **neighborhood** of a set Y of nodes are those nodes not in Y with at least one edge connecting them to at least one node in Y. We denote such a neighborhood by $Y.\eta$, that is $Y.\eta = \{z \notin Y | \exists y \in Y, (y, z) \in E\}$. We use this somewhat unusual suffix notation because we regard η as a set-valued operator acting on the set Y. By the **region** dominated by Y, denoted $Y.\rho$, we mean $Y.\rho = Y.\eta \cup Y$.

In our treatment of network structure, we will make use of the **neighborhood closure** operator, denoted by φ [15]. For all $Y \subseteq N$, this is defined to be $Y.\varphi = \{z \in Y.\rho : \{z\}.\rho \subseteq Y.\rho\}$ which is computationally equivalent to $Y.\varphi = Y \cup \{z \in Y.\eta : \{z\}.\eta \subseteq Y.\rho\}$. Readily $Y \subseteq Y.\varphi \subseteq Y.\rho$. Recall that a closure operator φ is one that satisfies the 3 properties: (C1) $Y \subseteq Y.\varphi$, (C2) $X \subseteq Y$ implies $X.\varphi \subseteq Y.\varphi$, and (C3) $Y.\varphi.\varphi = Y.\varphi$.

Because the structure of large networks can be so difficult to comprehend, it is natural to seek techniques for reducing their size, while still preserving certain essential properties [1,7], often by selective sampling [11]. Our approach is somewhat different. We view "structure" through the lens of neighborhood closure, which we then use to find the unique irreducible sub-network $\mathcal{I} \subseteq \mathcal{N}$.

A graph, or network, is said to be **irreducible** if every singleton subset $\{y\}$ is closed. A node z is **subsumed** by a node y if $\{z\}.\varphi \subseteq \{y\}.\varphi$. Since in this case, z contributes very little to our understanding of the closure structure of $\mathcal{N}$, its removal will result in little loss of information.

Proposition 1. *Let y subsume z and let $\sigma(x, y)$ denote a shortest path between x and y. If $z \neq y, \in \sigma(x, y)$, then there exists $\sigma'(x, y)$ such that $z \notin \sigma'$*

Proof. If not, we may assume without loss of generality that z is adjacent to y in σ. But, then $\sigma(x, z), x \notin \{y\}.\eta$ implies that $\{z\}.\varphi \not\subseteq \{y\}.\varphi$. (Also proven in [16].)   □

In other words, z can be removed from $\mathcal{N}$ with the certainty that if there was a path from some node x to y through z, there will still exist a path of equal length from x to y after z's removal. Such subsumed nodes can be iteratively removed from $\mathcal{N}$ without changing connectivity. This iterative reduction process we denote by ω.

Operationally, it is easiest to search the neighborhood $\{y\}.\eta$ of each node y, and test whether $\{z\}.\eta \subseteq \{y\}.\rho$ as shown in the code fragment of Figure 1. This code is then iterated until there are no more subsumable nodes. Let $y.\beta$ denote the set of nodes subsumed directly, or indirectly, by y. In a sense these subsumed nodes **belong** to y. Let $\tau(y)$ denote $|y.\beta|$. Since every node subsumes itself, $\tau(y) \geq 1$. In our implementation of this code, we also increment $\tau(y)$ by $\tau(z)$ every time node z is subsumed by y. So, $\tau(y) = |y.\beta|$. Consequently, $\sum_{y \in \mathcal{N}.\omega} \tau(y) = n = |\mathcal{N}|$.

```
for_each y in N
  {
  for_each z in y.nbhd
    {
    if (z.nbhd contained_in y.region
      {        // z is subsumed by y
      for_each x in z.nbhd
        remove edge (x, z)
      remove z from network
      }
    }
  }
```

Fig. 1. Key loop in reduction process, ω

Before considering the behavior of ω, we want to establish a few formal properties of the reduced network.

Proposition 2. *Let $\mathcal{N}$ be a finite network and let $\mathcal{I} = \mathcal{N}.\omega$ be a reduced version, then $\mathcal{I}$ is irreducible.*

Proof. Suppose $\{y\}$ in $\mathcal{I}$ is not closed. Then $\exists z \in \{y\}.\varphi_\eta$ implying $z.\rho \subseteq \{y\}.\rho$ or that z is subsumed by y contradicting termination of the reduction code. □

Two graphs, or networks, $\mathcal{N} = (N, E)$ and $\mathcal{N}' = (N', E')$ are said to be **isomorphic**, or $\mathcal{N} \cong \mathcal{N}'$, if there exists a bijection, $i : N \to N'$ such that for all $x, y \in N$, $(i(x), i(y)) \in E'$ if and only if $(x, y) \in E$. That is, the mapping i precisely preserves the edge structure, or equivalently its neighborhood structure. Thus, $i(y) \in i(x).\eta'$ if and only if $y \in x.\eta$.[1]

The order in which nodes, or more accurately the singleton subsets, of $\mathcal{N}$ are encountered can alter which points are subsumed and subsequently deleted. Nevertheless, we show below that the reduced graph $\mathcal{I} = \mathcal{N}.\omega$ will be unique, upto isomorphism.

Proposition 3. *Let $\mathcal{I} = \mathcal{N}.\omega$ and $\mathcal{I}' = \mathcal{N}.\omega'$ be irreducible subsets of a finite network $\mathcal{N}$, then $\mathcal{I} \cong \mathcal{I}'$.*

Proof. Let $y_0 \in \mathcal{I}$, $y_0 \notin \mathcal{I}'$. Then y_0 is subsumed by some point y_1 in $\mathcal{I}'$ and $y_1 \notin \mathcal{I}$ else because $y_0.\rho \subseteq y_1.\rho$ implies $y_0 \in \{y_1\}.\varphi$ so $\mathcal{I}$ would not be irreducible.

Similarly, since $y_1 \in \mathcal{I}'$ and $y_1 \notin \mathcal{I}$, there exists $y_2 \in \mathcal{I}$ such that y_1 is subsumed by y_2. Now we have two possible cases; either $y_2 = y_0$, or not.

Suppose $y_2 = y_0$ (which is most often the case), then $y_0.\rho \subseteq y_1.\rho$ and $y_1.\rho \subseteq y_0.\rho$ or $y_0.\eta = y_1.\eta$. Hence $i(y_0) = y_1$ is part of the desired isometry, i.

---

[1] Note that $i : N \to N'$ is a normal single-valued function on N, so we use traditional prefix notation. We reserve suffix notation for set-valued operators/functions.

Now suppose $y_2 \neq y_0$. There exists $y_3 \neq y_1 \in \mathcal{I}'$ such that $y_2.\rho \subseteq y_3.\rho$, and so forth. Since $\mathcal{I}$ is finite this construction must halt with some y_n. The points $\{y_0, y_1, y_2, \ldots y_n\}$ constitute a complete graph Y_n with $\{y_i\}.\rho = Y_n.\rho$, for $i \in [0, n]$. In any reduction all $y_i \in Y_n$ reduce to a single point. All possibilities lead to mutually isomorphic maps.   □

We call this unique subgraph, the **irreducible spine** of $\mathcal{N}$. In [12], Lin, Soulignac and Szwarcfiter, speak of a "*dismantling*" of a graph G as a graph H obtained by removing one dominated vertex of G, until no more dominated vertices remain"; and similarly conclude that "all dismantlings of G are isomorphic". This is precisely the process we have been describing.

For the remainder of this paper we will use a single example to illustrate our approach to describing network structure. In [14] Mark Newman describes a 379 node network in which each node corresponds to an individual engaged in network research, with an edge between nodes if the two individuals have co-authored a paper. The reader is encouraged to view an annotated version at www.umich.edu/~mejn/centrality.[2]

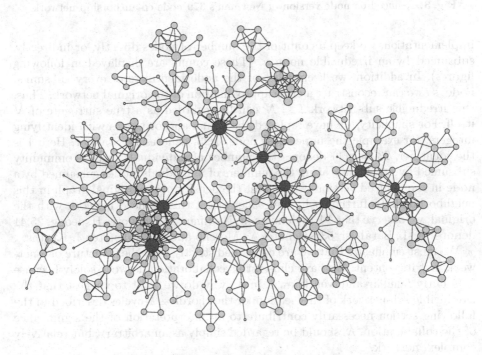

Fig. 2. 379 node collaboration network

As described in [16], we used the code of Figure 1 to reduce the 379 Newman collaboration network to the 65 node irreducible spine shown in Figure 3. In our

[2] Similar "collaboration" networks can be found in Stanford Large Network Database.

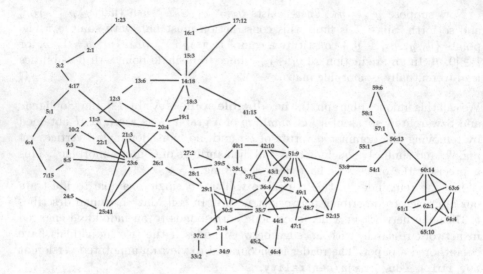

Fig. 3. Reduced 65 node version of Newman's 379 node co-authorship network

implementation, we keep a count of the number of nodes directly, or indirectly, subsumed by an irreducible node y. (These counts are displayed in following figures). In addition, we keep a list of the node identifiers of every subsumed node, so we can reconstruct a close approximation of the original network. Thus, this irreducible sub-network $\mathcal{I} \subseteq \mathcal{N}$ can be regarded as a true surrogate of $\mathcal{N}$ itself. For simplicity, we have replaced the actual author names with identifying integers; for example the uppermost node, 1:23, denotes D. Stauffer. Here 1 is the identifier, $23 = \tau(1)$ denotes the number of individuals in the community subsumed by 1.[3] By indicating the numbers of individuals/nodes subsumed by a node in the reduced version, we suggest the density of the original graph in this neighborhood. To further help the reader orient this reduced network with the original, we observe that 14:18 denotes M. Newman, 23:6 denotes H. Jeong, 25:41 denotes A.-L. Barabasi, 53:8 denotes Y. Moreno and 60:14 denotes J. Kurths.

We must emphasize that we are concerned strictly with the structure of a network, not its content. We have chosen this collaboration network solely because it is fairly familiar and well known. In no way do we want to suggest that the irreducible sub-network of this section, or the chordless k-cycles described in the following section necessarily contribute to an interpretation of the significance of the collaboration. $\mathcal{N}$ should be regarded simply as an arbitrary, but relatively complex, network.

What is the computational cost of reducing such a network to its irreducible spine?

---

[3] Because there is considerable randomness in the reduction process, several individuals other than Stauffer could have been chosen to represent this community. Still, the resulting graph would have been isomorphic to Figure 3.

The dominant cost is the loop in Figure 1 over all n nodes of N. So, it is at least $O(n)$. Then we have the embedded loop `for_each z in y.nbhd`. First, we assume that the degree $\delta(y)$ of each node is bounded (typically the case in large networks), thus its behavior will still be linear. In our implementation, all sets are represented by bit strings, with each bit denoting an element; set operators are thus logical bit operations. There is no need to loop over the elements of a set. Consequently, set operations such as union, intersection, or containment testing, are $O(1)$. In this case, the entire loop will still be $O(n)$.

However, the loop of Figure 1 must be iterated until no more nodes are subsumed. It is not hard to create networks in which only one node is subsumed on each iteration; Figure 4 is a simple example, if nodes are encountered in subscript order. So worst case behavior is $O(n^2)$.

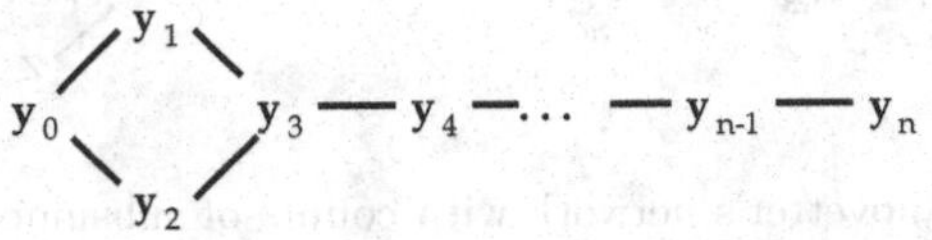

Fig. 4. Reduction, ω, has $O(n^2)$ behavior

The analysis above assumed that the degree of all nodes was bounded. Suppose not; suppose $\delta(y) \to n = |N|$. In this case the node y will subsume many nodes, thereby bounding the number of necessary iterations. We have no formal proof for this last assertion, but it appears to be true.

Using their H-graph structure, Lin, Soulignac and Szwarcfiter, show that the cost to dismantle a network is $O(n + \alpha m)$ where α denotes the arboricity of N [12]. Experimentally, our reduction, ω, of the Newman collaboration graph to its irreducible spine shown in Figure 3 required 5 iterations, with the last over the remaining 65 nodes to verify irreducibility. Reduction of a 4,764 node network depicting Norwegian corporate directorships [17] and a 5,242 node networks from the Stanford Large network database, to their 228, and 1,469, node irreducible spines respectively took 5 and 6 iterations. In practice, network reduction appears to be nearly linear.

3   Chordless k-Cycles

A **cycle** is a closed, simple path [2,9]. A cycle $C =< y_1, y_2, \ldots, y_k, y_1 >$ has length k. For each node $y_i \in C$, $|\{y_i\}.\eta| \geq 2$. The irreducible spine of Figure 3 has an abundance of cycles and no nodes x with $|\{x\}.\eta| = 1$.

A **chord** in a cycle is an edge/connection $(y_i, y_j) \in E$ where $j \neq i \pm 1$ (or $i = 1, j = k - 1$). A cycle C is **chordless** if it has no chords.[4]

---

[4] In [15,16], the author mistakenly used the term "fundamental cycle" for the chordless cycles that will be explored in this section.

It is the thesis of this paper that these chordless k-cycles provide a valuable characterization of the structure of a network. As a small example, consider Figure 5 from Granovetter's 1973 article on "weak ties" [8], which has been redrawn so as to emphasize the chordless 14-cycle. The nodes X, Y, Z represent

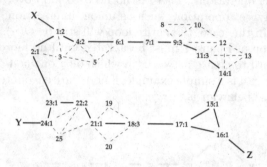

Fig. 5. Granovetter's network with counts of subsumed nodes

other portions of the network. Readily, describing this subset as a 14-cycle with 9 pendant nodes is an appropriate characterization. Our goal with this example is simply to show that these kinds of chordless cycles arise naturally in the literature and in real life.

The following proposition characterizes the structure of irreducible spines.

Proposition 4. *Let $\mathcal{N}$ be a finite network with $\mathcal{I} = \mathcal{N}.\omega$ being an irreducible version. If $y \in \mathcal{I}$ is not an isolated point then either*

(1) there exists a chordless k-cycle C, $k \geq 4$ such that $y \in C$, or

(2) there exist chordless k-cycles C_1, C_2 each of length ≥ 4 with $x \in C_1$ $z \in C_2$ and y lies on a path from x to z.

Proof. (1) Let $y_1 \in N_{\mathcal{I}}$. Since y_1 is not isolated, let $y_0 \in y_1.\eta$, so $(y_0, y_1) \in E$. With out loss of generality, we may assume $y_0 \in C_1$ a cycle of length ≥ 4. Since y_1 is not subsumed by y_0, $\exists y_2 \in y_1.\eta$, $y_2 \notin y_0.\eta$, and since y_2 is not subsumed by y_1, $\exists y_3 \in y_2.\eta$, $y_3 \notin y_1.\eta$. Since $y_2 \notin y_0.\eta$, $y_3 \neq y_0$.

Suppose $y_3 \in y_0.\eta$, then $< y_0, y_1, y_2, y_3, y_0 >$ constitutes a k-cycle $k \geq 4$, and we are done.

Suppose $y_3 \notin y_0.\eta$. We repeat the same path extension. $y_3.\eta \nsubseteq y_2.\eta$ implies $\exists y_4 \in y_3.\eta$, $y_4 \notin y_2.\eta$. If $y_4 \in y_0.\eta$ or $y_4 \in y_1.\eta$, we have the desired cycle. If not $\exists y_5, \ldots$ and so forth. Because $\mathcal{N}$ is finite, this path extension must terminate with $y_k \in y_i.\eta$, where $0 \leq i \leq n - 3$, $n = |N|$. Let $x = y_0$, $z = y_k$.

(2) follows naturally.   □

The points of those chordal subgraphs still remaining in Figure 5 such as the triangle $< 15, 16, 17 >$, are all elements of other chordless cycles as predicted by Proposition 4.

3.1   Centers and Centrality

A central quest in the analysis of social networks is the identification of its "important" nodes. In social networks, "importance" may be defined with respect to the path structure [5].

Let $\sigma(s,t)$ denote a **shortest path** between s and t, and let $d(s,t)$ denote its length, or **distance** between s and t. Those nodes $C_C = \{y \in \mathcal{N}\}$ for which $\delta(y) = \sum_{s \neq y} d(s,y)$ is *minimal* have traditionally been called the **center** of $\mathcal{N}$ [9], they are "closest" to all other nodes. It is well known that this subset of nodes must be edge connected. One may assume that these nodes in the "center" of a network are "important" nodes.

Alternatively, one may consider those nodes which "connect" many other nodes, or clusters of nodes, to be the "important" ones. Let $\sigma_{st}(y)$ denote the number of shortest paths $\sigma(s,t)$ containing y; then those nodes y for which $\sigma_{st}(y)$ is *maximal* are those nodes that are involved in the most connections. Let $C_B = \{y \in \mathcal{N}\}$, for which $\sigma_{st}(y)$ is maximal. This is sometime called "betweenness centrality" [3,5]. (Note: traditionally, centrality measures are normalized to range between 0 and 1, but we will not need this for this paper.)

In the following sequence we want to show that nodes with minimal distance and maximal betweenness measures will be found in the irreducible spine $\mathcal{I}$. This is non-trivial because it need not always be true. One problem is that, we may have several isomorphic spines, $\mathcal{I}_1, \ldots, \mathcal{I}_k$, so we can only assert that $C_C \cap \mathcal{I}_j$ and $C_B \cap \mathcal{I}_j$ are non-empty for all $1 \leq j \leq k$. Second, there exist pathological cases where the centers are disjoint from $\mathcal{I}$. The network of Figure 4, in Section 2, is an example. If $n = 8$ then $C_C = C_B = y_4$ because $18 = \delta(y_4) < \delta(y_3) = \delta(y_5) = 19$, and $24 = \sigma_{st}(y_4) > \sigma_{st}(y_3) = \sigma_{st}(y_5) = 23$. But, $y_4 \notin \mathcal{I}$. The conditions of Proposition 7 will ensure this cannot happen. We can assume $\mathcal{I}$ is connected, else we are considering one of its connected components.

Lemma 1. *Let $y \in \mathcal{I}$ and let z "belong" to y, i.e. $z \in y.\beta$. There exists a shortest path sequence $< y_0, \ldots, y_k >$ such that*

(a) $y_0 = y$,

(b) $y_k = z$, and

(c) $y_i.\eta \subseteq y.\rho = y_i.\eta \cup y_i$, $1 \leq i \leq k$.

Proof. This is a formal property of the subsumption process.   □

This sequence need not correspond to the sequence in which nodes are actually subsumed.

Lemma 2. *Let $y \in \mathcal{I}$, with $z \in y.\beta$ and let $\sigma(s,z)$ be a shortest path where $s \notin y.\beta$. Then there exists a shortest path $\sigma(s,z) = < s, \ldots, y_0, \ldots, y_i, z >$.*

Proof. Suppose $\sigma(s,z) = < s, \ldots, v, z >$. Since $z \in y.\beta$, $\exists i, z.\eta \subseteq y_i.\eta \cup y_i$. Now $v \in y_i.\eta \cup y_i$ hence $\sigma(s,z) = < s, \ldots, y_i, z >$ is also a shortest path. Iterate this construction for $k = i - 1, \ldots, 0$. This is also a corollary statement to Proposition 1.   □

Lemma 3. *Let $y \in \mathcal{I}$ and let $z \in y.\beta, z \notin y.\eta$. If $s \notin y.\beta$ then $d(s,z) \geq d(s,y) + 1$.*

Proof. By Lemma 2, $\exists y_k, k \geq 1$ such that $z \in y_k.\eta$ and $\sigma(s,z) = < s, \ldots, y_0, \ldots, y_i, z >$ is a shortest path. Readily $d(s,z) = d(s,y) + i \geq d(s,y) + 1$. $\qquad\square$

Proposition 5. *Let $y \in \mathcal{I}$ with $z \in y.\beta$. If $z \in y.\eta$ then*
   (a) For all s, t, $\sigma_{st}(y) \geq \sigma_{st}(z)$
   (b) $\delta(y) \leq \delta(z)$

Proof. (a) Since $z \in y.\eta$, and by Lemma 2, $i = 0$, for all shortest paths through z, there exists a shortest path through y.
(b) Readily, $z \in y.\eta$ and $z.\eta \subseteq y.\rho$ implies $d(s,y) \leq d(s,z)$ for all $s \neq y, z$. $\qquad\square$

In this case, z may, or may not, also be in an alternate spine $\mathcal{I}'$. Hence equality is possible in both (a) and (b).

Proposition 6. *Let $y \in \mathcal{I}$ with $z \in y.\beta$. Let $\sum_{x \in \mathcal{I}, x \neq y} \tau(x) \geq \tau(y)$ and let $\sum_{x \in \mathcal{I}, x \in y.\eta} \tau(x) \geq \tau(y)$. If $z \notin y.\eta$ then*
   (a) For all s, t, $\sigma_{st}(y) > \sigma_{st}(z)$
   (b) $\delta(y) < \delta(z)$.

Proof. (a) If $s \in y.\beta$ and $t \notin y.\beta$, then Lemma 2 establishes that $\sigma_{st}(y) \geq \sigma_{st}(z)$. Now suppose that $t \in y.\beta$, then $\sigma(s,t)$ through z need not imply a shortest path $\sigma(s,t)$ through y. The maximal possible number of such shortest paths occurs when $y.\beta - y$ is a star graph, such as shown in Figure 3.1. Let $k = \tau(y) - 2$.

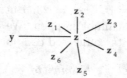

$\exists\, C(k, 2) = k \cdot (k-1)/2$ shortest paths $\sigma(s,t)$ through z with $s, t \neq z$, and k more with $t = z$.

Finally, assume $s, t \notin y.\beta$. Let $n = \sum_{x \in \mathcal{I}, x \in y.\eta} \tau(x) \geq \tau(y)$, a condition of this proposition. This ensures that $\exists\, C(n, 2) + n$ shortest paths through y avoiding z. Since $n > k$, $\sigma_{st}(y) > \sigma_{st}(z)$.
(b) $\delta(y) = \sum_{s \in y.\beta} d(s,y) + \sum_{t \notin y.\beta} d(t,y)$ and similarly $\delta(z) = \sum_{s \in y.\beta} d(s,z) + \sum_{t \notin y.\beta} d(t,z)$. Let $k = d(y,z)$, $k \geq 2$. $\sum_{s \in y.\beta} d(s,y) < \sum_{s \in y.\beta} d(s,z) + k \cdot \tau(y)$. $\sum_{t \notin y.\beta} d(t,y) < \sum_{t \in y.\beta} d(t,z) - k \cdot |t \notin y.\beta|$. So, provided $\sum_{x \in y.\beta, x \neq y} \tau(x) = |t \notin y.\beta| > \tau(y)$, we have $\delta(y) \leq \delta(z)$. $\qquad\square$

Proposition 7. *Let $\mathcal{I}$ be an irreducible spine of a network $\mathcal{N}$ with centers C_C and C_B.*
If for all $y \in \mathcal{I}$, $\sum_{x \in \mathcal{I}, x \neq y} \tau(x) \geq \tau(y)$ and $\sum_{x \in \mathcal{I}, x \in y.\eta} \tau(x) \geq \tau(y)$ then there exist $x_i \in \mathcal{I}$ and $y_j \in \mathcal{I}$ such that $x_i \cap C_C$ and $y_j \cap C_B \neq \emptyset$.
Moreover, $C_C \subseteq \cup_i(x_i.\eta)$ and $C_B \subseteq \cup_j(y_j.\eta)$.

Proof. We compare $y \in \mathcal{I}$ with any $z \in y.\beta$. The first assertion is just a corollary of propositions 5, where $z \in y.\eta$, and 6, where $z \notin y.\eta$.

The second assertion follows because the inequalities of Proposition 6 are all strict.   □

The conditions of Proposition 7 (and Proposition 6) are sufficient to eliminate pathological situations such as Figure 4; but are by no means necesary. In practice, one really only needs that $\mathcal{I}$ be sufficiently large, and that its subsumed sub-graphs not be too unbalanced.

3.2   Estimation of Other Network Properties

The performance of many important network analysis programs is of order $O(n^k)$, where $k > 1$. They execute much faster on a small network such as the irreducible spine rather than the network itself. Using $\mathcal{I}$ one can often approximate the value with considerable accuracy. We illustrate by calculating the diameter using Figure 3. Recall that the **diameter** of a network is the maximal shortest path between any two points. In [4], the cost to find a diameter using the Floyd-Warshall algorithm is $O(n^3)$. This can be reduced to $O(n^2 \log n)$ by Johnson's algorithm, but we know of no better exact solutions. In Figure 3 we can do this by hand.[5]

Readily, node 65, in the lower right hand corner is an extreme node. Expanding out by shortest paths, one finds that node 6 on the left edge is at distance 13, that is $d(6, 65) = 13$, and this is maximal *in this irreducible spine*. The center of this subgraph will be nodes at distance 6 or 7 from both extremes. These are nodes 35, 48 and 51, which are necessarily connected in $\mathcal{I}$. Using Proposition 7 we can assume that at least one of these is in the actual center of $\mathcal{N}$, and that C_C is contained in its neighborhood.

We continue our estimation of the diameter by considering the subsumed portions of the network. What is the nature of the suppressed portions of the network?

Let $y \in \mathcal{I}$, $y.\beta$ is a chordal subgraph, where a subgraph is said to be **chordal** if it has no chordless cycles of length ≥ 4. Chordal graphs are mathematically quite interesting and have been well studied [2,10,13]. Succinctly, they can be regarded as tree-like assemblages of complete graphs; they can be generated by a simple context-free graph-grammar. In effect, they are pendant tree-like structures that are attached to the irreducible spine, $\mathcal{I}$, at one (or two adjacent) nodes. Thus β is a set-valued operator that associates a pendant tree of complete graphs with y.

Readily, the diameter, $diam_n$ of a chordal graph on n points satisfies $1 \leq diam_n \leq n - 1$, with the lower bound occurring if $\{y\}.\beta = K_n$, and upper bound when $\{y\}.\beta$ is linear. In lieu of a better expectation, we will estimate the diameter of a pendant chordal graph $\{y\}.\beta$ of n nodes to be $n/2$. (A much better expectation could be made if both the number of nodes, and number of edges, were recorded in the reduction process, ω. This would not be hard.)

---

[5] This ability is an artifact of this graph structure and not generally feasible.

With this expected value, we can estimate the length of a maximal shortest path (u, v) in $\mathcal{N}$ *through* nodes 6 and 65 to be $d(u, 6) + d(6, 65) + d(65, v)$ or $4/2 = 2 + 13 + 5 = 10/2$, or $d(u, v) = 20$, where $u \in \{6\}.\beta$ and $v \in \{65\}.\beta$.

However, this (u, v) path does not appear to actually be the longest path (*i.e.* diameter). For the adjacent node 7, $\{7\}.\beta = 15$. So for $u \in \{7\}.\beta$ we estimate $d(u, v)$ to be $7.5 + 12 + 5 = 24.5$. In fact, the real diameter is $d(u, 7) + d(7, 65) + d(65, v) = 5 + 12 + 5 = 22$.

A similar process can be used to count triangles in the network [18].

3.3 Network Signatures

If we count the cycles in the reduced Newman collaboration graph of Figure 3, we get the following enumeration. This distribution of chordless cycle lengths may serve as a kind of spectral analysis, or "signature" of the network. Much more research is needed to determine the value of these signatures for discriminating between networks. For example, at SocInfo 2012 in Lausanne, Switzerland, it was suggested that $CC = \sum_k k \times n_k/|n|$, where n_k is the number of k cycles, might serve as a measure of connective complexity.

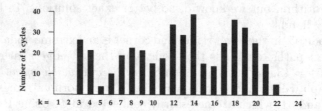

Fig. 6. Distribution of k-cycles in Figure 3

As we see, there are still 26 triangles in this reduction; such graphs are not "triangle-free". However, it is the 5 chordless cycles of maximum length that are of most interest. We might call them "major cycles". One of them is: < 4, 5, 6, 7, 8, 9, 10, 11, 12, 13, 14, 41, 51, 49, 36, 37, 38, 39, 28, 29, 26, 21, 4 >. This one has been emboldened in Figure 7 where we emphasize this 22-cycle of maximal length, while suppressing other aspects of this network.

The reduction process, ω, retains only nodes on, or between, chordless k-cycles, $k \geq 4$ (Proposition 4). It eliminates the "chordal" subgraphs of $\mathcal{N}$. It can be argued that by removing the chordal portions of a network $\mathcal{N}$, ω is only deleting well understood sub-sections that can be reasonably well simulated and "re-attached" to the irreducible spine. Just retaining the size of these subsumed subgraphs permits calculation of certain global attributes, such as diameter and centrality, as described earlier.

Looking at Figure 7 we see a similar process taking place. The entire subgraph consisting of nodes $\{54, 55, \ldots, 65\}$ is another pendant portion which will be ignored if one concentrates solely on the longest k-cycles. The edge/connection

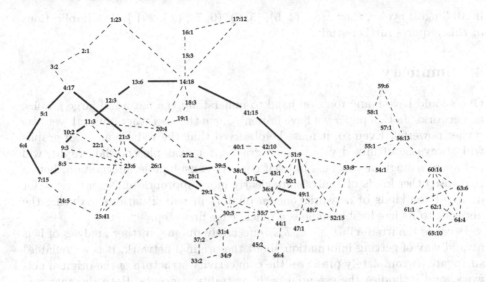

Fig. 7. A maximal chordless cycle in the reduced Newman graph of Figure 3

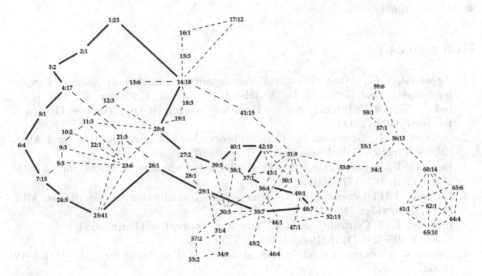

Fig. 8. Another maximal chordless cycle in the reduced Newman graph of Figure 3

$(56, 60) \in E$ is retained, solely because it connects the two 4-cycles among nodes $\{60, \ldots, 65\}$ to the main body, as described in Proposition 4.

The 22-cycle shown in Figure 7 is only one of five longest chordless cycles; a second is shown in Figure 8. As can be seen, it involves other paths.

If the five longest k-cycles are intersected, we discover that 10 nodes occur in all. They are $\{4, 5, 6, 7, 14, 26, 29, 36, 39, 49\}$. And, four connections appear in all longest cycles, they are { $(4, 5)$, $(5, 6)$, $(6, 7)$, $(26, 29)$ }. The implications of this requires further study.

4   Summary

One should have many tools on hand to understand the nature of large graphs, or networks. In this paper we have presented one that is rather unusual, yet also rather powerful. Even so, it must be observed that the reduction, ω, of graphs will always be of mixed value. Some graphs, for example chordal graphs, will reduce to a single node. This in itself conveys considerable information, but in this case other kinds of analysis are clearly more appropriate. Nevertheless, for many of the kinds of networks one encounters in real situations, reducing the network to its irreducible spine is a quick, easy first step.

Because the irreducible spine, $\mathcal{I}$, is effectively unique, further analysis of it is a valid way of getting information about the original network. It is a "reliable" surrogate. It completely preserves the connectivity structure of the original network, and embodies the essential path centrality concepts. Both distance and betweeness centers are contained within it. Consequently, this kind of analysis with respect to closed sets can provide valuable insights into the nature, and the structure, of networks.

References

1. Aggarwal, C.C., Wang, H.: On Dimensionality Reduction of Massive Graphs for Indexing and Retrieval. In: Abiteboul, S., Böhem, K., Koch, C., Tan, K.-L. (eds.) IEEE, 27th Intern. Conf. on Data Engineering (ICDE), Hanover Germany, pp. 1091–1102 (2011)
2. Agnarsson, G., Greenlaw, R.: Graph Theory: Modeling, Applications and Algorithms. Prentice Hall, Upper Saddle River (2007)
3. Brandes, U.: A Faster Algorithm for Betweeness Centrality. J. Mathematical Sociology 25(2), 163–177 (2001)
4. Cormen, T.H., Leiserson, C.E., Rivest, R.L.: Introduction to Algorithms. MIT Press, Cambridge (1996)
5. Freeman, L.C.: Centrality in Social Networks, Conceptual Clarification. Social Networks 1, 215–239 (1978/1979)
6. Linton, C.: Freeman. Visualizing Social Networks. J. of Social Structure 1(1), 1–19 (2000)
7. Gilbert, A.C., Levchenko, K.: Compressing Network Graphs. In: Proc. LinkKDD 2004, Seattle, WA (August 2004)
8. Granovetter, M.S.: The Strength of Weak Ties. Amer. J. of Sociology 78(6), 1360–1380 (1973)
9. Harary, F.: Graph Theory. Addison-Wesley (1969)
10. Jacobson, M.S., Peters, K.: Chordal graphs and upper irredundance, upper domination and independence. Discrete Mathematics 86(1-3), 59–69 (1990)
11. Leskovec, J., Faloutsos, C.: Sampling from Large Graphs. In: 12th Intern. Conf. on Knowledge Discovery and Data Mining, KDD 2006, Philadelphia, PA, pp. 631–636 (2006)

12. Lin, M.C., Soulignac, F.J., Szwarcfiter, J.L.: Arboricity, h-Index, and Dynamic Algorithms. arXiv:1005.2211v1, pp. 1–19 (May 2010)
13. McKee, T.A.: How Chordal Graphs Work. Bulletin of the ICA 9, 27–39 (1993)
14. Newman, M.E.J.: Finding community structure in networks using the eigenvectors of matrices. Phys. Rev. E 74(036104), 1–22 (2006)
15. Pfaltz, J.L.: Mathematical Continuity in Dynamic Social Networks. In: Datta, A., Shulman, S., Zheng, B., Lin, S.-D., Sun, A., Lim, E.-P. (eds.) SocInfo 2011. LNCS, vol. 6984, pp. 36–50. Springer, Heidelberg (2011)
16. Pfaltz, J.L.: Finding the Mule in the Network. In: Alhajj, R., Werner, B. (eds.) Intern. Conf. on Advances in Social Network Analysis and Mining, ASONAM 2012, Istanbul, Turkey, pp. 667–672 (August 2012)
17. Seierstad, C., Opsahl, T.: For the few not the many? The effects of affirmative action on presence, prominence, and social capital of female directors in Norway. Scandinavian J. of Management 27, 44–54
18. Tsourakakis, C.E., Drineas, P., Michelakis, E., Koutis, I., Faloutos, C.: Spectral counting of triangles via element-wise sparsification and triangle-based link recommendation. Soc. Network Analysis and Mining 1(2), 75–81 (2011)

SocWeb: Efficient Monitoring of Social Network Activities*

Fotis Psallidas[1], Alexandros Ntoulas[2,3], and Alex Delis[3]

[1] Columbia University, New York, NY 10027
[2] Zynga, San Francisco, CA 94103
[3] Univ. of Athens, Athens, 15784, Greece
fotis@cs.columbia.edu, {antoulas,ad}@di.uoa.gr

Abstract. Although the extraction of facts and aggregated information from individual *Online Social Networks* (*OSN*s) has been extensively studied in the last few years, cross–social media–content examination has received limited attention. Such content examination involving multiple *OSN*s gains significance as a way to either help us verify unconfirmed-thus-far evidence or expand our understanding about occurring events. Driven by the emerging requirement that future applications shall engage multiple sources, we present the architecture of a distributed crawler which harnesses information from multiple *OSN*s. We demonstrate that contemporary *OSN*s feature similar, if not identical, baseline structures. To this end, we propose an extensible model termed *SocWeb* that articulates the essential structural elements of *OSN*s in wide use today. To accurately capture features required for cross-social media analyses, *SocWeb* exploits intra-connections and forms an "*amalgamated*" *OSN*. We introduce a flexible *API* that enables applications to effectively communicate with designated *OSN* providers and discuss key design choices for our distributed crawler. Our approach helps attain diverse qualitative and quantitative performance criteria including freshness of facts, scalability, quality of fetched data and robustness. We report on a cross-social media analysis compiled using our extensible *SocWeb*-based crawler in the presence of *Facebook* and *Youtube*.

1   Introduction

The unprecedented growth rate of *Online Social Networks* (*OSN*s) both in terms of size and quality poses multiple research challenges. As individuals flock, the respective *OSN* volume is constantly increasing. Regarding quality, users often discuss about aspects of their daily life, thus making *OSN*s a source of information that is valuable in many different areas of interest. Among those, detection of events [4,24,11], identification of trends [5,1,11], announcement of news, detection of communities [2,17], sentiment analysis, and location tracking [23] have been in the epicenter of attention. All of the above point into an ever–increasing need to better understand both the exhibited behavior and its development by either individuals or groups of users. In doing so, numerous forms of social awareness are being developed [19].

* This work was supported by PIRG06-GA-2009-256603.

X. Lin et al. (Eds.): WISE 2013, Part II, LNCS 8181, pp. 118–136, 2013.
© Springer-Verlag Berlin Heidelberg 2013

The typical process to unveil and further analyze underlying patterns is that given a single stream of social data, a *Complex Event Processing (CEP)* mechanism [12,10,26] is deployed to identify trends and formations of interest in an on-line fashion. Although, a number of studies have been conducted mining data streams emanating from a single social source, there is great interest in attaining cross-social media analyses involving multiple streams from different *OSN*s. Meaningful such aggregation of information can certainly lead to improved fact verification and enhance trend establishment [4]. Consequently, data originated from multiple *OSN*s should not be considered disjointly but rather should be co-developed and co-referenced. In turn, *CEP*-engines should blend social streams from multiple *OSN*s to benefit from their inter-connections. For instance, let an individual A be a user in 2 of the most widely-used *OSN*s: *Facebook* and *Twitter*. A well-known challenge where social media content can assist a great deal is the tracking of the movement and the identification of the current location of A. In [23], it is argued that the current location of A can be inferred using information about her friends. If we limit ourselves by extracting information from a a specific social network say *Twitter*, we are unable to correlate data potentially available from both accounts regarding the location of A. In this respect, we lose vital information for the location tracking task. Furthermore, information about an event detected in an online stream can be extended or cross-validated using other *OSN*s through query formulation strategies [4]. In this context, there is a pressing need to re-consider and benefit from intra-*OSN* relations and produce novel types of analysis and applications in numerous fields including news, events, polls, ads, marketing, games, information tracking, and intra-social awareness.

Obtaining meaningful data simultaneously from multiple *OSN*s is however not a trivial path to follow. Conventional crawling approaches for the "open" Web proposed in the last decade fall short in fetching effectively from multiple *OSN*s; such methods include (1) BFS crawling [18,16], (2) contiguous crawling, (3) focused crawling [22], and (4) random walking [3,14]. Furthermore, content found on database systems rather than on web servers, also known as "deep/hidden", is often crawled (or searched) by filling forms using appropriate keywords [20,15]. Given the steadily increasing volume of data and the inherent physical-network limitations, the distribution and/or the parallelization of crawling have been proposed as an effective means to realize crawling [25,9].

*OSN*s raise different crawling challenges that cannot be captured by state-of-the-art web crawling techniques. By nature, traditional Web data have two salient features that facilitate access: they are available freely via web-servers or supporting databases and they can be fetched in a straightforward manner. Social network providers allow only "subscribed" applications to fetch their data using exclusively provided *API*s. However, the fetched data is considered private with high sensitivity and heterogeneity. The popular *Facebook* and *Twitter* impose strict limitations and regulations on use of their data.[1] Conventional web crawling techniques that strive to obtain data from *OSN*s are very likely to deviate from the legal limitations and provider regulations imposed. Moreover, using the *API*s provided by individual *OSN*s comes at a high productivity cost. Such

---

[1] See for example the discussion on rate limits here:
https://dev.twitter.com/docs/faq

API calls can be parameterized in a multitude of ways requiring so user sophistication. Matters are also inherently more complex when multiple social networks using diverse structures and building elements are involved in the crawling as there is always need to properly disambiguate the returned results. *SN* providers also frequently impose constraints in terms of response time. Provider resources available for responding to application queries remain limited making those applications prone to network and processing bottlenecks. Indeed, several approaches have applied traditional crawling techniques to fetch data from specific *OSN*s encountering some of the above problems [6,13,7]. For instance, [6] uses BFS and uniform sampling crawling to gather data for *Facebook*'s analysis purposes; the study reports resource limitations, privacy restrictions and *API* misbehavior by both the provider and its users.

To alleviate the aforementioned problems and driven by the motivation to describe and seamlessly access multiple *OSN*s in an *amalgamated* form, we propose an extensible model *SocWeb*. *SocWeb* maps the structures of the underline *OSN*s, helps capture their relationships and ultimately offers the basis to develop a versatile distributed crawler/fetcher. *SocWeb* leverages two fundamental concepts that we introduce: the model and its requisite generic social network *API* or *SNAPI*. *SocWeb*'s programmatic interfaces intend on addressing issues encountered by standard web crawling techniques. The design choices of *SNAPI* adhere to the principle of appropriately abstracting the procedure of connecting one or more applications to desired social network providers by minimizing the effort required and complying with the norms and regulations imposed by the providers.

The contributions of our work are:

- We present and describe an "amalgamated" *OSN*, based on the observation that the underlying structures of *OSN*s remains similar, if not identical.
- We present the *SNAPI* that enables simultaneous connection to one or more generic applications with specific social network providers.
- We outline the design choices of our distributed crawler for automated content extraction from multiple *OSN*s.
- We evaluate qualitatively and quantitatively our system and provide characteristics of the data retrieved as a proof of concept that our system can efficiently monitor data from different *OSN*s.

2   Representing *OSN*s in SocWeb

At a high level, an *OSN* s is typically represented by a directed graph $G_s(V_s, E_s)$ where vertices V_s correspond to objects (e.g. users, photos, comments) in the social network and edges E_s (which can be potentially labeled) correspond to relationships between those objects (e.g. user A posted photo p_1). Since our goal is to efficiently monitor a variety of *OSN*s, we need to extend this definition to include a set of *OSN*s.

To this end, we employ the following definitions for the building blocks (vertices, edges) of an *OSN*.

Object Types (OT). In our *SocWeb Model* we define two different basic types of objects (vertices), *primitive* and *composite*. More specifically a vertex $v \in V_s$ of *OSN* s is:

- *Primitive, iff* $d_{out}(v) = 0$ and $d_{in}(v) \geq 1$, or
- *Composite, iff* $d_{out}(v) \geq 1$

where $d_{in}(v)$, $d_{out}(v)$ are the in- and out-degree of v respectively. Intuitively, the *primitive* vertices define the boundary of a given *OSN*, while composite vertices may link to either primitive or composite vertices thus playing the role of both forming and populating a social network.[2]

In some cases, a given *OSN* may specialize the types of its objects. For example, Facebook has object types such as user, post, album and photo. Note that these types correspond to composite vertices. Every social network has to also define types for primitive nodes. Types that correspond to primitive nodes are, for example, integers, strings and timestamps. Furthermore, each vertex has a unique object type. We denote the object type of a vertex v as $V_{OT}(v)$.

Link Types (LT). Following the OT definition, we also define two kinds of basic link types between two vertices v_1, v_2:

- *P-link, iff* $(v_1, v_2) \in E_s$ and $Composite(v_1)$ and $Primitive(v_2)$
- *C-link, iff* $(v_1, v_2) \in E_s$ and $Composite(v_1)$ and $Composite(v_2)$
  Similar to the object types, some social networks may further specialize their link types. For example, a Youtube user may be linked to a post using a 'posted', 'liked' or 'disliked' link. We denote the link type of an edge (v_1, v_2) as $E_{LT}(v_1, v_2)$. We should note here that not all objects can be linked with any link type. For example, it may not make sense to connect two users with a link denoted as 'uploaded'.
- *Collections and Colinks.* One characteristic of *OSN*s is that an object can link to collections of objects of the same object and association type. For instance, a Facebook user can be associated with several photo albums using an 'uploaded' link type. Instead of referring to each of the albums as a different connection we encapsulate the photo albums to form an album collection and use a single link from the objects. In the *SocWeb Model*, we call such types of links Co-links and we formally define them as: $Colink(v, \mathbf{v}) = \{v_1, v_2, \ldots, v_n\}$, where $\mathbf{v} = (v_1, v_2, \ldots, v_n)$ is an object representing the collection of objects $v_1, \ldots, v_n$ with the same type, and v is the object linking to the collection. Note that we create collections of objects based on the link type rather that the object type. An object may be connected to objects of the same type but the connection between them has a different meaning. For instance, a Facebook user can be connected to posts either because she liked or posted or commented on them.
- *Intra-OSN edges and S-links.* By using the definitions above, we can describe a graph that represents a single *OSN* s. Since our goal is to monitor a set of *OSN*s, we need a way to represent the interconnections among them. One straightforward approach would be to consider the union of the set of *OSN*s, assuming that their graphs are disconnected. By following this approach, however, we are not able to take advantage of the fact that a user may have accounts in different *OSN*s and

---

[2] Note that if $d_{out}(v) = 0$ and $d_{in}(v) = 0$ then v is an isolated vertex. Such vertices are difficult to discover as there are no links to them and are very unlikely to appear in a real *OSN*. For simplicity, in our model, we assume that we do not have such nodes.

attribute her objects and links to the same person. Being able to identify the same person across OSNs is of great importance for several different scenarios such as opinion mining, personalization and ad targeting.

To this end, we also define a special kind of cross-OSN links. More specifically, for two different social networks s_1, and s_2, we consider the additional set of intra-OSN edges E_{s_1,s_2}, which is the set of directed edges (v_k, v_l) where $v_k \in V_{s_1}$ and $v_l \in V_{s_2}$. For these intra-OSN edges, we define the following types:

- $SP\text{-}link(v_1, v_2)$ iff $(v_1, v_2) \in E_{s_1,s_2} \wedge Primitive(v_2) \wedge Composite(v_1)$
- $SC\text{-}link(v_1, v_2)$ iff $(v_1, v_2) \in E_{s_1,s_2} \wedge Composite(v_2) \wedge Composite(v_1)$
- $SCo\text{-}link(v_1, \mathbf{v})$ iff $v_1 \in V_{s_1} \wedge v_2 \in V_{s_2} \wedge s_1 \neq s_2 \wedge Colink(v_1, \mathbf{v})$

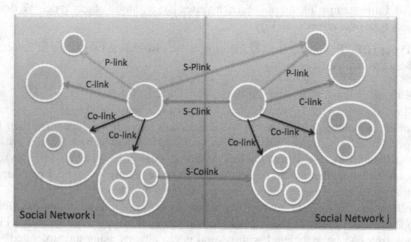

Fig. 1. Multi-social graph example with all possible basic object and link types of *SocWeb Model*. Nodes in blue and green refer to primitive and composite objects respectively.

To illustrate the abstraction that *SocWeb Model* offers, we show 2 OSNs in Figure 1 with all the possible basic object and link types. These types can be further specialized based on the description of each social network.

3   Accessing OSNs through a Generic API

Most of the social networks today provide an Application Programming Interface (API) to their data in order to allow developers to build applications. In almost all cases, the API provides a way for authorizing an application or a user to access data on the OSN. There are typically two levels of authorization: (a) acquiring the minimum credentials that each user/application needs to connect to the OSN, and (b) potentially selecting additional credentials that are useful to a given application (e.g. to access the birth dates of a user's friends).

Since our goal is to access multiple OSNs at the same time, we need to be able to create and maintain such multiple credentials and interact with a variety of APIs. This, however, is a challenging problem for the following reasons:

- Given the number of available *OSN*s there is also a large number of available APIs with multiple versions available. For instance, Twitter supports a REST API and a Streaming API and the REST API has multiple versions.
- Each API supports a large number of calls, parametrizable in a multitude of ways.
- Each *OSN* provides a diversity of encoding formats that can differ from one *OSN* to another. For example, language encodings are typically one of UTF-8, ISO-8859-1 or 1-3byte Unicode sequences(e.g "\u00ed").
- There is a large number of different ways to exchange information with an OSN, for example JSON, XML, KML, RSS (2.0) or ATOM.
- Different *OSN*s use different authorization procedures. Most of them rely on the standard OAuth protocol, which has a multitude of versions. Additionally, different platforms (Web, desktop, mobile) require different authorization procedures within the same *OSN*.
- Each *OSN* enforces its own limitations to the amount of requests that are allowed. Such limitations can be enforced by IP, by application, by user or by authorized/non-authorized calls within a given time period (typically per hour or per day). This implies that we need to be aware that an application limit has been reached when accessing information in an *OSN* and back off if necessary.
- Each *OSN* has its own set of error codes. Hence, we need to take appropriate action which may be different per *OSN*.

To alleviate these problems we propose a generic Social Network API (SNAPI) for arbitrary *OSN*s that is based on the *SocWeb Model* discussed in Section 2. We proceed by introducing the socWebObject data structure that is central to our system and we then define the SNAPI, whose purpose is to interact with arbitrary *OSN*s using the corresponding APIs through the use of socWebObjects.

3.1 The socWebObject Data Structure

Based on our discussion in Section 2, each node of an *OSN* graph has a set of p-links, c-links, co-links, sp-links, sc-links and sco-links associated with it. Additionally, each node has a unique identifier, a unique object type and belongs to a single *OSN* and may have one or more nodes pointing to it. To this end, in our system we use the following data structure to represent a SocWebObject:

```
socWebObject(id, type, plinks, clinks, colinks, splinks, sclinks,
scolinks)
```

where p-links, c-links, sp-links, sc-links are maps to other socWebObjects based on the association type and co-links and sco-links are maps to an array of socWebObjects. For instance, let obj be a socWebObject representing a Facebook user, then we can reference the user name as obj.plinks["username"], the hometown as obj.clinks["hometown"] and the *i*-th post of the user as obj.colinks["posts"][i]. Additionally, the hometown is another socWebObject and we can reference its longitude as obj.clinks["hometown"].plinks["longitude"].

Table 1. SNAPI Calls. $Def(s)$ is the definition of *OSN* s, i.e. the set of nodes and different types of links of s. linktype[] represents a subset of the different kinds of links p-links, c-links, etc.

Call
initialize($\bigcup_{i=1}^{N} Def(s_i)$, app_id[])
apply_definition($Def(s)$)
apply_credentials(user_id, app_id, credentials)
apply_constraints(constraint[])
get_object(object_id, type, linktype[])
get_links(object_id, type, linktype[])

3.2   A Generic Social Network API (SNAPI)

In our system, we provide access to the different *OSN*s by implementing wrappers for the different API requests together with the most popular parametrization options. Overall, our generic API consists of the set of calls as shown in 1, which we discuss in more detail.

SNAPI Initialization. SNAPI is initialized by providing the definition (i.e. set of nodes and links with their types) of the *OSN*s that we are interested in following to the initialize() call together with the application ids that are authorized to access information in the *OSN*s. In the cases where we discover a new *OSN*, or we decide to change accessing data through a different application, we can do so on-the-fly through the apply_definition() call.

SNAPI Authorization. For authorization purposes SNAPI provides the apply_credentials() call which uses the login and password provided to connect and acquire the necessary credentials. In most cases, the credentials are access tokens returned by the OAuth protocol.

SNAPI Constraints. In order to make our system's monitoring capabilities more flexible, we have provided SNAPI with the capability to specify a set of constraints that can be enforced during the monitoring process. More specifically, for each object type we can define rules for each one of its connected object types (i.e. p-,c-,co-,sp-,sc-,sco-links). For instance, suppose we want to retrieve the location of a tweet only if its longitude is within a specific range. We can express this rule as: tweet.clinks[``location''].plinks[``longitude''] $\geq$ minimum_longitude && tweet.clinks[``location''].plinks [``longitude''] $\leq$ maximum_longitude Providing this capability allows us to only monitor parts of an *OSN* that we are interested in thus saving resources. In our current implementation, we only allow simple boolean constraints to be provided as input to SNAPI. We plan to implement these constraints in the form of a full-scale Complex Event Processing system [12,10,26] in the future.

SNAPI Fetching of Data. We provide two different calls for fetching within *SocWeb*, fetching of objects and fetching of links, through the get_object() and get_links() calls respectively. For both calls, we need to provide an object type together with its id which uniquely identifies an object within a given *OSN*. Such an id is typically created by the *OSN* and found by following the different kinds of links during monitoring.[3] We can also specify the kinds of links that we are interested in (c-links, p-links, etc.) through the `linktype[]` parameter. In the case of `get_object()`, the system proceeds iteratively by fetching the object's links and applying the provided constraints. If, when retrieving an object, we cannot immediately decide whether it satisfies a constraint (e.g. we have not discovered a property such as longitude that was specified in a constraint), we place it in a queue which we periodically clean in order to keep only objects that satisfy the given constraints.

Efficiently fetching the data from a set of *OSN*s is a challenging task which we discuss next.

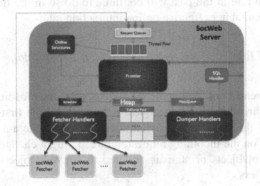

Fig. 2. SocWebFS Architecture

4   Efficient Monitoring of *OSN*s: The SocWeb Fetcher System

We have so far discussed how we represent a set of *OSN*s within SocWeb as well as its generic API that enables us to access the *OSN*s. Given the enormous size and update rates of information in the *OSN*s we need to find ways of monitoring this information in an efficient way. Not all users post information at the same time, or at the same rate. If we blindly start downloading everything our system comes across, we may end up with redundant information. In addition, since most of the *OSN*s impose limitations to the amount of objects that we can download within a given period of time, making smart decisions of what to download and how often is of paramount importance.

In this section we discuss our design around a fetcher system (the SocWebFS) which is built with these constraints in mind and can make decisions on-the-fly on what to download at a given time. Our system consists of a central master server that coordinates a set of fetchers. The server keeps track of statistics on previously downloaded objects and makes estimations on what to download next. We present an overview of the whole system in Figure 2 and we describe each of the components in the following subsections.

---

[3] Of course, there could be a conflict where a given object id may correspond to objects in more than one *OSN*s. We handle this case internally in our system by having an additional *OSN* id attached to the object id.

4.1   SocWebFS Server

This is essentially the brain of the system which decides which object to fetch, when to fetch and how often. The SocWebFS Server comprises seven modules: (a) the Request Queue-Thread Pool, (b) the Frontier Queue, (c) the Scheduler, (d) the Fetcher Handlers, (e) the ToDump Pool, (g) the Dumper Handlers. As is typical in these kinds of architectures, these modules operate in a pipeline fashion, sharing global online data structures in order to coordinate.

Online Structures. This module is responsible for maintaining statistics regarding the online activity of the users. These statistics are useful in deciding which users to prioritize when downloading objects. The intuition is that a user who has been posting information at a high rate in the past will continue to do so in the future. To this end, for each user we maintain her effective online interaction rate:

$$EIR(u) = \frac{1}{7} \cdot \#Requests(u) \ in \ the \ last \ 7 \ days$$

Request Queue-Thread Pool. To handle the requests for downloading within our system we maintain a thread pool. Each request is handled by the first available thread, while the number of threads is tuned dynamically based on the amount of incoming requests. Depending on the incoming request, a thread, might change the online structures, update socWebObjects or statistics related to socWebObjects or spawn/delete fetcher/dumper handlers.

Frontier Queue. This is the most important part in the system as it decides on the prioritization of which objects to download at any given moment. In its simplest form [9], the first item from the queue is fetched and then it is placed back in queue to be refreshed again later. This operation can be performed in a variety of ways in order to optimize for freshness or age of the objects [8], optimize for bandwidth or cost [20], or adhere to politeness policies.

In our implementation, the frontier module consists of 2 sub-modules: a set of F front FIFO queues, that guarantee prioritization of socWebObjects to download, and a set of Q back queues that guarantee polite behavior of the fetcher (i.e. ensuring that we are not downloading too fast from a given OSN) by monitoring download rates from the OSNs. Within the front queues we prioritize the objects based on a set of metrics:

– **User-based.** In this case, we take into account the user's interaction rates with the OSN. The higher the EIR for a given user, the higher priority her objects are given in the queue.
– **Change-Rate-based.** Since the objects change periodically, we need a way to keep track of which ones are more likely to change in order to prioritize them first. We consider an object to have changed if any of its c-linked objects has changed. To this end, we use a change rate metric for each object:

$$CR(obj, link) = \frac{\#Changed \ c\text{-}links}{t_l - t_s}$$

where t_l, t_s are the timestamps of the last and first changed c-link objects respectively. This definition of average change rate works well in most cases, but our system is flexible to employ different and more sophisticated change rate approaches (e.g. fitting a probabilistic distribution to the changes to estimate the rate).

– **Importance-based.** Depending on the social network schema, it is sometimes the case that some objects are more important than others. For example, when a user updates their location may be more important than a new comment that she just posted.

To capture this difference in importance among the objects, we employ an importance metric $I(obj)$ for each object. $I(obj)$ can be static per object type (e.g. location is 10 times more important than comment) or can be dynamically adjusted. In our implementation, we followed a dynamic approach and we define the importance of each object as:

$$I(obj) = \frac{1}{n} \sum_{u \in Users} EIR(u) \cdot App(obj, u)$$

This is essentially the average EIR of all the users associated with the given object weighted by an application-specific weight App. For example, if we were using SocWeb to implement a search engine $App(obj, u)$ could be the number of times that obj was returned as a result to the user u. In this case, the more times the user sees obj the higher the weight.

Based on these metrics, all the objects of the frontier are given a priority and are placed in a queue to be scheduled for fetching.

Scheduler. The goal of the scheduler component is distribute the prioritized objects from the frontier across several fetcher handlers. The scheduler has three main goals:

– To balance the total fetching workload across the several SocWebFS Fetcher nodes by deciding when and where to send an object for fetching. To this end, the Scheduler maintains histograms per machine to estimate the amount of time needed for each machine to perform each request. Given a socWebObject and its change rate per link as computed in the change rate level, we estimate the amount of requests that have to be performed to fetch an object and we pick the machine with the smallest load to handle the fetching.

– To ensure that fetching of an object will not exceed the limitations (e.g. IP, time, API) posed by the *OSN*s. If the scheduler estimates that fetching of an object may potentially exceed one or more limitations imposed by the *OSN*, it postpones its fetching for later and periodically repeats this estimation.

– To ensure fault tolerance by guaranteeing that fetches that received an error or time out will be considered for fetching again in the near future. To this end, each object gets a unique session id (sid) that uniquely identifies the transaction. This sid together with the initial request timestamp t_s are used to detect whether we have waited sufficient time before we consider the fetching of the given object as timed-out. If the object request has timed out, we place the object back to the frontier to be fetched again later. The process of resending an object back to the frontier is performed only a fixed number of times per object (set to 3 in our system).

Fetcher Handlers. The fetcher handlers are responsible for the communication of SocWebFS Server with the SocWebFS Fetchers. Each SocWebFS Fetcher corresponds to a single Fetcher Handler that serves as middleware for the communication of the SocWebFS Fetcher with Server's resources. Upon retrieving a set of requests the handler serializes them and sends them to the fetchers for processing. The handler also updates the initial request timestamps t_s in the frontier.

SocWebFS Fetchers. The goal of a SocWebFS fetcher is to retrieve a single object that is assigned to it by its corresponding handler. Each fetcher uses the SNAPI that we described in Section 3 which is initialized at the startup of the system. After authentication, the fetchers operate in four states: (a) connect, which initializes the connection to the *OSN*, (b) wait for requests to be assigned, where the fetcher blocks and awaits yet-to-be-fetched objects from the server, (c) fetching, which requests objects from the *OSN*, and (d) upload, which uploads the fetched objects in a bulk mode back to the SocWebFS server. When a SocWebFS fetcher has finished fetching of objects it issues an upload request that returns a subset or all of the fetched objects along with statistics(amount of requests, time per request, limitations reached). Next, Fetcher handler has to send the statistics to the Scheduler, check whether the fetching of each object was valid or resulted in an error, and append the fetched object to the Dump Pool.

Dumper Handlers and Dump Pool. Each Fetcher Handler has a corresponding Dumper Handler which takes on the task of saving the retrieved objects to disk and appending newly found objects to the frontier. To enable the communication of those components, we use a Dump Pool which is essentially implemented as a set of queues, with each queue corresponding to a pair of Fetcher-Dumper Handlers. In this case, the Fetcher acts as a producer and the Dumper as a consumer. We handle updates of objects by maintaining multiple versions of objects, but keeping only those versions that are linked by other objects.

4.2   Privacy Considerations

We have discussed the overall architecture of our SocWeb system which enables us to download and store objects from a set of *OSN*s locally. Our system is flexible enough to handle different *OSN*s with different APIs and limitations on the data that we can store.

However, given the sensitive nature of some of the data in the *OSN*s we may need to enforce additional limitations in certain cases. For example, certain *OSN*s pose limitations on whether the data collected can be at all stored on disk or can only be used on-the-fly. To this end, SocWeb provides subscriptions of external applications and Dumper Handlers instead of storing the data locally. The only indirect requirement that we have for the external handlers is that they are capable of consuming the data at the rate that the fetcher retrieves them from the *OSN*s. If the external rate is slower, SocWeb drops some of the objects to match the external consuming rate.

In addition to storing the data, there are also challenges in enforcing access-level constraints to the data. For example, consider the scenario when users A and B are both friends with user C but user A can see C's birthdate but B cannot because C specified so. In this case, users A and B have different access permissions to user C's p-link objects. This scenario may happen for all different kinds of links that we defined in Section 2. To solve this problem, we additionally employ a per-user storage space where we keep, for a given user, the conflicting parts of an object that are different from the global version of the object in the system. In this way, we can enable applications using the data that SocWeb collected to adhere to the privacy of the data because of the different user access levels.

5 Evaluation

Our main objective in evaluating our architecture and the *SNAPI* interface is to establish the utility of our approach in terms of a number of key characteristics.

- *Robustness:* to avoid traps, *SNAPI* can be parameterized with constraints for cycle detection during the fetching process. Furthermore, Dumper Handlers determine whether an object may be eligible for (re-)fetching. Application dependent traps are also detected through parameterization of these handlers.
- *Politeness:* each *OSN* poses its own limitations in terms of calls per-application, per-*IP* and/or per-user. Our *Scheduler* and *SocWebFS*-fetchers offer compliance with imposed retrieval rates as designated by *OSN*s. The above can also dynamically re-set limitations that change on the fly. *SNAPI* also determines whether heavy workload crawling activities are in place and makes use the respective *Facebook/Youtube* streaming *API* to better facilitate fetching of objects.
- *Distributeness:* our proposed architecture is based on the single-*SocWebFS*–server and multiple-concurrent-clients model, all operating in star-like fashion. The design warrants for fault tolerance (*Frontier*), workload balance (*Scheduler*), elasticity (*Scheduler*) and redundancy manipulation (*Dumper Handlers*). Even in the case of a software crash, the *SocWebFS*–server loses no data as the crawlers will await for the main server to become alive anew. The state of the server is maintained as expressed by the *Frontier* and *MetaQueue* structures is maintained by the back-end database.
- *Quality:* the relevance of retrieved data is of paramount importance to users and their applications. Our policies that differentiate between data of "interest" and "no interest" and so assist in achieving per-user extraction quality characteristics.
- *Scalability and High Performance:* *SocWebFS* is able to scale up (or down), in the presence of more (less) machines and/or bandwidth changes. To examine the scalability of our system we conducted corresponding experiments.
- *Extensibility:* The modularity introduced in the design of *SocWebFS* allows several levels of extensibility. Application dependent policies are introduced in the form of constraints to parameterize *SocWebFS* components (*SNAPI*, *Scheduler*, *Dumper Handlers*). These policies help the proposed model to render a simple yet powerful abstraction for multi-social networks description.

By and large, the above characteristics are those that have been used over time to ascertain effectiveness in Web crawling [18,21,25]. We also treat carefully the trade-off between performance and maintenance of up to date information (i.e., freshness). As an object may receive repetitive requests for either edited or deleted distributed content, this will inevitably lead to performance degradation. *Scheduler*'s design uses the object change-rate as an estimator for changes expected in the future and offers accurate quantitative information regarding this issues helping attain a good balance in the trade-off at hand.

We should also point out that we have designed our systems so as to be able to handle diverse types of retrieved objects, fetching protocols, arbitrary social networks as well as multiple data formats such as *XML, JSON*, UTF8-encoding, etc.

5.1 Evaluation Approach and Settings

We experimented with *SocWebFS* using 2 popular *OSN*s: *Facebook* and *YouTube*. Our evaluation is weaved around 2 experimental use-cases that illustrate not only the use of our system but also its compliance of our prototype with the aforementioned behavioral and performance characteristics. In the first use-case, we exclusively use *Facebook* to demonstrate the applicability and value of *SocWebFS* in the presence of a single *OSN*. The second scenario uses both *Facebook* and *YouTube* networks and explores the interconnections formed as soon as users deposit videos on one and proceed with respective posts on *Facebook*. We refer to the first use-case as *Facebook Spider* and to the second as *Aggregated Social World*.

For both experiments, we employed a private laboratory cloud made up of physical machines featuring Intel(R) Xeon(R) CPU X3220 processors at 2.40GHz, 8GB of RAM all connected through a 1GBps Ethernet switch. We used 7 virtual machines (VMs) from this cloud with each VM featuring a 2GB main memory running a client module (i.e., *SocWebCrawler*). One of these VM servers also undertook the role of the *gateway* as all *SocWebCrawlers* would issue requests to *OSN*s through this gateway using the *NAT* protocol.

5.2 Facebook Spider

In this use-case, we employ *SocWeb* to play the role of a social *spider* for *Facebook*: given an initial number of objects-nodes in the *OSN* graph, we intend to retrieve objects by exploiting adjacent links to already visited nodes. Social graphs are however inherently dynamic. Thus, we will have to continually monitor already visited nodes for new, deleted and/or updated adjacent links.

Before we start working with *SocWeb*, we are first required to formally define an abstraction of the candidate *OSN* to be crawled as Section 2 outlines. Most of the definition effort here is directed towards designating the specific objects and link type of interest than the entire network. In this use-case, we create an abstraction of *Facebook* that consists of users, albums and posts as our key-interest object types. Each user is linked with her posts, albums and friends (*co-links*), his hometown, school, work (*c-links*) as well as his first name, surname, birthday, last post's and album's creation time (*p-links*). Each post and album is connected to the number of likes and comments received (*p-links*). We also consider albums to be connected with their photos (*co-link*).

During the first time of crawling of a user we want to download the full list of her posts, albums and friends and her *c-* and *p-links*. We decide to re-introduce a user to the *Frontier* module, if and only if her amount of posts is more than a threshold [4]; this threshold can be perceived as a constraint of significance (i.e., user["posts"]["likes_cnt"] > 1000). We also append to the *Frontier* her friends and corresponding albums if they were uploaded during the last week with respect to the time of the their crawling. For albums, we also require to be of significant interest (i.e., user["albums"]["comments_cnt"]>1000). Once an object has been fetched, its monitoring commences with regard to changes in her links (i.e., user["posts"]["created_time"] > user["last_post_time"]). If we are required to crawl a friend of a user, we follow the same procedure but we don't further crawl the friends of friends. If *SocWeb* determines to re-crawl some specific object we consider only content created after the last time it was crawled.

The last action during *SocWeb* initiation is to acquire a set of nodes in order to start crawling. Here, this initial set of nodes consists of users whose profiles are publicly available. A large database containing such names found at drupal.org/project/namedb, contains 129,036 names listed in alphabetical order. We then use *Facebook-API* to obtain user–ids with similar names. We were able to retrieve 25,556,963 user–ids from which we considered only 881,549 users who maintain entirely public profiles.

We focused our crawling in the period of July 26th, 2011, to August 19th, 2011. Our assumption was that a good number of crawled users in this period were on vacation. In this regard, they were presumably uploading live content of their daily activities in a bursty manner throughout the day. Also, live content regarding vacation is invariably deemed attractive for followers (i.e., friends, subscribers etc.) to either "like" or comment on. Table 2 depicts the overall outcome of this specific crawling exercise that yielded a sizeable dataset of 2,437 albums and 75,370,629 posts. Apart from the live content, we also crawled their full history.

Table 2. *Facebook* Dataset

Names	Ids Found	Public Users	Albums	Posts
129,036	25,556,963	881,549	2,437	75,370,629

Figure 3 presents the distribution of the posts made by users in this use-case. This distribution has a mean of 85.98, standard deviation of 182.48, skewness of 4.10, kyrtosis of 23.79 and follows a power-law. Evidently, the rates at which users generate new content vary significantly. The overwhelming majority of users does not create much content and they infrequently use *Facebook*; some may even have deactivated their accounts. On the other hand, a small portion of users produce content at a high-rate. The *Frontier* component of *SocWeb* in conjunction with the *Scheduler* can effectively monitoring the change–rate metric in order quantify content differentiations. Responding to such observations, users with high content creation rates are placed to high–priority queues.

To further quantify the effectiveness of the change-rate metric, we measure its fit on the retrieved dataset of Table 2. In our setup, we sort the retrieved posts in ascending creation time. Then, for each user, we capture the change rate of his/her posts at any

---

[4] Manually set to 1,000 for this experiment.

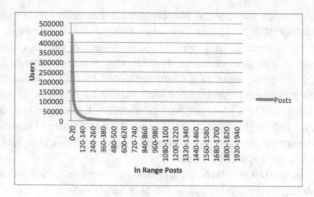

Fig. 3. Histogram of posts made by users

point in time while processing the sorted dataset in a streaming fashion. As soon as a new user post shows up in the stream, we use the current change–rate to measure whether it could accurately predict or not the creation time of the new post. This strategy is reflected in the decision making process of our *SocWebCrawlers*. In this context, we are only interested in establishing mis-predictions as yield of our under-estimation of the creation times of posts. Over-estimating the creation time of posts, while it is considered a misbehavior in general, in our setup is partially alleviated by deciding a time window of maximum expected change (i.e. the *Scheduler* decides to re-crawl an object if this time window has elapsed). In this use-case, we have empirically set this time window to 1 month. In Figure 4a, we show the distribution of posts and the fraction of mis-predictions yielded by the change-rate metric per month, beginning between April 2005 and some time in August 2012. An interesting property of this distribution is the growth of *Facebook* in terms of posts per month. Thus, the choice of the scaling factor of the change-rate metric introduced in Section 4 can adequately fit this kind of increasing *OSN* behavior. Finally, our analysis shows that the amount of mis-predictions as a percentage of posts on a monthly basis remains significantly low as Figure 4b indicates.

5.3   Aggregated Social World

Figure 5 shows how *Facebook* users create a post on which they upload content from *Youtube*. The objective of this use-case is to demonstrate that *SocWeb* readily facilitates fetching *Youtube* information pertinent to *Facebook* posts. Offering aggregate information has been successfully employed in event processing before [4]. When more multiple *OSN*s are present, information aggregation calls for communication across social networks, a function that *SocWeb* can readily offer. Here, we also report on *SocWeb* performance as the number of *SocWebCrawlers* increases and discuss the quality of retrieved data.

We initiate *SocWeb* by following the same steps as in the *Facebook Spider* case; object and link types of interest are shown in Figure 5. By exploiting *SocWeb Model*'s semantics, we formally define a *Facebook* post to be connected to *Youtube* video(s) as *sc-link(s)*. Further, we let *SNAPI* trigger requests via *Youtube*'s *API* depending on

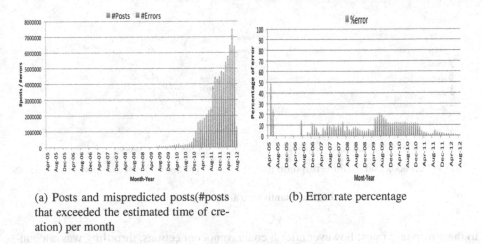

(a) Posts and mispredicted posts(#posts that exceeded the estimated time of creation) per month

(b) Error rate percentage

Fig. 4. *Facebook*'s growth and *SocWeb*'s adaptivity

Fig. 5. An example of intra-social connections

the content of *Facebook* post by also applying constraints (i.e. post[``type''] = ``video" && is_youtube_video(post[``link''])). As social networks inherently display power-law distributions (i.e. Figure 1), we select 10,000 users from our first experiment, whose posts follow the same skewed distribution, as node-seeds to commence crawling.

Every object that undergoes crawling spends time passing through the various *SocWeb* components including the fetcher handlers, clients, dump pool and finally the dumpers. The expended time for such trips largely depends on the workload of *SocWeb* and the available clients to perform the crawling step. The major overhead though comes from the time required to fetch each object; in our experiments, we establish that an average 84% of the time goes towards fetching objects when all VMs are in use. Moreover, social network providers penalize the concurrent access of their graphs. Thus, incrementing the number of *SocWebCrawlers* employed doesn't necessarily reciprocate in terms of the volume of data retrieved. In this regard, we report on the scalability of *SocWeb* in terms of data returned to the *SocWeb* server from the *SocWebCrawlers* in the unit of time (bytes per second-*bps*) as well as the average *bps* retrieved from each social network provider as a function of the *SocWebCrawlers* employed.

Figure 6a shows the average *bps* rates obtained. *Youtube*'s response rate is uniformly higher than that of *Facebook* although no major optimization (i.e., batch requests) was included in our implementation to the *Facebook*'s *API*. Both providers show consistency

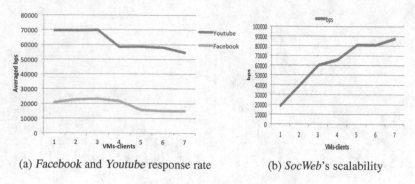

(a) *Facebook* and *Youtube* response rate   (b) *SocWeb*'s scalability

Fig. 6. Quantitative analysis of scale

to their response rates; however after 3 concurrent connections, throttling was encountered. Moreover, when the number of *SocWebCrawlers* increased, the *OSN*-provider limitations became apparently severe.

Figure 6b shows the average *bps*-rate extracted by *SocWebCrawlers* as the number of deployed *SocWebCrawlers* increases. The *OSN*s-imposed limitations for more than 3 concurrent connections affect the obtained *bps*-rates. In our experiment, no NAT-pipe saturation occurred and contention was sufficiently low; as far as the degradation of the stream rate (*bps*) achieved in the dumpers was only 2% of the average stream rate of *SocWebCrawlers*.

It is also worth pointing out some limitations we encountered as *OSN*s apparently impose constraints on (possibly) *IP*s, *Application ID*s and/or volume of data fetched. In the early stages of our experimentation, *SocWeb* faced an outage of more than 1 day due to the above limitations whose nature appeared to be dynamic. By taking into account the history log, we were able to guide our *Scheduler* to a more productive fetching cycle by following a more polite etiquette and building on our experience regarding the specific times of the day in which we could launch more voluminous crawling activities.

6   Conclusions

In this paper we present *SocWeb*, a distributed crawling system that helps monitor multiple *OSN*s simultaneously. We introduced the *SocWeb Model* to formally define not only every participating *OSN* bit also existing and developing intra-connections among them. We discuss problems that emerge when applications communicate with with social network providers in the presence of multiple *OSN*s and suggest a generic API, *SNAPI*, to alleviate them. Using the semantics of *SocWeb Model* and the *SNAPI* we outline the key design choices for our monitoring system based on *SocWebCrawlers*. We demonstrate the utility of *SocWeb* while experimenting with *Facebook* and *Youtube* and working on two use-cases.

References

1. Asur, S., Huberman, B.A., Szabo, G., Wang, C.: Trends in Social Media. In: 5th Int. AAAI Conf. on Weblogs and Social Media, Barcelona, Spain (February 2011)
2. Backstrom, L., Huttenlocher, D., Kleinberg, J., Lan, X.: Group Formation in Large Social Networks: Membership, Growth, and Evolution. In: Proc. of the 12th ACM SIGKDD Conf., Philadelphia, PA (October 2006)
3. Bar-Yossef, Z., Berg, A., Chien, S., Fakcharoenphol, J., Weitz, D.: Approximating Aggregate Queries about Web Pages via Random Walks. In: Proc. of 26th Int. VLDB Conf., Seoul, Korea, pp. 535–544 (September 2006)
4. Becker, H., Iter, D., Naaman, M., Gravano, L.: Identifying Content for Planned Events across Social Media Sites. In: Proc. of 5th ACM Int. Conf. on WSDM, Seattle, WA (February 2012)
5. Budak, C., Agrawal, D., El Abbadi, A.: Structural Trend Analysis for Online Social Networks. Proc. of the VLDB Edowment 4(10), 646–656 (2011)
6. Catanese, S.A., De Meo, P., Ferrara, E., Fiumara, G., Provetti, A.: Crawling facebook for social network analysis purposes. In: Proc. of the Int. Conf. on Web Intelligence, Mining and Semantics (WIMS 2011), Songdal, Norway (May 2011)
7. Chau, D.H., Pandit, S., Wang, S., Faloutsos, C.: Parallel crawling for online social networks. In: Proc. of the 16th Int. Conf. on WWW, Banff, Canada, pp. 1283–1284 (May 2007)
8. Cho, J., Garcia-Molina, H.: Synchronizing a database to improve freshness. In: Proc. of the 2000 ACM SIGMOD Conf., Dallas, TX, pp. 117–128 (May 2000)
9. Cho, J., Garcia-Molina, H.: Parallel Crawlers. In: Proc. of the 11th Int. Conf. on WWW, Honolulu, HI, pp. 124–135 (May 2002)
10. Rundensteiner, E.A., Wang, D., Ellison, R.T.: Active Complex Event Processing Over Event Streams. Proc. of the VLDB Endow 4(10), 634–645 (2011)
11. Dou, W., Wang, K., Ribarsky, W., Zhou, M.: Event Detection in Social Media Data. In: IEEE VisWeek Workshop on Interactive Visual Text Analytics, Seattle, WA (October 2012)
12. Ali, M.H., et al.: Microsoft CEP Server and Online Behavioral Targeting. Proc. of the VLDB Endow. 2(2), 1558–1561 (2009)
13. Gjoka, M., Kurant, M., Butts, C.T., Markopoulou, A.: Walking in Facebook: A Case Study of Unbiased Sampling of OSNs. In: Proc. of the 29th INFOCOM Conf., San Diego, CA (March 2010)
14. Henzinger, M.R., Heydon, A., Mitzenmacher, M., Najork, M.: On Near-uniform URL Sampling. In: Proc. of the 9th Int WWW Conf., Amsterdam, The Netherlands (May 2000)
15. Ipeirotis, P.G., Agichtein, E., Jain, P., Gravano, L.: To search or to crawl?: Towards a query optimizer for text-centric tasks. In: Proc. of the ACM SIGMOD Cong., Chicago, IL, pp. 265–276 (June 2006)
16. Kahle, B.: Preserving the Internet. In: Scientific American. Nature Publishing Group (March 1997), www.sciamdigital.com
17. Leskovec, J., Lang, K.J., Mahoney, M.: Empirical Comparison of Algorithms for Network Community Detection. In: Proc. of the 19th Int. Conf. on WWW, Raleigh, NC, pp. 631–640 (April 2010)
18. Manning, C.D., Raghavan, P., Schutze, H.: Introduction to Information Retrieval. Cambridge University Press, New York (2008)
19. Naaman, M., Boase, J., Lai, C.-H.: Is It Really About Me?: Message Content in Social Awareness Streams. In: Proc. of ACM Conf. on Computer Supported Cooperative Work (CSCW 2010), Savannah, GA, pp. 189–192 (February 2010)
20. Ntoulas, A., Zerfos, P., Cho, J.: Downloading Textual Hidden Web Content Through Keyword Queries. In: Proc. of the 5th ACM/IEEE JCDL Conf., Denver, CO (June 2005)
21. Rabinovitch, M., Spatscheck, O.: Web Crawling and Replication. Addison Wesley (2001)

22. Punera, K., Chakrabarti, S., Subramanyam, M.: Accelerated focused crawling through on-line relevance feedback. In: Proc. of the 2002 ACM WWW Conf., Honolulu, Hawaii, USA, pp. 148–159 (2002)
23. Sadilek, A., Kautz, H., Bigham, J.P.: Finding your Friends and Following Them to Where You Are. In: Proc. of the 5th ACM Int. Conf. on WSDM, Seattle, WA, pp. 723–732 (February 2012)
24. Sakaki, T., Okazaki, M., Matsuo, Y.: Earthquake Shakes Twitter Users: Real-time Event Detection by Social Sensors. In: Proc. of the 19th Int. Conf. on WWW, Raleigh, NC, pp. 851–860 (April 2010)
25. Shkapenyuk, V., Suel, T.: Design and Implementation of a High-performance Distributed Web Crawler. In: Proc. of the 18th IEEE ICDE Conf., San Jose, CA, pp. 357–368 (February 2002)
26. Wu, E., Diao, Y., Rizvi, S.: High-Performance Complex Event Processing Over Streams. In: Proc. of the 2006 ACM SIGMOD Conf., Chicago, IL, pp. 407–418 (June 2006)

A Multiple Feature Integration Model
to Infer Occupation from Social Media Records

Xiang Wang, Lele Yu, Junjie Yao, and Bin Cui

Department of Computer Science
Key Lab of High Confidence Software Technologies (Ministry of Education)
Peking University
{kingflying.pkueecs,yulele214}@gmail.com,
{junjie.yao,bin.cui}@pku.edu.cn

Abstract. With the rapid development of more and more social media applications, lots of users are connected with friends and their daily life and opinions are recorded. Social media provides us an unprecedented way to collect and analyze billions of users' information. Proper user attribute identification or profile inference becomes more and more attractive and feasible. However, the flourishing social records also pose great challenge in effective feature selection and integration for user profile inference. This is mainly caused by the text sparsity and complex community structures.

In this paper, we propose a comprehensive framework to infer user's occupation from his/her social activities recorded in micro-blog message streams. A multi-source integrated classification model is set up with some fine selected features. We first identify some beneficial basic content features, and then we proceed to tailor a community discovery based latent dimension solution to extract community features.

Extensive empirical studies are conducted on a large real micro-blog dataset. Not only we demonstrate the integrated model shows advantages over several baseline methods, but also we verify the effect of homophily in users' interaction records. The different effects of heterogeneous interactive networks are also revealed.

Keywords: User Profile Modeling, Occupation Inference, Feature Selection, Heterogeneous Network, Micro-blog.

1 Introduction

In the recent fast availability of Web 2.0 and social network services, social media has become more and more popular among the world and has already exerted great influence on billions of users' ordinary life. Based on recent statistics, Facebook has more than one billion registered users and enjoys 660 million active users every day. Twitter has around 500 millions registered users and several hundreds of million messages are posted everyday[1].

---

[1] http://goo.gl/90KjQ,
http://blog.twitter.com/2013/03/celebrating-twitter7.html

X. Lin et al. (Eds.): WISE 2013, Part II, LNCS 8181, pp. 137–150, 2013.

The explosive development of social media has brought great opportunities to many fields. Billions of users' life are recorded. Enterprise and research areas face an unprecedent opportunity to extract and analyze users' taste, interest and other information. Since usually little information is explicitly provided by users themselves due to privacy or other concerns, automatic user attribute inference is required to infer the missing attributes of users like gender, age and interest. Through accurate attribute inference, personalized and targeting services, such as product and content recommendation, can be improved to each individual user. Occupation inference can also be introduced to adjust the advertising and user profile modeling.

Besides these great opportunities, challenges also exist. The main challenge is caused by the complicated nature of social media: the extreme rich features of data, low quality of social content and complex user interaction network within them.

Recently, several works have focused on the task of user attribute inference. For example, [16] implemented a classification method to infer three characteristics of users: political affiliation, ethnicity identification and affiliation to Starbucks. [12] used a community detection method to identify the department and college affiliation of undergraduate students. Inference of gender, age, location and other attributes were discussed in [18,10,15,1]. A common assumption to infer user attribute is homophily [11]. Homophily indicates that similar users tend to interact with each other. User attribute inference can be resolved through the information of similar users.

However, there are still many unsolved challenges to infer user attribute on social media. First, user representation is difficult because we need to extract proper features from lots of noisy data, and different features should be used for different inference tasks. A flexible way to integrate features is also valuable so that user representation can quickly be achieved for different tasks.

Another challenge is information rich heterogeneous networks. Heterogeneous networks are network systems consisting of multiple object types and multiple link types. In social media sites, users can be connected through friendship, co-discussion and mention activities. [8] discussed knowledge about such networks is often hidden in massive links. [20] put forward a concept called meta-path to encode the different relationships in heterogeneous networks to cluster objects under the limited guidance of users. However, many facets of heterogeneous networks, such as unstructured data and cyber-physical networks, are still untouched.

In this paper, we propose a multi-source integration model to infer the occupation of users on social media sites. We carry out a comprehensive feature analysis on a large real dataset, and identify language behaviors of users in different occupation categories are very different. Besides, we propose a latent network factor, i.e., *latent social dimension* to capture the community structure of users. To integrate these features, we represent each user as a feature vector and utilize the supervised machine learning classification framework to infer user's occupation.

Analyses and experiments are conducted on Sina Weibo[2], the largest microblog platform in China. Comprehensive results demonstrate the significant advantage of our proposed model. The contributions of this paper can be summarized as follows:

1. We systematically analyze the feature representation of users and dive into the network structure to capture users' latent community affiliations.
2. We propose a multi-source supervised classification framework combined with both content-based and community-based features.
3. We conduct several experiments and the results show the good performance of both content and community features, and especially the community ones. Besides, we validate homophily assumption in this user inference task.

The rest of this paper is organized as follows. In Section 2 we discuss related work. Section 3 introduces the problem definition and approach framework. Section 4 details the feature selection and engineering work in the inference model. Section 5 presents experiments and evaluations on a real large dataset and finally we conclude this work in Section 6.

2   Related Work

Research in this paper is related to several areas. Here we briefly review the corresponding literature.

User Profiling: Works in this field focus on expertise modeling, influence inference, and interest extraction.[22] proposed a model to propagate interests of an item among users via their friendships. [10] put forward an unified discriminative influence probabilistic model to identify users' locations. [3] measured user's influence from in-degree, retweets, mentions, topics and time respectively. Another common method to infer user profiling is collective classification [19]. The idea of collective classification is to infer user attribute using neighbors' information. Normally a relational classifier is constructed based on the relational features of labeled data, and then an iterative process is required to infer the unlabeled data. However, the main drawback of collective inference is that it only considers the direct neighbors of users and the interactions between indirectly linked users are ignored. Besides, collective classification fails to capture the presence of underlying factors that actually influence user's behaviors.

Community Detection: Community detection has been a trending topic for a long time. Traditional community detection algorithm uses closeness metric, by adding edges into an empty network one by one. However, to cut the hierarchical tree and determine the final network community, manual division is required. [6] put forward edge betweenness metric to divide community. This method removes edges with larger betweenness from the original network, which is opposite to closeness method. [14] proposed modularity metric to identify community. Larger modularity means that there is larger number of intra-community

---

[2] http://weibo.com/

edges than inter-community edges. Many previous work used community metric to infer user attribute. [12] used a greedy algorithm to maximize a new evaluation metric called normalized conductance, which measures the quality of a single community, to detect communities, and then assigned an identical attribute value to users in the same community. The disadvantage of this method is that using labeled data in community to infer the unlabeled data involves too much noise and can't capture the interactions between communities.

3   Problem Formulation and Approach Framework

In this section, we first define the problem of user occupation inference and then introduce our multi-source integration framework to infer users' occupation information.

Data Scenario: Dataset used in this work is based on one of largest micro-blog platforms–Sina Weibo[3]. Users can post, re-tweet and comment messages. At the same time, they can follow other users. Everyday, hundreds of millions messages are posted and spread in this social media site[4]. Sina Weibo has already labeled a small subset of its users and categorize them into 12 occupation classes, such as entertainment, media and government[5]. We use the open API provided by Sina Weibo to crawl these users' data and get about ten thousand accounts. Profiles, tweets, tags, friend and follower lists are collected. After removing some low active users, we select 65828 accounts for later empirical study. Here, we can not only identify the difference in users' language behaviors from their posted messages but also catch the strength variety of users' interactions by utilizing the community structure.

In Table 1, we provide the occupation distribution of this dataset. We find that media accounts for the largest proportion(26%), followed by entertainment class (18%). The percentage of public welfare is smallest, which is only 1%.

Table 1. Occupation Distribution of Verified Users on Sina Weibo

Transport	Government	Finance	Electronic	Public welfare	Education
1.9%	14.4%	8.9%	2.8%	1.2%	10.0%

Estate	Media	Service	Entertainment	Others	Medical
8.8%	26.1%	3.7%	17.9%	3.0%	1.3%

To make our problem clear and unambiguous, we give the formal definition of it in the following.

Definition 1. *There are K occupation labels $\kappa = \{c_1, \ldots, c_K\}$. Given a social network $G = (V, E, Y)$ where V is the set of user vertices, E is the set of connection edges and Y is the set of users' occupation labels. $y_i \in Y$ and $y_i \in \kappa$*

---

[3] http://weibo.com

[4] http://www.36kr.com/p/201443.html

[5] Verified Account: http://verified.weibo.com/

represents the occupation label of user vertex i, and we have already labeled the occupation labels Y_{know} of some vertices V_{know}. The occupation inference task is aiming to select the occupation labels Y_{unknow} for the remaing vertices V_{unknow}.

Multi-source User Occupation Inference Model: To solve this problem, we utilize both the content and network features, and then transform this problem into a machine learning classification task. Figure 1 is the framework of our multi-source inference model. We can divide this approach into two stages:

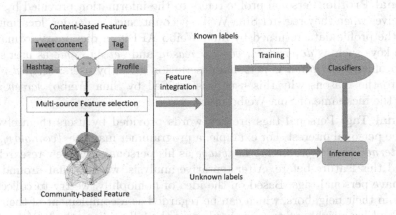

Fig. 1. Multi-source User Occupation Inference Model

1. *Feature Selection and Integration*: We integrate features from two categories. One is content feature, including tweet content, hashtag, tag and profile. Actually, we also investigate some other possible features, like location, temporal pattern of behaviors and linguistic characteristics, but these features don't perform well in our task. Another class of features is community, i.e., latent social dimension used in this work. We use it to capture the latent community affiliations of users so that the global network information can be utilized.

   After this selection, we then represent each user as a feature vector, which combines these two types of social media features. New features can be easily added into our model .

2. *Model Prediction*: After feature extraction and user representation, we choose a de facto supervised machine learning classifier to infer user's occupation. Common classifiers include *naive bayes, decision tree, support vector machine* and *logistic regression*. Theoretically, any classifier is adaptive in this case. However, the actual utility needs to be verified through experiments. As to the latent social dimension, it is very important for classifier to choose effective and significant community dimensions so that the inference can be optimized. We also conduct comparisons between different classifiers in section 5.

4   Feature Selection

In this section, we introduce the features used in occupation inference work. Sina Weibo is a rich social media platform, with a large variety of user generated content and multiple types of user interactions.

4.1   Content Feature

Personal Profile: Personal profile refers to the information provided by users themselves when they register Sina Weibo account, such as gender, location. Because the profile data returned by Sina Weibo API is in dictionary format, we choose keys such as *description*, *verified_reason* and *screen_name* as user's personal profile. *Description* is personal description given by users; *verified_reason* refers to the reasons why this user was verified by Sina Weibo; *screen_name* means the nickname of Sina Weibo users.

Personal Tag: Personal tags are key words provided by users themselves to describe personal interest. For example, a programmer may use *Technology, Mobile Internet* and *Programming Language* as his personal tags. Few research has touched this feature before. After tentative analysis, we find that around 76% users have personal tags. Based on the idea of homophily, we try to collect the tags from their neighbors, which can be regarded as a complement of their own tags. To utilize neighbors' tags, we implement the following method. We combine user's own tags with tags of his top-k most similar neighbors based on similarity measure. We use the Jaccard Similarity Coefficient[6] to measure the similarity of users. After joining neighbors' tags, we find that nearly every user has at least one tag and the average number of tags of users are between 10 and 20 based on the choice of k.

It should be noted that neighbors here refer to bi-follower friends. Bi-follower friends imply two users following each other. Friends or neighbors mentioned in this paper refer to bi-follower friends by default. Bi-follower friends indicate stronger relationship than one-directional relationship and thus can filter much noise.

Hashtag: A hashtag is a word or phrase fixed between the symbol #. It serves as a symbol to integrate similar tweets. The usage of hashtags is related to the occupations of users. For example, users from public welfare may concern hashtags like *Beijing Rescue Team* and *Social Public Welfare* while users of IT companies may be interested in *Iphone5* and *Google I/O Conference*. After extracting hashtags from tweets, we implement word filtering strategy based on word frequency and represent users as hashtag vectors.

Tweet Content: Sina Weibo allows users to post tweets within 140 words. It is intuitive to observe that users of different occupations often post tweets which are different in content. For example, a property developer may be accustomed to use words like *housing, bank, inflation,* while a famous singer may like using words

---
[6] http://en.wikipedia.org/wiki/Jaccard_index

such as *singing, composer, popularity*. However, it is still difficult to represent tweet content due to the following causes. First, the distribution of tweets is unbalanced. The number of tweets posted by most users is comparatively small, while only a small fraction of users post many tweets, which follows the Power-Law distribution. Another challenge is noise, which is caused by user's arbitrary writing habits and ordinary user's low quality background.

In order to solve the content representation problem, many methods have been proposed. Probabilistic Latent Semantic Analysis (PLSA) was introduced in [9] to project similar words into a latent dimension. Its disadvantage is the single topic assumption of each document. Latent Dirichlet Allocation(LDA) [2] was later proposed to solve the disadvantage of PLSA, and it allows multiple topics in each document, which is considered as state-of-the-art method. In this paper, we treat the tweets of a user as a document and use LDA to represent each user as a probabilistic distribution among topics. This method could identify user's latent topic distribution from their posted content.

4.2 Community Feature

Here we continue to discuss extraction method of community feature. There are a variety of heterogeneous networks on Sina Weibo, such as friendship, retweet and mention network(using @). One important phenomenon in social network is community structure. Here we set up a new community feature, i.e., latent social dimension, based on community structure of users to infer their occupations.

Latent Social Dimension: [21] presented a new way to utilize community structure, which is called latent social dimension. Actually, latent social dimension represents the affiliation of users to different communities. Figure 2 is a toy example to illustrate latent social dimension. There are three communities and five users in this graph. One user can affiliate to multiple communities and the thickness of lines between users and communities indicates the strength of affiliation. The mathematical format of latent social dimension is actually a vector, with each dimension corresponding to each community. Take User A as an example. User A can be represented as a vector$< 0.3, 0.2, 0.1 >$. 0.3 means the strength of affiliation between community 1 and user A is 0.3.

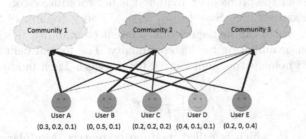

Fig. 2. A user and community interactive graph to illustrate latent social dimension

The advantage of latent social dimension compared to collective classification is that latent social dimension can capture the interactions between users from the whole network while collective classification can only infer user attribute with neighbors' help.

It is intuitive to conclude that users of same occupation are more likely to connect with each other in networks and thus are more likely to form community structure. Users of different occupations tend to form different communities, and thus their latent social dimensions are different. We will validate this assumption in section 5.

Extraction of Latent Dimension: This task is based on community detection algorithm. Traditional graph partition-based algorithm [7] aims to minimize the number of edges between communities. [13] points out that minimizing the number of edges between communities is not a good metric because it tends to divide most nodes into one community. A better metric to divide communities is modularity. Modularity is defined as number of intra-edges in our target network minus number of intra-edges in a comparable random network. Intra-edges means edges inside communities, not between communities. Community detection task is to maximize modularity function. The advantage of modularity-based algorithm is that it can find the communities which naturally exist in the network, without the need to pre-assign the number of communities to be detected.

To simplify our explanation, we make some definitions. We assume that the number of users in the network is n and the number of edges is m. We first define adjacent matrix $\mathbf{A}$: $A_{ij} = 1$ if there is an edge between node i and node j; $A_{ij} = 0$ otherwise. Here we ignore the direction of the graph, that is $A_{ij} = A_{ji}$ and we also don't consider the weight of edge. Next, we define modularity:

$$Q = \frac{1}{2m} \sum_{ij} [A_{ij} - P_{ij}] \delta(g_i, g_j) \tag{1}$$

where g_i represents the community of node i. δ is a function. $\delta(r, s) = 1$ if $r = s$ and $\delta(r, s) = 0$ if $r \neq s$. P_{ij} represents probability that there is an edge between node i and node j in a random network. For convenience, we choose P_{ij} as:

$$P_{ij} = \frac{k_i k_j}{2m} \tag{2}$$

where k_i indicates the degree of node i and can be calculated as $k_i = \sum_j A_{ij}$.

Now we consider the problem of dividing the network into c communities. We first define index matrix $\mathbf{S}$: $\mathbf{S} = (s_1|s_2|\ldots|s_c)$. Every column of $\mathbf{S}$ is a index vector of 0 or 1, which can be regarded as a latent community. 0 or 1 represents the disaffiliation or affiliation to this community. The formal definition of $\mathbf{S}$ is: $S_{ij} = 1$ if node i belongs to community j; 0 otherwise. Then modularity can be revised as:

$$Q = \frac{1}{2m} Tr(S^T B S) \tag{3}$$

where $B = A - P$, which is called modularity matrix. Modularity matrix $\mathbf{B}$ is a real symmetric matrix and its function is the same as that of Laplacian

Matrix in standard spectral partitioning. We decompose $\mathbf{B}$ as $B = UDU^T$, where $U = (u_1|u_2|\ldots)$ is a matrix made up of the eigenvectors of $\mathbf{B}$ and $\mathbf{D}$ is a diagonal matrix made up of eigenvalues of $\mathbf{B}$. Then we can revise modularity as:

$$Q = \frac{1}{2m} \sum_{j=1}^{n} \sum_{k=1}^{c} \beta_j (u_j^T s_k)^2 \qquad (4)$$

According to [17], when the column vectors of $\mathbf{S}$ are proportional to the leading eigenvectors of $\mathbf{B}$, modularity can be maximized. To avoid the problem of 0 or 1 in $\mathbf{S}$, we relax $\mathbf{S}$ to be continuous. In this case, when $\mathbf{S}$ is made up of the top-c eigenvectors of $\mathbf{B}$, modularity can be maximized theoretically. We should note that the number of communities, c, is uncertain. We need to choose proper c to maximize modularity in practice. According to equation 4, only when β_j is positive can it have positive effect on modularity. As a result, the maximum of c will not exceed the number of positive eigenvalues of $\mathbf{B}$.

With the discussion of the above features, we can enrich the classification framework introduced in Section 3. Empirical study will be presented in the next Section.

5   Experiments

In this section, we report our evaluation experiments on Sina Weibo. We first compare results of different inference models and different classifiers. Then we validate our homophily assumption from two aspects and finally dive into the heterogeneous networks characters.

5.1   Baseline and Evaluation Metrics

To demonstrate the improvement of the proposed model, we select the following baselines:

- *Weighted Random Model(WRand)*: This model ignores any content and network information and simply classifies user to a random occupation with the probability proportional to the percentage of that occupation.
- *Majority Model(Majority)*: This model also ignores any content and network information. Users are classified into the same occupation label which accounts for the largest proportion of all the occupations.
- *Content-based Model*: This model considers the content-based feature. We classify this model into 4 parts: tweet content model*(LDA)*, hashtag model *(Hashtag)*, tag model*(Tag)* and profile model*(Profile)*.
- *Community-based Model(Cmty)*: This model only contains the latent social dimension feature.
- *Combined Model*: This model contains both content and community features. We divide this model into 4 parts: tweet content and community*(LDA-Cmty)*, hashtag and community*(Hashtag-Cmty)*, tag and community*(Tag-Cmty)* and finally profile and community*(Profile-Cmty)*.

We choose common evaluation metrics to evaluate our model. They are *Precision*, *Recall* and *F-measure* respectively.

5.2   Classifier Choice

Figure 3 shows the performance of different classifiers using community feature. Here we take *logistic regression* and *support vector machine*(SVM) for example. From the result, we can find that though the precisions of these two classifiers are almost the same, SVM performs much worse than logistic regression in recall, more than 10% lower, which finally results in the poor f-measure for SVM.

This study indicates that regarding the dimension selection for the community feature, logistic regression performs much better than SVM.

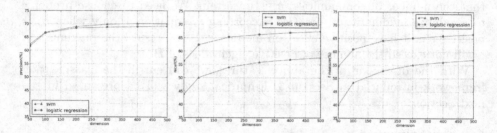

Fig. 3. Comparison of Different Classifiers

5.3   Inference Performance

Figure 4 shows the results of different inference models. To let different models comparable, we set the dimension of *Hashtag, Tag, Profile, Cmty* to all 500. From the results, we observe that *WRand* and *Majority* perform very poor because they do not consider any content and community information. For the other four models based on content, we find that the performance of *LDA, Tag* and *Profile* are almost the same, i.e., about 60% for all metrics while the performance of *Hahstag* is comparably poor, with f-measure just 24%. For *Cmty* model, it outperforms all the content-based models, with nearly 70% precision, recall and f-measure. Thus we can conclude that the community feature, i.e., latent social dimension, performs better than content-based features in our occupation inference task. The benefit of this finding is that we can infer user's occupation just based on network structure, without incorporating what he tweets or what he re-tweets. This is especially important for users who seldom post any tweets or make any comments.

5.4   Homophily Characters

In Section 4, we discuss the homophily hypothesis, i.e., users of same occupation are more likely to gather together and establish connections, thus forming significant community structure. We verify this assumption in this section from the following two aspects.

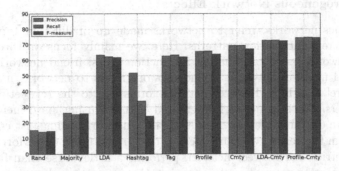

Fig. 4. Comparison of Different Inference Models(Logistic Regression Classifier)

- *Verification 1*: [4] points out that when the value of modularity in community discovery is larger than 0.3, an obvious community structure can be observed in the network. Based on this , we calculate the modularity of friend network and we get 0.59 which is much larger than 0.3. Thus, we are sure to conclude that a significant community structure exists in friend network.
- *Verification 2*: After verifying significant community structure in friend network, we analyze whether there is a dominant occupation in these communities. Figure 5 explains the occupation distribution of *top-10* communities detected from friend network. The red color indicates dominant occupation in this community. We find that there is always a dominant occupation in top-10 communities. This also indicates that users of the same occupation are more likely to connect with each other and form community structure.

Community size	Education	Service	Estate	Entertainment	Media	Medical	Government	Public welfare	Others	Electronic	Transport	Finance
1823	0.095	0.049	0.012	0.252	0.188	0.023	0.017	0.015	0.052	0.042	0.010	0.245
1319	0.019	0.033	0.595	0.020	0.047	0.014	0.014	0.011	0.020	0.087	0.006	0.134
984	0.023	0.447	0.027	0.034	0.061	0.012	0.045	0.021	0.066	0.075	0.149	0.039
927	0.068	0.037	0.013	0.024	0.035	0.011	0.040	0.634	0.071	0.033	0.008	0.027
810	0.075	0.101	0.017	0.060	0.064	0.044	0.017	0.002	0.072	0.475	0.021	0.049
715	0.022	0.076	0.001	0.006	0.027	0.580	0.069	0.048	0.091	0.052	0.014	0.015
693	0.017	0.014	0.001	0.003	0.087	0.007	0.694	0.006	0.088	0.020	0.046	0.016
644	0.012	0.022	0.003	0.042	0.054	0.009	0.002	0.006	0.017	0.030	0.764	0.039
485	0.501	0.019	0.008	0.016	0.031	0.019	0.153	0.023	0.167	0.049	0.012	0.002
336	0.042	0.021	0.015	0.036	0.054	0.074	0.083	0.024	0.196	0.009	0.435	0.012
270	0.089	0.063	0.004	0.052	0.085	0.537	0.004	0.019	0.019	0.067	0.015	0.048
250	0.172	0.020	0.012	0.012	0.080	0.028	0.128	0.004	0.396	0.036	0.044	0.068
239	0.025	0.100	0.004	0.548	0.151	0.013	0.038	0.013	0.038	0.029	0.004	0.038
100	0.060	0.080	0.000	0.000	0.090	0.050	0.110	0.010	0.480	0.080	0.010	0.030
100	0.040	0.030	0.060	0.170	0.080	0.000	0.100	0.030	0.390	0.050	0.020	0.030

Fig. 5. The Occupation Distribution of top 10 Communities Detected(Friend Network)

5.5 Heterogeneous Network Effects

Heterogeneous networks, refer to networks made up of different types of objects or different interaction patterns. Here, we mainly focus on three heterogeneous networks based on different interaction types: friend network, retweet network and mention network. Friend network refers to networks made up of bi-follower relationships. Retweet network forms due to the retweeting actions between users. Mention network is created by users using @ to mention each other. Figure 6 is the inference result of three networks using just community feature. From the result, we can conclude that friend network performs best in all three metrics and retweet network performs a little worse than friend network while mention network performs worst. One interesting phenomenon is that even though both retweet network and mention network are created due to user interaction behaviors, the performances of them are quite different.

To investigate the reason, we conduct a tentative analysis of three networks. From Table 2, we find that the statistical features of friend network and retweet network are almost the same while mention network is much sparser, with more than 10000 users of degree zero, which might be a reason for the poor performance of mention network. [5] points out that in social network, the mention function(@) often plays a role to connect users having different or even opposed opinions and behaviors, which results in the poor homophily phenomenon in mention network.

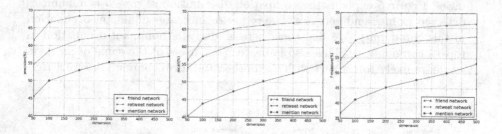

Fig. 6. Inference Performance of Different Heterogeneous Networks

Table 2. Statistical Characters of Heterogeneous Networks

	Friend Network	Retweet Network	Mention Network
Links	2770378	2745765	776177
Average degree	42	41	12
Number of nodes with degree of zero	618	3714	10811

6   Conclusions

In this paper, we propose a multi-source model to infer users' occupation categories on micro-blog platforms. We utilize both users' posted content feature and their interaction community features in this model. The content feature includes tweet content, hashtag, personal tag and personal profile. For the network feature, we propose to use latent social dimension, in order to better identify user's latent affiliation. Then we model this inference task as a supervised classification problem and introduce the manually labeled source to train a inference model.

We systemically analyze the data characters on a real large micro-blog(Sina Weibo) dataset and demonstrate the advantage of proposed approach. At the same time, we also reveal the patterns of different user interaction communities and homophily phenomenon among users of the same occupation category.

With the continuous growth of social media services, effective user profile extraction and user interest modeling become more and more important. Occupation inference model proposed in this paper has several promising future direction. For example, we can extract users' occupation evolution, profile variety and interest distribution among different groups.

Acknowledgements. This research was supported by the National Natural Science foundation of China under Grant No. 61272155 and 61073019.

References

1. Backstrom, L., Sun, E., Marlow, C.: Find me if you can: improving geographical prediction with social and spatial proximity. In: Proceedings of the 19th International Conference on World Wide Web, pp. 61–70. ACM (2010)
2. Blei, D.M., Ng, A.Y., Jordan, M.I.: Latent dirichlet allocation. The Journal of Machine Learning Research 3, 993–1022 (2003)
3. Cha, M., Haddadi, H., Benevenuto, F., Gummadi, K.P.: Measuring user influence in twitter: The million follower fallacy. In: 4th International AAAI Conference on Weblogs and Social Media (ICWSM), vol. 14, page 8 (2010)
4. Clauset, A., Newman, M.E., Moore, C.: Finding community structure in very large networks. Physical Review E 70(6), 066111 (2004)
5. Conover, M.D., Ratkiewicz, J., Francisco, M., Gonçalves, B., Flammini, A., Menczer, F.: Political polarization on twitter. In: Proc. 5th Intl. Conference on Weblogs and Social Media (2011)
6. Girvan, M., Newman, M.E.: Community structure in social and biological networks. Proceedings of the National Academy of Sciences 99(12), 7821–7826 (2002)
7. Hagen, L., Kahng, A.B.: New spectral methods for ratio cut partitioning and clustering. IEEE Transactions on Computer-Aided Design of Integrated Circuits and Systems 11(9), 1074–1085 (1992)
8. Han, J.: Mining heterogeneous information networks by exploring the power of links. In: Gama, J., Costa, V.S., Jorge, A.M., Brazdil, P.B. (eds.) DS 2009. LNCS, vol. 5808, pp. 13–30. Springer, Heidelberg (2009)
9. Hofmann, T.: Probabilistic latent semantic analysis. In: Proceedings of the Fifteenth Conference on Uncertainty in Artificial Intelligence, pp. 289–296. Morgan Kaufmann Publishers Inc. (1999)

10. Li, R., Wang, S., Deng, H., Wang, R., Chang, K.C.-C.: Towards social user profiling: unified and discriminative influence model for inferring home locations. In: Proceedings of the 18th ACM SIGKDD International Conference on Knowledge Discovery and Data Mining, pp. 1023–1031. ACM (2012)

11. McPherson, M., Smith-Lovin, L., Cook, J.M.: Birds of a feather: Homophily in social networks. Annual Review of Sociology, 415–444 (2001)

12. Mislove, A., Viswanath, B., Gummadi, K.P., Druschel, P.: You are who you know: inferring user profiles in online social networks. In: Proceedings of the Third ACM International Conference on Web Search and Data Mining, pp. 251–260. ACM (2010)

13. Newman, M.E.: Finding community structure in networks using the eigenvectors of matrices. Physical Review E 74(3), 036104 (2006)

14. Newman, M.E.: Modularity and community structure in networks. Proceedings of the National Academy of Sciences 103(23), 8577–8582 (2006)

15. Otterbacher, J.: Inferring gender of movie reviewers: exploiting writing style, content and metadata. In: Proceedings of the 19th ACM International Conference on Information and Knowledge Management, pp. 369–378. ACM (2010)

16. Pennacchiotti, M., Popescu, A.-M.: Democrats, republicans and starbucks afficionados: user classification in twitter. In: Proceedings of the 17th ACM SIGKDD International Conference on Knowledge Discovery and Data Mining, pp. 430–438. ACM (2011)

17. Pothen, A., Simon, H.D., Liou, K.-P.: Partitioning sparse matrices with eigenvectors of graphs. SIAM Journal on Matrix Analysis and Applications 11(3), 430–452 (1990)

18. Rao, D., Yarowsky, D., Shreevats, A., Gupta, M.: Classifying latent user attributes in twitter. In: Proceedings of the 2nd International Workshop on Search and Mining User-Generated Contents, pp. 37–44. ACM (2010)

19. Sen, P., Namata, G., Bilgic, M., Getoor, L., Galligher, B., Eliassi-Rad, T.: Collective classification in network data. AI Magazine 29(3), 93 (2008)

20. Sun, Y., Norick, B., Han, J., Yan, X., Yu, P.S., Yu, X.: Integrating meta-path selection with user-guided object clustering in heterogeneous information networks. In: Proceedings of the 18th ACM SIGKDD International Conference on Knowledge Discovery and Data Mining, pp. 1348–1356. ACM (2012)

21. Tang, L., Liu, H.: Relational learning via latent social dimensions. In: Proceedings of the 15th ACM SIGKDD International Conference on Knowledge Discovery and Data Mining, pp. 817–826. ACM (2009)

22. Yang, S.-H., Long, B., Smola, A., Sadagopan, N., Zheng, Z., Zha, H.: Like like alike: joint friendship and interest propagation in social networks. In: Proceedings of the 20th International Conference on World Wide Web, pp. 537–546. ACM (2011)

Recommending Interesting Landmarks
Based on Geo-tags from Photo Sharing Sites

Jinpeng Chen[1], Zhenyu Wu[1], Hongbo Gao[1], Changjie Zhang[1],
Xuejun Cao[1], and Deyi Li[1,2]

[1] State Key Laboratory of Software Development Environment, BeiHang University
[2] Institute of Electronic System Engineering

Abstract. In this paper, we aim to explore interesting landmark recommendations based on geo-tagged photos for each user. Meanwhile, we also try to answer such a question, i.e., when we want to go sightseeing in a large city such as Beijing, where should we go? To achieve our goal, first, we present a data field clustering method (*DFCM*). By using *DFCM*, we can cluster a large-scale geo-tagged web photo collection into groups (or landmarks) by location. And then, we provide more friendly and comprehensive overviews for each landmark. Subsequently, we model the users' dynamical behaviors using the fusion user similarity, which not only captures the overview semantic similarity, but also extract the trajectory similarity and the landmark trajectory similarity. Finally, we propose a personalized landmark recommendation algorithm based on the fusion user similarity. Experimental results show that our proposed approach can obtain a better performance than several state-of-the-art methods.

1 Introduction

In location recommendation, there have already been a reasonable amount of researches. In [12], authors proposed an approach to find like-minded users at different locations. To model the users' similarity, in [8], authors proposed a novel approach for recommending potential friends based on users' semantic trajectories. In [6], an approach which mines the similarity of people's trajectories based on location histories was proposed. However, our presented users' similarity measure not only considers the similarity of people's trajectories but also fuses the semantic similarity of landmark overviews. In [10], they provided a novel collaborative filtering (*CF*) approach to train a location and activity recommender. The *GM-FCF* system [14] directly made location aware recommendations to users using a novel combination of social relations and geographic information. In [13], they put forward a framework that encompasses new techniques for extracting semantically meaningful geographical locations from the proliferation of GPS data. In [11], they provided a novel category-regularized matrix factorization approach (*CRMF*) to recommend landmarks to individual users based on both user-landmark preference information and category-based landmark similarity. Our work aim is similar to [11], which focus on personalized landmark recommendation based on geo-tagged photos. However, our recommendation method is different from [11]. In this work, we regard *CRMF* as one of the baselines.

X. Lin et al. (Eds.): WISE 2013, Part II, LNCS 8181, pp. 151–159, 2013.
© Springer-Verlag Berlin Heidelberg 2013

In this paper, we face the three challenges: (1) How to organize a large collection of photos with all those kinds of information? (2) How to model the users' dynamical behaviors based on geo-tagged photos? (3) How to generate personalized landmark recommendations based on geo-tagged photos for each user? In order to handle these challenges, we present different strategies, which are introduced in other sections.

2   Detect Landmarks and Generate Landmark Overviews

2.1   Detect Landmarks

We intuit that interesting landmarks attract more visitors, so there are more geo-tagged photos taken in it. In order to detect landmarks of users' interests in the geographic space, we can cluster geo-tagged photos. But cluster results are influenced by granularity of the location. For instance the larger the extent of a place, the longer the distance the activities occur in [1]. To accommodate variable granularity, we propose a data field clustering method, which is a density-based clustering method initially developed to cluster point objects. We consider the Mean Shift (*MS*) [2], which has been shown effective for spatial data clustering in previous work [3, 4], as a baseline method in order to verify the effectiveness of our proposed algorithm. We elaborate the notion of data field [7] prior to introducing the data field clustering method.

Data Field. Inspired by the knowledge of physical fields, we introduce the interaction of particles and their field into the data space. Given a dataset containing n objects in space $\Omega \subseteq R^P$, i.e., $D = \{x_1, x_2, \ldots x_n\}$, where $x_i = (x_{i1}, x_{i2}, \ldots x_{ip})$, $i = 1, \ldots n$. Each data object can be considered as a mass point or nucleon with a certain field around it and the interaction of all data objects will form a data field through space.

Because Gaussian function has good mathematic properties, in this work we adopt Gaussian function to define the potential at any point x as,

$$\varphi(x) = \sum_{i=1}^n \varphi_i(x) = \sum_{i=1}^n (m_i \times \exp{(-(\frac{||x-x_i||}{\sigma})^2)}) \tag{1}$$

where x is the GPS coordinates, $||x - x_i||$ is the distance between object x_i and x, m_i is the mass of object x_i, and σ is the influence factor that indicates the range of interaction. In this work, we assume each data object x is supposed to be equal in mass and meet a normalization condition $\sum_{i=1}^n m_i = 1$.

Given a data set in space, the distribution of the associated data field is primarily determined by the influence factor σ once the form of potential function is fixed. In order to optimize σ, Shannon entropy principle is used as Equation 2.

$$min\ H = min_\sigma(-\sum_{i=1}^n \frac{\varphi_i}{z} log\left(\frac{\varphi_i}{z}\right)) \tag{2}$$

where $Z = \sum_{i=1}^n \varphi_i$ is a normalized factor.

Clustering Method Based on Data Field. According to the definition of the data field, we propose the data field clustering method (*DFCM*, in Algorithm 1) which clusters the points in the data space based on the strength of interaction of objects.

The idea of this clustering method is to optimally select the influence factor σ for generation of potential field distribution in the first step. Thereafter, the data objects contained in each equipotential line/surface are treated as a natural cluster, and the nested structures consisting of different equipotential lines/surface are treated as the cluster spectrum. Thus, the clustering at different hierarchies is realized.

The local maximum points could be regarded as "virtual field sources", and all the data objects are convergent by self-organization due to the attraction by their own "virtual field sources". Thus, the local maximum points can be regarded as cluster centers, and the initial partition is formed. To obtain the clustering at different hierarchies, the initial clusters are combined based on regular saddle point iteration between two local maximum points.

In order to find local maximum points and saddle points in the potential field distribution, the algorithm first searches all the critical points satisfying $\nabla \varphi(x) = 0$. Thereafter, it classifies the critical points according to the eigenvalues of the Hessse matrix $\nabla^2 \varphi(x)$. For a given critical point x, let $l_1 < l_2 \ldots < l_d$ be the d eigenvalues of the Hesse matrix, where $d >= 2$ is the dimension of the space. If $l_d < 0$, x is the local maximum point in the potential field distribution; if $l_1 > 0$, x is a local minimum point in the potential field distribution; if $l_1, l_2, \ldots, l_d \neq 0$ and the number of positive eigenvalues and the number of negative eigenvalues are both bigger than 1, the point x is the saddle point in the potential field.

Algorithm 1. DFCM

Input: $D = \{x_1, x_2, \ldots x_n\}$, sample number n_{sample}, noise threshold ξ
Output: the hierarchical partition $\{\Pi_1, \Pi_2, \ldots \Pi_k\}$
Steps: Select n_{sample} samples randomly to construct the sample data set *SampleSet*.
$\sigma = Optimal\_Sigma(SampleSet)$ //Optimization of the influence factor σ
$Map = CreatMap(D, \sigma)$ //Apply grid partition on the space, construct an index tree
$CriticalPoints = Search\_CriticalPoints(Map, \sigma)$ //Search in the topological critical points
Set *MaxP* as the set of local maximum points and *SadP* as the set of saddle points.
//Initially divide the data according to the set of local maximum points
$\Pi_1 = Initialization\_Partition(Map, D, MaxP, \sigma, \xi)$
//Combine the initial clusters iteratively according to the set of saddle points
$[\Pi_1, \Pi_2, \ldots \Pi_k] = Saddle\_Merge(Map, \Pi_1, MaxPoints, SadP, \sigma, \xi)$

2.2   Generate Landmark Overviews

Once the previous clustering work has been done, we regard each cluster as a landmark. Next, we will add an overview into each landmark. The work by Q. Hao et al. [5] proposed a landmark overview generation approach. They showed that the method was very effective, which provided an informative overview for a given location via mining location-representative tags. In this work, we adopt their approach to generate landmark overviews. As shown in the Table 1, it presents an example of a landmark (*Tsinghua*) overview. The first entity of each row represents a ranking topic of an overview, and the following entities represent ranking sub-topics.

Table 1. An overview of *Tsinghua*, three topics and three sub-topics are selected for displaying

Ranking Overview	Topic	Sub-topic		
		1	2	3
1	tsinghua university	students	classroom	auditorium
2	old summer palace	qing dynasty	lotus	imperial gardens
3	peking university	weiming lake	baya pagoda	campus

3   Recommending Interesting Landmarks

3.1   Preliminary

Definition 1 GPS Point (or GPS Coordinate): A GPS point p is a four-tuple: $<x, y, t, m>$, where x and y are Euclidean coordinates, t is the timestamp and m is the times a user visiting this GPS point (we can call it mass or score of the GPS point).

Definition 2 Landmark (or Hot Spot): A landmark L consists of a group of consecutive GPS points $P = \{p_1, p_2, ..., p_n\}$. Formally, we denote a virtual landmark center as $c = (x, y, ta, tl, lm, ov)$, where $c.x = \sum_{i=1}^{n} p_i.x/|P|$ (3), $c.y = \sum_{i=1}^{n} p_i.y/|P|$ (4), respectively stands for the average latitude and longitude of the collection P, $ta = p_1.t$ and $tl = p_n.t$ represent user's arrival and leaving time on L, $lm = \sum_{i=1}^{n} p_i.m$ represents the times a user visiting this landmark (we can call it mass or score of the landmark) and ov represents overviews of this landmark L.

Definition 3 Trajectory: A trajectory with a score Tr of a user is a sequence of GPS points based on a certain threshold ΔT. Here, $Tr = p_1 \rightarrow p_2 \rightarrow ... \rightarrow p_n$, where $p_i \in P$, $p_{i+1}.t > p_i.t$, $p_{i+1}.t - p_i.t < \Delta T$ ($1 \leq i \leq n$) and the score is times a user passing this trajectory. In addition, we also introduce another two definitions associated with trajectory, namely, landmark trajectory and M-length trajectory. A landmark trajectory with a score Ltr of a user is a sequence of landmark based on a certain threshold ΔT. If the node numbers in a trajectory is M, we call this trajectory M-length trajectory.

3.2   User Similarity Exploration

In this section, we detail the processes of user similarity exploration.

Location History Extraction. We construct two location histories: users' trajectories and user's landmark trajectories. With users' travel experiences and the interests of locations, we can calculate a classical score for each GPS point (or each landmark) and each trajectory (or each landmark trajectory) within the given geospatial region.

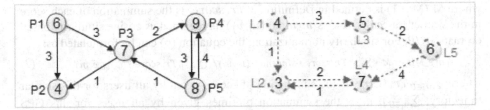

Fig. 1. Trajectory graph and landmark trajectory graph

As shown in the Figure 1, we present two graphs: the graph of GPS points in the left part and the graph of landmarks in the right part. These two graphs contain all trajectories and all landmark trajectories of a user respectively. For the graph of GPS points in the left part of the Figure 1, the graph nodes (*p1, p2, p3, p4* and *p5*) stand for GPS points, and the graph edges denote users' trajectories among them. Take 2-length trajectory (*p1* → *p2*) as an example, the number shown on the node (6 and 4) and the edge (3) represents the score of the GPS point and the trajectory, respectively. For the graph of landmarks, we have the similar analysis.

For similarity normalization (in next section), we construct user-GPS point matrix *UP* and user-landmark matrix *UL*. The row of *UP* (or *UL*) represents GPS-point (or landmark). The column of *UP* (or *UL*) represents user. Each entry in the *UP* (or *UL*) records how many times a user visit the GPS-point (or landmark).

Overview Semantic Similarity. In section 2.2, we introduce an overview consisted of topics and sub-topics. For calculating the overview semantic similarity, we construct user-topic matrix *UT* and user-sub-topic matrix *UST*. Each entry in the *UT* or *UST* denotes whether a user add tags to pictures with this topic or sub-topics. Here, 1 denotes a user uses it, while 0 expresses a user does not use this topic or sub-topic.

Based on these two matrixes, we can learn users' profiles utilizing user similarities, which are reflected via considering whether users add the same tags (topics or sub-topics). In other words, if two users are more similar to each other, they are likely to add more similar tags. Here, we can obtain two user similarities: user-topic matrix similarity($sim\_T(u_i, u_j)$) and user-sub-topic similarity ($sim\_ST(u_i, u_j)$). Note that, we use cosine similarity to compute $sim\_T(u_i, u_j)$ and $sim\_ST(u_i, u_j)$. The overview semantic similarity is formalized as:

$$sim\_view(u_i, u_j) = \beta_1\, sim\_T(u_i, u_j) + \beta_2\, sim\_ST(u_i, u_j) \tag{5}$$

where $\beta_1, \beta_2 > 0$, $\beta_1 + \beta_2 = 1$, $u_i, u_j \in U$, $U = \{u_1, u_2, ..., u_n\}$ is a set of users.

Trajectory Similarity. In section Location History Extraction, the trajectory graph and the matrix *UP* have been constructed. We adopt the similar sequence matching method proposed by Li Q.[6] in order to find the similar trajectories for each user-pair. The retrieved similar trajectories are used to calculate an overall similarity score for each user-pair. When computing the score, we take into account two factors: the length of a similar trajectory with weight (that is, M-length trajectory) and the mass of a node in this trajectory. So the score an M-length trajectory obtains can be formulated as:

$$score\_Mlt = M * \sum_{j=1}^{M-1} Tr.score * sim\_mass(u_i, u_j) \tag{6}$$

where M (M > 1) is defined in Definition 3, $Tr.score$ is the summation of each score in the similar trajectory, and $sim\_mass(u_i, u_j)$ is regarded as cosine similarity based on matrix UP. For similarity normalization, the equation (6) can be formulated as:

$$score\_Mlt = M * \sum_{j=1}^{M-1} Tr.score * sim\_mass(u_i, u_j) / \left(\sum_{u_i \in U} Tr.score * \sum_{u_i \in U} p.m \right) \qquad (7)$$

where $\sum_{u_i \in U} Tr.score$ is the summation of scores given by all users for this similar trajectory, $\sum_{u_i \in U} p.m$ is the summation of times given by all users for this GPS point.

As shown in equation (8), the trajectory similarity of two users is measured based on all the M-length similar trajectories. Here, n is the number of similar trajectories of two users. $score\_Mlt_i$ is the score of the i-th similar trajectory, which can be calculated according to equation (7). N_1 and N_2 denotes the number of the GPS points of two users, respectively.

$$sim\_Tra = \frac{1}{N_1 N_2} \sum_{i=1}^{n} score\_Mlt_i \qquad (8)$$

Landmark Trajectory Similarity. Like trajectory similarity in section Trajectory Similarity, the landmark trajectory similarity of two users can be formulated as:

$$sim\_LTra = \frac{1}{N_1 N_2} \sum_{i=1}^{n} score\_Mllt_i \qquad (9)$$

Note that, the difference between equation (8) with equation (9) is that we need replace trajectory and GPS point with landmark trajectory and landmark respectively.

3.3   Landmark Recommendation

Incorporating the three similarity measures, we propose the user similarity fusion method, which is demonstrated in equation (10).

$$sim\_Fusion(u_i, u_j) = \gamma_1 sim\_view(u_i, u_j) + \gamma_2 sim\_Tra + \gamma_3 sim\_LTra \qquad (10)$$

where $\gamma_1, \gamma_2, \gamma_3 > 0 \; and \; \gamma_1 + \gamma_2 + \gamma_3 = 1$.

Based on the user similarity fusion, we propose a landmark-based CF strategy.

$$\widehat{score(u, L)} = \overline{score_u} + \frac{\sum_{v \in U_k} sim\_Fusion(u_i, u_j)(score(v, L) - \overline{score_v})}{\sum_{v \in U_k} sim\_Fusion(u_i, u_j)} \qquad (11)$$

where $\overline{score_u}$ and $\overline{score_v}$ represents average times a user u and v visiting L respectively, $score(v, L)$ represents times a user v visiting L and U_k is the k most-similar neighbors of u.

4   Experiments

4.1   Data Set and Settings

The photos for our experiments were collected from the datasets of geo-tagged photos available on Flickr using the site's public API. These photos we crawled meet such requirements: they were taken in Beijing and the upload time is between 4th, January, 2005 and 10th, February, 2012.These photos collection contains 533,594 unique photos associated with 2,760,614 textual tags and taken by 16,196 unique users.

The data set are split into training and test sets. Additionally, we have further split the training data to validation data to optimize the parameters β_1, β_2, γ_1, γ_2, γ_3 and k, the neighborhood size. We have varied k from 10-60 by an interval of 10 and the other five parameters from 0 to 1 by an interval of 0.1. Using the validation data, we have found the best β_1, β_2, γ_1, γ_2, γ_3 to be 0.8, 0.2, 0.2, 0.3, 0.5 and k to be 20.

Our algorithm computes a ranking score for each candidate landmark (i.e., one user has not visited) and returns the top-K highest ranked landmarks as recommendation to a targeted user. To evaluate the prediction accuracy, we focus on how many locations previously removed in the preprocessing step re-appear in the recommended results. Therefore, we apply four popular performance metrics, namely, Mean Average Precision(*MAP*), *Precision@K*, *Recall@K* and *nDCG* (normalized Discounted Cumulative Gain) [9], to capture the performance of our proposed algorithm.

4.2   Experimental Results

In this work,we employ five baselines for comparison: *MeanShift (MS) + sim_ Fusion* ($sF(\gamma_1 = 0.2, \gamma_2 = 0.3, \gamma_3 = 0.5)$), *DFCM + sF($\gamma_1 = 1$)*(or *DFCM+sV*), *DFCM+ sF ($\gamma_2 = 1$)*(or *DFCM+sT*), *DFCM+ sF($\gamma_3 = 1$)* (or *DFCM+sL*) and *CRMF*. Here, *sF* represents the user similarity fusion, *sim_Fusion*. *sV*, *sT* and *sL* represents the overview semantic similarity (*sim_View*), the trajectory similarity (*sim_Tra*) and the landmark trajectory similarity (*sim_LTra*), respectively. In order to simplify the expression, *sF($\gamma_1 = 0.2, \gamma_2 = 0.3, \gamma_3 = 0.5$)* is expressed as *sF*.

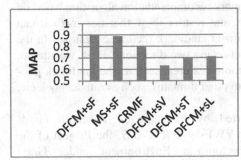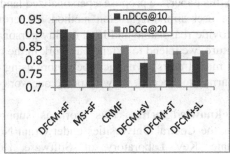

Fig. 2. Comparision of *MAP* and *nDCG* among different methods

Figure 2 shows the comparision of *MAP* and *nDCG* between our approach (*DFCM+sF*) and baselines. Our proposed method (*DFCM+sF*) is slightly superior to (*MS+sF*) and significantly superior to other methods. Using *nDCG*, Figure 2 further differentiates our approach from baselines. Obviously, (*DFCM+sF*) leads the performance in both *nDCG@10* and *nDCG@20* among these methods. Moreover, (*DFCM+sF*) better improves the performance comparing with (*MS+sF*) and *CRMF*. The possible reason for these is: 1) *CRMF* only captures the similarity of category-based and user-landmark preference (a.k.a trajectory similarity); 2) $(DFCM+sF)$ not only captures the overview semantic similarity, but also extracts the trajectory similarity and the landmark trajectory similarity; 3) $(DFCM+sV)$, $(DFCM+sT)$ and $(DFCM+sL)$ only consider one of the three similarities.

In Figure 3, we can observe that, according to precision, our proposed method $(DFCM+sF)$ slightly outperforms $(MS+sF)$ and significantly outperforms other methods. The explanation for this is similar to ones for Figure 2. Note that, when the number of the recommended landmarks reaches 10, the optimal value of precision is obtained. Because of the space limit, we have the similar analysis for recall.

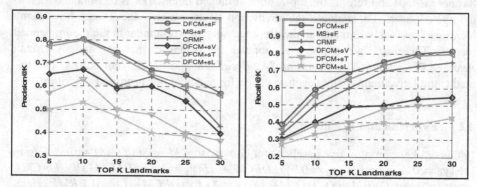

Fig. 3. Precision and Recall changing over the number of recommended landmarks

5   Conclusion

In this paper, we propose a personalized landmark recommendation algorithm based on the user similarity fusion method. Experimental results show that our method can provide reasonable and high quality personalized landmark recommendations. In the future, we intend to extend our work in the following two directions. First, we attempt to model users' dynamical behaviors using more features, such as the landmark popularity etc. Second, we will spread our work to other domains, such as music, book etc.

Acknowledgments. This work is supported by Fundamental Research Funds for the Central Universities under Grant No.YWF-13-T-RSC-077, the Project of the State Key Laboratory of Software Development Environment under Grant No.SKLSDE-2013ZX-26.

References

1. Deng, D.P., Chuang, T.R., Lemmens, R.: Conceptualization of Place via Spatial Clustering and Co-occurrence Analysis. In: Proceedings of the International Workshop on Location Based Social Networks, pp. 49–55 (2009)
2. Fukunaga, K., Hostetler, L.: The estimation of the gradient of a density function, with applications in pattern recognition. IEEE Trans. Information Theory 21(1), 32–40 (1975)

3. Cao, L., Luo, J., Gallagher, A., Jin, X., Han, J., Huang, T.S.: A worldwide tourism recommendation system based on geotagged web photos. In: ICASSP, pp. 2274–2277 (2010)
4. Crandall, D.J., Backstrom, L., Huttenlocher, D., Kleinberg, J.: Mapping the world's photos. In: Proceedings of the 18th International Conference on World Wide Web Conference, WWW 2009, pp. 761–770. ACM, New York (2009)
5. Hao, Q., Cai, R., et al.: Generating Location Overviews with Images and Tags by Mining User-Generated Travelogues. In: ACM MM (2009)
6. Li, Q., Zheng, Y., Xie, X., Chen, Y., Liu, W., Ma, W.: Mining user similarity based on location history. In: Proc. of the 16th Intl. Conf. on Advances in Geographic Info. System, pp. 298–307. ACM Press, Santa Ana (2008)
7. Gan, W., Li, D., Wang, J.: An Hierarchical Clustering Method Based on Data Field. Acta Electron-ICA Sinica 34(2) (2006)
8. Ying, J.J.-C., Lu, E.H.-C., Lee, W.-C., Weng, T.-C., Tseng, V.S.: Mining user similarity from semantic trajectorie. In: LBSN, pp. 19–26 (2010)
9. Manning, C.D., Raghavan, P., Schtze, H.: Introduction to Information Retrieval. Cambridge University Press (2008)
10. Zheng, V.W., Zheng, Y., Xie, X., Yang, Q.: Collaborative location and activity recommendations with gps history data. In: WWW, pp. 1029–1038 (2010)
11. Shi, Y., Serdyukov, P., Hanjalic, A., Larson, M.: Personalized landmark recommendation based on geotags from photo sharing sites. In: ICWSM 2011, pp. 622–625. AAAI (2011)
12. Zheng, V.W., Cao, B., Zheng, Y., Xie, X., Yang, Q.: Collaborative filtering meets mobile recommendation: A user-centered approach. In: AAAI (2010)
13. Cao, X., Cong, G., Jensen, C.: Mining Significant Semantic Locations From GPS Data. In: VLDB, pp. 1009–1020 (2010)
14. Ye, M., Yin, P., Lee, W.-C.: Location Recommendation in Location-based Social Networks. In: GIS, pp. 458–461 (2010)

Topical Discussions on Unstructured Microblogs: Analysis from a Geographical Perspective

Seema Nagar, Kanika Narang, Sameep Mehta,
L.V. Subramaniam, and Kuntal Dey

IBM Research India, New Delhi and Bangalore, India
{senagar3,kaninara,sameepmehta,lvsubram,kuntadey}@in.ibm.com

Abstract. Social networks today have emerged as hotbeds of online user conversations. Social microblog sites like Twitter have become favorite portals for users to discuss and express opinions on events and topics. Established event detection techniques on microblog streams today are capable of detecting events early in their lifecycle, amidst the volumes of user message exchanges. Techniques have been proposed in literature to identify topical conversations on microblogging portals comprising of unstructured data with no explicit discussion thread, distinguishing such conversations from isolated expressions of topical interests. However, evolutions of discussion topics have not been studied in a geographical context before. In the current work, we identify and characterize topical discussions at different geographical granularities, such as countries and cities. We observe geographical localization of evolution of topical discussions. Experimental results suggest that these discussion threads tend to evolve more strongly over geographically finer granularities: they evolve more at city levels compared to country levels, and more at country levels compared to globally. Our algorithm to find geographical evolution of discussion sequences and the derived insights can be used for information spread analyses and related applications on microblogging networks.

1 Introduction

The online social networks and social media today have emerged as the hottest hubs of user generated content. Social networks like Facebook and Google Plus, microblogging platforms like Twitter and image and video sharing networks like Pinterest, Flickr and Youtube are some of the largest repositories of content generated by users of these online networks. These digital media networks today serve as some of the leading message exchange platforms across users.

Detecting and analyzing user interests and opinions with respect to given events on such online networks has emerged as a topic of research interest. Research has shown that microblogging platforms like Twitter act as online news media rather than only social message exchange platform ([12]). A Twitter message reaches to followers of the person sending the tweet. A person receiving a message can *retweet* it to spread the message further forward to her followers. [12] shows that a tweet can, on an average, be expected to reach around

X. Lin et al. (Eds.): WISE 2013, Part II, LNCS 8181, pp. 160–173, 2013.

1,000 people irrespective of the number of followers of the tweet originator. This indicates, contemporary information diffuses significantly over Twitter.

Enormous content, generated by multiple millions of active users, makes Twitter rich in information and immense in information entropy. Conversations on this microblogging platform moves from one set of topics to another. Events detection on Twitter has been addressed under different constraints and settings ([19], [21], [22]). Discovering discussion threads, in absence of apriori corpus for contemporary events, has been addressed in [16]. At the same time, it is interesting to note that discussions on social media cross physical boundaries imposed by cities, countries and continents. While [10] attempts to explore the problem of modeling geographical topical patterns on Twitter using a sparse generative model, it does not address the geographical granularity of evolution of geographical topical discussions. In the online world, where location does not limit participation, geographically characterizing topical discussion threads on microblogging platforms without explicit discussion threads like Twitter is a challenging task. The different information dimensions (such as spatial, temporal and topical), as well as the information diffusion phenomenon seen in Twitter, makes such characterization an interesting research problem with application potentials.

In this work, we attempt to discover and characterize the localization and geographical evolution of discussion topics on microblogging platforms in form of threads, using Twitter data for our experiments. We add knowledge from the geographical data available on tweets and Twitter user profiles. We assign locations to topic clusters with simple probabilistic measures, and establish cross-cluster relationships by capturing the geographical overlaps of pairs of clusters. We also establish extended semantic links using external data sources such as contemporary online news media.

We use Brazil Flood (2010) and London Olympics (2012) Twitter datasets for experimentation. In our experiments, we create topic clusters using tweets to identify topics in which each tweet belongs to a topic. We connect the clusters in three different ways to construct three different graphs - an extended semantic graph ([16]) using externally available semantic data sources, a location graph using geographical data as found on user profiles and tweet locations, and a temporal graph using the timing of the tweets. We detect discussion topics and find them to be geographically evolving. We characterize the temporal evolution of these discussions. We use normalized mutual information based analyses to assess the goodness of our insights.

We observe that discussion threads emerging and evolving on microblog platforms tend to evolve more strongly on geographically finer granularities compared to coarser ones. Specifically, we find that topical discussions evolve more at city levels compared to country levels, and more at country levels compared to globally, which is a novel observation based upon our study of existing literature. Since the datasets used for experiment, namely Brazil Flood (2010) and London Olympics (2012), are matters of national an international interest for all practical purposes, this observation is surprising.

The discovered geographical characteristics of evolution of the discussions can be applied for commercial applications such as marketing campaigns and promotions. As a specific example of application of our work, marketing campaigns that depend upon spread of information over discussions on microblog networks would want to focus more on city levels compared to country levels. Our work can be further used for geographically targeted information spreading, collecting and analyses applications, and information diffusion studies, using microblog networks as data source.

In summary, the contributions of our work are the following.

1. We propose a methodology to study geographical characteristics of evolution of topical discussions on microblogging platforms at different granularities.
2. We empirically relate the evolution of discussion threads to geographical locations.
3. We further show that the discussion threads on microblogging platforms tend to evolve more strongly over geographically finer granularities compared to coarser ones.
4. We demonstrate our findings on microblogging data for 2 real events.

The rest of the paper is organized as follows. In Section 2 we describe the problem setting and our approach. We detail our experiments and observations in Section 3. We investigate the related literature in Section 4. Finally, we conclude in Section 5.

2   Algorithm

In this section, we describe the problem settings and propose the solution approach that we follow.

2.1   Problem Settings

Objective. The objective of our work is to identify and characterize topical discussion sequences based upon geographical (locational) attributes, and derive insights about the evolution of discussions from one topic to another over geographical locations of different granularities.

In order to meet our objective, there are a number of technical challenges to be overcome. We list the set of challenges below.

– First of all, detection of the topic-based clusters is required for the event. This will identify the independent set of discussions happening around a given event.
– We now need to establish different types of relationships across these clusters. One can think of the clusters as graph vertices and relationships as graph edges. In our setting, we need to identify geographical discussion threads, which implies we need to establish three core relationships: namely, contextual semantic, geographical and temporal relationships. Please note that the temporal edges will help us identify the evolution of discussions.

– Finally, we need to be able to characterize the geographical evolution of microblog discussion threads, and provide a quantitative assessment and a qualitative understanding of the localizations granularities of discussion topics.

2.2 Overview of Relationships

We now provide an overview of the contextual semantic, geographical and temporal relationships and their extraction algorithms. To make this article self contained, we present an overview of these relationships.

Contextual Semantic Relationships. This relationship is extremely useful but challenging to establish. We motivate the need for such relationship by a simple example. Consider two events with associated keywords $E_1 = \{$damage, flood, death, toll$\}$ and $E_2 = \{$flood, relief, shelter$\}$. Now, lets pick one word from each set *damage* and *relief*. One cannot establish any of the widely accepted relationships like synonymns, antonymns, hypernymns, hyponyms *etc.* when the words are taken in isolation. However, coupled with prior knowledge about the larger event *flood*, the words can be semantically related. In essence (with abuse of notation and terminology), damage and relief are independent variables without extra information, however, they are related given *flood*. Therefore, we would like to add the semantic edge between these events. Following the approach of [16], we use external corpus, namely contemporary online news, to extract and quantify such semantic relationships. In general, the corpus could be news stories, articles or books.

Geographical Relationships. This relationship associates an event cluster, comprising of multiple tweets, with one or more geographical locations. Since each event may comprise of users belonging to and tweets originating from multiple locations, we associate a *belongingness* value to each location that a cluster is associated with. We propose two different measures to compute belongingness, computed upon the number of users and the number of tweets belonging to a given locaiton respectively. We subsequently relate the event clusters by combining these geographical belongingness values using a well-known fuzzy technique ([24]). We conduct our experiments with two different granularities of geographical locations: cities and countries.

Temporal Relationships. Allen [3] presents an exhaustive list of temporal relationships which can exist given two time periods (events in our case). The relationships include *overlap*: partof event A and event B co-occur, *meets*: event A starts as soon as event B stops, *disjoint*: event A and event B share no common time point.

2.3 Our Approach

As part of our overall solution, we generate an event topic graph $\mathcal{G} = \{\mathcal{E}, \{R\}\}$, in which $\mathcal{E}$ represents the event topics (topical clusters), and act as the vertices of the graph. The set $\{R\}$ represents the relationship edges between the clusters, and are formed from each of semantic, temporal and geographical perspectives. These different types of edges, along with the structural patterns they form, are analyzed experimentally to obtain insights and meet our objective.

Topic-Based Cluster Creation

An event topic E^i is represented as $\{(K_1^i, K_2^i, \ldots, K_n^i), (L_1^i, L_2^i, \ldots, L_m^i), [T_s^i, T_e^i]\}$, where K^i denotes the set of keywords extracted from the tweets which form the event E^i, L^i is set of locations in event cluster and T^i is time period of the event. We use existing established methods for computing K, L and T. K contains the *idf* vector and proper nouns (extracted by PoS tagging) from the tweets. L is generated using the location from tweet locations and user profiles, and uses Standford's NLP Toolkit and associated Named Entity Recognizer. T denotes the time of first and last tweet in the event topic cluster. Since, event topic extraction is not the focus of our work, we use a well-known online clustering algorithm ([22]) to generate event topics from streaming Tweet data, thereby creating clusters of tweets forming event topics.

Extracting the Relationships

We now extract the three core relationships, namely semantic, geographical and temporal relationships, across the event topic cluster vertices.

Contextual Semantic Relationship Extraction. We establish contextual semantic relationships across event clusters following the *extended semantic edge* detection algorithm proposed by [16]. We generate $|K^i| \times |K^j|$ pairs of keywords per cluster pair which need to be evaluated for contextual semantic relationship. To avoid similar keywords that would skew the results, we prune pairs which are related semantically (synonyms, antonyms, hypernymns and hyponynms). Since POS tagging is done on the tweets in the event, we also remove pairs where one of the words is a Proper Noun or Active Verb. We use the well-established Lin's method ([14]) to compute similarity scores of K^i and K^j, using the feature vector built into the Wordnet lexical database. We retain a pair of words if the similarity score S_{ij} is lesser than a desirable similarity threshold S, and prune the pair otherwise.

For sake of completeness, please note that Lin's measure of similarity between a pair of words $w1$ and $w2$ is defined as: $sim(w1, w2) = \frac{2I(F(w1) \cap F(w2))}{I(F(w1)) + I(F(w2))}$, where $F(w)$ is the set of features of a word w, and $I(S)$ is the amount of information contained in a set of features S. Assuming that features are independent of one another, $I(S) = -\sum_{f \in S} log(P(f))$, where $P(f)$ is the probability of feature f.

Now, all the concept pairs from K^i and K^j, after the above pruning, are chosen from the concepts present in clusters i and j. Using external contemporary relevant corpus such as Google News, pairwise *coupling* scores of these concepts are now measured as $(C(K_l^i, D_t), C(K_m^j, D_t))$, where D_t denotes the set of documents in which the concepts K^i and K^j co-occur, and $C(K_l^i, D_t)$ gives the *tf-idf* score of word K^i in document D_t. Finally, these scores are combined across all pairs of concepts from all the clusters, computed as $\frac{\sum_{K^i, K^j} Coupling}{(|K^i| \times |K^j| - w_{ij})}$, where for a given pair of event clusters E^i and E^j, w_{ij} pairs of keywords were pruned and the rest were retained. The final scores are ranked in descending order and top N% are selected based on user preference or can be pruned based on threshold.

Geographical Relationship Extraction. We extract geographical relationships at two different granularities, countries and cities. Given a location L^i and an event topic cluster E^i, $L^i \in E^i$ iff $(\exists M)$, a microblog post, made from a location in L^i, such that $M \in E^i$. Please note that with this definition, a location could simultaneously be a part of multiple clusters.

As noted earlier, every event topic cluster has a vector $L^i = (L^i_1, L^i_2, \ldots, L^i_m)$ associated. In the first step of our two-step process of extracting geographical relationships, we assign a *belongingness* value to each location that a cluster is associated with, thereby forming a belongingness value vector. In the next step, we quantify the strength of geographical relationships across each pair of clusters, by combining the belongingness value vectors of each cluster to the geographies, using a well-known fuzzy technique.

Step 1: Computing belongingness - In the first step, we augment the L^i vector to the $\hat{L}^i$ vector, in which each element $\hat{L^i}_c$ belonging to $\hat{L}^i$ is assigned its belongingness value $\hat{B^i}_c$, in which c denotes the index of an individual location (such as a city or country) within the location vectors L^i and $\hat{L}^i$, and $1 \leq c \leq m$. We use two different measures, namely the number of distinct people belonging to each location and the number of tweets originating from each location, to assign belongingness values to each belongingness vector $\hat{B}^i$.

If there are a total of $|\hat{S^i}_p|$ distinct people from location $\hat{L^i}_p$ in event cluster E^i comprising of tweets from $|S^i_p|$ distinct people in total in the cluster, then the belongingness value of $\hat{L^i}_c$ to E^i, by the measure of number of distinct people, is calculated as: $\hat{B^i}_c = \frac{|\hat{S^i}_p|}{|S^i_p|}$. The belongingness value of $\hat{L^i}_c$ to E^i, by the measure of number of tweets (instead of number of people) from this location, namely $|\hat{S^i}_t|$, and the total number of tweets, namely $|S^i_t|$, is calculated similarly as $\hat{B^i}_c = \frac{|\hat{S^i}_t|}{|S^i_t|}$.

Step 2: Establishing geographical relationships - We subsequently attempt to establish geographical relationships across event topic clusters. Since a cluster belongs to multiple geographical locations with a certain probability (belongingness), we use a well-known fuzzy technique to determine the strength of their relationships ([24]). The similarity-based weight W_{ij} of a given pair of event topic clusters, namely E_i and E_j, having belongingness value vectors $\hat{B}^i$ and $\hat{B}^j$ respectively, is given by measuring similarity between fuzzy sets ([24]):

$$W_{ij} = \frac{\hat{B^i}_c.\hat{B^j}_d}{max(\hat{B^i}_c.\hat{B^i}_c, \hat{B^j}_d.\hat{B^j}_d)} \tag{1}$$

Please note that, we consider all $1 \leq c \leq m$ and $1 \leq d \leq m$. Also, $\hat{B^i}_c.\hat{B^j}_d \neq 0$ iff $c \neq d$, which ensures that the set similarity formula can be applied on our vectors. For our experiments, we treat the operator dot (.) as the multiplication operator.

Temporal Relationship Extraction. The third kind of relationship we attempt to extract is temporal relationship. We look at two kinds of temporal relationships. (a) We draw a temporal edge from event E^i to event E^j if E^i ended within a threshold time gap before E^j started. For experimentation, we restrict the maximum time gap limit to a realistic 2 days. This follows from the assumption that on microblogging services like Twitter, a discussion thread will not last longer than this. This thresholding also prevents the occurrence of spurious edges across different clusters. It captures the *meets* and *disjoint* relationships described by [3]. We call this a *follows* temporal relationship. (b) We draw a temporal edge from event E^i to event E^j if E^i started before E^j started, and ended after the start but before the end of E^j. This captures the *overlaps* relationship described by [3]. Please note that unlike the undirected semantic and social relationship edges, a temporal relationship edge is directed but unweighted. The direction is from the event with the earlier starting time to the one with the later starting time.

Identifying and Characterizing Discussions

Finally, after establishing the relationships, we attempt to identify discussions and characterize those geographically.

Identifying Discussion Sequences. A discussion sequence graph is defined as, a directed acyclic graph (DAG) of topics that are related using the semantic edges obtained by our earlier semantic relationship extraction process, where the relationships are established over time. Intuitively, a discussion sequence captures the topical evolution of discussions over time. We identify discussion sequences using the logical intersection (AND) of the relationship set of the undirected semantic and the directed temporal graphs, with the directions of the latter preserved in the output. So, the discussion sequence DAG $\mathcal{G}_{DS}$ is formed as: $\mathcal{G}_{DS} = \{\mathcal{E}, \{\{R_T\} \cap \{R_S\}\}\}$, where the set $\{R_T\}$ represents the edge set of the directed temporal graph and the set $\{R_S\}$ represents the edge set of the undirected semantic graph.

Identifying Geographical Discussion Threads. A geographical discussion thread is defined as a DAG of discussion sequences that are further connected by geographical similarity of participation strengths. The similarity is computed in terms of people or tweets, as constructed by the process of extracting geographical relationships. Thus, a geographical discussion thread is identified by performing a logical intersection (AND) of the relationship set of the DAG $\mathcal{G}_{DS}$ with the undirected geographical relationship graph, resulting in a DAG. The retained discussion sequences capture the geographical evolution of discussion topics around events on microblogs over time, and hence, are identified as **geographical discussion threads**. The corresponding graph $\mathcal{G}_{GDS}$ is formed as: $\mathcal{G}_{GDS} = \{\mathcal{E}, \{\{R_{DS}\} \cap \{R_G\}\}\}$, where the set $\{R_{DS}\}$ represents the geographical relationship set.

Measuring the Goodness of our Approach. We quantify the evolution of topically evolving discussions along the geographical axis by measuring mutual information found by inspecting link structures. We discover BGLL ([4])

communities that maximizes modularity ([5], [8]) independently on the semantic and geographical relationships, and compute the normalized mutual information (NMI: [6]) across these two sets of communities. We independently discover BGLL communities on the discussion sequences and geographical discussion threads, and compute the NMI across these two. A favorable comparative value of these two sets of NMI measures, observed in our experiments, indicates the goodness of our method.

For sake of completeness, please note that, the mutual information of X and Y, namely $I(X, Y)$, is given by: $I(X, Y) = \sum_{y \in Y} \sum_{x \in X} p(x, y) log(\frac{p(x,y)}{p(x)p(y)})$. If H(X) and H(Y) denote the marginal entropies of X and Y respectively, then the NMI measures are given by $C_{XY} = \frac{I(X,Y)}{H(Y)}$ and $C_{YX} = \frac{I(X,Y)}{H(X)}$ respectively.

In our experiments, we observe lower NMI values in the first case, and much higher NMI values in the second. The improved NMI qualitatively indicates the characteristic of discussion sequences to tend to evolve geographically, in spite of low structural overlap of the base semantic and geographical graphs. The number and strength of geographical discussion threads observed experimentally, indicates the soundness of our approach.

3   Experiments

We now conduct our experiments, using Twitter data, following the approach illustrated in Section 2.

3.1   Constructing the Baseline Graphs

For our experiments, we collected the Twitter data for Brazil Flood (2010) and London Olympics (2012), two high-trending events that had received significant user traction on Twitter. For Brazil Flood, we used (*Brazil* OR *flood*) as the target keyword for data collection, implying that any tweet collected will have at least one of these terms in it. For London Olympics, we used (*London* OR *Olympics*). While collecting the tweet data from Twitter, we also collected the profile location of each user tweeting at least once around any of these events. Given the sparse nature of the location attribute in the Twitter posts, we use the profile locations to assign a user, and the tweets made by the user in absence of tweet location attribute, to a corresponding geographical location. Table 1 shows the basic statistics of the datasets.

Table 1. The columns in the table show a) keywords used to search Twitter to collect the dataset, b) dates for which the data was collected, c) number of tweets collected, d) number of clusters and e) number of contemporary external news documents collected

Dataset	Keywords	Timespan	Tweets	Clusters	News Docs
Flood	Brazil, Flood	14 Jan - 28 Jan'10	335,704	196	149
Olympics	London, Olympic	27 Jun - 13 Aug'12	2,319,519	299	1,186

Event topic cluster detection not being the focus of our work, we used an existing online clustering algorithm to form the clusters from the tweets. As outlined earlier, we now establish semantic, geographical and temporal relationships across the clusters. We use Google News (http://news.google.com), a news text data source, to establish semantic links, selecting news documents contemporary to the event using appropriate date ranges. The search results had 1,186 and 149 unique news documents for London Olympics and Brazil Flood respectively. These documents served as the external contemporary relevant corpus for our experiments. As described in Section 2, we deduplicate concepts to avoid repetitions using WordNet and use the *tf-idf* sum of all pairs of selected words across the given pair of clusters as edge weights, using the method of [16]. We use Lin's ([14]) method of computing WordNet similarity, and use 0.4 as the threshold value in our experiments.

We form geographical links based upon the similarities of each given pair of clusters, as described in Section 2, at country and city levels. The weight of the link is determined by Equation 1. We further form two kinds of temporal links - namely *follows* (set union of *meets* and *disjoint*), and *overlaps*. This completes the process of construction of the baseline graphs.

3.2 Identifying Structural Overlap of Baseline Graphs

We now look at the inherent structural overlap of the basic semantic and geographical graphs. We use the well-known NMI ([6]) measure to investigate structural similarities of these two graphs. Table 2 shows low NMI values for each comparison of the semantic and geographical graphs, implying significant structural differences between the two, and showing that these two graphs are inherently different. This inherent difference makes the existence and evolution of geographical discussion threads (that we find subsequently) surprising.

Table 2. NMI across top 10%, 20%, 30%, 40% and 50% of edges retained in semantic and geographical graphs in the Brazil Flood and London Olympics data sets

Geo-gra- nularity	Measurement method	NMI (Brazil Flood)					NMI (London Olympics)				
		10%	20%	30%	40%	50%	10%	20%	30%	40%	50%
Country level	People	0.032	0.025	0.069	0.074	0.070	0	0.033	0.056	0.071	0.061
	Tweet	0.009	0.010	0.074	0.064	0.057	0	0.062	0.054	0.064	0.057
City level	People	0	0.018	0.007	0.033	0.021	0	0.011	0.014	0.022	0.027
	Tweet	0	0.001	0.030	0.023	0.008	0	0.011	0.026	0.026	0.026

3.3 Identifying Discussion Sequences

We find discussion sequences from extended semantic graphs and temporal graphs, as described in Section 2, using the method of [16] and Lin's WordNet similarity measure. We perform edge set intersection (logical AND) of the semantic and temporal graphs, retaining the directions observed in the temporal graphs. For manual inspection, we pick the induced subgraph formed by 30 event topic clusters at random from the the semantic graph, and build the discussion sequence graph around it by taking a logical intersection with the temporal graph. We then visually inspect the resulting discussion sequence graph (showing vertices with at least one connecting edge within this set), as shown in Figure 1. Clearly, many discussion sequences prominently appear in the visualization.

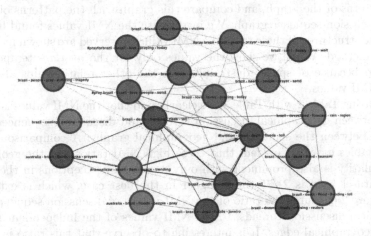

Fig. 1. Discussion sequences created from *Lin's semantic* and *temporal overlaps* relationships for Brazil Flood: vertices of the same color belong to one community (thicker edges have higher weights)

3.4 Identifying and Characterizing Geographical Discussion Threads

We now attempt to identify geographical correlations of the discussion sequences, and characterize the identified geographical discussion threads. We identify the geographical discussion threads by taking edge set intersection (logical AND) of the geographical relationships and discussion sequence edges.

In order to determine whether the geographical discussion thread graphs thus found is quantitatively substantial and qualitatively meaningful, we look at some of the basic attributes of these graphs. The Brazil Flood geographical discussion thread graph comprises of 169 vertices and 3,013 edges, and the corresponding London Olympics graph comprises of 254 vertices and 6,183 edges. Retaining the top 50% edges of these graphs result in retaining 162 and 229 vertices respectively, while dropping the remaining vertices.

Considering all the paths across each of the geographical discussion threads with the *overlaps* temporal relationship, we find an average path length, indicating the average number of clusters participating in a geographical discussion thread, to be 7.16, for the Brazil Flood data set for the people-based input geographical relationship graph. The maximum path length we observe in this graph is 13. These are significant for a 15-day timespan (ref: Table 1). The tweet-based graph for Brazil Flood has an average path length of 9.53 and maximum path length of 18, which is even higher. For London Olympics, the average and maximum path lengths are 9.58 and 17 for the people-measure graph and 9.6 and 17 respectively. All these values indicate substantial presence of geographical discussion threads among these events.

We characterize the geographical discussion threads by observing the structural patterns of the graph, and compare this graph with the patterns observed in the discussion sequence graph. We investigate the NMI values found by comparing the structures of these two. The NMI values observed are shown in Table 3 for each method. While we show the values only for the *overlaps* temporal relationship because of space constraints, but the other relationships also yield similar NMI values.

Comparing Table 2 with Table 3, we clearly find that the NMI values is always higher between geographical discussion threads and discussion sequences, compared to between the semantic and geographical graphs. A comparison across the two tables makes it evident that these high NMI ratios and the geographical granularity is also prominent with only some stray exceptions in the Brazil Flood dataset, and is 21.75 times higher in the best case, which is extremely high. Figure 2 captures the ratio of the NMI values of discussion sequences and geographical discussion threads, to the NMI values of the independent semantic and geographical edges. It is interesting to observe that this ratio is always higher than unity for London Olympics, and more than 3 times higher in many of the cases, the highest being as much as 7.0769.

These high ratios of NMI, observed across Table 2 with Table 3, affirm that our observations are not accidental, but consistent across data sets over multiple events and geographies.

One can qualitatively think of the low NMI between geographical and semantic graphs to indicate isolated discussions to be happening randomly over different geographies. However, the (often much) higher NMI observed between discussion sequences and geographical discussion threads affirms our belief that discussions on microblogging platforms like Twitter evolve strongly along geographical regions. Further, the ratios of NMI at the city-level geographical granularity are much higher than country-level granularity, and as high as 21.75 in the best case. This indicates a stronger topical similarity for tightly bound geographical locations. Figure 3 shows some randomly picked geographical discussion sequences at a country level for visual inspection. The figure makes it evident that, most of the discussion sequences that emerge, evolve over well-bounded geographical axes.

4   Related Work

Significant research has been conducted on content analysis of information discussed on social media sites [12]. Grinev et al. [9] demonstrate TweetSieve, a system that obtains news on any given subject by sifting through the Twitter stream. Along similar lines, Twinner by Abrol et al. Abrol and Khan [1] identify news content of a query by taking into account the geographic location and

Table 3. NMI across top 10%, 20%, 30%, 40% and 50% of discussion sequences and geographical discussion threads in Brazil Flood and London Olympics data sets

Geo-gra-nularity	Measurement method	NMI (Brazil Flood)					NMI (London Olympics)				
		10%	20%	30%	40%	50%	10%	20%	30%	40%	50%
Country level	People	0.071	0.035	0.069	0.019	0.078	0.148	0.092	0.161	0.102	0.143
	Tweet	0.069	0.020	0.065	0.062	0.109	0.157	0.150	0.171	0.075	0.162
City level	People	0.102	0.052	0.072	0.120	0.151	0.055	0.072	0.047	0.036	0.070
	Tweet	0.026	0.017	0.187	0.291	0.175	0.066	0.051	0.084	0.078	0.184

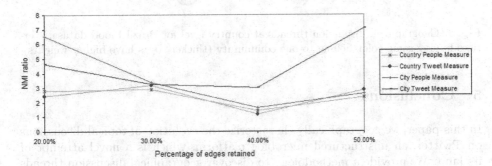

Fig. 2. Ratio of NMI of geographical discussion threads and discussion sequences, to NMI of independent semantic and geographical graphs, for London Olympics data set: all ratios are well over unity, and the highest is 7.0769

the time of query. Nagar et al. [15] demonstrate how content flow occurs during natural disasters.

Several ways to cluster social content have been studied. There has been work on clustering based on links between the users by doing agglomerative clustering, min-cut based graph partitioning, centrality based and Clique percolation methods ([7], [18]). Other approaches consider only the semantic content of the social interactions for the clustering [23]. More recently there has been work on combining both the links and the content for doing the clustering ([17], [20]). In [16] relationships between clusters are determined based on semantic, linkage and temporal information.

Doing location analysis on Twitter text has been studied in the past [1]. The location of a particular news event is determined by extracting the location information from the tweet messages [2]. Geo-social event detection [13] has been used to contrast crowd behavior in different geographies. [10] models geographical topics in Twitter stream, but does not attempt to study topical discussions at different geographical granularities. The impact of location on spreading hashtags has been studied recently by [11]. In this paper we use topical semantics, location and time as the dimensions to find relationships between Tweet clusters, and study the formation, evolution and granularity of propagation of topical discussions from a geographical perspective.

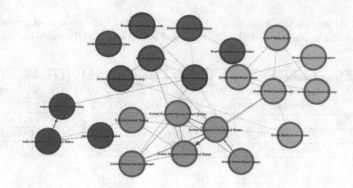

Fig. 3. Geographical discussion threads at country level for Brazil Flood dataset: vertices having same color belong to one community (thicker edges have higher weights)

5   Conclusions

In this paper, we geographically characterize the evolution of topical discussions on Twitter, an unstructured microblog platform, which is a novel attempt of its kind. We provide a methodology to discover geographical discussion threads from microblogs. We show that on microblogging platforms like Twitter, topical discussion threads evolve over well-defined geographical boundaries. Experimentally, we use 2 real event microblogging datasets, over two different granularities (country and city levels), to confirm our findings. We observe that discussion threads evolve within well-defined geographical boundaries, the most prominent granularity of development being at city levels, followed by country levels.

Using microblogs are the data sources, this work can be used for settings such as geographically fine-tuned information spreading and diffusion studies, and geographical information gathering and analyses applications. Further, our work is relevant for commercial applications such as geographically focused campaigns and marketing promotions.

Since discussions on Twitter also evolve socially ([16]), hence, as future research work, it will be interesting to investigate the combined impact of social relationships and geographical characteristics on information spreading and topical discussions. A comparative study of the impact of social and geographical relationships will also be of significant academic interest.

References

1. Abrol, S., Khan, L.: Twinner: Understanding news queries with geo-content using twitter. In: Proceedings of the GIS (2010)
2. Agarwal, P., Vaithiyanathan, R., Sharma, S., Shroff, G.: Catching the Long-Tail: Extracting Local News Events from Twitter. In: Proceedings of ICWSM (2012)
3. Allen, J.F.: Maintaining Knowledge about Temporal Intervals. Communications of the ACM (1983)

4. Blondel, V.D., Guillaume, J.L., Lambiotte, R., Lefebvre, E.: Fast unfolding of communities in large networks. J. Stat. Mech., P10008 (2008)
5. Clauset, A., Newman, M.E.J., Moore, C.: Finding community structure in very large networks. Phys. Rev. E. 70(066111) (2004)
6. Coombs, C.H., Dawes, R.M., Tversky, A.: Mathematical psychology: An elementary introduction. Prentice-Hall, Englewood Cliffs (1970)
7. Fortunato, S., Barthelemy, M.: Resolution limit in community detection. Proceedings of the National Academy of Sciences 104(1), 36–41 (2007)
8. Girvan, M., Newman, M.E.J.: Community structure in social and biological networks. Proceedings of the National Academy of Sciences 99(7821) (2002)
9. Grinev, M., Grineva, M., Boldakov, A., Novak, L., Syssoev, A., Lizorkin, D.: Tweetsieve: Sifting microblogging stream for events of user interest. In: Proceedings of the SIGIR (2009)
10. Hong, L., Ahmed, A., Gurumurthy, S., Smola, A.J., Tsioutsiouliklis, K.: Discovering geographical topics in the twitter stream. In: WWW (2012)
11. Kamath, K.Y., Caverlee, J., Lee, K., Cheng, Z.: Spatio-temporal dynamics of online memes: a study of geo-tagged tweets. In: WWW (2013)
12. Kwak, H., Lee, C., Park, H., Moon, S.: What is Twitter, a Social Media or a News Media. In: Proceedings of the WWW (2010)
13. Lee, R., Sumiya, K.: Measuring geographical regularities of crowd behaviors for Twitter-based geo-social event detection. In: Proceedings of the WWW (2010)
14. Lin, D.: An information-theoretic definition of similarity. In: Proceedings of the International Conference on Machine Learning (1998)
15. Nagar, S., Seth, A., Joshi, A.: Characterization of Social Media Response to Natural Disasters. In: Proceedings of the WWW (2012)
16. Narang, K., Nagar, S., Mehta, S., Subramaniam, L.V., Dey, K.: Discovery and analysis of evolving topical social discussions on unstructured microblogs. In: European Conference on Information Retrieval (2013)
17. Pathak, N., DeLong, C., Banerjee, A., Erickson, K.: Social topics models for community extraction. In: Proceedings of the 2nd SNA-KDD Workshop (2008)
18. Porter, M.A., Onnela, J.P., Mucha, P.J.: Communities in networks. Notices of the American Mathematical Society 56(9), 1082–1097 (2009)
19. Ritter, A., Mausam, E.O., Clark, S.: Open domain event extraction from Twitter. In: ACM SIGKDD (2012)
20. Sachan, M., Contractor, D., Faruquie, T.A., Subramaniam, L.V.: Using content and interactions for discovering communities in social networks. In: Proceedings of the WWW (2012)
21. Sakaki, T., Okazaki, M., Matsuo, Y.: Earthquake shakes Twitter users: Real-time event detection by social sensors. In: Proceedings of the WWW (2010)
22. Weng, J., Lee, B.S.: Event detection in Twitter. In: ICSWM - Proceedings of the AAAI Conference on Weblogs and Social Media (2011)
23. Zhou, D., Manavoglu, E., Li, J., Giles, C.L., Zha, H.: Probabilistic models for discovering e-communities. In: Proceedings of the WWW (2006)
24. Mitrovic, Z., Rusov, S.: Mitrovic Z., Rusov S.: Z similarity measure among fuzzy sets. FME Transactions 34(2) (2006)

Discovering Correlated Entities from News Archives

Lili Yang[1], Chunping Li[1], Qiang Ding[2], and Li Li[2]

[1] Tsinghua National Laboratory for Information Science and Technology,
School of Software, Tsinghua University, Beijing 100084, China
`yangll11@mails.tsinghua.edu.cn`, `cli@tsinghua.edu.cn`
[2] Shannon Lab, Huawei Technologies Co. Ltd., Beijing 100095, China
`{q.ding,jollylili.li}@huawei.com`

Abstract. Most textual documents contain references to real-word entities such as *people, locations* and *organizations*. The understanding of their correlations is behind many applications including social relationship construction platform and major search engines, etc. This paper aims to discover entity correlations from news archives by means of the proposed hierarchical Entity Topic Model (hETM). hETM is a semantic-based analysis model which follows the gist of probabilistic topic models and in which a directed acyclic graph (DAG) is leveraged to capture arbitrary topic correlations. Entity extraction is taken as a preprocessing step of our model and we then employ different generative processes for ordinary words and entities. The discovering of entity correlations is achieved via the analysis of the dependencies between entities and their associated topics as well as topic correlations. We evaluate the approach upon BBC news dataset and results demonstrate the higher quality of discovered entity correlations compared with existing methods.

Keywords: Entity, Correlation, Topic Model, News.

1 Introduction

A large percentage of the textual documents published on the Internet are associated with entities: news articles often mention people's names and organizations, scientific papers usually contain authors and technical terms, and advertisements are generally associated with company names. To discover the correlations of these entities is of interest on account of its importance in plenty of applications including intelligent search engines and social relationship construction platform. For example, Google[1] returns a list of "Related Searches" when a user inputs an entity as the query to the search engine. Renlifang[2] extracts *People Names* from millions of web pages on the Internet and calculates correlations between these people to form a people relationship network.

---

[1] `http://www.google.com/`
[2] `http://renlifang.msra.cn/`

X. Lin et al. (Eds.): WISE 2013, Part II, LNCS 8181, pp. 174–187, 2013.
© Springer-Verlag Berlin Heidelberg 2013

A direct way to measure entity correlations is based on their co-occurrence in a document. For entities that often co-occur, we think they are closely correlated and for those seldom co-occur, their correlation is relatively weak. A big shortage of this method is that for those never co-appeared entities, we can only treat them as uncorrelated but as a matter of fact, they may have potential relationships as was implicated by "Six Degrees of Separation" [1,2]. Another drawback is that this method only leverages entity frequencies without taking their semantic relations into consideration.

Probabilistic topic models [3,4] have made it possible for mining entity correlations upon their semantics and also could relieve the aforementioned obstacle that unable to mine correlations between entities never co-appear. The marrow of these models lies in the power of clustering semantically similar terms into a topic. Hence, closely related entities are prone to be assigned to the same topic and their correlations could be measured by probabilities they belong to associated topics. [5] proposed several statistical entity-topic models which intersect topic modeling and entity modeling to get dependencies between entities and topics to measure entity correlations. [6] used a multi-type topic model to directly capture correlations between an arbitrary number of *who-entities* and *where-entities* mentioned in each document. Although these models could capture entity correlations in consideration of their semantic relations, they still have some limitations. We may notice that in spite of the low probability to be associated with same topics, some entities may still have correlations with each other: *Oxford* and *Ballet West*, the former is a famous university in England and the latter is an American ballet company and hence the two entities are prone to be associated with different topics. But both of their associated topics are probable to be related with topic *children education*, therefore *Oxford* and *Ballet West* could be treated as correlated to some extent. Existing entity-correlation discovering methods based on topic models fail to capture this kind of correlation because they treat all discovered topics as irrelevant.

In this paper, we propose a novel topic model based approach for discovering correlated entities. Our intuition is that if two topics are highly related, entities assigned to them are possibly to be correlated with each other. So we present a hierarchical Entity Topic Model (hETM) to acquire dependencies between entities and topics as well as topic correlations. Then the correlation between a pair of entities could be measured by not only their probabilities in the same topic but also topic correlations if they are associated with different topics. Evaluation of the effectiveness of our proposed approach focuses mainly on two aspects: missing entity prediction and entity co-participance prediction. The former is to predict missing entities based on obtained entity correlations and the latter is to predict whether two entities will co-participate in a future event. We did experiments on BBC news dataset and compared our approach with some existing models. The results demonstrate the advantage of the proposed approach.

The remainder of this paper is organized as follows. Section 2 introduces the background and related works. In Section 3 we present our proposed approach in details. In Section 4, we show experimental results on real world dataset. We have the concluding remarks in Section 5.

2   Related Work

In this section, we briefly introduce some related works with respect to entity mining with topic models as well as researches about correlated topic models.

The recent years have witnessed a surge of interests in entity mining with topic models. The first branch to be mentioned is entity resolution. [7] developed an unsupervised model inspired by LDA and exploited collaborative group structure for making resolution decisions. [8] proposed a LDA-dual model that involved both author names and words and emphasized the use of the global information of the corpus obtained after learning of the model. [9] used Wikipedia's category hierarchy as the topic hierarchy to disambiguate entities with hierarchical topic models and [10] tackled the problem of identity resolution with a hierarchal generative nonparametric model which integrated both topic and co-author information.

Another entity mining technique via topic models is entity recognition. [11] and [12] proposed approaches to recognize entities in web search queries. [11] addressed the problem by employing a weakly supervised learning which considered contexts of a named entity as words of a document and classes of the named entity as topics and [12] solved the limitation that lack of context information in short queries by utilizing the search session information before the query as its context. [13] conducted entity mining by using click-through data collected on search engine and then employed a topic model which was learned by weak supervision. These topic model based entity mining approaches demonstrated promising results and showed great potential of topic models in this area.

Correlated topic models are related to our approach and some models have been presented in past years. [14] applied a logistic normal distribution to capture pair-wise topic correlations and [15] used a Dirichlet-Tree prior over the topic proportions to simulate the intimacy between topics. PAM, which was proposed in [16], is able to capture arbitrary topic correlations with a pre-defined DAG where each internal node represents a multinomial distribution over their children. Furthermore, [17] introduced hLDA which uses a nested Chinese Restaurant Process (nCRP) to form the prior of a topic tree and each document could choose only one path from the tree while hPAM [18], which is a combination of both PAM and hLDA, allows all words to choose any path and level from a pre-defined DAG.

Our proposed hETM is inspired by hPAM to learn topics and topic correlations. The main advantage of hETM over all the aforementioned correlated topic models is that it models entities explicitly after we extract them from a text corpus by employing different generative processes for entities and words. This facilitates the discovery of entity correlations and it also can make predictions about missing entities and entity co-participance in future event.

3 Entity Relationship Modeling

In this section, we present our proposed model for discovering correlated entities. We start with the introduction of hPAM, based on which our approach is inspired. We then turn to the details of our proposed hETM, as well as how to utilize it to derive entity correlations.

3.1 Hierarchical Pachinko Allocation Model

hPAM is a flexible framework for hierarchical topic modeling. By manually designing a DAG beforehand and allowing each node be associated with a distribution over vocabulary, this model is able to learn a hierarchical topic structure exactly as the defined DAG, with upper nodes representing more general topics (super-topics) and lower nodes representing specific topics (sub-topics). Each super-topic is associated with a dirichlet distribution over sub-topics. Topic correlations are indicated by probabilities that super-topics generate sub-topics. For example, super-topic *London Olympic* has high probabilities to generate sub-topics *Torch Relay* and *Ticket Sales*, then we think *London Olympic* is highly correlated with both *Torch Relay* and *Ticket Sales*.

A fully connected four-level structure (see Fig.1(a)) is frequently adopted in previous researches. This structure consists of one root topic at the top, T super-topics at the second level, t sub-topics at the third level and words at the bottom where the root is connected to all super-topics, super-topics are fully connected to sub-topics and sub-topics are fully connected to words. Circles in Fig.1(a) represent topics and solid black boxes are the word distributions associated with each topic. The graphic model of this four-level hPAM is shown in Fig.1(b) and the generative process is as follows:

- For each of the $1 + T + t$ topics k
    Draw word distribution of the topic $\phi_k \sim Dirichlet(\beta)$
- For each of the M documents m
    1. Draw a distribution over super-topics $\theta_{rm} \sim Dirichlet(\alpha_r)$
    2. For each of the T super-topics, draw distributions over sub-topics $\theta_{sm} \sim Dirichlet(\alpha_s)$
- For each of the N_m words w_{mn} in document m
    1. Draw a super-topic $z_{rmn} \sim Multinomial(\theta_{rm})$
    2. Draw a sub-topic $z_{smn} \sim Multinomial(\theta_{z_{rmn}m})$
    3. Draw a level $l_{mn} \sim Multinomial(\tau_{z_{rmn}z_{smn}})$
        If $l_{mn} = 0$, draw word $w_{mn} \sim \phi_0$
        If $l_{mn} = 1$, draw word $w_{mn} \sim \phi_{z_{rmn}}$
        If $l_{mn} = 2$, draw word $w_{mn} \sim \phi_{z_{smn}}$

As we have noticed in the generative process, the generation of a word needs to choose a path from root to bottom first and then choose a level to determine which topic the word will be generated from. Correlations among super-topics and sub-topics are captured by the dirichlet distribution parameters associated

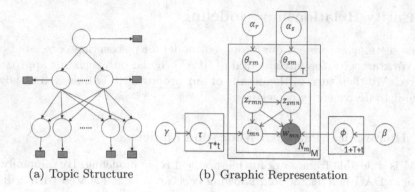

(a) Topic Structure   (b) Graphic Representation

Fig. 1. Four-level hPAM

with each super-topic hence the updating of these parameters is extremely important. Assuming α_{xy} represents the parameter of super-topic x to generate sub-topic y, it is updated in each Gibbs Sampling iteration according to the following rules proposed by [16]:

$$mean_{xy} = \frac{1}{N_x} \times \sum_d \frac{n_{xy}^{(d)}}{n_x^{(d)}}, \quad var_{xy} = \frac{1}{N_x} \times \sum_d (\frac{n_{xy}^{(d)}}{n_x^{(d)}} - mean_{xy})^2$$

$$m_{xy} = \frac{mean_{xy} \times (1 - mean_{xy})}{var_{xy}} - 1, \quad \alpha_{xy} = \frac{mean_{xy}}{exp(\frac{\sum_y log(m_{xy})}{t-1})} \quad (1)$$

where N_x is the total number of documents with non-zero counts of super-topic x, $n_x^{(d)}$ is the number of occurrences of super-topic x in document d, $n_{xy}^{(d)}$ is the number of times that sub-topic y is sampled from its parent x in document d. Note that smoothing is important to estimate these parameters. Even when sub-topic y does not generated from super-topic x in one iteration, it still gets some probability in subsequent iterations.

3.2   Hierarchical Entity Topic Model

As the objective of this paper is to model entity correlations, we made a modification for the aforementioned hPAM so that we can model entities explicitly. For ordinary words, the generative process follows the instructions of hPAM, which is to say, to choose a path from the pre-defined DAG first and then choose a level to determine which topic the word is generated from. For the generation of entities, we assume a multinomial distribution over paths and levels that are chosen by words in the same document. This assumption is based on the observation that words and entities are often focused on similar topics in a document. In another way, if more words in a document are focused on a topic, more entities in the same document will also focus on the same topic. In this way, although the modeling of words and entities are separate, there's strong coupling between them. Since both words and entities have probabilities to be assigned to any topic at any level, it is obvious that each topic is consisted of a distribution over words and a distribution over entities.

As the same as we have mentioned in hPAM, our model also has the ability to acquire topic hierarchies and their correlations which are extremely valuable in calculating entity correlations later. Our model also adopts the fully-connected four-level tree structure to model topic hierarchy and correlations. The graphic model is shown in Fig.2 and the generative process is as follows:

- For each of the $1 + T + t$ topics k:
  1. Draw word distribution $\phi_k \sim Dirichlet(\beta)$
  2. Draw entity distribution $\tilde{\phi}_k \sim Dirichlet(\tilde{\beta})$
- For each of the M documents m
  1. Draw a distribution over super-topics $\theta_{rm} \sim Dirichlet(\alpha_r)$
  2. For each of the T super-topics, draw distributions over sub-topics $\theta_{sm} \sim Dirichlet(\alpha_s)$
- For each of the N_m words w_{mn} in document m
  1. Draw a super-topic $z_{rmn} \sim Multinomial(\theta_{rm})$
  2. Draw a sub-topic $z_{smn} \sim Multinomial(\theta_{z_{rmn}m})$
  3. Draw a level $l_{mn} \sim Multinomial(\tau_{z_{rmn}z_{smn}})$
     If $l_{mn} = 0$, draw word $w_{mn} \sim \phi_0$
     If $l_{mn} = 1$, draw word $w_{mn} \sim \phi_{z_{rmn}}$
     If $l_{mn} = 2$, draw word $w_{mn} \sim \phi_{z_{smn}}$
- For each of the E_m entities e_{mn} in document m
     Draw path and level $< \tilde{z}_{rmn}, \tilde{z}_{smn}, \tilde{l}_{mn} > \sim Multinomial(C_{z_{rm}z_{sm}l_m})$
     If $\tilde{l}_{mn} = 0$, draw entity $e_{mn} \sim \tilde{\phi}_0$
     If $\tilde{l}_{mn} = 1$, draw entity $e_{mn} \sim \tilde{\phi}_{\tilde{z}_{rmn}}$
     If $\tilde{l}_{mn} = 2$, draw entity $e_{mn} \sim \tilde{\phi}_{\tilde{z}_{smn}}$

where $C_{z_{rm}z_{sm}l_m}$ is number of words in document m assigned to super-topic z_r, sub-topic z_s and level l.

The Gibbs Sampling approach is applied to perform the parameter estimation. In our model, the super-topic z_{rw}, sub-topic z_{sw} and level l_w are needed to be sampled for every word w. The sampling process follows the formula (2):

$$p(z_{rw} = x, z_{sw} = y, l_w = l | \bullet) \propto \frac{N_x^{(d)} + \alpha_x}{N^{(d)} + \sum_{x=1}^{T} \alpha_x} \frac{N_{xy}^{(d)} + \alpha_{xy}}{\sum_{y=1}^{t} N_{xy}^{(d)} + \sum_{y=1}^{t} \alpha_{xy}} \qquad (2)$$

$$\frac{N_{xy}^l + \gamma}{\sum_{l=1}^{3} N_{xy}^l + 3\gamma} \frac{N_{kw} + \beta}{\sum_w N_{kw} + V\beta}$$

where N_{kw} is the times word w assigned to topic k, N_{xy}^l is the number of times super-topic x and sub-topic y lead to level l, $N^{(d)}$ is the number of total words in document d and other parameters are the same as mentioned before.

For entities, the super topic $\tilde{z}_{re}$, sub topic $\tilde{z}_{se}$ and level $\tilde{l}_e$ also need to be sampled. A multinomial distribution parameterized by the fraction of choices

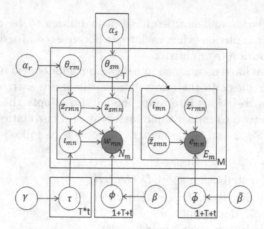

Fig. 2. Graphic Representation of hETM

of words in the same document is applied and the sampling process for entities follows the formula (3):

$$p(\widetilde{z}_{re} = x, \widetilde{z}_{se} = y, \widetilde{l}_e = l|\bullet) \propto \frac{N_{xyl}^{(d)}}{N^{(d)}} \frac{N_{ke} + \widetilde{\beta}}{\sum_e N_{ke} + E\widetilde{\beta}} \tag{3}$$

where N_{ke} is the times entity e assigned to topic k, $N_{xyl}^{(d)}$ is the number of words assigned to super-topic x, sub-topic y and level l in document d, E is the total number of entities in the text corpus.

Dirichlet parameters associated with super-topics are calculated similar to (1). It is important to note that when counting the number of occurrences of super-topics and sub-topics, we have to include the entity part. Yet those dirichlet parameters could only indicate correlations between super-topics and sub-topics. What we need is correlations between any topic pairs to model entity correlations. Suppose that x represents the super-topic and y represents the sub-topic, we have

$$p(y|x) = \frac{\alpha_{xy}}{\sum_{y_i} \alpha_{xy_i}}, \quad p(x|y) = \frac{\alpha_{xy}}{\sum_{x_i} \alpha_{x_iy}} \tag{4}$$

$$p(x_2|x_1) = \sum_{y_i} p(x_2|y_i)p(y_i|x_1), \quad p(y_2|y_1) = \sum_{x_i} p(y_2|x_i)p(x_i|y_1)$$

where x_1 and x_2 are instances of the super-topic and y_1 and y_2 are instances of the sub-topic. After we get the correlations between any topic pairs, we can obtain correlations between any pair of entities by formula (5) and (6):

$$Affinity(e_i, e_j) = \frac{p(e_i|e_j) + p(e_j|e_i)}{2} \tag{5}$$

$$p(e_i|e_j) \propto \sum_t [p(e_i|t)p(t|e_j) + \sum_{t'} p(e_i|t')p(t'|t)p(t|e_j)] \tag{6}$$

where t' represents set of topics that strongly correlated with topic t. From (6), we are aware that correlations between a pair of entities are contributed by two aspects: the first part is how likely they both appear in the same topic and the second part is how likely they appear in two different but correlated topics. We select the strongly correlated topics t' according to the conditional probability $p(t'|t)$ and we'll show in Section 4 that the choice of different number of t' will have some impact on experimental results.

4   Experiments and Analysis

In this section, we conduct experiments on real-word dataset to demonstrate the effectiveness of our proposed model. We took CI-LDA, corrLDA and corrLDA2 used in [5] as well as LDA to be our base line models. We treated entities as the same as ordinary words in LDA because this model doesn't distinguish entities and words. For hETM, we applied a fully connected four-level DAG structure with 30 super-topics and 60 sub-topics and initialized hyper-parameters as $\alpha_r = \alpha_s = \gamma = 0.1, \beta = 0.02$. For the base line models, we set their topic numbers to be 91(the same as total topic numbers of hETM) and empirical values of $\alpha = 50/K, \beta = 0.01$ were determined as their hyper-parameters. We ran 500 Gibbs Sampling iterations to fit those models as we found 500 iterations to be sufficient for convergence by monitoring the perplexity every 50 iterations.

The time complexity of hETM in Gibbs Sampling procedure is $O(TtlNI)$, where T, t and l are the number of super-topics, sub-topics and levels in hETM. N is number of words in the corpus and I represent for number of Gibbs Sampling iterations. Compared to LDA, whose time complexity is $O(NKI)$, where $K = T + t + 1$,we notice hETM takes more time to fit due to the hierarchy of learned topics.

4.1   Dataset

News of BBC for different countries, e.g, United Kingdom (UK), United States (US) and China (CN) from January 1st, 2012 to May 17th, 2012 were collected in advance. After preprocessing, we got 3000 documents altogether. We extracted entities in this corpus by making use of an open source tool Zemanta[3] , which is a high performance content analyzer capable of extracting entities and linking them to Wikipedia entries. We removed stop words as well as words that occur less than 3 times in the corpus and we also removed entities that appear in less than 5 documents. Statistics of BBC dataset after preprocessing is shown in Table 1.

4.2   Topic and Entity Correlation Learning

To investigate the quality of topics discovered by hETM, we show 4 topics in Table 2 with the most likely 10 words and their probabilities above as well as 6 entities and their probabilities below. The titles of topics were labeled by

---

[3] http://www.zemanta.com/

Table 1. Statistics of BBC Dateset

Document Count	3000
Unique Words	17742
Unique Entities	2439
Total Words	558656
Total Entities	172934

ourselves. As we can see, for each topic, the entity probabilities seem to be higher than ordinary word probabilities, which accords with what we observe in Table 1 that each entity appears at an average of 70.9 times in the corpus while each word appears only 31.5 times on average. Besides, we can notice the entity distribution contains profound meanings which could give great help for the understanding of topics. This informs us the separate modeling for entities and words is senseful for topic modeling.

For all of the discovered topics, we calculated correlations between any two topics according to Equation (4) and the correlations between entity pairs according to Formula (5) and (6). For the strongly correlated entity pairs, we connected them with a line and the length of the line indicates the degree of their correlation. Shorter lines represent for more closely relatedness while longer lines mean not so closely related. We show entities related with *Afghanistan, Economy, Hacking* and *Health* in Fig. 3. For the entity *Afghanistan*, the highly related entities include places of Afghanistan (LAshkar Gah, Helmand, etc), military base (Camp Bastion) and the Secretary of State for Defence (Philip Hammond). For the entity *Hacking*, the strongly correlated entities are people related to News of the World phone-hacking scandal, including victim (Milly Dowler), barrister (Tom Crone) and so on. We did a thorough observation for all the correlated entities in this figure by means of finding entities' corresponding entries in Wikipedia and confirmed the rationality of the connection.

4.3 Missing Entity Prediction

The task of missing entity prediction is that given a news article with one entity missing, we are trying to predict what the missing entity is. An example of this task is shown in Table 3. The first row is a fragment excerpted from a news article with one missing entity filled with "***" instead. The second row contains other entities contained in this news fragment and the third row contains top 9 possible entities we predicted. We mark the hit entity with a "*" at the top right.

In this task, four base models were fitted to be compared with our approach. We carried out the experiment in two ways: (1) Use only other entities and the obtained entity correlations. (2) Use other entities and all words in the article as context information without leverage the entity correlations.

We denote the missing entity as e_x and other entities as e_i, if we only use other entities and their correlations to predict the missing entity, the probability of

Table 2. Examples of topics discovered by hETM

Titanic		Olympic		Business		Traffic	
cruise	0.016632	athlete	0.015753	deal	0.012193	transport	0.016259
board	0.011929	event	0.014559	based	0.011531	train	0.012441
coast	0.011257	team	0.014321	overseas	0.010471	route	0.012318
april	0.009073	part	0.012968	international	0.010338	network	0.011702
sank	0.006889	organiser	0.007877	group	0.010338	line	0.010101
survived	0.006889	day	0.006365	share	0.007555	lane	0.008254
hit	0.006721	world	0.006365	largest	0.006760	journey	0.007884
class	0.006385	start	0.006047	operation	0.006363	project	0.007761
night	0.006050	venue	0.005490	brand	0.006098	miles	0.006653
lifeboat	0.005714	race	0.005490	buy	0.006098	speed	0.006529
Ship	0.032254	Game	0.046540	Company	0.024915	London	0.025866
The Titanic	0.019488	Olympic	0.030152	Business	0.014049	Road	0.012688
Passengers	0.015120	London	0.025617	Firm	0.014049	Traffic	0.010963
Crew	0.011425	Sport	0.011298	Market	0.009676	HS2	0.009855
Boat	0.008401	Stadium	0.007161	Chinese	0.007820	Travel	0.008500
Atlantic	0.007225	Football	0.005729	Investment	0.006628	Passengers	0.007884

* Upper part of each topic shows the most likely words with their probabilities and lower part shows the most likely entities with their probabilities. The title of each topic is label by ourselves.

Table 3. Missing Entity Prediction Example

"there will be a need for more planes to come from the developing economies but the government should be looking at the number of short-haul flights in the south east, particularly ***," he said. "if passengers could be persuaded to use high-speed rail there would not be so many of those short-haul flights and those slots could be made available."

Contained Entities: Economies, Government, Passengers, High-speed rail

Predicted Entities: UK, Heathrow*, Government, Passengers, Security, Green, EU, BAA, London

e_x appears in this news article could be estimated as $p(e_x) \propto \sum_i p(e_x|e_i)$ where $p(e_x|e_i)$ in our model is estimated by Formula (6) and we set the number of t' to be 2. In other base models, it is estimated as $p(e_x|e_i) = \sum_t p(e_x|t)p(t|e_i)$. If we use other entities and all words as context information to predict the missing entity without taking entity correlations into consideration, the probability of entity e_x could be estimated as $p(e_x|d) = \sum_t p(e_x|t)p(t|d)$ where the $p(t|d)$ of test articles is obtained by resampling using learned word distribution $p(w|t)$ and entity distribution $p(e|t)$ from the training set.

In this experiment, we randomly chose 200 articles to be our test set and learned topics and topic correlations from other 2800 articles. For each of these

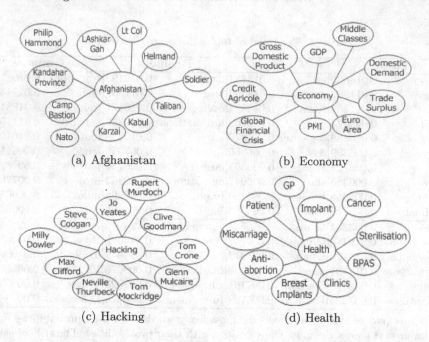

(a) Afghanistan   (b) Economy

(c) Hacking   (d) Health

Fig. 3. Entities related with (a)Afghanistan (b) Economy (c) Hacking (d) Health. Shorter lines represent for more closely relatedness while longer lines mean not so closely related.

test articles, we removed an entity randomly and then made the predicted entity list according to $p(e_x)$. The rank of each hit entity was recorded to calculate metrics to evaluate the performance. The metrics include *Mean Reciprocal Rank (MRR)* and *Average Rank (AveRank)* of the 200 hit entities for the test news articles. The results are shown in Table 4.

From the results, we can see missing entity prediction using entity correlations could achieve better results than using other entities and words as contexts with respect to both the MRR and Average Rank metrics in all of the five models with only one exception: the corrLDA model about the Average Rank metric. It indicates that the entity correlations extracted by the topic model based approaches are effective. Furthermore, as we can see in Table 4, although the LDA model performs well with regard to MRR metric, it's not satisfactory at the Average Rank metric and corrLDA model could perform the best with regard to Average Rank metric yet not good at the MRR metric. Instead, our proposed model hETM does not perform the best using neither the MRR nor the Average Rank metric, it is only a little worse than the optimum. Most important of all, it achieves preferable results with regard to both of these two metrics. We ran the experiment again by choosing another 200 articles randomly as test set and it demonstrated similar results. It proves that our approach is superior and stable at missing entity predictions.

Table 4. Missing Entity Prediction

	Only Entity		Words+Entity	
	MRR	AveRank	MRR	AveRank
LDA	0.138	151.06	0.122	182.95
hETM	0.128	148.17	0.116	156.67
CI-LDA	0.117	199.88	0.092	209.69
corrLDA	0.115	146.70	0.109	145.53
corrLDA2	0.107	167.51	0.111	180.0

Table 5. Entity Co-participance Prediction

	Accuracy(%)
LDA	57.12
hETM	65.26
CI-LDA	58.19
corrLDA	59.95
corrLDA2	64.86

4.4 Entity Co-participance Prediction

We further carried out experiments on entity co-participance prediction. The objective of this experiment is to predict whether two entities would co-appear in the future event based on their correlations acquired in the learning step. We divided our BBC news dataset into two parts: news from January to April was treated as training set and that of May as test set. Entity pairs did not co-appear in training set but co-appeared in the test set were first selected as *"true pairs"*. *"false pairs"* are those never co-appeared in neither training set nor test set. In order to control the quantity, we randomly permutated the second entity of each true pair to generate a false pair and as a result, we got 15824 true pairs in total and the same number of false pairs.

In this experiment, LDA, CI-LDA, corrLDA and corrLDA2 were still the base models to be compared with. We shuffled the 31648 entity pairs and calculated the entity correlations for each pair. For our model, the correlation calculation followed (5) and (6) and we set the number of t' to be 4. For other models, the entity correlations were calculated as $Affinity(e_i, e_j) = (p(e_i|e_j) + p(e_j|e_i))/2$ and $p(e_i|e_j) = \sum_t p(e_i|t)p(t|e_j)$. We saved the median of the affinity value of all these pairs and predicted pairs whose affinity above the median as true and below the median as false. The prediction accuracy is shown in Table 5.

In table 5, we notice that our model could gain the highest accuracy by 14.1% improvement over LDA and 8.9% improvement over corrLDA. Compared with Table 4, we find although LDA and corrLDA perform well at missing entity prediction, they are not good at entity co-participance prediction while corrLDA2 is weak at missing entity prediction but shows good ability at entity co-participance prediction. However, our proposed hETM proves to be outstanding in both experiments.

4.5 Impact of Number of Correlated Topics

As we can see in Formula (6), when we calculate the correlations of entity pairs, there exists a parameter t' indicating strong correlated topic set used to calculate entity correlations. In the experiment of missing entity prediction, we set number of correlated topics to be 2 and in entity co-participance prediction, we set it as 4. We evaluated its impact under different settings of topic structures using the

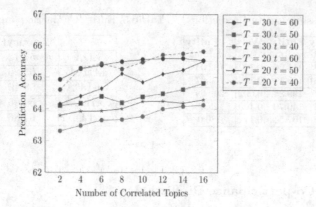

Fig. 4. Entity co-participance prediction accuracy with different choices of number of correlated topics under different topic settings

entity co-participance prediction experiment. The prediction accuracy is shown in Fig.4. We set the number of t' from 2 two 16 and T represents for the number of super-topics and t is the number of sub-topics in this experiment.

In Fig.4, we can see the prediction accuracy tends to increase when number of t' becomes larger in almost all the topic settings because bigger value means to bound more topics and thus will strengthen the relatedness between correlated entity pairs. Besides, we noticed that topic numbers and structures plays a significant role in this experiment as we can see that when $T = 30$ $t = 60$ and $T = 20$ $t = 40$ achieve high accuracy than other circumstances. These results show that setting the appropriate topic structures as well as a good choice of number of t' can yield better performance.

5   Conclusion

In this paper we present a novel statistical topic model based approach to model the entity correlations by not only considering entity-topic dependencies but also cooperating with topic-topic correlations. We propose the hETM model inspired by Hierarchical Pachinko Allocation Model (hPAM) to acquire topic correlations and at the same time, the generative processes of entities and words are set to be separated. To evaluate the impact of considering the topic correlations when measuring entity correlation, we have done a series of experiments including missing entity prediction and entity co-participance prediction and the results demonstrate its effectiveness.

Our future work will make use of the more detailed information of entities we can get from hETM (for example, a generalized entity or a specific entity) to acquire more rational entity correlations. We are also interested in investigating more intensive nonparametric approaches instead of defining the DAG structure manually to learn more suitable topic correlations for different datasets.

References

1. Elmacioglu, E., Lee, D.: On six degrees of separation in DBLP-DB and more. SIGMOD Record 34(2) (June 2005)
2. Kleinfeld, J.: Could it be a big world after all? the "six degrees of separation". Myth. Society (2002)
3. Blei, D., Ng, A., Jordan, M., Lafferty, J.: Latent dirichlet allocation. Journal of Machine Learning Research 3(993-1022) (2003)
4. Teh, Y.W., Jordan, M.I., Beal, M.J., Blei, D.M.: Hierarchical Dirichlet processes. Technical Report, Department of Statistics, UC Berkeley (2004)
5. Newman, D., Chemudugunta, C., Smyth, P.: Statistical entity-topic models. In: Proceedings of the 12th ACM SIGKDD International Conference on Knowledge Discovery and Data Mining (2006)
6. Shiozaki, H., Eguchi, K., Ohkawa, T.: Entity Network Prediction Using Multi-type Topic Models. IEICE-Transactions on Information and Systems E91-D(11), 2589–2598 (2008)
7. Bhattacharya, I., Getoor, L.: A latent dirichlet model for unsupervised entity resolution. In: Sixth SIAM Conference on Data Mining (2006)
8. Shu, L., Long, B., Meng, W.: A Latent Topic Model for Complete Entity Resolution. In: Proceedings of the 2009 IEEE International Conference on Data Engineering, pp. 880–891 (2009)
9. Kataria, S.S., Kumar, K.S., Rastogi, R.R., et al.: Entity disambiguation with hierarchical topic models. In: Proceedings of the 17th ACM SIGKDD International Conference on Knowledge Discovery and Data Mining (2011)
10. Dai, A.M., Storkey, A.J.: The grouped author-topic model for unsupervised entity resolution. In: Honkela, T. (ed.) ICANN 2011, Part I. LNCS, vol. 6791, pp. 241–249. Springer, Heidelberg (2011)
11. Guo, J., Xu, G., Cheng, X., Li, H.: Named entity recognition in query. In: Proceedings of the 32nd International ACM SIGIR Conference on Research and Development in Information Retrieval (2009)
12. Du, J., Zhang, Z., Yan, J., et al.: Using search session context for named entity recognition in query. In: Proceedings of the 33rd International ACM SIGIR Conference on Research and Development in Information Retrieval (2010)
13. Xu, G., Yang, S.-H., Li, H.: Named entity mining from click-through data using weakly supervised latent dirichlet allocation. In: Proceedings of the 15th ACM SIGKDD International Conference on Knowledge Discovery and Data Mining (2009)
14. Blei, D., Lafferty, J.: A correlated topic model of Science. The Annals of Applied Statistics 1(1), 17–35 (2007)
15. Tam, Y.-C., Schultz, T.: Correlated latent semantic model for unsupervised LM adaptation. In: IEEE Intl. Conf. on Acoustics, Speech and Signal Processing, pp. 41–44 (2007)
16. Li, W., McCallum, A.: Pachinko allocation: DAG-structured mixture models of topic correlations. In: Proceedings of the 23rd International Conference on Machine Learning, pp. 577–584 (2006)
17. Blei, D., Griffiths, T., Jordan, M., Tenenbaum, J.: Hierarchical topic models and the nested Chinese restaurant process. In: Advances in Neural Information Processing Systems 16. MIT Press, Cambridge (2004)
18. Mimno, D., Li, W., McCallum, A.: Mixtures of hierarchical topics with Pachinko allocation. In: Proceedings of the 24th International Conference on Machine Learning, pp. 633–640 (2007)

High Quality Microblog Extraction
Based on Multiple Features Fusion
and Time-Frequency Transformation*

Min Peng, Jiajia Huang, Hui Fu, Jiahui Zhu, Li Zhou, Yanxiang He, and Fei Li

Computer School, Wuhan University, Wuhan, China
{pengm,huangjj,fuhui,zhujiahui,zhouli_88,yxhe}@whu.edu.cn,
kevin.lifei@gmail.com

Abstract. Online social media exhibits massive social event relevant messages. Some of them contain useful and meaningful information, while others might not worth reading. In this paper, for a given social event, we focus on extracting high quality information from massive social media messages, since the extracted information has valuable textual content, and is widely propagated and posted by authority. We propose an extraction framework to get high quality information by considering different features globally in social media. Specially, in order to reduce computing time and improve extraction precision, some important social media features are employed and transformed into wavelet domain and fused further, to get a weighted ensemble value. A large scale of Sina microblog dataset is used to evaluate the framework's performance. Experimental results show that the proposed framework is effective to extract high quality information.

Keywords: Information Extraction, Feature Fusion, K-Dimensions Feature, Wavelet Coefficient, EM Algorithm.

1 Introduction

With Web 2.0 technologies prevailing, more and more people like to share news and express opinions on various social events through social media platforms such as *Twitter*, *Facebook* and *Blogger*. According to statistics, over 200 million tweets are generated on Twitter per day [17]. Social media has developed into a platform for sharing and collecting real-world event information.

Social media information always relate to certain social events, so they can be classified into kinds of topic sets. For a topic set, qualities of messages posted by different users are quite different. Some describe a detailed process of the event or express poster's opinion clearly, while others contain incomprehensible text, Internet catchphrase or even nonsense advertisements. The former have more

* This material is based on the work supported by National Science Foundation of China (NSFC) under Award 61070083 as well as the Key Technologies R&D Program of Wuhan under Award 201210421135.

X. Lin et al. (Eds.): WISE 2013, Part II, LNCS 8181, pp. 188–201, 2013.

valuable information, while the latter may not. From the other point of view, people like to copy similar content from authority users and post it on their own homepages. That means, for a social event, the topic set exists massively redundant messages. However, in many situations, end users want to catch a rough sketch of an event in a short time via a few of selected and prioritized information. In this paper, we address this problem by extracting high quality information from a topic set which contains massive event-relevant information.

Here, high quality information in a topic set is defined as a small subset of messages, which contains five characteristics, i.e., 1) Relates to the topic, 2) Describes the event or expresses the opinion very clearly, 3) Generates widely attention among users, 4) Posted by influential people and 5) With lower redundant contents. For a certain topic, high quality information extracting is a significant application in emergency response, viral marketing, disease outbreaks, social management, social event predicting, etc.

However, there are some challenges for extracting the high quality information from an event set. We list some as follows:

- *How to effective extract high quality information with lower redundant content and more abundant information?*
- *How to measure the composite impacts of content, user authority and information propagation (such as comment or forward messages) when extracting high quality information?*
- *How to extract high quality information efficiently when topic set is very large?*

In this study, to meet the above challenges, we build a high quality information extraction framework for microblog data. Specifically, based on statistics analysis (section 3), microblog information is verified to have high content redundancy, while the redundant contents have significant differences in other features such as *comment number* and *forward number*. Hence, in this paper, the proposed framework will consider multiple features such as comment number, forward number, URL, textual content and *follower number*. We address this task with three steps. First, construct a K-dimensions feature matrix as an extraction basis matrix. Second, transform the K-dimensions feature into time-frequency domain to reduce computing time and improve extraction quality. Third, estimate each feature's contribution to the information quality based on EM algorithm.

The main contributions of this paper are: First, define a high quality social media information extraction problem and detail the task. Then, reveal the internal relationships among different features through statistic analysis. At last, propose a novel extraction framework to reduce the content redundancy among high quality information.

Since there is no existing dataset directly fitting for the evaluation of the proposed method here, the framework is evaluated by downloading dataset (300,000 micrioblogs comprised with eight hot events) from Sina Microblog, which is the most popular microblogging website in China. There are several key insights in our experimental results. (1) The proposed framework could extract high quality information with lower content redundancy. (2) Based on time-frequency transformation, time consuming reduces to almost an half. (3) Based on performance evaluation, the experimental result based on our framework is better than several classic and traditional methods.

2   Related Work

During the last couple of years, many research in the field of social media focused on studying information extraction [3, 8] and summarization [11, 21] or otherwise presentation of Twitter event content [4]. Traditional approaches mainly consider textual content of information. These methods extract topic relevant information based on LDA model [5], TFIDF model [11] or other topic models [12, 18]. Most of the above mentioned works only refer to one or two social media feature, such as poster influence [5, 9].

However, social media information contains not only media content such as texts, pictures and videos, but also additional social attributions such as comment number and forward number, as well as authority attributions such as influence of poster. It may get totally different results between two messages with same content. Therefore, besides content features, other variety features, such as poster influence, tags, URL and posting time also have significant values for improving extraction precision. For example, [13] identified events by combining a variety of social media features based on a few domain-specific similarity metrics and a weighted ensemble clustering. [2, 10] proposed a sampling method with compress sampling idea by considering some features.

Many existing social media information extraction and sampling approaches haven't considered time consuming of the extracting processes, others haven't analyze some of the features which were really important for information extraction. In this paper, we try to model the relationships among more relevant and necessary features (Section 4). Besides, to reduce the time consumption and the complexity of the computing, we transform the features into time-frequency domain (Section 5). Through spaces transformation, we utilize EM algorithm to get a weight ensemble score to produce high quality information.

3   Problem Definition

In this section, we formally define the problem and propose a framework of K-dimensions feature fusion based on time-frequency transformation (KD-FF_TF).

We start by assuming that there is a set of social media messages Γ that relates to a certain event. These messages can be from *Twitter*, *Facebook* or other platforms which concludes social structure, textual content and poster authority or other features. Now we want to extract a subset of high quality information for representing a rough sketch of this event, which makes people have a concise knowledge on the entire process of the event and people's opinions. Therefore, the task of high quality microblog extracting can be defined on the basis of different features.

For the event messages set Γ, each item (e.g.,message) $d_i \in \Gamma$ $(i = 1, 2, ..., N)$ has K feature scores of the form $F_k = \{f_{i1}, f_{i2}, ..., f_{iK}\}$. For the N items, we construct a K-dimensions feature matrix $F = \{f_{ij}\}, (i = 1, 2, ..., N; j = 1, 2, ..., K)$. Our goal is to extract high quality information by fusing all these features with matching weights and get the top-M items. This task involves four steps: (1) features analysis and K-dimensions feature matrix F quantification, (2) time-frequency domain transformation of each feature vector F_k $(k = 1, 2, ..., K)$, (3)time-frequency domain coefficients fusion based on EM algorithm, (4) new signal recomposing and high quality microblog information producing.

4   Analysis and Measure of K-Dimensions Feature Matrix

In this section, we analyze some relevant features to detect the internal relationships among them and then quantize them to generate a K-dimensions feature matrix.

4.1   Feature Analysis

First, the mutual effects of features are explained from some statistical properties.

In this paper, three types of attributions are regarded as important features. These attributions are important in many kinds of social media analysis tasks, such as user influence analysis [6], predication based on content [14] or event information extractions [8, 10]. These attributions are listed in Table 1.

Table 1. Five most relevant features

Attribution	Feature
Social Attribution	Comment / Forward / URL number
Content Attribution	Topic relevance
Authority Attribution	Follower number

Via several statistical analyzing over eight experimental datasets (detail described in 6.1), we get the following observations.

1. There are thousands of URLs in an event based microblog set. Per URL's occurrence frequency is no more than 5%. 20% of items own at least one URL. Besides, per URL's occurrence times satisfies power-law distribution. Detail statics results are shown in Fig. 1.
2. 10% to 25% items contain the most high frequency words or phrases (e.g., topic words or phrases).
3. Based on content calculation with TFIDF, 30% to 60% of items are highly similar to others.

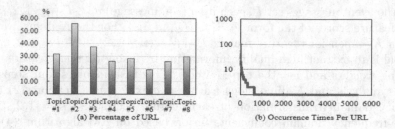

Fig. 1. (a) Percentage of items own at least one URL over eight sets; (b) Power-law distribution of per URL occurrence times over Topic #8

4. Large comment number usually correlates to large forward number. Large comment or forward number usually correlates to large follower number. The rate of *comment number/ forward number* of more than 99% items is less than 5. And, the rate of *follower number/(comment+forwar) number* of more than 50% items is less than 5.
5. There is a big difference among comment or forward number of different items with same content score. The largest comment number may be up to thousands while the smallest is 0. So as to the follower number.

From the above observations, we draw the following conclusions.

- URL is an important feature in social media information and satisfies power-law distribution (see observation (1)).Descriptions of an event or user's opinions are expressed as short texts in social media, detailed information usually is linked by URLs to videos or news websites.
- Event oriented information set has massive redundant textual content (see observations (2, 3)). Therefore, it is hard to extract high quality information effectively only based on content feature.
- Other features could be treated as supporting parts in content-based high quality information extracting (see observation 5). That means, when redundant information cannot be identified via content feature, other features are useful indicators for extracting.
- There are intricate relationships among all features. Each one is mutually affected by others (see observations (4, 5)). That is to say, the high quality information extraction task can't be well accomplished only based on content extraction method [3, 15].

Therefore, in the task of high quality information extraction defined in this paper, we construct a K-dimensions feature Matrix F by considering all of these features.

4.2 Measure of Matrix F

Next, three methods are developed to quantize the item score in the occurrence of each feature.

Method 1: score(f_{ik})=$logN_{ik}$, where N_{ik} is the ith item's number of kth feature, $k = \{comment, forward, follower\}$

Method 2: score(f_{ik})=$log(\sum_{URL_j \in d_i} w_j)$, where w_j is the jth URL's frequency, $k = \{URL\}$.

Method 3: score $(f_{ik}) = \sum_{(term_j \in d_i)} t_{ij}$, where t_{ij} is the $tf * log(idf + 1)$ value of jth term when occurring in i^{th} item.

Based on the three methods, a quantized score of item i in the occurrence of feature k is generated. By normalization the scores into a uniform interval, we generate a K-dimensions feature matrix $F \in R^{N \times K}$. Each element in matrix F denotes an observed value of ith item in the occurrence of the kth feature.

5 K-Dimensions Feature Fusion Based on Time-Frequency Domain

This section introduces the K-dimensions feature fusion framework based on time-frequency domain transformation.

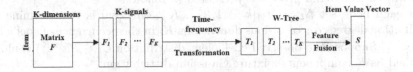

Fig. 2. K-dimensions feature fusion framework: Time-frequency transformation and fusion of K-dimensions feature

Set K-dimensions feature matrix as $F \in R^{N \times K}$, each dimension vector $F_k \in R^{N \times 1}(k = 1, 2, ..., K)$ is regarded as a one-dimension signal (items in F is ordered by posted time). In order to effective fuse these features, each feature is transformed into wavelet domain to get wavelet trees $T_k(k = 1, 2, .., K)$, the transformation process is described in Section 5.1. Section 5.2 discusses the K-dimension feature fusion algorithm based on wavelet domain. Entire process is shown in Fig. 2.

5.1   Wavelet Transformation

In this paper, the K-dimensions feature are transformed into time-frequency domain via wavelet transformation [1]. Wavelet is mathematical function that is very useful in representing data due to various properties like compact support, vanishing moments, dilating relations, etc. [1]. Suppose wavelet basis function is h, fast wavelet transformation of each dimension k results in a wavelet tree T_k with L coefficient vector nodes. Each node $t_{kl}(l = 1, 2, ..., L)$ has a value vector $W_{kl}(\in R^{n^l \times 1})$ to denote a set of corresponding coefficients (n^l denotes the number of coefficients in l^{th} node of T_k, the lth nodes in all the T_k have a same n^l). Here, the basis function h could be Haar wavelet, Daubechies$N$ wavelet or other orthogonal wavelet [1].

All these trees $T_k(k = 1, 2, .., K)$ have a homogeneous structure. Thus, we get a coefficient set $T = \{W_{kl}|k = 1, ..., K; l = 1, 2, ..., L\}$. To facilitate the following discussion, all the coefficients W_{kl} in set Γ are also expressed as $C = \{C_l|l = 1, 2, ..L\}$, where $C_l \in R^{n^l \times K}$ and $C_{lk} = W_{kl}$.

5.2   Feature Fusion in Wavelet Domain

As each feature vector is transformed into wavelet domain within L nodes, we need to fuse K-dimension features through fusing each group of K coefficient vectors of lth nodes in all Tree $T_k(k = 1, 2, ..., K)$ respectively. Then, with L fused coefficient vectors $C_l^*(l = 1, 2, ..., L)$, we reconstruct a new signal $S(\in R^{N \times 1})$, where each element in S is a comprehensive value of each item.

First, we analyze K-dimensions feature fusion with K coefficient vectors $C_{lk}(k = 1, 2, ..., K)$ of lth nodes in Tree T_k. Here, EM algorithm is employed to estimate *contribution degree* of each coefficient vector C_{lk}. EM algorithm is a classical method in parameter evaluation when its distribution contains latent variables [7]. The detail process is presented as follows.

For each node set $N_l = \{t_{kl}|k = 1, 2, ..., K\}$, our goal is to determine the contribution degrees α_{lk} of various dimensions K_l in the occurrence of coefficient matrix C_l. Suppose the observed coefficient distribution $C_l(\in R^{n^l \times K})$ could be described as K components mixture Gaussian distribution.

Thus, based on K components mixture model, probability of occurrence of a fused coefficient c_{li} can be written as:

$$f(c_{li}) = \sum_{k=1}^{K} \alpha_{lk} f_{lk}(c_{lki}) \tag{1}$$

where $\alpha_{lk} \geq 0, \sum_{k=1}^{K} \alpha_{lk} = 1$. $f_{lk}(c_{lki})$ denotes a probability density function of the kth component distribution $N(\mu_{lk}, \sigma_{lk}^2)$. α_{lk} denotes a contribution degree of the $k^{th}(k = 1, 2, ..., K)$ component in the lth node set N_l.

Set $\Phi_l = \{\Theta_l; \Lambda_l\} = \{\alpha_{l1}, ..., \alpha_{lK}; \mu_{l1}, ..., \mu_{lK}, \sigma_{l1}^2, ..., \sigma_{lK}^2\}$ as the parameters vector of mixture distribution. Λ_l denotes a vector of the mixture model's distributions parameters of the lth node set.

Then, formula (1) could be expressed with parameters Φ_l as follows (formula (2))

$$f(c_{li}|\Phi_l) = \sum_{k=1}^{K} \alpha_{lk} f_{lk}(c_{lki}, \mu_{lk}, \sigma_{lk}^2) = \sum_{k=1}^{K} \alpha_{lk} \frac{1}{\sqrt{2\pi\sigma_{lk}^2}} e^{-\frac{(c_{lki}-\mu_{lk})^2}{2\sigma_{lk}^2}} \tag{2}$$

Hence, the likelihood function over the entire l^{th} node set N_l is given as:

$$L(\Phi_l) = \prod_{i=1}^{n^l} f(c_{li}|\Phi_{lK}) = \prod_{i=1}^{n^l} \sum_{k=1}^{K} \alpha_{lk} \frac{1}{\sqrt{2\pi\sigma_{lk}^2}} e^{-\frac{(c_{lki}-\mu_{lk})^2}{2\sigma_{lk}^2}} \tag{3}$$

In order to employ maximum likelihood method to estimate the parameter vector Φ_l, formula (3) is therefore given by:

$$l(\Phi_l) = ln(L(\Phi_l)) = \sum_{i=1}^{n^l} ln(\sum_{k=1}^{K} \alpha_{lk} \frac{1}{\sqrt{2\pi\sigma_{lk}^2}} e^{-\frac{(c_{lki}-\mu_{lk})^2}{2\sigma_{lk}^2}}) \qquad (4)$$

In this paper, EM algorithm [7] is used as an iterative procedure for maximizing $l(\Phi_l)$. The EM algorithm of estimating vector Φ_l is presented as Algorithm 1.

Data: Node set N_l, coefficient matrix $C_l = \{C_{lk}, k = 1, 2, ..., K\}$

1 Initialize: assign a group of initial vaules to parameter vector Φ_l
$$\Phi_l^{(0)} = \{\alpha_{l1}^{(0)}, ..., \alpha_{lK}^{(0)}; \mu_{l1}^{(0)}, ..., \mu_{lK}^{(0)}, \sigma_{l1}^{2(0)}, ..., \sigma_{lK}^{2(0)}\};$$

2 **repeat**

3 | Calculate posterior probability w_{lki} of each element in C_l with formula (5);

$$w_{lki}^{(s)} = \frac{\alpha_{lk}^{(s-1)} f_{lk}(c_{lki}, \mu_{lk}, \sigma_{lk}^2)}{\sum_{t=1}^{K} \alpha_{lt}^{(s-1)} f_{lt}(c_{lti}, \mu_{lt}, \sum_{lt}^2)} \qquad (5)$$

Estimate new values for vector $\Phi_l^{(s)}$ by taking partial derivatives of formula (4) for each parameter in Φ_l;

4 **until** $|l^{(s)}\Phi_l - l^{(s-1)}\Phi_l| \le \varepsilon$;

5 Return mutual concentrations $\Theta_l = \{\alpha_{l1}, ..., \alpha_{lK}\}$.

Algorithm 1. EM estimation of parameter vector Φ_l

Based on Algorithm 1, a group of optimum parameters $\Theta_l = \{\alpha_{l1}, ..., \alpha_{lK}\}$ is produced to fuse K-dimension coefficients matrix C_l into 1-dimension coefficient vector C_l^* by linearly weighting with Θ_l as follows:

$$c_{li}^* = \sum_{k=1}^{K} \alpha_{lk} c_{lki} \qquad (6)$$

Each node set $N_l(l = 1, 2, ..., L)$ gets its 1-dimension coefficient vector $C_l^* = \{c_{li}^*|i = 1, 2, ..., n^l\}$.

Then, a new signal S is reconstructed with coefficient set $C^* = \{C_l^*|l = 1, 2, ..., L\}$ via wavelet inverse transformation. Each element s_i in S denotes a finial comprehensive value of item i. Entire process of K-dimensions feature fusion based on wavelet transformation is presented as Algorithm 2.

Since the information items N is usually quite large and goes into ten thousands, scalability of KD-FF_TF algorithm is important. Computing wavelet transform takes $O(N * j)$ for each dimension, where j is the wavelet decomposition level. Besides, estimating parameter vector Θ_l takes $O(s^l * K^3 * n^l)$ for each node set N_l, where s^l is iteration times. As each Θ_l is estimated independently, the EM estimation process could be parallel executed for all node sets $N_l(l =$

Data: K-dimensions feature matrix $F \in R^{N \times K}$
1 **for** $k=1$ to K do **do**
2   T_k = wavelet transformation $h(F_k)$;
3 **end**
4 **for** $l=1$ to L do **do**
5   Estimate fusion parameter Θ_l of the lth coefficient C_l with Algorithm 1;
6   Calculate the fused coefficient C_l^* with formula (6);
7 **end**
8 S=Inverse wavelet transformation $h(C^*)$;
9 **return** items new signal S

Algorithm 2. K-dimensions feature fusion based on time-frequency domain (KD-FF_TF)

$1, 2, ..., L$). So the EM estimation process takes only $\max(O(s^l * K^3 * n^l))$. Therefore, the overall complexity of KD-FF_TF algorithm is $O(N * j) + \max(O(s^l * K^3 * n^l))$. However, max $(n^l) \approx \frac{N}{2}$, the decomposition level $j \in [2, 8]$, while $s^l \in [30, 50]$, so $O(N * j)$ is much smaller than $\max(O(s^l * K^3 * n^l))$. Therefore the complexity of KD-FF_TF algorithm approximates to $\max(O(s^l * K^3 * n^l))$.

We address two problems in wavelet transformation adopting. First, Mallat algorithm [16] is used to carry out wavelet transformation. In addition,two-solution property of the Discrete Daubechies7 Wavelet (DW) and Haar Wavelet (HW) [1] is adopted as basis function with 5 levels decomposition respectively. Hence, based on the two wavelet basis functions, we get two methods: KD-FF_DW and KD-FF_HW.

6   Experimental Evaluation

6.1   Dataset and Setup

To evaluate the proposed high-quality information extraction framework, we utilized microblogs about social event posted in Sina Microblog. Via a crawler program, we collected mocroblogging data from January to May in 2013. The data is consisted of eight hot social events happened among those months and is classified into eight datasets. Each event dataset contains 10,000 to 100,000 relevant original microblogs and a similar number of involved users. The description of the datasets is shown in Table 2.

Table 2. Microblogs number and involved users number

Topic	#1	#2	#3	#4	#5	#6	#7	#8	Average
Microblog	73,337	9,603	26,519	78,814	101,978	32,613	19,282	20,421	45,320
User	59,257	8,571	20,072	68,760	87,577	30,201	17,176	18,670	38,785

Keywords of Topic #1 to Topic #8 are: *Two Sessions, Chinese First Lady Liyuan Peng, New Real Estate Policy, H7N9 Bird Flu, Ya'an Earthquake, Beijing Mist, Dead Pig* and *Funeral of Mrs. Thatcher.*

In this paper, several classic or basic techniques are employed to compare and evaluate the proposed extraction methods (as KD-FF_DW and KD-FF_HW in the rest of the paper). First of all, we use two current state-of-art microblog extraction methods, which were also used in [2] as baselines. One is Most Recent Tweets [2, 4] (as MR) method, and the other is the Most Tweeted URL [2] based (as MTU) method. Then, we use TFIDF method, which is often used as microblog summarization and extraction [3, 4, 9] baseline method. At last, we use a variant of our method: EM estimation only based on K-dimensions (as KD) feature matrix F. That is to say, the matrix F isn't transformed into time-frequency domain.

6.2 Experimental Results

Effect of Reducing Content Redundancy. We collected information redundancy statistics of top-M items results from different methods. As the most important feature of high quality information is textual content, we calculate variance ($variance = avg(\sum (s_i - avg(s_i)^2)))$ of top-M items, where s_i is the content score of the ith item. Higher variance denotes lower content redundancy. Take Topic #1, Topic #2, Topic #5 and Topic #8 as the examples, the results are shown in Fig. 3 (the higher is better).

From Fig.3, the best performance comes from our proposed methods. Their variance are higher than others. That means, our framework is effective in reducing redundant content. Especially, high quality information set produced

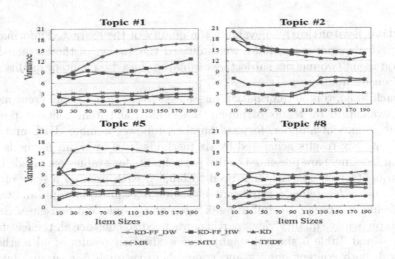

Fig. 3. Comparison of proposed methods against baselines. Results are shown for variance of top-M items with M ranging from 10 to 200 over four topics.

by KD-FF_DW method always has the least redundancy, namely, this set has the most abundance information. In addition, KD method also has well performance, because EM algorithm can estimate each features contribution degree reasonablely. Furthermore, when $M = 30$, our methods always have much higher variance value than KD method. The performances of the other three methods are bad and instable. Especially, the MTU results are always very poor when $M < 30$, because many highly similar microblogs are linked with a same high-occurrence URL. These microblogs are usually forwarded from an authority user.

Comparison of Time Consumption. In the second set of experiments, we compare time consumption of our method against KD method. As analyzed in section 5, computing complexity of an iteration of KD method is $O(K^3 * N)$. Table 3 presents the running time for the two methods respectively.

Table 3. Running time(sec.) of our framework and KD method over eight topic sets

Topic	#1	#2	#3	#4	#5	#6	#7	#8
KD-FF_TF	0.868	0.123	0.261	0.917	3.128	0.410	0.235	0.251
KD	1.940	0.225	0.679	2.00	5.317	0.826	0.484	0.562

It is easy to see that time consumption of our framework (e.g., KD-FF_DW and KD-FF_HD, their time consumption of an iteration have very slight difference.) is about half of the KD method. It is consistent with our theoretical analysis. As time consumption is a big deal in processing big data, our method shows its advantage in massive scale of data extraction.

Subjective Evaluation. To evaluate the quality of the extracted information of our methods against baselines, we evaluated the quality of the extracted information along two metric: subjective evaluation based on content quality and comprehensive quality.

For each topic, top-30 information messages are extracted by different methods for evaluation by teams of human assessors. Each item contains posters nickname, comment number, forward number, follower number, URL and textual content. Six results generated by six methods (two from our methods and four from baselines) are presented to five participants for evaluation. Grade from content quality view, participants need to labeled each item on a scale of 1-5, where a score of 5 signifies textual content strong relevant to its topic and worth reading while a score of 1 signifies weak relevant to topic and nonsense. Grade from comprehensive quality view, participants need to consider all the features.

Table 4 and Table 5 show the subjective evaluation results of all methods. In general, both content quality and comprehensive quality generated by our two methods and KD method are much better than the rest. It indicates the necessity of considering multiple features in extracting salient information.

Table 4. Content quality evaluation of different methods over eight event sets

Topic	#1	#2	#3	#4	#5	#6	#7	#8
KD-FF_DW	3.12	**3.53**	**3.37**	3.43	2.73	3.41	4.02	**3.55**
KD-FF_HW	3.43	3.42	3.22	3.50	2.93	3.47	**4.33**	3.17
KD	**3.76**	3.50	2.96	**3.66**	**2.97**	**3.6**	3.87	3.20
MTU	2.57	2.13	2.58	3.46	2.33	3.55	2.69	3.17
MR	2.59	2.28	2.51	2.72	2.70	2.74	3.29	2.47
TFIDF	2.51	2.52	2.55	3.74	2.32	3.13	3.17	3.35

Table 5. Comprehensive quality evaluation of different methods over eight event sets

Topic	#1	#2	#3	#4	#5	#6	#7	#8
KD-FF_DW	3.41	3.59	3.30	**4.03**	3.19	3.50	**4.00**	**3.50**
KD-FF_HW	3.93	3.58	**3.38**	3.92	**3.68**	3.75	3.93	3.40
KD	**4.02**	**3.62**	3.07	3.99	3.35	**3.86**	3.76	3.32
MTU	2.08	1.91	1.83	3.36	1.80	3.26	2.24	2.63
MR	2.13	2.10	1.93	2.59	2.08	2.42	2.64	1.90
TFIDF	2.40	2.19	2.01	3.65	1.73	2.92	2.45	2.69

Note that KD method and our methods have the mostly similar results, and it's hard to judge which is better directly. Therefore, *mean* and *variance* of eight scores over two kinds of metrics are calculated and presented in Table 6.

Table 6. Statics of scores comparison with different methods

	Content quality		Comprehensive quality	
	Mean	Variance	Mean	Variance
KD-FF_DW	3.395	0.136	3.565	0.081
KD-FF_HW	3.434	0.168	**3.696**	**0.052**
KD	**3.44**	**0.124**	3.624	0.119

As for the statics of scores, KD method performance is slightly better at content quality metric while KD-FF_HW method is better at comprehensive quality metric. Again note the mean of content quality for KD-FF_HW (3.434) is very close to KD (3.44). In a word, our framework doesn't have a significant advantage over KD method.

Here, we list several possible interpretations of why the subjective evaluation results are not very satisfactory. First, items with high content quality could acquire high score. But in terms of the whole high quality information set, the result set generated by KD method has more redundancy items with high content quality (testified by Fig. 3). So its whole content quality evaluation is better than ours. But these items have significantly difference on other feature such as comment/forward/follower number. So the comprehensive evaluation scores of them are amended. Second, in terms of content quality evaluation, there are only subtle differences between two methods. It may relate to some randomness of participant grading. Take Topic #6 for example, when set M as 15, we find result produced from KD-FF_HW has the maxima comprehensive score (3.92) while from KD method has a second score (3.64).

In a word, from the human evaluation, we can conclude that multiple features fusion is important to extract high quality information, and time-frequency transformation can improve the comprehensive quality of extracted information.

7 Discussion and Conclusion

In this paper, we proposed a high quality information extraction framework based on the idea of multiple features fusion. First, we formulated a high quality information extraction problem and analyze the internal relationships among different features. Then, we developed a framework of K-dimensions feature fusion based on time-frequency transformation (KD-FF_TF). Finally, along with a host of baselines, we evaluated the information generated by our framework on the basis of Sina Microblog datasets.

With sets of eight hot events containing 36,000 microblog messages and 31,000 involved users, we evaluated the proposed framework through content redundancy, time consuming and subjective evaluation. The proposed KD-FF_TF (both KD-FF_DW and KD-FF_HW) achieves better performance in reducing redundant content and cutting down almost half of the computing time compared with KD method. In addition, subjective evaluation through scored by human assessors, the proposed methods (e.g., KD-FF_DW and KD-FF_HW) get an improvement over some of the state-of-art methods.

Above all, our work provides the means to study salient content selection and presentation by extracting high quality information from microblogging data. There are still a number of additional interesting directions for future work on high quality information extraction. For example, how do other inexplicit features such as comment content and user tags impact on high quality extraction? Other future direction includes high quality extraction based on both multiple topics and multiple features.

Acknowledgments. This material is based on the work supported by National Science Foundation of China (NSFC) under Award 61070083 as well as the Key Technologies R&D Program of Wuhan under Award 201210421135. The authors are very grateful for these generous supports.

References

1. Daubechies, I.: Ten Lectures on Wavelets. Philadelphia: Society for Industrial and Applied Mathematics (1992)
2. Agichtein, E., Castillo, C., Donato, D., Gionis, A., Mishne, G.: Finding High-quality Content in Social Media. In: The International Conference on Web Search and Web Data Mining, pp. 183–194. ACM Press, New York (2008)

3. Becker, H., Naaman, M., Gravano, L.: Selecting Quality Twitter Content for Events. In: The Fifth International AAAI Conference on Weblogs and Social Media. AAAI Press, Barcelona (2011)
4. Ramage, D., Dumais, S., Liebling, D.: Characterizing Microblogs with Topic Models. In: International AAAI Conference on Weblogs and Social Media, pp. 130–137. AAAI Press, Washington (2010)
5. Xia, W., He, Y., Tian, Y., Chen, Q., Lin, L.: Feature Expansion for Microblogging Text Based on Latent Dirichlet Allocation with User Feature. In: 2011 6th IEEE Joint International Information Technology and Artificial Intelligence Conference, pp. 228–232. IEEE Press, Chongqing (2011)
6. Weng, J., Lim, E.P., Jiang, J., He, Q.: Twitterrank: Finding Topic-sensitive Influential Twitterers. In: Proceedings of the Third ACM International Conference on Web Search and Data Mining, pp. 261–270. ACM Press, New York (2010)
7. Dempster, A.P., Laird, N.M., Rubin, D.B.: Maximum Likelihood from Incomplete Data via the EM Algorithm. J. Roy. Stat. Soc. Series B (Methodological), 1–38 (1977)
8. De Choudhury, M., Counts, S., Czerwinski, M.: Find Me the Right Content! Diversity-Based Sampling of Social Media Spaces for Topic-Centric Search. In: the 5th International AAAI Conference on Weblogs and Social Media. AAAI Press, Barcelona (2011)
9. Vosecky, J., Leung, K.W.-T., Ng, W.: Searching for Quality Microblog Posts: Filtering and Ranking Based on Content Analysis and Implicit Links. In: Lee, S.-g., Peng, Z., Zhou, X., Moon, Y.-S., Unland, R., Yoo, J. (eds.) DASFAA 2012, Part I. LNCS, vol. 7238, pp. 397–413. Springer, Heidelberg (2012)
10. Lin, Y.R., Candan, K.S., Sundaram, H., Xie, L.: SCENT: Scalable Compressed Monitoring of Evolving Multirelational Social Networks. J. TOMCCAP 7(1), 29 (2011)
11. Sharifi, B., Hutton, M.A., Kalita, J.K.: Experiments in Microblog Summarization. In: 2010 IEEE Second International Conference on Social Computing (SocialCom), pp. 49–56. IEEE Press, Minneapolis (2010)
12. Becker, H., Naaman, M., Gravano, L.: Event Identification in Social Media. In: The ACM SIGMOD Workshop on the Web and Databases. ACM Press, Rhode Island (2009)
13. Harabagiu, S.M., Hickl, A.: Relevance Modeling for Microblog Summarization. In: The Fifth International AAAI Conference on Weblogs and Social Media. AAAI Press, Barcelona (2011)
14. Asur, S., Huberman, B.A.: Predicting the Future with Social Media. In: 2010 IEEE/WIC/ACM International Conference on Web Intelligence and Intelligent Agent Technology, pp. 492–499. IEEE Press, Toronto (2010)
15. Becker, H., Naaman, M., Gravano, L.: Selecting Quality Twitter Content for Events. In: The Fifth International AAAI Conference on Weblogs and Social Media. AAAI Press, Barcelona (2011)
16. Mallat, S.G.: A Theory for Multiresolution Signal Decomposition: The Wavelet Representation. J. IEEE T. Pattern Anal. 11(7), 674–693 (1989)
17. TwitterEngineering: 200 million tweets per day, http://blog.twitter.com/2011/06/200-million-tweets-per-day.html
18. Sharifi, B., Hutton, M.A., Kalita, J.: Summarizing Microblogs Automatically. In: Human Language Technologies: The 2010 Annual Conference of the North American Chapter of the Association for Computational Linguistics, pp. 685–688. ACL Press, Los Angeles (2010)

Exploiting Structural Similarity
for Automatic Information Extraction from Lists

Dat T. Huynh[1], Jiajie Xu[2], Shazia Sadiq[1], and Xiaofang Zhou[1,2]

[1] School of Information Technology and Electrical Engineering,
University of Queensland, Australia
[2] School of Computer Science and Technology, Soochow University, China
{tandat.huynh,uqssadiq,zxf}@uq.edu.au, xujj@suda.edu.cn

Abstract. In this paper, we propose a novel technique to reduce dependency on knowledge base for ONDUX, the current state-of-art method for information extraction by text segmentation. While the existing approach mainly relies on high overlapping between pre-existing data and input lists to build an extraction model, our approach exploits structural similarity of text segments in the sequences of a list to align them into groups to achieve effectiveness with low dependency on pre-existing data. Firstly, a structural similarity measure between text segments is proposed and combined with content similarity to assess how likely two text segments in a list should be aligned in the same group. Then we devise a data shifting-alignment technique in which positional information and the similarity scores are employed to cluster text segments into groups before their labels are revised by an HMM-based graphical model. The experimental results on different datasets demonstrate the ability of our method to extract information from lists with high performance and less dependence on knowledge base than the current state-of-art method.

1 Introduction

A large number of web pages contain information about entities in the form of lists in which data values are organised in textual sources, e.g. postal addresses, advertisings, references. Nevertheless, how to extract information of the data values from such lists is still a challenge and an important research problem which has been addressed in recent studies ([18], [16], [5]).

In the literature, the problem of information extraction from a list of field values is addressed as information extraction by text segmentation (IETS) in which information of entities is organised in implicit semi-structured records ([16]). Since field values are listed in a textual representation in the implicit semi-structured forms, traditional wrapper-based methods ([6], [3]) which depend on HTML tags cannot be applied for the textual inputs which are formatted differently in HTML. A dominant approach for this problem is the deployment of statistical methods, such as Hidden Markov Model (HMM) ([4]) and Conditional Random Fields (CRFs) ([10]) to extract information. However, obtaining a large amount of training data, which includes the associations between text segments

X. Lin et al. (Eds.): WISE 2013, Part II, LNCS 8181, pp. 202–215, 2013.

with their corresponding labels, to build an extraction model requires a lot of laborious work and may be very expensive or even unfeasible in some situations.

Later, some studies proposed the usage of pre-existing datasets to alleviate the need for manually labeled training data ([2], [12], [18]). However, as the training dataset is directly built from a database, those methods made an assumption about the overlapping of format-related features between field values in knowledge base and input texts. Moreover, those methods have low performance in running time because they execute inference step and training step for each time they perform a new extraction ([18], [5]).

Recently, authors in [5] have proposed ONDUX, an on-demand unsupervised method, to overcome those drawbacks. The authors exploited high overlapping between a knowledge base and an input list to label text segments and use the labels to build an HMM-based graphical model to revise the results. An implicit assumption in their work is that the majority of labels which are correctly assigned in matching step can help to build a graphical model to rectify incorrect and mismatched ones.

Our work is an improvement of ONDUX when we keep its good features but we reduce the dependency on the overlapping between a knowledge base and input lists in building its extraction model. Our idea is to exploit the structural similarity between field values in different sequences within an input list to align them into groups or columns before we revise their labels by using a graphical model. To realise the idea, we firstly propose a novel structural similarity measure to evaluate how likely two segments should be aligned into the same group. Different from traditional similarity measures which focus on the contents of input strings, our proposed structural similarity measure is defined by exploiting robust features on structures of two text segments within a list. Moreover, we devise a data shifting-alignment technique to group similar text segments into clusters or groups by using our proposed structural similarity. Our proposed technique exploits positional information of text segments to combine with structural similarity to discover the repeated patterns among portions of strings within a list to group text segments in alignment process. The labels of text segments in the groups are then exploited to build a graphical model and revise the results for extracting information. We conduct an extensive experimental study with real datasets on the web. The experiments show that our extraction method is robust and performs well with high precision and recall and less dependent on knowledge base than existing study ([5]).

The remaining sections of this paper are organised as follows. Section 2 discusses related studies in the literature. Next, we present the phases of our method in section 3. Our proposed structural similarity as well as data shifting-alignment technique will be described in detail in this section. Then experimental results are analysed in section 4. Eventually, section 5 concludes the paper and suggests some future work.

2   Related Work

IETS is the process of converting an unstructured document which contains implicit records into structured form by splitting the document or a list into substrings which contain data values ([16]). The dominant approach to segment texts in an input list to extract field values in the literature is the application of machine learning with two different techniques for generating training data. The first technique, which is called supervised approach, builds a training data set manually by human ([17], [8], [4], [10], [14], [12]). Meanwhile, the second technique exploits existing data in a knowledge base or reference tables to build extraction models automatically ([2], [12], [18], [5]).

The studies of [17] and [8] can be considered as the first studies addressing this problem in the literature. In their work, an HMM for recognising the field values in an input text was constructed from a provided training dataset. Later, this approach was extended in the system DATAMOLD ([4]), in which each state of an external HMM contains an internal HMM for recognising the value of each attribute. Later, CRFs ([10]) was proposed as an alternative model for HMM to solve the task of labelling texts. CRFs-based methods are proven to outperform all previous learning-based methods in both theory and experimental evaluations for the problem of sequence labelling ([14], [12], [16]). Although the quality of extraction results of HMM and CRFs are good, those supervised methods require to have a large amount of manually training data to build an extraction model.

Meanwhile, the general idea of unsupervised approach is to exploit a pre-existing data source to build a statistical extraction model. The study of [2] followed this idea and proposed an unsupervised method with HMM. Authors in [18] proposed a similar technique but adapted the idea to CRFs. In both methods, training data is built automatically by directly concatenating field values in a reference table. They assume that the field values in an input text are in single order and their features of field values in a reference tables and input texts are similar. According to the experiments of [5], performance of [18] is low when testing data and referent table come from different sources. Moreover, each time those methods perform a new extraction, they need to infer the oder of field values to build training data and construct a new extraction model.

Recently, authors in [5] have proposed ONDUX, an on-demand unsupervised approach for the problem of IETS. In their method, a knowledge base is employed to label text segments via some attribute matching functions of common terms between text segments in an input text and field values in the knowledge base. The labels are then used by a reinforcement phase to build a graphical model to verify and potentially correct the labels which were generated by matching phase. Therefore, the high overlapping between knowledge base and the source list must be maintained in their work so that a graphical model can be generated correctly.

We argue that building such a good knowledge base is a formidable task when the method is applied to extract information from any arbitrary target list in any domain on the web. In this paper, we propose a novel technique to reduce the dependency on knowledge base. The text segments are firstly aligned into groups by using their structural similarity before we revise their labels for information extraction by using a graphical model. In next section, we will present our proposed method in detail.

3 Proposed Method

3.1 Similarity Measures for Text Segments

In order to measure how likely two text segments should be put in the same group, we propose the structural similarity between field values in different sequences of an input list. We identify common features sharing in field values belonging to the same group in a list and use them to define the measure of our proposed structural similarity as the follows.

Structural Similarity. This similarity describes the way field values or text segments display in a textual list. We firstly consider the features which provide general information of the strings which form two text segments v_1 and v_2. Those features include: (1) number of letters, (2) percentage of lower case letters, (3) percentage of upper case letters, and (4) percentage of digits in a value. For each feature f_i, we compute a numeric value for the feature and the similarity of v_1 and v_2 on a feature is defined as in equation 1.

$$sim_{f_i}(v_1, v_2) = 1 - \frac{|a_i - b_i|}{max(a_i, b_i)} \qquad (1)$$

where a_i and b_i are numeric values for a particular feature of values v_1 and v_2.

Beside the general characteristics of strings as above, we also consider the similarity on the organisation of tokens in a string. We observe that two field values or text segments belonging to the same group in an input list often share similar format or representation style. For an instance, person names (e.g "D T Huynh") often start with some abbreviations, which include capital letters, followed by a word which represents a family name. Based on the observation, we devise a format-related similarity of two values v_1 and v_2. We firstly define a set of masks to represent the tokens in the textual string of each value. We tokenise the strings v_1 and v_2 and encode them by using *symbol masks*. For example, the value "D T Huynh" is encoded as "[A-Z] [A-Z] [A-Z][a-z]+", where the mask [A-Z] represents an uppercase letter, the mask [a-z]+ represents a consecutive string of one or many lowercase letter. The similarity between two values v_1 and v_2 is computed based on those sequences of masks. This idea is actually adapted from the study of [4] which employs *symbol masks* to capture the format of values in an inner HMM. However, their work utilised a training dataset to build an HMM-based statistical model to capture the format of strings. Meanwhile, our study utilises the concept *symbol masks* to define our structural similarity measure.

Given two values v_1 and v_2, we encode v_1 and v_2 by using the symbol masks as above to obtain two sequences of masks for v_1 and v_2. The distance measure between two masks of v_1 and v_2 is defined as the minimum number of insertions, deletions or substitutions to transfer from this mask to the other one. Then, we apply dynamic programming ([11]) to find the minimum number of operations. Formally, the format-related similarity between two values v_1 and v_2 is illustrated as the equation 2.

$$sim_F(v_1, v_2) = 1 - \frac{dist(m_{v_1}, m_{v_2})}{|m_{v_1}| + |m_{v_2}|} \tag{2}$$

where m_{v_i} is the sequence of masks encoded for the value v_i, $|m_{v_i}|$ is the number of masks in the sequence, and $dist$ is an edit distance function between two sequences of masks.

Eventually, the structural similarity $sim_S(v_1, v_2)$ is the average of all feature similarities (equation 3).

$$sim_S(v_1, v_2) = \sum_i^n \alpha_i \times sim_{f_i}(v_1, v_2)) \tag{3}$$

where n is the number of features of v_1 and v_2 and f_i is a particular feature.

Content Similarity. Beside structural similarity, we also consider the similarity of the contents of two text segments. If two text segments or field values share some certain keywords, it will be high possibility that they belong to the same concept and therefore they should be align in the same group. For example, the segments "IEEE Transaction on Knowledge and Data Engineering" and "IEEE Transaction on Multimedia" share some common keywords such as "IEEE" and "Transaction" and both of them are the names of the same concept "journals". Therefore, the content-related similarity measure sim_C can be defined by using a string similarity function between two text segments. A large number of approximately string matching techniques have been proposed in the literature. Popular measures include edit distance functions ([7]), Jaccard coefficient, Cosine similarity measure in information retrieval ([15]), and their extensions to utilise *q-grams* instead of words ([9]). In our work, we adapt the idea of using q-grams in [9] to define the content-related similarity measure. As formulated in equation 4, it is the Jaccard similarity between two sets of q-grams of the text values v_1 and v_2.

$$sim_C(v_1, v_2) = \frac{|qg(v_1) \cap qg(v_2)|}{|qg(v_1) \cup qg(v_2)|} \tag{4}$$

where $qg(v_i)$ is the set of q-grams associated with the text segment v_i.

Knowledge Base Support. Intuitively, if two field values v_1 and v_2 co-occur in an attribute of a table or a knowledge base, they should belong to the same group. Since the labels of text segments are assigned in the matching step, two segments in different sequences of an input list with the same label should be grouped into the same cluster. We set the similarity $sim_K(v_1, v_2)$ between two values v_1 and v_2 to one if they co-occur in an attribute of a knowledge base or zero otherwise.

Given two field values v_1 and v_2, the similarity score between v_1 and v_2 is defined by the weighted sum of the similarity scores of all features between them. We combine them together to define a score function as in the equation 5.

$$sim(v_1, v_2) = w_1 * sim_C(v_1, v_2) + w_2 * sim_S(v_1, v_2) + w_3 * sim_K(v_1, v_2) \quad (5)$$

where sim_i's are the similarity between v_1 and v_2 on their different features; the weights w_i's are real numbers in $[0, 1]$ and their total is one.

3.2 Data Shifting-Alignment Technique

The purpose of data alignment phase is to align similar text segments in different sequences into groups by using their similarity scores. It can be seen that text segments in the same position in different sequences of a list usually belong to the same concept and can be clustered in a group. However, this assumption is not always correct because some field values in a sequence can be missed or the number of field values are different in sequences. Therefore, we cannot simply use only positional information of text segments to cluster text segments in an input list into groups.

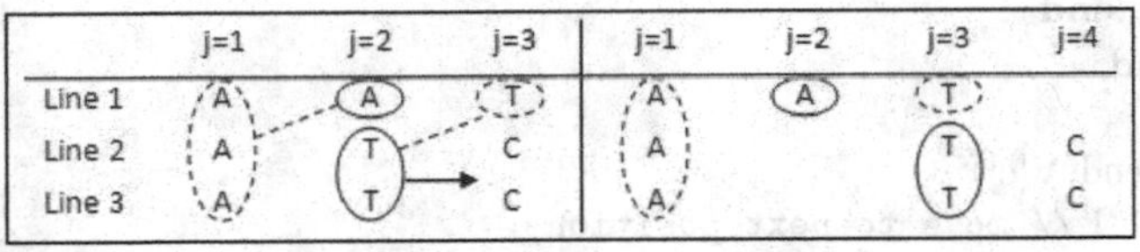

Fig. 1. A demonstration of data shifting-alignment technique

An example of the problem can be illustrated in Figure 1. In the example, each letter 'A', 'T', and 'C' accordingly stands for an author name, a paper title and a conference name in bibliography domain. Because there are differences between the number of field values (e.g author names) in different lines of the input list, the text segments in the position two of the sequences are clustered into two groups 'A' and 'T'. Therefore, we need to have a method to move the group 'T' to the next position so that the text segments in the group can be aligned with the text segments in the next positions.

We propose a shifting-alignment technique to overcome this problem in data alignment phase. We firstly exploit positional information of text segments and their similarity scores to cluster them into groups. If there is only one group returned from the procedure, this means that all text segments will belong to the same group. In contrast, if a list of groups is returned, we keep one group in the list to obtain a group and perform a shifting step to move the remaining groups to the next position and the alignment process will be repeated for the next positions to align the text segments in those groups.

Algorithm 1 illustrates steps in our alignment phase. In order to cluster similar text segments in a list into groups, we adapt the idea of agglomerative clustering

---

Algorithm 1. Data alignment

Input: A two-dimensional list of segments in input list L
Output: A set of groups R

```
1  j = 0
2  while true do
3  |    G_j = [ ]
4  |    for i in range(0, |L|) do
5  |    |    G_j.append(L[i][j])
6  |    end
7  |    if G_j is empty then
8  |    |    break loop
9  |    end
10 |    V = CLUSTERING(G_j)
11 |    c = 0
12 |    if |V| > 1 then
13 |    |    c = CHOOSE-GROUP(V, j, R)
14 |    |    for k in range(0, |V|) do
15 |    |    |    if k != c then
16 |    |    |    |    for L[i][j] in V[k] do
17 |    |    |    |    |    insert NIL at position j of L[i]
18 |    |    |    |    end
19 |    |    |    end
20 |    |    end
21 |    end
22 |    R.append(V[c])
23 |    j = j + 1 // move to next position
24 end
25 return R
```

Algorithm 2. Choosing a group for data alignment

Input: A list of groups V at the position j; R: preceding groups of V in a list
Output: Index of a chosen group in V

```
1  S = [ ]
2  for i in range(0, |L|) do
3  |    for y in range(j+1, |L[i]|) do
4  |    |    S.append(L[i][y])
5  |    end
6  end
7  score = 0
8  for k in range(0, |V|) do
9  |    sim_S = similarity(V[k], S)
10 |    sim_P = similarity(V[k], R[j-1])
11 |    if sim_P - sim_S > score then
12 |    |    score = sim_P - sim_S; c = k
13 |    end
14 end
15 return c
```

algorithm to cluster text segments into groups by their structural similarity. In the algorithm, the shifting step is performed by inserting a NIL value into a position in a list (line 17). Moreover, this clustering step (line 10) in algorithm 1 is illustrated in detail as follows. Initially, each group of G contains a segment in V. Then we merge any two groups which have the highest similarity and the similarity is greater than a threshold θ_s. This process is repeated until we cannot find any two groups whose similarity above the threshold. After we cluster the text segments, we obtain a set of groups V and each group contains a list of elements of the same concepts.

In order to choose a group in a list of groups in the alignment results, we employ some heuristics on the groups in a list. Firstly, given a group V_c in a list of groups, if there is one or many other groups in the next positions which are similar to the group V_c, the group V_c should be moved to the next position so that it can be aligned with the other groups. For example, let's consider the group 'T' at the position two in Figure 1. Because there is a group 'T' at position three which is similar to the current group 'T' at position two, the group 'T' at position two should be moved to the next position so that it can be aligned with the group 'T' in the next position in the next processing step. Secondly, if a group V_c is similar to a group in the previous position in input list, the group V_c should be kept in the position to be aligned into a group and other groups in the list should be moved to new positions. For example, the group 'A' at position two in Figure 1 is similar to the group 'A' in the previous position. Therefore, we should keep the group 'A' in alignment results and move the group 'T' to the next position so that it can be aligned with other text segments.

Based on the observations on a group in a list, we define a score for a group V_c which is the least similar to the following groups and most similar to the preceding groups in a list V. It is computed by the subtraction of the similarity between the group V_c and preceding groups and the similarity between V_c and the succeeding groups. The group V_c with the highest score in the list of groups V will be chosen to form a group and then the remaining groups in the list V will be shifted to the next positions in alignment process. Formally, the choice of group V_c from a list of group V can be described in the equation 6.

$$V_c = argmax_{v \in V}(sim_P(v) - sim_S(v)) \tag{6}$$

where $sim_P(v)$ and $sim_S(v)$ are respectively the similarity between a group v and preceding and succeeding groups. The procedure to choose a group from a list is described in the algorithm 2.

3.3   Information Extraction Steps

The process of extracting information from an input list can be divided into three phases: text-blocking and matching, data alignment, and refinement phase.

Text-Blocking and Matching Phase. In this phase, we firstly split the sequences of an input list into text segments by using punctuation and match them with field values in a knowledge base to assign labels. Then we merge any two segments which co-occur in the same field in the knowledge base. We reuse the matching score defined in the study of [5] to compute the matching scores between a field value and an attribute A_j via a fitness function as in equation 7.

$$M(s, A_j) = \frac{\sum_{w \in s} fitness(w, A_j)}{|s|} \tag{7}$$

The fitness scores are computed for all tokens w in the query string s and the label A_j and it is defined as in the equation 8.

$$fitness(w, A_j) = \frac{freq(w, A_j)}{freq(w)} \times \frac{freq(w, A_j)}{freq_{max}(A_j)} \tag{8}$$

where $freq(w, A_j)$ is the number of values of the label A_j containing the token w, $freq(w)$ is the total number of instance values in the knowledge base containing the token w, and $freq_{max}(A_j)$ is the highest frequency of any token in the instance values of the label A_j.

Data Alignment Phase. After text-blocking phase, each sequence in an input list is split into a set of text segments. In data alignment phase, the text segments in different sequences of an input list are aligned into groups according to their similarity scores. We exploit the data shifting-alignment technique which has been presented in section 3.2 to cluster the text segments into groups. Finally, the text segments in the same groups are assigned the same labels and they are used to build a graphical model to revise the results in a final refinement phase.

Refinement Phase. We employ a positional and sequential model (PSM) proposed in [5] to revise the labels of unmatched segments and rectify mismatched ones in this phase. A PSM is defined by the following three components: (1) A set of states $T = \{begin, t_1, t_2, ..., t_N, end\}$ where each state t_i represents a label of a text segment; (2) A matrix A where each element a_{ij} is the probability of making a transition from state i to state j. Each element a_{ij} in the matrix A is defined as the equation 9; (3) A matrix P where the entry p_{ik} denotes the probability of the label t_i appearing in the position k-th in an input list. Formally, p_{ik} is defined as in the equation 10.

$$a_{ij} = \frac{Number\ of\ transitions\ from\ state\ t_i\ to\ state\ t_j}{Total\ number\ of\ transitions\ out\ of\ state\ t_i} \tag{9}$$

$$p_{ik} = \frac{Number\ of\ observations\ of\ t_i\ in\ k}{Total\ number\ of\ segments\ in\ k} \tag{10}$$

To compute the probability to have a label t for a text segment, matching score, sequential and positional model score are combined by using Bayesian disjunctive operator, also known as Nosiy-OR-Gate ([13]), as in equation 11.

$$sim(s, t) = 1 - (1 - M(s, t)) \times (1 - a_{ij}) \times (1 - p_{ik}) \tag{11}$$

where $M(s,t)$ is a matching score between a segment s and a label t, which is defined as in the equation 7; i is the index of the label t in a list of labels T, j is the index of the label of the next segment of s; and k is the position of s in an input sequence. The value of a_{ij} and p_{jk} are accordingly defined by sequential model as in the equation 9 and positional model in equation 10.

4   Experiments

4.1   Experimental Setup

In the experiments, our priority is to choose the public datasets which were used in previous work or available on the Internet. In the domain *Addresses*, we utilise *BigBook* and *Restaurants* dataset from RISE repository ([1]) and then manually label field values in the datasets. These datasets were introduced in the experiments of previous studies ([18], [5]). Next, we employ journal references in *PersonalBib* and *Cora* dataset used in the studies of [5] and [14] as data source and testing dataset in bibliographic data domain. To evaluate the affect of the overlapping terms between knowledge base and an input list and compare with ONDUX, we also use *BigBook* as in the experiments of [5]. Detailed information about the number of records and attributes in each dataset is described in table 1. In our experiments, we assign equal weights to features in similarity measures and set the threshold $\theta_s = 0.3$ for alignment algorithm.

Table 1. Domains, data sources, and datasets used in the experiments

Domain	Data source	Attributes	Records	Testing dataset	Attributes	Text Inputs
Address	*BigBook*	5	2,000	*BigBook*	5	2,000
Address	*BigBook*	5	2,000	*Restaurants*	4	250
Bibliography	*PersonalBib*	7	395	*Cora*	6	150

We utilise well known precision, recall and F1 measure in information extraction. We denote A_i as a referent set and B_i as testing results to be compared with A_i. The precision (P_i), recall (R_i) and F1 measure (F_i) are accordingly defined as in equation 12.

$$P_i = \frac{|A_i \cap B_i|}{|B_i|}; R_i = \frac{|A_i \cap B_i|}{|A_i|}; F_i = \frac{2 \times P_i \times R_i}{P_i + R_i} \qquad (12)$$

4.2   Extraction Quality

Firstly we demonstrate the overlapping between data source and testing datasets as in table 2. The column "%Same" in the table shows that a large percentage of text segments in testing data are found in the values of the same attribute in a corresponding knowledge base or data source. To demonstrate the effectiveness of our method in whole process of information extraction, we analyse the performance obtained after each step on each testing data and compare the experimental results with previous study. In tables 3, 4, and 5, the columns "ONDUX", "Matching", "Alignment", and "Refinement" accordingly describe the results of ONDUX and the steps of our method on testing datasets. In general, when there is a high overlapping between knowledge base and datasets, the results of both our method and ONDUX are extremely high for all attributes. We note that the experiments are conducted with whole knowledge base obtained from data sources. The good performance is obtained due to the high overlapping between knowledge base and testing lists. However, our method gives better results in some cases on *Restaurants* and *Cora* datasets. It can be explained that our method exploits structural information of text segments to group similar text segments together. Therefore, more text segments are labeled in different sequences of an input list. Due to this, the graphical model built from the labels of text segments in different rows can capture more statistical information of the transitions of labels in the input text.

Table 2. The overlapping between data source and testing dataset

Source	Dataset	%Same	%Unknown
BigBook	*BigBook*	93.36%	5.46%
BigBook	*Restaurants*	91.78%	7.21%
PersonalBib	*Cora*	76.17%	18.67%

Table 3. Experimental results on *BigBook* dataset using *BigBook* source

Field	ONDUX	Matching	Alignment	Refinement
Name	0.996	0.802	0.974	0.996
Street	0.995	0.814	0.924	0.995
City	0.996	0.922	0.937	0.996
State	1.000	0.894	0.963	1.000
Phone	1.000	0.936	0.971	1.000
Average	0.997	0.874	0.954	0.997

Table 4. Experimental results on *Restaurants* dataset using *BigBook* source

Field	ONDUX	Matching	Alignment	Refinement
Name	0.958	0.618	0.778	0.975
Street	0.980	0.907	0.924	0.984
City	0.986	0.739	0.937	0.987
Phone	0.992	0.981	0.981	0.992
Average	0.979	0.811	0.905	0.985

4.3   Impact of Previously Known Data to Extraction Results

In this section, we illustrate the advantage of our proposed method as compared to ONDUX, the state-of-the-art method for IETS. We analyse the impact of the

Table 5. Experimental results on *Cora* dataset using *PersonalBib* source

Field	ONDUX	Matching	Alignment	Refinement
Author	0.905	0.708	0.835	0.912
Title	0.822	0.788	0.796	0.823
BookTitle	0.846	0.797	0.825	0.857
Pages	0.849	0.762	0.878	0.921
Year	0.912	0.775	0.827	0.926
Volume	0.940	0.872	0.907	0.942
Average	0.879	0.784	0.845	0.897

overlapping terms between knowledge base and an input list on the quality of
information extraction of both methods. Similar to the study of [5], the experiments are performed on *BigBook* dataset which contains about 4,000 entries
coming from the *RISE* repository [1]. We randomly use 2,000 records as testing
data and 2,000 remaining records as knowledge base in the experiments. Then we
vary the number of shared terms between the knowledge base and testing data.
We repeat the experiment five times and compute the average F1-measure of all
field values in each step. Figure 2 represents the experimental results when we
vary the number of shared terms between knowledge base and the input list in
the experiments from 50 to 1,000 terms. In general, when the number of shared
terms approximately approaches 1,000, both methods reach similar extraction
quality. However, when the number of overlapping terms are not large enough,
the performance of ONDUX drops dramatically. It can be observed in Figure 2
that the F-measure values obtained by ONDUX are quite low when the number
of common terms are less than 200. In other words, ONDUX is quite dependent
on the overlapping between knowledge base and input list. Meanwhile, our proposed method still keep good performance with more than 77% of F1-measure.

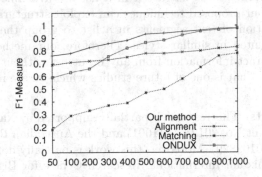

Fig. 2. F-Measure obtained when varying shared terms for *BigBook* dataset

Those experimental results are expected because ONDUX only exploits the overlapping terms to obtain the statistics about the structure of the testing list. Once overlapping terms are not large enough, ONDUX cannot build a good statistical model to revise the results which were generated by their matching step. Meanwhile, our proposed method can exploit the structural information of text segments within a list to cluster them into groups. Then, the group can be labelled with only some overlapping data with knowledge base. As a result, the alignment step in our method can help to increase the number of assigned labels to obtain statistical information about the structure of the input list to revise the results in refinement phase. In practice, the requirements of high overlapping between a knowledge base and input lists could not be obtained all the times, especially when we perform an extraction on an arbitrary list. Therefore, we can conclude that our method is more robust than ONDUX in terms of less dependency on the overlapping between knowledge base and input lists.

5   Conclusion

We have presented our approach to reduce the dependency on knowledge base for information extraction by text segmentation from textual lists. In our approach, structural similarity is exploited to group similar text segments into clusters before we revise the labels by a graphical model. In order to implement the idea, we propose a structural similarity to measure the likelihood of two text segments being similar. Moreover, we devise a shifting-alignment technique, in which positional information of text segments is combined with a shifting technique to cluster data into groups. We have experimented our proposed method on the datasets in different domains and the results show that our techniques can extract information from lists with high performance and less dependence on knowledge base than ONDUX, the current state-of-art study.

Currently, both our method and ONDUX still make an assumption that the attributes of the values to be extracted in a list are available in a knowledge base. Meanwhile, our proposed techniques can exploit structural similarity and positional information of text segments in a list to group them into columns if they are represented in similar styles. Therefore, our method can open an opportunity to extract information from any input list without the assumption of knowledge base. That is one of future studies which we are investigating.

Acknowledgments. This research is partially supported by National 863 High-tech Program (Grant No. 2012AA011001) and the Australian Research Council (Grants No. DP120102829). Moreover, this work is partially done when the authors visited Sa-Shixuan International Research Centre for Big Data Management and Analysis hosted in Renmin University of China. This center is partially funded by the Chinese National 111 Project "Attracting International Talents in Data Engineering and Knowledge Engineering Research".

References

1. Rise - a repository of online information sources used in information extraction tasks (1998), http://www.isi.edu/info-agents/rise/index.html
2. Agichtein, E., Ganti, V.: Mining reference tables for automatic text segmentation. In: Proceedings of the Tenth ACM SIGKDD Conference, pp. 20–29 (2004)
3. Arasu, A., Garcia-Molina, H.: Extracting structured data from web pages. In: Proceedings of the 2003 ACM SIGMOD, pp. 337–348 (2003)
4. Borkar, V., Deshmukh, K., Sarawagi, S.: Automatic segmentation of text into structured records. In: Proceedings of ACM SIGMOD, pp. 175–186 (2001)
5. Cortez, E., da Silva, A.S., Gonçalves, M.A., de Moura, E.S.: Ondux: on-demand unsupervised learning for information extraction. In: Proceedings of the 2010 ACM SIGMOD, pp. 807–818 (2010)
6. Crescenzi, V., Mecca, G., Merialdo, P.: Roadrunner: Towards automatic data extraction from large web sites. In: Proceedings of the 27th International Conference on Very Large Data bases, pp. 109–118 (2001)
7. Elmagarmid, A.K., Ipeirotis, P.G., Verykios, V.S.: Duplicate record detection: A survey. IEEE TKDE 19, 1–16 (2007)
8. Freitag, D., McCallum, A.: Information extraction with hmm structures learned by stochastic optimization. In: Proceedings of the Seventeenth National Conference on Artificial Intelligence and Twelfth Conference on Innovative Applications of Artificial Intelligence, pp. 584–589. AAAI Press (2000)
9. Gravano, L., Ipeirotis, P.G., Koudas, N., Srivastava, D.: Text joins in an rdbms for web data integration. In: Proceedings of the 12th International Conference on World Wide Web, pp. 90–101. ACM (2003)
10. Lafferty, J.D., McCallum, A., Pereira, F.C.N.: Conditional random fields: Probabilistic models for segmenting and labeling sequence data. In: Proceedings of the 18th International Conference on Machine Learning, pp. 282–289 (2001)
11. Levenshtein, V.: Binary codes capable of correcting deletions, insertions, and reversals. Soviet Physics Doklady 10(8), 707–710 (1966)
12. Mansuri, I.R., Sarawagi, S.: Integrating unstructured data into relational databases. In: Proceedings of the 22nd ICDE, pp. 29–40 (2006)
13. Pearl, J.: Probabilistic reasoning in intelligent systems: networks of plausible inference. Morgan Kaufmann Publishers Inc., San Francisco (1988)
14. Peng, F., McCallum, A.: Information extraction from research papers using crfs. Information Processing and Management 42, 963–979 (2006)
15. Salton, G., McGill, M.J.: Introduction to Modern Information Retrieval. McGraw-Hill, Inc., New York (1986)
16. Sarawagi, S.: Information extraction. Foundation and Trends in Databases 1(3), 261–377 (2008)
17. Seymore, K., Mccallum, A., Rosenfeld, R.: Learning hidden markov model structure for information extraction. In: AAAI 1999 Workshop on Machine Learning for Information Extraction, pp. 37–42 (1999)
18. Zhao, C., Mahmud, J., Ramakrishnan, I.V.: Exploiting structured reference data for unsupervised text segmentation with conditional random fields. In: Proceedings of the SIAM International Conference on Data Mining, pp. 420–431 (2008)

A Minwise Hashing Method for Addressing Relationship Extraction from Text

David S. Batista, Rui Silva, Bruno Martins, and Mário J. Silva

Instituto Superior Técnico and INESC-ID, Lisboa, Portugal
{dsbatista,rui.teixeira.silva,bruno.g.martins,
mario.gaspar.silva}@ist.utl.pt

Abstract. Relationship extraction concerns with the detection and classification of semantic relationships between entities mentioned in a collection of textual documents. This paper proposes a simple and on-line approach for addressing the automated extraction of semantic relations, based on the idea of nearest neighbor classification, and leveraging a minwise hashing method for measuring similarity between relationship instances. Experiments with three different datasets that are commonly used for benchmarking relationship extraction methods show promising results, both in terms of classification performance and scalability.

Keywords: Text Mining, Relationship Extraction, Minwise Hashing.

1 Introduction

The task of relationship extraction concerns with the detection and classification of semantic relationships between entities mentioned in a collection of textual documents. Popular application domains include the detection of gene-disease relationships or protein-protein interactions in biomedical literature [4,11,19], the detection of associations between named entities referenced in news or web corpora (e.g., birthplace relations between persons and locations, or affiliation relations between persons and organizations) [3,6], or the detection of relations between pairs of nominals in general [9].

Over the years, multiple approaches have been proposed to address relationship extraction [9,11,17]. Rule-based methods employ a number of linguistic rules to capture relation patterns. Feature-based methods, on the other hand, transform the text into a large amount of linguistic features (e.g., lexical, syntactic and semantic features), later capturing the similarity between these feature vectors through traditional supervised learning methods. Recent developments have mainly relied on kernel-based learning approaches, either exploring kernels for representing sequences [4], in an attempt to capture sequential patterns within sentences, or kernels specific for trees or graph structures in general, to learn features related to parse tree structures [3,13]. Kernel methods are better than feature-based methods at circumventing data sparseness issues and at exploring very large feature spaces, but nonetheless they are also computationally demanding. Whenever one needs to address real-world problems involving hundreds of

X. Lin et al. (Eds.): WISE 2013, Part II, LNCS 8181, pp. 216–230, 2013.
© Springer-Verlag Berlin Heidelberg 2013

relationship classes as expressed on large amounts of textual data, and when dealing with large training datasets, scalability becomes an issue.

In this paper we explore the automated extraction of semantic relations, based on nearest neighbor classification. To make the nearest neighbor search computationally feasible, we leverage an efficient method based on minwise hashing and on Locality-Sensitive Hashing (LSH) [2,15]. Experiments with three different collections that are commonly used for benchmarking relationship extraction methods, namely the dataset from the SemEval 2010 task on multi-way classification of semantic relations between pairs of nominals [9], the Wikipedia relation extraction dataset created by Culotta et al. [6], and the AImed dataset of human protein interactions [4], showed good results. We specifically tested different configurations of the proposed method, varying the minwise hashing signatures and the number of considered nearest neighbors. Our best results correspond to a macro-averaged F1 score of 0.69 on the SemEval dataset, a macro-averaged F1 score of 0.43 on Wikipedia, and an F1 score of 0.52 on AImed. These values come close to the state-of-the-art results reported for these datasets, and we argue that the method has advantages in simplicity and scalability.

Section 2 of the paper presents related work. Section 3 details the proposed method, describing the considered representation for the relation instances, and presenting the minwise hashing approach that was used. Section 4 presents the experimental evaluation of the proposed method. Finally, Section 5 summarizes our conclusions, and outlines possible directions for future work.

2   Related Work

Extracting semantic relations between nominal expressions (i.e, named entities like persons, locations or organizations) in natural language text is a crucial step towards document understanding, with many practical applications. Several authors have addressed the problem, for instance by formulating it as a binary classification task (i.e., classifying candidate instances of binary relations, between pairs of nominal expressions, as either related or not). Relevant previous approaches include those that adopt feature-based supervised learning methods [10,21], or kernel-based methods [17,18] to perform relation extraction. The major advantage of kernel methods is that they allow one to explore a large (often exponential or, in some cases, infinite) feature space in polynomial computational time, without the need to explicitly represent the features. Nonetheless, kernel methods are still highly demanding in terms of computational requirements, whenever one needs to manipulate large training data sets. The main reason for this lays in the fact that kernel methods, even if relying only on very simple kernels, are typically used together with models and learning algorithms such as Support Vector Machines (SVM), where training involves a quadratic programming optimization problem and is typically performed off-line. Moreover, given that SVMs can only directly address binary classification problems, it is necessary to train several classifiers (i.e., in a one-versus-one or a one-versus-all strategy) to address multi-class relation extraction tasks.

Given a set of positive and negative binary relation examples, feature-based methods start by extracting syntactic and semantic features from the text, using them as cues for deciding whether the entities in a sentence are related or not. Syntactic features extracted from the sentences include (i) the entities themselves, (ii) the semantic types of the entities, (iii) the word sequence between the entities, (iv) the number of words between the entities, and (v) the path in a parse-tree containing the two entities. Semantic features can for instance include the path between the two entities in a dependency tree. The features are presented to a classifier in the form of a feature vector, which then decides on the relation class. Previous works have explored different types of supervised learning algorithms and different feature sets [10,21].

Feature-based methods have the limitation of involving heuristic choices, with features being selected on a trial-and-error basis, to maximize performance. To remedy the problem of selecting a suitable set of features, specialized kernels have been designed for relationship extraction. They leverage rich representations of the input data, exploring the input representations exhaustively, in an implicit manner and conceptually in a higher dimensional space.

For instance, Bunescu and Mooney presented a generalized subsequence kernel that works with sparse sequences, containing combinations of words and parts-of-speech (POS) tags to capture the word-context around the nominal expressions [4]. Three subsequence kernels are used to compute the similarity between sequences (i.e., between relation instances) at the word level, namely comparing sequences of words occurring (i) before and between, (ii) in the middle, and (iii) between and after the nominal expressions. A combined kernel is simply the sum of all three sub-kernels. The authors evaluated their approach on the task of extracting protein interactions from MEDLINE abstracts contained in the AImed corpus, concluding that subsequence kernels in conjunction with SVMs improve both precision and recall, when compared to a rule based system. Bunescu and Mooney also argued that, with this approach, augmenting the word sequences with POS tags and entity types can lead to better results than those obtained with the dependency tree kernel by Culotta and Sorensen [7].

Zelenko et al. described a relation extraction approach, based on SVMs or Voted Perceptrons, which uses a tree kernel defined over a shallow parse tree representation of the sentences [18]. The kernel is designed to compute the similarity between two entity-augmented shallow parse tree structures, in terms of a weighted sum of the number of subtrees that are common between the two shallow parse trees. These authors evaluated their approach on the task of extracting *person-affiliation* and *organization-location* relations from text, noticing that the proposed method is vulnerable to unrecoverable parsing errors.

Culotta and Sorensen described a slightly generalized version of the previous kernel, based on dependency trees, in which a bag-of-words kernel is also used to compensate for errors in syntactic analysis [7]. Every node of the dependency tree contains rich information like word identity, POS and generalized-POS tags, phrase types (noun-phrase, verb-phrase, etc.), or entity types. Using a rich structured representation can lead to performance gains, when compared to bag-of-words approaches. A further extension is proposed by Zhao and Grishman, using composite kernels to integrate information from different syntactic sources [20]. They incorporate tokenization, parsing, and dependency analysis, so that processing errors occurring at one level may be overcome by information from other levels. Airola et al. introduced the all-dependency-paths kernel [1]. They use a representation based on a weighted directed graph that consists of two unconnected subgraphs, one representing the dependency structure of the sentence, and the other representing the sequential ordering of the words. Bunescu and Mooney presented yet another alternative approach which uses information concentrated in the shortest path, at a dependency tree, between the two entities [3]. These authors argue that the shortest path between the two nominals encodes sufficient information to infer the semantic relation between them.

Recent studies continue to explore combinations or extensions of the previously described kernel methods [11,13]. However, most proposals have been evaluated on different data sets, making it difficult to assess which is better.

3   Relationship Extraction as Similarity Search

The proposed approach for classifying pairs of substrings (e.g., pairs of nominal expressions) from a given sentence, according to the semantic relation that the sentence expresses over these substrings, is based on the idea of finding the most similar relation instances from a given database of examples. A representation for each candidate binary relation, obtained from the sentence where the pairs of substrings co-occur, is first generated. We then assign the relationship type, to the instance being classified, according to the most frequent/similar type at the top k most similar relation instances, gathered from the database of examples. This procedure essentially corresponds to a weighted kNN classifier, where each example instance has a weight corresponding to its similarity towards the instance being classified, and where the more similar instances have therefore a higher vote, in the classification, than the ones that are more dissimilar.

The considered representation for each binary relation instance is essentially based on character quadgrams, specifically considering (i) the character quadgrams occurring between the two operands that constitute the binary relation (i.e., between the two substrings corresponding to the nominals that are related), (ii) the quadgrams occurring in a window of three tokens before the first operand and between operands, and (iii) the quadgrams occurring between operands and in a window of three tokens after the second operand. Consider, for instance, the following sentence, where the related nominals are in bold: *the **micropump** is fabricated by **anisotropic etching**, considering orientation.* The substring (i) would correspond to *is fabricated by the*, while substring (ii) would correspond to *the micropump is fabricated by*, and substring (iii) would correspond to *is fabricated by anisotropic etching, considering orientation.*

This representation follows the observation that a relation between two entities is generally expressed using only words that appear in one of three basic

Table 1. Frequent relational patterns associated to different relation types

Dataset	Relationship	Relational Patterns
SemEval	cause-effect	caused by; have; caused; result in; triggered by;
	entity-origin	have; come from; be from; run away from; arrive from;
Wikipedia	member-of	play for; have also; serve in; elect to; serve as;
	award	nominate for; award; win; receive; run;
AImed	related	bind to; bind; have; interact with; do
	not-related	bind to; bind; have; do; may

patterns, namely before-between (i.e., tokens before and between the two entities involved in the relation), between (i.e., only tokens between the two entities), and between-after (i.e., tokens between and after the two entities) [4]. Besides character quadgrams, we also use prepositions, verb forms in the past participle tense, the infinitive forms of verbs, and a relational pattern corresponding to a verb, followed by nouns, adjectives, or adverbs, and ending with a preposition. The previous relational pattern is inspired by one of the features used in ReVerb, an unsupervised open-domain relation extraction system [8]. The morphological information is extracted from the three same textual windows considered for the quadgrams (i.e., the substrings before-between, between, and between-after), with the help of the MorphAdorner NLP package[1].

We experimented with other representations for the relation instances, using different textual windows and different n-gram sizes, as well as with n-grams of tokens, after normalizing the text through lowercasing, lemmatization and/or WordNet-based generalization operations. However, the features described before achieve the best trade-off between accuracy and computational performance.

Each quadgram/token, at each of the three groups (i.e., before-between, between, and between-after), is assigned to a globally unique identifier. The similarity between two instances is measured through the Jaccard similarity coefficient between each set of globally unique identifiers.

For illustration purposes, Table 1 shows the top 5 most frequent relational patterns for two different relations in each of the three datasets that were used in our experiments – see Section 4. It is important to notice that in the case of the AImed dataset, similar relational patterns are extracted for both classes. We noticed that in instances from the *not-related* class, it is often the case that patterns such as *bind to* are preceded by different kinds of negation patterns.

The naive approach to finding the most similar pairs of relation instances, in a database of size N, involves computing all N^2 pairwise similarities, which quickly becomes a bottleneck for large N. Even if the task is parallelizable, overcoming this complexity is necessary to achieve good scalability, and it is therefore highly important to devise appropriate pre-processing operations that facilitate relatedness computations on the fly. In our case, this is done by approximating the Jaccard similarity coefficient through a minwise hashing (i.e., min-hash) procedure,

---

[1] http://morphadorner.northwestern.edu/

later leveraging a Locality-Sensitive Hashing (LSH) method for rapidly finding the kNN most similar relation instances. We therefore have that, contrary to traditional kNN classifiers, where training takes virtually zero computation time (i.e., it just involves storing the example instances) but classification is highly demanding, our approach involves a LSH method for indexing the training instances, which allows classification to be made efficiently, since we only have to measure similarity towards a small set of candidates. Similarly to other kNN classifiers, our approach remains relatively simple, performs the learning in an on-line fashion, and can directly address multi-class problems.

The min-hash technique was first introduced in the seminal work of Broder, where it was successfully applied to duplicate Web page detection [2]. Given a vocabulary Ω of size D (the set of all representative elements occurring in a collection of relation instances) and two sets, S_1 and S_2, where $S_1, S_2 \subseteq \Omega = \{1, 2, \ldots, D\}$, we have that the Jaccard similarity coefficient, between the sets of elements, is given by the ratio of the size of the intersection between S_1 and S_2, to the size of the union of both datasets:

$$J(S_1, S_2) = \frac{|S_1 \cap S_2|}{|S_1 \cup S_2|} = \frac{|S_1 \cap S_2|}{|S_1| + |S_2| - |S1 \cap S_2|} \tag{1}$$

The two sets are more similar when their Jaccard index is closer to one, and more dissimilar when their Jaccard index is closer to zero. In large datasets, efficiently computing the set sizes is challenging, given that the total number of possible elements is huge. However, suppose a random permutation π is performed on the ordering that is considered for the elements in the vocabulary Ω, i.e., suppose $\pi : \Omega \longrightarrow \Omega$. An elementary probability argument shows that the Jaccard coefficient can be estimated from the probability of the first (i.e., the minimum) values of the random permutation π, for sets S_1 and S_2, being equal, given that the Jaccard coefficient is the number of common elements to both sets over the number of elements that exist in at least one of the sets.

$$\Pr\left(\min(\pi(S_1)) = \min(\pi(S_2))\right) = \frac{|S_1 \cap S_2|}{|S_1 \cup S_2|} = J(S_1, S_2) \tag{2}$$

After the creation of k minwise independent permutations (i.e., $\pi_1, \pi_2, \ldots, \pi_k$) one can efficiently estimate $J(S_1, S_2)$, without bias, as a binomial distribution:

$$\hat{J}(S_1, S_2) = \frac{1}{k} \sum_{j=1}^{k} \text{one–if–true}(\min(\pi_k(S_1)) = \min(\pi_k(S_2))) \tag{3}$$

$$\text{Variance}(\hat{J}(S_1, S_2)) = \frac{1}{k} \hat{J}(S_1, S_2) \left(1 - \hat{J}(S_1, S_2)\right) \tag{4}$$

Equations 3 and 4 show, respectively, the expected value of the binomial distribution used for estimating the Jaccard coefficient from the k random permutations, and the corresponding variance, which decreases for larger values of k. The function one–if–true() in Equation 3 returns one if the two sets share the same minimum value, and zero otherwise.

In the implementation of the minwise hashing scheme, each of the independent permutations is a hashed value, in our case taking 32 bits of storage. Each of the k independent permutations is associated to a polynomial hash function $h^k(x)$ that maps the members of Ω to distinct values. For any set S we take the k values of $h^k_{min}(S)$, i.e. the member of S with the minimum value of $h^k(x)$. The set of k values is referred to as the min-hash signature of an instance.

Efficient nearest neighbor search is implemented through a Locality-Sensitive Hashing (LSH) technique, that leverages the min-hash signatures to compress the relation instance representations into small signatures (i.e., to generate small signatures from the set of all character quadgrams, prepositions, verb forms in the past participle tense, infinitive forms of verbs, and relational patterns occurring before-between, between, and between-after the relation operands), at the same time preserving the expected similarity of any pair of instances. This technique uses L different hash tables (i.e., L different MapDB[2] persistent storage units), each corresponding to an n-tuple from the min-hash signatures, that we here refer to as a band. At classification time, we compute a min-hash signature for a given target instance and then consider any pair that hashed to the same bucket, for any of the min-hash bands, to be a candidate pair. We check only the candidate pairs for similarity, using the complete min-hash signatures to approximate the Jaccard similarity coefficient. This way, we avoid the pair-wise similarity comparisons against all example instances, thus performing the kNN classification in a efficient manner. Chapter 3 of the book by Rajaraman and Ullman details the use of minwise hashing with LSH techniques, in applications related to finding similar items [15].

A complete outline for the proposed method is therefore as follows: First, sets of character quadgrams, prepositions, verb forms in the past participle tense, infinitive forms of verbs, and relational patterns, are extracted from the substrings that occur before-between, between, and between-after the relation operands, for each possible relation at each sentence from a given database of examples. Then, a list of min-hashes are extracted from the sets generated in the first step. The min-hashes are split into bands, and hashed into the L different hash tables. At classification, when checking if a given binary relation, as expressed in some sentence, is being described, we start by also extracting the same set of features, from the substrings that occur before-between, between, and between-after two named entities to be considered the relation operands. A min-hash signature is then generated from this set of features. Relation instances, from the collection of examples, with at least one identical LSH band, are considered as candidates, and their Jaccard similarity coefficient towards the target instance is then estimated, from the available min-hashes. The candidates are sorted according to their similarity towards the target instance, and the most relevant relationship type, computed with basis on the weighted votes from the top kNN most similar instances, is returned as the identified semantic relationship type.

---

[2] http://www.mapdb.org/

Table 2. Statistical characterization of the considered datasets

| | SemEval | | Wikipedia | | AImed |
	Train	Test	Train	Test	Data
# Sentences	8,000	2,717	2,199	926	2,202
# Terms	137,593	46,873	49,721	20,656	75,878
# Relation classes	19	19	47	47	2
# Relation instances (except *not-related/other*)	6,590	2,263	15,963	6,386	1,000
# Nominals	16,001	5,434	5,468	2,258	4,084
Avg. sentence length (terms)	119.8	119.4	177.2	172.8	184.2
StDev. sentence length (terms)	45.0	44.4	104.5	100.1	98
Avg. instances/class	421	143	295.6	135.9	1,961.5
StDev. instances/class	317.5	105.5	1707.3	728.2	1,372.5
Max. instances/class (except *not-related/other*)	844	22	268	113	1,000
Min. instances/class	1	1	1	1	1,000

4 Experimental Validation

The minwise hashing method proposed semantic relation extraction was evaluated with three different document collections, from different domains, which are commonly used as benchmarks for this problem.

A statistical characterization of the three datasets used in our experiments is given in Table 2. The SemEval dataset[3] consists of 10,717 sentences, annotated according to 19 possible classes between two nominals in each sentence (9 non-symmetric relations types, such as *cause-effect* or *instrument-agency*, plus another label for denoting that no relationship is being expressed). The dataset is relatively balanced between the classes, and is split into training and testing subsets, with 8,000 instances for training and 2,717 for testing.

The Wikipedia dataset[4] consists of paragraphs from 441 Wikipedia pages, containing annotations for 4,681 relation mentions of 53 different relation types like *job-title*, *birth-place*, or *political-affiliation*. The dataset comes split into training and testing subsets, with about 70% of the paragraphs for testing, and the remaining 30% for training. In the Wikipedia dataset, the distribution of the examples per class is highly skewed: *job-title* is the most frequent relation (379 instances), whereas *grandmother* and *discovered* have just one example in the dataset. Moreover, although the full dataset contains annotations to the 53 different semantic relationship types, only 46 types are included in both the training and testing subsets, and still, of these 46 relation types, 14 of them have less than 10 examples. We therefore measured our results over a subset of 46 relationship types. In the experiments with this Wikipedia dataset, we only considered the problem of predicting the relationship type (i.e., classifying according to one of the 46 semantic relationship types, or as *other*), and not according to the direction of the relation. Of all the three datasets, this was the one that least fitted our general approach for modeling the relationship extraction task, and significant adaptations had to be made.

---

[3] http://semeval2.fbk.eu/semeval2.php?location=tasks&taskid=11
[4] http://cs.neiu.edu/~culotta/data/wikipedia.html

Table 3. Results obtained with different configurations of the proposed method

Dataset	Min Hash	1 kNN			3 kNN			5 kNN			7 kNN		
		P	R	F1	P	R	F1	P	R	F1	P	R	F1
	200/25	0.662	0.622	0.641	0.683	0.642	0.662	0.698	0.652	0.674	0.698	0.637	0.666
	200/50	0.662	0.621	0.640	0.683	0.643	0.662	0.698	0.651	0.673	0.698	0.636	0.666
	400/25	0.664	0.636	0.650	0.685	0.668	0.676	0.708	0.672	0.690	0.691	0.667	0.679
SemEval	400/50	0.663	0.635	0.649	0.684	0.664	0.674	**0.708**	0.674	**0.690**	0.694	0.670	0.682
(18 classes)	600/25	0.657	0.631	0.644	0.677	0.660	0.669	0.697	0.674	0.685	0.695	0.660	0.677
	600/50	0.657	0.631	0.644	0.676	0.658	0.667	0.699	**0.678**	0.688	0.694	0.664	0.678
	800/25	0.654	0.630	0.642	0.675	0.656	0.665	0.694	0.662	0.678	0.696	0.658	0.677
	800/50	0.654	0.632	0.643	0.677	0.658	0.667	0.698	0.665	0.681	0.696	0.658	0.676
	200/25	0.410	0.336	0.369	0.434	0.335	0.378	0.439	0.310	0.363	0.489	0.323	0.389
	200/50	0.409	0.336	0.369	0.435	0.336	0.379	0.440	0.310	0.364	0.489	0.321	0.387
	400/25	0.453	0.350	0.394	0.472	0.354	0.405	0.507	0.348	0.413	0.485	0.323	0.388
Wikipedia	400/50	0.450	0.349	0.393	0.468	0.354	0.403	0.503	0.350	0.412	0.509	0.328	0.399
(46 classes)	600/25	0.419	0.344	0.378	0.439	0.352	0.391	0.492	0.364	0.419	0.522	**0.365**	**0.430**
	600/50	0.419	0.343	0.377	0.444	0.354	0.394	0.485	0.353	0.408	**0.532**	0.353	0.425
	800/20	0.416	0.344	0.377	0.431	0.348	0.385	0.493	0.351	0.410	0.513	0.343	0.411
	800/50	0.419	0.345	0.378	0.433	0.350	0.387	0.515	0.346	0.414	0.517	0.338	0.409
	200/25	0.405	0.545	0.465	0.430	0.509	0.466	0.480	0.484	0.482	0.507	0.460	0.482
	200/50	0.405	0.545	0.465	0.430	0.509	0.466	0.480	0.484	0.482	0.507	0.460	0.482
	400/25	0.420	0.589	0.491	0.451	0.554	0.497	0.481	0.524	0.501	0.516	0.502	0.509
AImed	400/50	0.420	0.588	0.490	0.455	0.561	0.502	0.484	0.529	0.505	**0.519**	0.505	0.512
(1 class)	600/25	0.409	0.605	0.488	0.445	0.571	0.500	0.475	0.529	0.500	0.511	0.513	0.512
	600/50	0.409	0.605	0.488	0.445	0.571	0.500	0.475	0.530	0.501	0.511	0.513	0.512
	800/25	0.416	0.613	0.496	0.453	0.595	0.514	0.481	0.547	0.512	0.490	0.512	0.501
	800/50	0.418	**0.614**	0.498	0.454	0.596	**0.515**	0.482	0.545	0.511	0.489	0.514	0.501

Finally, the AImed corpus[5] consists of 225 MEDLINE abstracts, 200 of which describing interactions between human proteins. There are 4,084 protein references and approximately 1,000 tagged interactions. In this data set there is no distinction between genes and proteins, and the relations are symmetric and of a single type. We relied on a 10-fold cross validation methodology in the experiments with AImed, using the same splits as Mooney and Bunescu [4].

Using the three different datasets, we experimented with different parameters for the minwise hashing-based scheme, namely by varying the number of nearest neighbors that was considered (1, 3, 5 or 7), the size of the min-hash signatures (200, 400, 600 or 800 integers) and the number of LSH bands (25 or 50 bands). Notice that when using b bands of r rows each, the probability that the signatures of two sets S_1 and S_2 agree in all the rows of at least one band, therefore becoming a candidate pair, is $1 - (1 - J(S_1, S_2)^r)^b$. With 50 bands and a min-hash signature of size 600, roughly one in 1000 pairs that are as high as 85% similar will fail to become a candidate pair through the LSH method, and thus be a false negative. With these parameters, instances with a similarity bellow 85% are also likely to be discarded through the LSH method, which can contribute to the confidence in a correct classification (i.e., we are also trading precision for recall, when selecting the minwise hashing parameters).

As evaluation measures, we used macro-averaged precision, recall and F1-scores over the relation labels, apart from the *not-related/other* labels. This corresponds to calculating macro-averaged scores over 18 classes in the case of the SemEval dataset, over 46 classes in the case of the Wikipedia dataset, and we

---

[5] ftp://ftp.cs.utexas.edu/pub/mooney/bio-data/

measured results over the single *is-related* class in the AImed dataset. Table 3 presents the obtained results, showing that using the 5 or the 7 nearest neighbors, instead of just the most similar example, results in an increased performance for the SemEval and Wikipedia datasets, while a better F1 score was obtained for the AImed dataset when considering the 3 nearest training examples.

Table 4 presents per-class results in the case of the SemEval dataset, considering the configuration that achieved the best performance in the results from Table 3 (i.e., a configuration using the 5 nearest neighbors, with a min-hash size of 400, and with 50 bands in the LSH method). Besides the results on the regular SemEval classification setting, involving relation types with direction, we also present results in a setting that ignores the relation directions (i.e., considering 8 different relationship classes) as well as individual results for the class corresponding to *other-relations*. The results show that some classes, such as *cause-effect* are relatively easy to classify, whereas classes such as *instrument-agency* are much harder. For the class corresponding to *entity-destination(e1,e2)*, the dataset only contains one instance for training and one instance for testing.

In terms of comparisons with the current state-of-the-art, the best participating system at the SemEval 2010 task achieved a performance of over 0.82 in terms of the F1 metric, whereas the second best system reported an F1 score of 0.77. The median F1 score was of 0.68, while our best F1 score was of 0.69. Participating systems used a variety of methods and resources. For instance the winning entry used Google's n-gram collection and an approach that splits the task into two classification steps (i.e., relationship type and direction), whereas the second best participant used Cyc as a semantic resource. Almost all participants used features derived from WordNet, Roget's Taxonomy, or Levin's verb classes. In our approach, we used a simpler set of features common to different domains and mostly language independent: we only need POS tags for computing some of the features, and POS tagging can be made efficiently and accurately for most languages [14]. Comparing with other approaches over the same datasets, our method is focused on addressing scalability, although still attaining competitive results in terms of accuracy.

In the specific case of the AImed dataset, previous studies have mostly compared different kernel-based methods, reporting F1 scores ranging from 0.26 to 0.60, with a common cross-validation methodology [1,17]. For instance the subsequence kernel from Bunescu and Mooney achieves an F1 score of 0.54 [4], while the all-dependency-paths kernel from Airola et al. achieves an F1 score of 0.56 [1]. Our min-hash approach has slightly inferior results, with an F1 score of 0.52, but we argue that our method has advantages over kernel-approaches in terms of simplicity, support for multi-class on-line learning, and scalability.

Our approach is significantly different from that of commonly used kernel-based methods, which involve two main components. First, the linguistic structures, to be used within the kernels, have to be generated. Dependency parsers can be about an order of magnitude faster than syntax parsers, on average taking 130 milliseconds per sentence on modern hardware. POS tagging is about 1.5 orders of magnitude faster than dependency parsing, on average taking 4

Table 4. Results obtained for different relation classes in the SemEval dataset

Relation	Direction	Instances (train/test)	Asymmetrical			Symmetrical		
			Precision	Recall	F1	Precision	Recall	F1
Cause-Effect	(e1,e2)	344/134	0.843	0.843	0.843	0.798	0.902	0.847
	(e2,e1)	659/194	0.735	0.902	0.810			
Component-Whole	(e1,e2)	470/162	0.572	0.759	0.653	0.628	0.670	0.648
	(e2,e1)	150/129	0.609	0.520	0.561			
Entity-Destination	(e1,e2)	844/291	0.744	0.911	0.819	0.747	0.901	0.817
	(e2,e1)	1/1	1.000	0.000	0.000			
Entity-Origin	(e1,e2)	568/211	0.789	0.815	0.802	0.756	0.795	0.775
	(e2,e1)	148/47	0.667	0.723	0.694			
Product-Producer	(e1,e2)	323/108	0.670	0.602	0.634	0.673	0.589	0.628
	(e2,e1)	394/123	0.654	0.569	0.609			
Member-Collection	(e1,e2)	78/32	0.778	0.438	0.560	0.767	0.777	0.772
	(e2,e1)	612/201	0.776	0.791	0.783			
Message-Topic	(e1,e2)	490/210	0.751	0.733	0.742	0.778	0.778	0.778
	(e2,e1)	144/51	0.750	0.706	0.727			
Content-Container	(e1,e2)	374/153	0.726	0.778	0.751	0.706	0.802	0.751
	(e2,e1)	166/39	0.627	0.821	0.711			
Instrument-Agency	(e1,e2)	97/22	0.429	0.545	0.480	0.605	0.667	0.634
	(e2,e1)	407/134	0.615	0.679	0.645			
Other	—	1410/454	—	—	—	0.442	0.293	0.352
Macro-average	—	—	0.708	0.674	0.690	0.718	0.764	0.740

milliseconds per sentence on modern hardware. Our method mostly relies on character quadgrams, although we also used POS tags for improving accuracy. Nonetheless, we avoid complex NLP operations associated with parsing the sentences. Second, we have that the substructures used by the kernels have to be determined, and the classifier has to be applied. Using the AImed dataset, Tikk et al. reported on times of approximately 66.4 and 10.8 seconds, for training and testing (i.e., without feature extraction) an SVM classifier using a shallow linguistic kernel that is essentially a simplified version of the subsequence kernel from Bunescu and Mooney [4], as well as times of approximately 4517.4 and 3.7 seconds from training and testing, using the all-paths-graph kernel [17]. In our experiments, which were executed on modern hardware and using a single core, it took us 172 seconds for processing all three stages (i.e., feature extraction, training and testing) with the SemEval 2010 dataset, when considering the 5 nearest neighbors, min-hash signatures of size 400, and 50 bands. Each fold from the AImed dataset takes around 161 seconds to process considering the 3 nearest neighbors, min-hash signatures of size 800 and 50 bands.

Although a direct comparison against the current stat-of-the-art methods, in terms of computational performance, cannot be made easily (i.e., this would require a common set of tools for performing feature extraction, as well as running the implementations of the different algorithms on the same hardware and over exactly the same datasets) we provide some detailed figures for the performance of our method. The charts on Figure 1 present the processing times, in seconds, that are separately involved in feature extraction from the training and testing data, training (i.e., data indexing) and classification, for the different settings represented in Table 3. In the case of the AImed dataset, the charts show the average time over the 10 different folds.

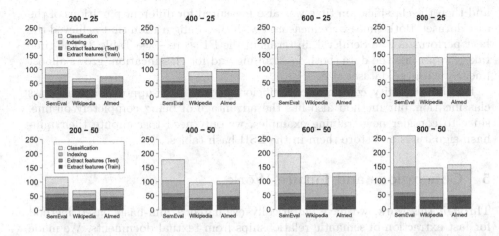

Fig. 1. Processing times, in seconds, for each dataset and configuration

The total processing time involved in each experiment naturally increases with the size of the dataset being considered. The time taken to extract features is independent of the LSH configuration being used, and the results indicate that these values represent a significant amount of the total processing time that is involved in each experiment. The results also show that the indexing times increase significantly as the size of the min-hash signatures gets larger, since more hash functions need to be computed, and more min-hash values have to be stored and compared. Augmenting the number of bands in turn increases the classification time, since the number of hash tables where we have to look for candidate instances, and possibly also the number of candidates increases.

Figure 2a shows how the proposed method performs, in terms of processing times, over the training phase and with an increasingly larger dataset, specifically when considering 25%, 50%, 75% and 100% of the SemEval dataset. In Figure 2b, the training was performed over the full SemEval training dataset,

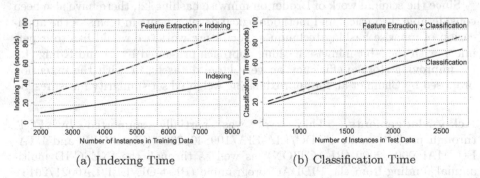

(a) Indexing Time   (b) Classification Time

Fig. 2. Processing times with different partitions of the SemEval dataset

and then the classification time was also measured for different partitions of the test dataset. Both times were measured with the configuration that achieved the best performance on SemEval, in terms of the F1 score – see Table 3. The time taken to process the data, both for training and for classification, grows almost linearly with the dataset size.

Besides simplicity, computational efficiency, and direct support for multi-class classification, our method also has the advantage of being completely on-line, since to consider new training examples, we only need to compute their min-hash signatures and store them in the LSH hash tables.

5   Conclusions and Future Work

Through this work, we have empirically evaluated a min-hash based method for fast extraction of semantic relationships from textual documents. We made experiments with well known datasets from three different application domains, showing that the task can be performed with high accuracy, using a simple on-line method based on similarity search, that is also computationally efficient.

Despite the interesting results, there are also opportunities for improvement. For instance, some of the state-of-the-art kernel methods for relation extraction also explore similarity between graph-based representations for the relation instances, derived from both lexical information and from constituency/dependency parsing [13]. Recent studies have proposed minwise hashing methods for comparing graphs [16], and it would be interesting to experiment with the application of these methods in the task of relationship extraction from textual documents, using richer graph-based representations for the relation instances. The proposed approach for measuring similarity between relation instances, based on the Jaccard similarity coefficient, could also be integrated into a general kernel-based framework (i.e., an SVM classifier using a kernel function based on the Jaccard similarity coefficient), and it would be interesting to see if supervised learning methods could be used to discover weights for properly combining different Jaccard similarity coefficients, computed for instance with basis on different sub-strings surrounding the relation operands.

Since the seminal work of Broder on minwise hashing [2], there have also been considerable theoretical and methodological developments in terms of the application of minwise hashing techniques. For future work, we would like to experiment with the b-bit minwise hashing approach presented by Li and König [12] for improving storage efficiency on very large datasets, or with the extension proposed by Chum et al. for approximating weighted set similarity measures [5].

Acknowledgements. This work was partially supported by FCT, through project grants PTDC/EIA-EIA/109840/2009 (SInteliGIS) and UTA-Est/MAI/0006/2009 (REACTION), as well as through the INESC-ID multi-annual funding from the PIDDAC programme (PEst-OE/EEI/LA0021/2013). The author David Batista was also supported through the PhD scholarship from FCT with reference SFRH/BD/70478/2010.

References

1. Airola, A., Pyysalo, S., Björne, J., Pahikkala, T., Ginter, F., Salakoski, T.: A graph kernel for protein-protein interaction extraction. In: Proceedings of the Workshop on Current Trends in Biomedical Natural Language Processing (2008)
2. Broder, A.: On the resemblance and containment of documents. In: Proceedings of the Conference on Compression and Complexity of Sequences (1997)
3. Bunescu, R., Mooney, R.: A shortest path dependency kernel for relation extraction. In: Proceedings of the Conference on Empirical Methods in Natural Language Processing (2005)
4. Bunescu, R., Mooney, R.: Subsequence kernels for relation extraction. In: Proceedings of the Conference on Neural Information Processing Systems (2006)
5. Chum, O., Philbin, J., Zisserman, A.: Near duplicate image detection: min-hash and tf-idf weighting. In: Proceedings of the British Machine Vision Conference (2008)
6. Culotta, A., McCallum, A., Betz, J.: Integrating probabilistic extraction models and data mining to discover relations and patterns in text. In: Proceedings of the Conference of the North American Chapter of the ACL (2006)
7. Culotta, A., Sorensen, J.: Dependency tree kernels for relation extraction. In: Proceedings of the Annual Meeting of the ACL (2004)
8. Fader, A., Soderland, S., Etzioni, O.: Identifying relations for open information extraction. In: Proceedings of the Conference on Empirical Methods in Natural Language Processing (2011)
9. Hendrickx, I., Kim, N., Kozareva, Z., Nakov, P., Séaghdha, D., Padó, S., Pennacchiotti, M., Romano, L., Szpakowicz, S.: Semeval-2010 task 8: Multi-way classification of semantic relations between pairs of nominals. In: Proceedings of the International Workshop on Semantic Evaluation (2010)
10. Kambhatla, N.: Combining lexical, syntactic, and semantic features with maximum entropy models for extracting relations. In: Proceedings of the Annual Meeting of the ACL (2004)
11. Kim, S., Yoon, J., Yang, J., Park, S.: Walk-weighted subsequence kernels for protein-protein interaction extraction. BMC Bioinformatics 11(107) (2010)
12. Li, P., König, C.: b-bit minwise hashing. In: Proceedings of the International Conference on World Wide Web (2010)
13. Nguyen, T.-V., Moschitti, A., Riccardi, G.: Convolution kernels on constituent, dependency and sequential structures for relation extraction. In: Proceedings of the Conference on Empirical Methods in Natural Language Processing (2009)
14. Petrov, S., Das, D., McDonald, R.T.: A universal part-of-speech tagset. In: Proceedings of the Conference on Language Resources and Evaluation (2012)
15. Rajaraman, A., Ullman, J.: Mining of massive datasets, ch. 3. Finding Similar Items. Cambridge University Press (2011)
16. Teixeira, C., Silva, A., Junior, W.: Min-hash fingerprints for graph kernels: A trade-off among accuracy, efficiency, and compression. Journal of Information and Data Management 3(3) (2012)
17. Tikk, D., Thomas, P., Palaga, P., Hakenberg, J., Leser, U.: A comprehensive benchmark of kernel methods to extract protein protein interactions from literature. PLoS Computational Biology 6(7) (2010)
18. Zelenko, D., Aone, C., Richardella, A.: Kernel methods for relation extraction. Journal of Machine Learning Research 3 (2003)

19. Zhang, Y., Lin, H., Yang, Z., Wang, J., Li, Y.: Hash subgraph pairwise kernel for protein-protein interaction extraction. IEEE/ACM Transactions on Computer Biology and Bioinformatics 9(4) (2012)
20. Zhao, S., Grishman, R.: Extracting relations with integrated information using kernel methods. In: Proceedings of the Annual Meeting of the ACL (2005)
21. Zhou, G., Zhang, M.: Extracting relation information from text documents by exploring various types of knowledge. Information Processing and Management 43(4) (2007)

Generating a Conceptual Representation
of a Legacy Web Application*

Roberto Rodríguez-Echeverría, Víctor M. Pavón, Fernando Macías,
José María Conejero, Pedro J. Clemente, and Fernando Sánchez-Figueroa

University of Extremadura (Spain),
Quercus Software Engineering Group
{rre,victorpavon,fernandomacias,chemacm,pjclemente,fernando}@unex.es
http://quercusseg.unex.es

Abstract. Web application (WA) development has been fueled by the
definition and evolution of web application frameworks since late 90's. In
parallel, Model Driven Web Engineering approaches have been defined
and successfully applied to reduce the effort of web application development and reuse, fostering the independence of the implementation technology. Although they pursue similar objectives, both approaches have
lived and evolved separately. The work presented herein tries to reduce
the gap between them by defining a model-driven reverse engineering
process to generate a conceptual representation from a framework-based
legacy web application. This work is part of a bigger project, named
MIGRARIA, whose main goal is to define a model-driven modernization process to obtain Rich Internet Aplications (RIAs) from legacy web
systems.

Keywords: Web Models Transformations, Reverse Engineering, Model-
Driven Web Engineering.

1 Introduction

The emergence of the software factory organizational model have favored the
massive adoption of development frameworks, well-known design patterns and
the definition of code conventions in order to organize the development of a
software product as a systematic factory process. Since late 90's, widespread
language-specific web development frameworks (e.g. Struts[1]) have supported
most of the actual development of WAs. These frameworks are often strongly
tied to the programming-language level, increasing the complexity of its maintenance and evolution. Thus, such processes have been performed in an ad-hoc
manner, resulting in very expensive and error-prone tasks.

---

* Work funded by Spanish Contract MIGRARIA - TIN2011-27340 at Ministerio de
Ciencia e Innovación and Gobierno de Extremadura (GR-10129) and European Regional Development Fund (ERDF).

[1] http://struts.apache.org/

X. Lin et al. (Eds.): WISE 2013, Part II, LNCS 8181, pp. 231–240, 2013.

At the same time, within an academic context, Model Driven Web Engineering (MDWE) approaches [8] have been defined to leverage model driven engineering methods and techniques in the development of WAs. MDE fosters the definition of languages in the problem space instead of the solution space. In the context of WA development, MDWE approaches propose different specification languages that share a main goal: independence of the implementation technology. The rapid evolution of web-related implementation technologies supposes an important issue in the development of WAs, but also in its maintenance and evolution. MDWE approaches allow defining WAs in a declarative way by means of conceptual representations and provide code generation engines to tackle the constant change of implementation technologies.

This work proposes an approach to approximate these two different worlds of web application development. A model-driven reverse engineering process is described. Such process defines the necessary activities, artifacts and tools to generate a conceptual representation from a legacy web application developed by means of a MVC²-based framework. As a model driven engineering process, abstraction, reusability, automatization and replicability are key factors to consider. In order to provide a detailed description, a concrete application scenario has been specified. In this case, on the one hand, it is considered that the legacy web application has been implemented by using a concrete set of frameworks (Struts v1.3 and Hibernate v3.6), conventions (naming, configuration, etc.) and design patterns (e.g. Data Access Object pattern). And, on the other hand, WebML [2] has been selected as target conceptual representation. A case study has been performed to validate the approach, the Conference Review System (CRS). Obviously, the proposed process is described at a conceptual level and different realizations are possible.

It also worths to mention that the work presented herein is part of a larger research project, called MIGRARIA³, where a systematic and semi-automatic process to modernize legacy non-model-based data-driven WAs into RIAs has been defined. One of the leading ideas of this project is to use model driven techniques and tools to deal with the complexity of extraction and interpretation processes. Inside this project and concerning the same motivation, some preliminary works have been already published [6,7]. The work presented herein supposes a major revision and extension of such works providing a comprehensive reverse engineering process not limited to the navigational concern.

The rest of the paper is structured as follows. In Section 2 an overview of the reverse engineering process is presented. Then, a detailed presentation of the generation of the MIGRARIA MVC model is performed in Section 3. The final transformation to WebML is introduced in Section 4. The related work is discussed in Section 5. And, finally, main conclusions and future work are outlined in Section 6.

---

² Model View Controller Pattern
³ http://www.unex.es/eweb/migraria

2   Reverse Engineering Process

The main goal of this work consists on the extraction of conceptual models from legacy WAs developed by means of web development frameworks based on the MVC pattern. A specific model driven reverse engineering process has been defined to fulfill that purpose. Common model-driven methods, techniques and tools are utilized to define such process in a systematic, replicable and reusable manner. Figure 1 describes the main steps of our model-driven reverse engineering process. Such process is organized as a sequence of three steps:

1. Technology-dependent model extraction. First, model representations of the software artifacts implementing the legacy WA are directly extracted by means of text to model transformations defined from the grammars of the programming languages involved to the metamodels of such grammars. The objective of this first step is to provide a comprehensive representation of the legacy WA in a format that can be uniformly processed by means of model driven techniques and tools. That way, reverse engineering techniques, as static analysis, can be performed as model queries or model transformations.
2. Conceptual MVC model generation. Within this work a specific language has been defined to generate a conceptual representation of a legacy MVC-based web application, the MIGRARIA MVC metamodel. This metamodel has been designed based on the MVC pattern that constitutes the conceptual foundation of a great number of web development frameworks. The objective of this second step is to generate a conceptual representation of the legacy WA based on MVC concepts and independent of implementation technology platform details, such as the concrete development frameworks or the programming languages used.
3. MVC to MDWE transformation. Most mature MDWE approaches provide engineers with friendly concrete syntax and sound tools to assist the development process. These features make them a very interesting target representation for our reverse engineering approach. The objective of this third step is to obtain a conceptual representation of the legacy WA in a concrete MDWE approach by defining a model to model transformation from the MVC model. Most MDWE approaches provide models to define presentation, navigation and data concerns in a conceptual point of view.

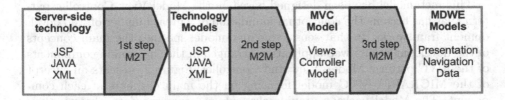

Fig. 1. Model-Driven Reverse Engineering

3   MVC Model Generation

In order to generate a conceptual MVC-based representation of the legacy system is necessary to define (1) a MVC metamodel for web applications and (2) the generation process by means of model-to-model transformation rules.

3.1   MIGRARIA MVC Metamodel

According to the goal of the MIGRARIA project, defining a modernization process framework for legacy web applications, a specific language has been defined to generate a conceptual representation of a legacy MVC-based web application, the MIGRARIA MVC metamodel.

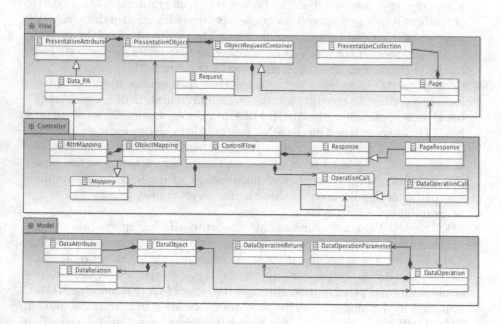

Fig. 2. MIGRARIA MVC metamodel overview

This metamodel has been designed based on the Model-View-Controller pattern that has become the conceptual foundation of a great number of web development frameworks. In that sense, this metamodel specifies the main concepts of the development of a web application arranged in the three main components of the MVC pattern: Model, View and Controller. Figure 2 presents an excerpt of the MIGRARIA MVC model focusing on the main elements of each component. The Model package provide elements to represent data objects, their attributes, their relationships and the operations defined over them. The View package provide elements to represent pages, as main containers, and presentation objects and requests, as main contents. Presentation objects, basically, have a set of attributes, can indicate data input or data output and can be presented individually or inside a collection. Meanwhile, requests are characterized by their parameters and path and define connection points with controller elements by means of the request-handler association. The Controller package provides elements to represent request handlers (*ControlFlow*), their mappings defined between presentation and data objects, their response defining a relationship with the target page element, and the sequence of operation calls performed to execute the requested action or to fetch the requested data. This metamodel is a comprehensive revision and extension of previous work [7].

3.2   MIGRARIA MVC Model Generation

The generation process is accomplished in three sequential steps, one for each package defined by MVC metamodel. First, all the interesting information related with data schema and data operations is collected in the model representing the Model component of the legacy system. Second, all the information concerning web pages composition and interaction is gathered in the model representing its View component. Those two models are generated independently of each other. And finally, all the information related to request handling and operation execution is collected by the model representing its Controller component. This last model plays a fundamental role and connects the former ones by defining mappings and relationships between them.

Figure 3 presents an example of a create operation. In this case, at the top, an excerpt of the author submission page from the CRS case study is shown that is basically composed by a HTML form allowing to specify the title, abstract, subjects and track of the paper to submit. As it can be observed, the form contains a pair of input text controls and a pair of select controls. The JSP producing the submission page is analyzed to get its presentation objects and requests. The product of that analysis is the page *PaperCreate* of the View model. This model element is composed of an input presentation object (HTML form) that contains a pair of data presentation attributes (input text) and a pair of dataset presentation attributes (select) representing corresponding form controls. Two presentation collections are also composing such view. Each one allows specifying the set of elements populating each dataset presentation attribute, i.e. the data for the select control forms. And the view also contains the specification of the request *new* that represents the submission of the form and references to the input presentation object *paperForm* by means of its *submit* attribute.

As a basic common pattern, at least, two different controllers are related to a concrete view, one populating its objects and collections and the other one handling the request generated from the view. In this case, the controller *PaperPopulateCreateAction* specifies the data operations called (*trackDAO.getAll* and *subjectDAO.getAll*) and the mappings defined to fetch the data for every dataset presentation attribute of the presentation object *paperForm*. The mappings define the relation between the controller instances storing the return of the operation calls (tracks, subjects) and the corresponding presentation collections defined in the view (*tracks* and subjects). Meanwhile, the second controller,

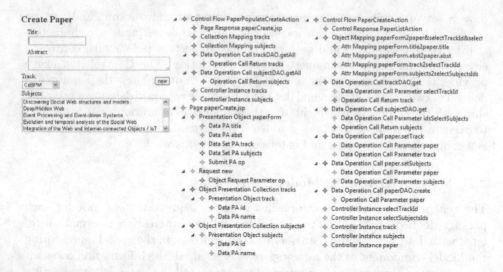

Fig. 3. MIGRARIA MVC model of paper creation (new submission)

PaperCreateAction, is responsible of handling the request *new*. A controller that receives a presentation object as input defines a controller instance to define the mapping. In the example, the controller instance *paper* is defined as an instance of the data object *Paper* and a mapping is established between that instance and the input presentation object *paperForm*. The mapping attributes allow defining finer grain mappings. In this case, on the one hand, the data presentation attributes of the input presentation object are mapped to the corresponding attributes of the data object *paper* referenced by the controller instance *paper*. And on the other hand, the dataset presentation attributes are both mapped to two different controller instances: one to store the track *id* selected and the other one to store the *ids* of the selected subjects. Those instances are used as parameters of data operation calls to fetch the data objects for those ids (*trackDAO.get* and *subjectDAO.get*). The returned data objects are stored on controller instances (*tracks* and *subjects*) and passed again as parameters of data operation calls to associate them with the instance *paper* representing the object in creation. Finally, the last data operation call of the sequence represents the creation of a new *Paper* object from the controller instance *paper* (*paperDAO.create*).

Additionally, controllers specify their response by means of a *response* element. This response element may have two different types, *page* or *control*, indicating whether the controller returns directly a view or delegates the control to another handler to generate the response view. Figure 3 presents both types of response. The page response is used by the controller that fetches the data to populate the *paperCreate* view, whilst the control response is used by the controller that handles the create operation to delegate the generation of the response view in another controller (*paperListAction*).

4   WebML Model Transformation

In this work, we have selected WebML as target MDWE approach because of its well-known syntax and professional development tools. For the sake of brevity, just an illustrative example of the generation of the hypertext model is presented in this work.

Figure 4 presents the elements generated to model the creation of a paper (a new submission) from the MVC model of Figure 3. The main structure of this model is conformed by the following elements:

- A *page*, named *paperCreate*, as main container, representing the author submission page.
- An *Entry Unit*, named *EU_Paper*, representing the HTML form to input the submission data.
- Two *Selector Units*, named *SU_Subjects* and *SU_Tracks*, representing the select controls of the HTML form.
- A *Create Unit*, named *CU_Paper*, representing the creation of a new data object paper.
- A *Connect Unit*, named *CU_PaperSubjects*, representing the association between the new paper and the selected conference subjects.
- And four links to connect every unit correspondingly.

As expected, the organization of these elements resemble the controller-view-controller pattern aforementioned. In this example, the controller responsible of populating the objects of the response view is transformed in the two selector units shown. Meanwhile the controller responsible of handling the request is modeled as the operation units presented.

A selector unit is then derived from the following pattern in a MVC model: an output object presentation collection used by a dataset presentation attribute of an input presentation object. In our example, the page *PaperCreate* contains the input presentation object *paperForm* with the dataset attribute *subjects* connected to the presentation collection *subjects*. Once identified the selector unit, it is necessary to find which data should be fetched by this unit. Such information may be extracted from the mapping defined in the controller between the collection and the instance storing the dataset resulting. That instance makes a reference to the data object from the Model defining its type. As a data object corresponds to a WebML entity, the selector unit entity is defined by the referenced data object.

Accordingly, an input presentation object is transformed into a entry unit. In the example, the entry unit *EU_Paper* is generated from the input presentation object *paperForm*. And it is located inside the WebML page generated from the MVC page *PaperCreate*. Every data presentation attribute of the input object is transformed in a *field* element of the entry unit. While every dataset presentation attribute generates a *selectionField* of the entry unit. In case the attribute *multiSelect* is set in a dataset attribute the generation process produces a *multiSelectionField* instead of a *selectionField*. The attributes of a MVC presentation object may indicate its value must be hidden, such information may be specified

in the fields of a WebML entry unit. Additionally, the requests defined over an
input presentation object are mapped to normal links connecting the entry unit
with the corresponding operation unit. And the parameter mapping is generated from the attribute mappings defined for the input object in the controller
handling the request. Figure 4 presents *EU_Paper* properties to show those attribute and request transformations. Additionally, a transport link is generated
between a selector unit and a *(multi)selectionField* of an entry unit, when the
aforementioned selector-unit pattern appears in the controller populating the
page containing the input presentation object (entry unit). Figure 4 presents
the properties of the transport link *SU_Subjects-EU_Paper*, illustrating how one
of the attributes of the presentation object (collection item) is marked as output
(*id*) and the other one as label.

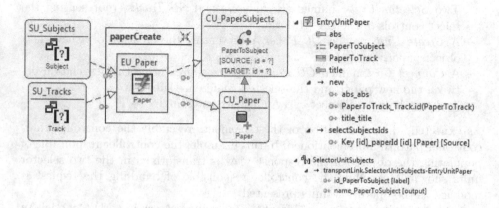

Fig. 4. WebML create operation

Concerning the WebML operation units, they are derived from the operation call sequence of a MVC controller. A MVC data operation call indicates a
reference to an operation defined within a concrete data object. The operation
type is used to identify the kind of operation unit to generate (create, update
or delete). And the data object indicates the entity of the operation unit. In
the example, the data operation call *paperDAO.create* is transformed into the
create unit *CU_Paper*. And an OK link is created to connect such operation
with the next one. In this case, the connect operation unit *CU_PaperSubjects*.
A connect operation unit is then derived from an operation call referencing a
data operation which represents a manytomany relation between two different
data objects. In the example, the data operation call *paperDAO.setSubjects* is
converted to the connect unit *CU_PaperSubjects*. The associated transport link
stems from the controller instance passed as parameter to that call operation.

Finally, it worths to mention that similar rules have been defined to generate
the other CRUD operations. Practically the complete set of WebML core units
are fully supported in this transformation process.

5   Related Work

Web Application information extraction has been traditionally performed by reverse engineering techniques [4]. Most approaches presented in that survey propose strategies of static analysis taking just web pages as input of the extraction process. They treat the legacy system as a black box and focus on analyzing its output (HTML pages). Meanwhile, the approach presented herein propose to extract the information concealed in the source code of the legacy system by means of static analysis. The majority of them are not conceived as model-driven approaches and pursue other aims that the generation of a conceptual representation.

[3][1] propose a model-driven process to generate a conceptual representation of a web application by using Ubiquitous Web Application (UWA) approach. But they also apply static analysis to the web application output not considering the system source code. That approach generates directly the final model without intermediate representations.

[5] proposes a reverse engineering process to obtain a WebML representation from a legacy PHP web shop application based on static code analysis. First, the source application is refactored to obtain a MVC version of it. Next, a code to model transformation into an intermediate model of the MVC web application is carried out. The last step is a model to model transformation from the the MVC model into a WebML model. Both approaches share the definition of a MVC metamodel and the transformation proccess to derive a WebML model from a MVC model. However, in our opinion, our MVC metamodel represents more accurately the general concepts defined by mainstream MVC-based web application frameworks, meanwhile the MVC metamodel proposed by the authors is closer to WebML schema. We then consider our approach provides a higher degree of reutilization.

6   Conclusions and Future Work

This work tackles an important part of the modernization process defined within the MIGRARIA project which faces the evolution of the presentation of a legacy web system towards a Web 2.0 new RIA client. In this paper we have focused on the reverse engineering process defined to extract and represent the relevant information from the legacy system. Its main activities have been detailed and organized into three main steps: i) the extraction of technology-dependent information from the legacy system; ii) the generation of a model conformed to our MVC metamodel and iii) the projection of this model towards a MDWE approach, in our case WebML. For the first step, the MoDisco discoverers have been extended to enable the extraction of MVC-frameworks specific semantics, e.g. Struts.The second step was accomplished by defining a new language, MIGRARIA MVC metamodel, instead of using one of the ripe existing MDWE approaches [8] cause most of them are designed to be used within a forward engineering process. They are conceived from a conceptual point of view which usually does not fit properly with the information extracted by means of a reverse engineering process. The projection to WebML is carried out by means of model-to-model transformations (MVC-to-WebML) that consider patterns identified in the MVC model to generate not only the navigational information of the system but also the existing CRUD operations into the WebML one. Then, the methods and tools of a concrete MDWE approach may leverage our latest modernization steps.

As main lines for future work we consider the following: (1) dealing with technology and code convention variability by means of transformation rules generation or adaptation techniques; (2) applying the approach to real legacy systems; and (3) improving the tool support of the reverse engineering process to simplify the engineer's decision making.

References

1. Bernardi, M.L., Di Lucca, G.A., Distante, D.: The RE-UWA approach to recover user centered conceptual models from Web applications. International Journal on Software Tools for Technology Transfer 11(6), 485–501 (2009)
2. Ceri, S., Fraternali, P., Bongio, A.: Web modeling language (webml): a modeling language for designing web sites. In: Proceedings of the 9th International World Wide Web Conference on Computer Networks: The International Journal of Computer and Telecommunications Netowrking, Amsterdam, The Netherlands, pp. 137–157. North-Holland Publishing Co. (2000)
3. Di Lucca, A.G., Distante, D., Bernardi, M.L.: Recovering Conceptual Models from Web Applications. In: Pierce, R., Stanrney, J. (eds.) SIGDOC 2006 Proceedings of the 24th Annual ACM International Conference on Design of Communication, pp. 113–120. ACM Press (2006)
4. Patel, R., Coenen, F., Martin, R., Archer, L.: Reverse Engineering of Web Applications: A Technical Review. Technical Report July 2007, University of Liverpool Department of Computer Science, Liverpool (2007)
5. Rieder, M.: From Legacy Web Applications to WebML models. A Framework-based Reverse Engineering Process. PhD thesis, TUW (2009)
6. Rodríguez-Echeverría, R., Conejero, J.M., Clemente, P.J., Villalobos, M.D., Sánchez-Figueroa, F.: Extracting navigational models from struts-based web applications. In: Brambilla, M., Tokuda, T., Tolksdorf, R. (eds.) ICWE 2012. LNCS, vol. 7387, pp. 419–426. Springer, Heidelberg (2012)
7. Rodríguez-Echeverría, R., Conejero, J.M., Clemente, P.J., Villalobos, M.D., Sánchez-Figueroa, F.: Generation of WebML Hypertext Models from Legacy Web Applications. In: 14th IEEE International Symposium on Web Systems Evolution, Riva del Garda, Italy, pp. 91–95. IEEE Computer Society (2012)
8. Rossi, G., Pastor, O., Schwabe, D., Olsina, L.: Web Engineering: Modelling and Implementing Web Applications. Human-Computer Interaction Series (October 2007)

AKMiner: Domain-Specific Knowledge Graph Mining
from Academic Literatures

Shanshan Huang and Xiaojun Wan[*]

Institute of Computer Science and Technology,
The MOE Key Laboratory of Computational Linguistics, Peking University,
Beijing 100871, China
{huangshanshan2010,wanxiaojun}@pku.edu.cn

Abstract. Existing academic search systems like Google Scholar usually return a long list of scientific articles for a given research domain or topic (e.g. "document summarization", "information extraction"), and users need to read volumes of articles to get some ideas of the research progress for a domain, which is very tedious and time-consuming. In this paper, we propose a novel system called AKMiner (Academic Knowledge Miner) to automatically mine useful knowledge from the articles in a specific domain, and then visually present the knowledge graph to users. Our system consists of two major components: a) the extraction module which extracts academic concepts and relations jointly based on Markov Logic Network, and b) the visualization module which generates knowledge graphs, including concept-cloud graphs and concept relation graphs. Experimental results demonstrate the effectiveness of each component of our proposed system.

Keywords: Knowledge graph generation, academic knowledge extraction, academic literature mining, Markov logic, AKMiner.

1   Introduction

Academic literatures offer scientific researchers knowledge about current academic progress as well as history in a specific research domain (e.g. "document summarization", "information extraction"). By reading scientific literatures, the beginners grasp basic knowledge of a research domain before in-depth study, and experienced researchers conveniently obtain the information of recent significant progress. However, relevant academic information is usually overloaded, and researchers often find an overwhelming number of publications of interests. Digital libraries offer various database querying tools, and Internet search companies have developed academic search engines. Typical academic search engines like *Google Scholar*, *Microsoft Academic Search* and *CiteSeer*, all achieve good retrieval performance.

However, most of the academic search systems simply return a list of relevant articles by matching keywords or author's information. Usually, there are quite a lot of articles for a specific topic, and it is hard for researchers to have a quick glimpse on the whole structure of knowledge, especially for the beginners.

---

[*] Corresponding author.

X. Lin et al. (Eds.): WISE 2013, Part II, LNCS 8181, pp. 241–255, 2013.

Another way to catch the development of a research domain or topic is to read review papers, such as survey papers published in ACM Computing Surveys every year. However, there are usually only a few high-quality review papers available for most research domains. Besides, the publishing cycle of review papers is relatively long, compared to fast updating of research achievements. Moreover, review papers are usually very long, and the knowledge is embedded in texts, which often fails to show the knowledge structure visually and vividly.

To help researchers relieve the burden of tedious paper reading, and acquire information about the recent achievements and developments quickly, we propose a novel system called AKMiner (Academic Knowledge Miner) to extract academic concepts and relations from academic literatures and generate knowledge graphs for a given research domain or topic. Hence, researchers can quickly get a basic vision of a research domain and learn the latest achievements and developments directly.

Our AKMiner system consists of two phases: a) extraction of academic concepts and relations, and b) academic knowledge graph generation. For the first step, Markov Logic Network (MLN) is applied to build a joint model for extracting academic concepts and their relations from literatures simultaneously. A concept cloud graph and a concept relation graph are then generated to visually present domain-specific concepts and relations, respectively. For concept cloud graph generation, the PageRank algorithm [19] is applied to calculate the importance scores of different concepts in a domain. The more important a concept is, the bigger it is displayed in the graphs. Experiments are conducted on datasets in four domains, and the training data and test data are from different domains. Evaluation results show that our proposed approach is more effective than several baseline methods (e.g. support vector machine (SVM), conditional random fields (CRF), C-Value) for knowledge extraction. Case studies show several good characteristics of the generated knowledge graphs.

The contributions of our study are summarized as follows:

- We propose a novel AKMiner system to mine knowledge graphs for a research domain, which can be used for enhancing existing academic search systems.
- We propose a joint model based on Markov Logic Network to extract academic concepts and their relations.

Experimental results on four datasets and cases of knowledge graphs show the effectiveness of our proposed system.

2   Related Works

2.1   Information Extraction

Information extraction (IE) techniques have been widely investigated for various purposes in the text mining and natural language processing fields. Typical IE tasks include named entity recognition (NER), relation detection and classification, and event detection and classification. State-of-the-art NER systems usually formulate NER as a sequence labeling problem, and employ various discriminative structured prediction models (e.g. hidden Markov model, maximum-entropy Markov model, CRF) to resolve it. Relation detection and classification aims to extract relations among entities and classify them into different categories. Many statistical techniques have been investigated to predict entity relations such as [32]. Recently, joint models have been proposed to extract entities and relations simultaneously [20], [21], which achieve superior performance to pipeline models. Event detection and classification aims to mine significant events and group them into relevant topics, benefiting more discussions and comparisons within and crossing topics.

Term extraction or terminology extraction is a subtask of information extraction, which aims to extract multi-word expressions from a large corpus. Different methodologies for automatic term extraction have been investigated, including linguistic, statistical and hybrid approaches [5]. Linguistic approaches basically identify terms by using some heuristic rules or patterns [8]. Statistical approaches usually rank candidate terms according to a criterion, which is able to distinguish among true and false terms and give higher confidence to the better terms [3], [7], [10]. Hybrid approaches usually combine linguistic and statistical approaches into a two-stage framework [7], [15]. Keyphrase extraction is a very similar task with term extraction. Most keyphrase extraction methods first extract candidate phrases with natural language processing techniques, and then rank the candidate phrases and select the final keyphrases with supervised or unsupervised algorithms [14], [18], [30].

2.2   Academic Literature Mining

In recent years, various text mining techniques have been investigated in the academic domain. Such techniques include metadata extraction [11], [13], paper summarization [1], [23], survey generation [2], [23], [31], literature search [9], [17], paper recommendation [28], and trend visualization [6], [25], [29]. In particular, [29] extracts elemental technologies through the structure of research paper's title and analyzes technical trend in any research fields.   [6] reports an effort to integrate statistics, text analytics, and visualization in a multiple coordinated window environment to support rapid understanding of scientific paper collections. [25] creates metro maps using metrics of influence, coverage, and connectivity for scientific literature, and shows the relations between papers. However, they all do not present fine-grained knowledge structure from paper content directly.

2.3   Markov Logic Networks

MLN is a powerful representation for statistical relational learning, and it has been applied in a few tasks in the area of information extraction. For example, [2] uses Markov Logic Network to extract database records from text or semi-structured sources. [26] proposes a system utilizing MLN for entity resolution, and [22] implements it into joint unsupervised coreference resolution. Besides, [27] uses MLN to discover social relationships in consumer photo collections.

3   Overview

3.1   System Overview

The purpose of our AKMiner system is to generate knowledge graphs for academic domains, which is useful for researchers to get useful information quickly and visually. Given a set of literatures in a specific domain, academic knowledge graphs can be built through the flow in Figure 1.

The framework of our proposed system consists of two main procedures: a) the extraction module which extracts academic concepts and relations, and b) the visualization module which visualizes knowledge graphs. Prior to the two main procedures, a preprocessing step is taken, as described in Section 6. To make inputs for MLN, we use a chunking tool to get noun phrases (NPs) from academic literatures and build the candidate set by

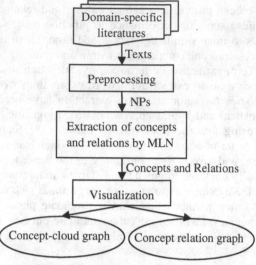

Fig. 1. Framework of AKMiner system

some preprocessing. Then, we extract useful academic concepts and relations among them with our proposed joint model. Finally, knowledge graphs, including concept-cloud graph and concept relation graph, are generated based on the extracted concepts and relations.

3.2   Definition of Concepts and Relations

In this section, we define academic concepts and relations used in our system. To describe an academic domain, we often present the main tasks in the domain, and summarize the main methods applied to solve the tasks. Hence, in this study, we focus on two kinds of academic concepts: *Task* and *Method*.

The *Task* concepts are specific problems to be solved in academic literatures, including all concepts related to tasks, subtasks, problems and projects, like "machine translation", "document summarization", "query-focused summarization", etc.

The *Method* concepts are defined as ways to solve specific *Tasks*, including all concepts describing algorithms, techniques, models, tools and so on, like "Markov logic", "CRFs", "heuristic-based algorithm", and "XML-based tool".

The relations extracted by our system include two kinds of relations: the relations between the two kinds of academic concepts (i.e. *Method-Task* relations), and the relations within the same kind of academic concepts (i.e. *Method-Method* relations and *Task-Task* relations). The first kind of relations are formed when a *Method* is applied to a referred *Task* (e.g., "extractive method" for "document summarization"). The latter kind of relations between *Methods* or between *Tasks* are formed by dependency, evolution and enhancements (e.g., "Markov model" and "hidden Markov model", "single document summarization" and "multi-document summarization").

4 Markov Logic Networks

Markov Logic Network (MLN) [24] is a probabilistic logic model which combines the idea of Markov network with first-order logic. A first-order knowledge base (KB) is a set of formulas in first-order logic, in which the predicates and functions are used to describe properties and relations among the objects. Particularly, a function (e.g., Anna = MotherOf(Bob)) represents a mapping from objects to objects, while a predicate represents relations among objects or some attributes (e.g., Friends(Jim, Bob)). In a first-order KB, if a case violates even one formula, it has zero probability.

The basic idea of Markov logic is to soften these hard constraints with associated weights, so that they can be violated with some penalty.

Formally, a Markov Logic Network L is a set of pairs (F_i, w_i), where F_i, is a formula in first-order logic, and w_i is the weight of formula F_i. A Markov Logic Network specifies a probability distribution over the set of possible worlds χ as follows,

$$P(X = x) = \frac{1}{Z} \cdot exp\left(\sum_i w_i \cdot n_i(x)\right)$$

where $n_i(x)$ is the number of true groundings of F_i in possible worlds χ, and Z is the normalization constant, given by $Z = \sum_{x \in \chi} exp(\sum_i w_i \cdot n_i(x))$.

5 MLN for Extraction of Concepts and Relations

In this section, we describe the construction of Markov Logic Network for concepts and relations extraction from academic literatures. Particularly, the attributes and relations of concepts are represented by predicates, and rules impose certain constraints over those predicates. We now describe our proposed approach in detail.

5.1 Concepts and Predicates

Our goal is to extract *Method* and *Task* concepts. For sake of predicates formulation, we add another kind of concepts, *Other* concept, not belonging to *Method* nor *Task*.

We define two kinds of predicates: query predicates and evidence predicates. The values of query predicates are unknown and need to be inferred. Here, the query predicates are the categories of concepts and the relations between them, including:

- **Method(concept)**: It indicates that the concept is a *Method*.
- **Task(concept)**: It indicates that the concept describes a *Task*.
- **Relation(concept₁,concept₂)**: It indicates that there is a relation between a *Method* and a *Task*. The first concept is a *Method* and the second one is a *Task*.
- **Relation_m(concept₁,concept₂)**: It indicates a relation between two *Methods*.
- **Relation_t(concept₁,concept₂)**: It indicates a relation between two *Tasks*.

The evidence predicates can be any features extracted from the inputs, and the values of them are already known. We represent the evidence predicates as follows:

- **Key_m(concept)**: A concept contains a keyword related to *Method* concept. For example, "graph-based method" contains *Method* related keyword "method".
- **Key_t(concept)**: A concept contains a keyword related to *Task* concept.

- **Key_m_outside(concept)**: A keyword related to *Method* is detected around the concept in a text. For example, in the text "... we apply CRF in sequence annotation ...", the keyword "apply" appears before concept "CRF".
- **Key_t_outside(concept)**: A keyword related to *Task* is detected around the concept in a text. For example, the keywords "task of" occurs before *Task* concept "paper summarization" in the text "... task of paper summarization ...".
- **Neighbor(concept$_1$,concept$_2$)**: Two concepts appearing closely are neighbors. We define "closely" within three sentences in our experiments.
- **Contain(concept$_1$,concept$_2$)**: This predicate is true only in two cases. One is that concept c_1 only contains one more suffix than concept c_2, such as "CRF model" and "CRF". Another case is that the last words of two concepts are the same, such as concept "single document summarization" and "document summarization".
- **Apposition(concept$_1$,concept$_2$)**: It indicates that the concepts are appositional in a text. We define the appositional format as "concept$_1$ and (or) concept$_2$".

Note that in our experiments, all the keywords are collected and summarized manually. We investigate through reading numerous articles and collect four lists of keywords for Key_m, Key_t, Key_m_outside and Key_t_outside, respectively. The keywords for Key_m/t only contain nouns (17 words for *Method*, and 10 for *Task*), such as "algorithm", "method", "model" for *Method*, and "project", "problem" for *Task*. The keywords for Key_m/t_outside contain nouns and verbs (60 for *Method* and 31 for *Task*), such as "propose", "present", "describe", etc. These keywords are independent of domains, and can be utilized in any domains.

5.2   Rules in MLN

Hard Rules
Hard rules describe the hard constraints that should always hold true. These rules are given a prior weight larger than other rules.

Non-overlapping Rules
We make the rule that three categories of concepts (*Task*, *Method* and *Other*) do not overlap. That is to say, a concept only belongs to one of the categories.

$$Task(c) \Rightarrow !Method(c) \wedge !Other(c) \tag{1}$$
$$Method(c) \Rightarrow !Task(c) \wedge !Other(c) \tag{2}$$
$$Other(c) \Rightarrow !Task(c) \wedge !Method(c) \tag{3}$$

Rules from Definition.
These rules are from the definition of three query predicates indicating relations.

$$Relation(c_1, c_2) \Rightarrow Method(c_1) \wedge Task(c_2) \tag{4}$$
$$Relation\_m(c_1, c_2) \Rightarrow Method(c_1) \wedge Method(c_2) \tag{5}$$
$$Relation\_t(c_1, c_2) \Rightarrow Task(c_1) \wedge Task(c_2) \tag{6}$$

Soft Rules
These rules describe constraints that we expect to be usually true, but not all the time.

Neighbor-based Rules

We assume that two neighbor concepts probably have a relation.

$$Neighbor(c_1, c_2) \wedge Method(c_1) \wedge Task(c_2) \Rightarrow Relation(c_1, c_2) \qquad (7)$$

$$Neighbor(c_1, c_2) \wedge Method(c_1) \wedge Method(c_2) \Rightarrow Relation\_m(c_1, c_2) \qquad (8)$$

$$Neighbor(c_1, c_2) \wedge Task(c_1) \wedge Task(c_2) \Rightarrow Relation\_t(c_1, c_2) \qquad (9)$$

Keyword-based Rules

We consider keywords as important clues to extract academic concepts. For example, a concept with the word "algorithm" as suffix is probably a *Method*, and a concept with "problem" as suffix is likely to be a *Task*. But sometimes this rule is not true. For instance, "efficient method" is not a useful concept. So keyword-based rules are soft.

$$Key\_m(c) \Rightarrow Method(c) \qquad (10)$$

$$Key\_t(c) \Rightarrow Task(c) \qquad (11)$$

In addition, the words around concept phrases also offer much useful information, such as the words "propose", "demonstrate", and "present". A concept with this kind of keywords appearing around probably belongs to the related concept category. The rules are as follows.

$$Key\_m\_outside(c) \Rightarrow Method(c) \qquad (12)$$

$$Key\_t\_outside(c) \Rightarrow Task(c) \qquad (13)$$

Containing-based Rules

From texts, we find that if the predicate *Contain* (c_1, c_2) is true, c_1 likely contains more modifiers than c_2. If c_1 is a *Task*, c_2 tends to be a *Task*, too. For example, *Contain("Spanisah to English MT", "MT") ∧ Task("Chinese to English MT") ⇒ Task("MT")*. In this case, c_1 helps to determine the category of c_2. But if c_1 is a *Method*, we cannot perform the inference. For example: *Contain("phrase-based MT", "MT") ∧ Method("phrase-based MT") ⇏ Method("MT")*.

However, if c_2 is a *Method* concept, c_1 is probably a *Method* concept. For instance, *Contain ("Hidden Markov model", "Markov model") ∧ Method ("Markov model") => Method ("Hidden Markov model")*. On the other hand, if c_2 is a *Task* concept, the inference is unreasonable. For example, *Contain ("phrase-based statistical machine translation", "machine translation") ∧ Task ("machine translation") ⇏ Task ("phrase-based statistical machine translation")*. In all, we have the formulas below.

$$Contain(c_1, c_2) \wedge Task(c_1) \Rightarrow Task(c_2) \qquad (14)$$

$$Contain(c_1, c_2) \wedge Method(c_2) \Rightarrow Method(c_1) \qquad (15)$$

Apposition Rules

Apposition information can also be used for extraction of concepts. Given the predicate *Apposition* (c_1, c_2), we add a rule that if two concepts are appositive, they likely belong to the same category. The rule is built below:

$$Apposition(c_1, c_2) \Rightarrow (Method(c_1) \wedge Method(c_2)) \vee$$
$$(Task(c_1) \wedge Task(c_2)) \vee (Other(c_1) \wedge Other(c_2)) \qquad (16)$$

Actually, keyword-based rules are the basic rules to recognize concepts, and containing-based and apposition rules help to detect some missing concepts (e.g. conditional random fields) and correct some wrong judgments, so that the concept extraction is more complete.

Transitivity Rules

Actually, some useful concept pairs may fail to be extracted, so we apply transitivity in relation inference. It is supposed that if c_1 and c_2 have a relation, while c_2 and c_3 have a relation, and then concept c_1 and c_3 potentially have a relation. However, the precondition is that they are in the same category. Relations between *Task* and *Method* are not assumed to have transitivity. The rules are represented below:

$$Relation\_m(c_1, c_2) \wedge Relation\_m(c_2, c_3) \Rightarrow Relation\_m(c_1, c_3) \qquad (17)$$

$$Relation\_t(c_1, c_2) \wedge Relation\_t(c_2, c_3) \Rightarrow Relation\_t(c_1, c_3) \qquad (18)$$

6   Empirical Evaluation

6.1   Evaluation Setup

Dataset and Evaluation Metrics

Our goal is to automatically extract useful knowledge from a set of literatures in a specific domain[1]. Since there exist many different research domains and new research domains emerge rapidly, it is impossible to label training data on each domain in practice. Therefore, experiments will be performed on a cross-domain setting, i.e. the training data and the test data are from different domains. To build the datasets, we collect 200 literature articles from the ACL corpus[2] on 4 domains in the field of natural language processing: "Statistical Machine Translation" (SMT), "Document Summarization" (DS), "Sentiment Analysis and Opinion Mining" (SAOM) and "Reference Resolution" (RR), and each domain contains 50 literature articles. We use articles from ACL corpus because they are well formatted, with pages about 7 to 10, and the text edition is conveniently available on the Internet. In our experiments, only the title, abstract, introduction and related work sections are extracted and used, because these sections have covered most concepts and relations in literature articles. Besides, the other text sections, such as approach and experiments, often contain much noise, like equations, figures and tables. Therefore, we only focus on partial texts of the literatures in this study. We read the texts and manually label the concepts and their relations. In our supervised experiments, we use the labeled data from three domains as training set, and use the labeled data from the remaining one domain as test set. Therefore, four-fold validation are conducted. We calculate the Precision (P), Recall (R) and F-measure (F) values to measure the performance of extractions.

Data Preprocessing

Considering that a concept is usually a noun phrase, we first use a noun phrase (NP) chunking tool - the StanfordNLP toolkit[3] to get all NPs from literature texts.

---

[1] The size of the literature set in a domain can range from several to thousands.
[2] http://www.acl.org/   The datasets used will be published on our website.
[3] http://nlp.stanford.edu/

The initial NP set contains many useless phrases, such as "we", "this paper", "future work", etc. So we build a simple filter to filter out some NPs by using several linguistic rules, and also collect a "stop words" list to abandon some useless NPs or some useless words in NPs, such as "some", "many", "efficient", "general", etc. In addition, too long or too short NPs are also excluded.

Baselines

To verify the performance of our proposed joint model, we develop three baseline methods (CRF model, SVM classifier, and C-Value method) for concept extraction and a baseline method (SVM classifier) for relation extraction.

The CRF model [16] can be used to extract academic concepts in literatures. To make training data for the CRF model, we search the labeled concepts in the literatures and mark the occurrences of the concepts. If one concept occurs in a literature, each word covered by the concept phrase will be labeled with relevant tags (T, M). The other words are marked with irrelevant tag (O). When two concept phrases are overlapping, such as "machine translation" and "statistical machine translation", we consider the longer one as the complete phrase. The features used in the CRF model include the current word, words around the current word, part of speech and keyword-based features. The lists of keywords are the same as they are for MLN. Besides, to guarantee the fairness of the comparison, NP features are also used in CRF, including a word's position in an NP (i.e. outside, beginning, middle and ending of an NP).

Support Vector Machine [4] is another popular method for information extraction, classifying NPs into three categories. The features used for SVM include prefix and suffix information of NP, and keyword-based features. The keyword-based features also include keywords inside and around concepts, which are used in MLN.

C-Value [10] is an unsupervised algorithm for multi-word terms recognition, and here we refer to concept phrases. It extracts phrases according to the frequency of occurrence of phrases, and enhances the common statistical measure of frequency by using linguistic information and combining statistical features of the candidate NPs.

For concept relation extraction, we use SVM as a baseline. Classifications are conducted on candidate concept pairs, which are combinations of two adjacent concepts. The candidate concept pairs are classified into two categories, having relations or not. The features for classification include phrases' length and position information, relation related keywords and the keywords' position information. Phrases' lengths are calculated by ignoring brackets and the context inside brackets. The position information includes positions in a sentence, the orders of phrase pairs, and the distances between phrases in pairs. Relation related keywords are also collected manually, including phrases like "based on", "enhancement", "developed on", etc.

Alchemy

We utilize the Alchemy system[4] in our experiments. Alchemy is an open-source package providing a series of algorithms for statistical relational learning and probabilistic logic inference, based on the Markov logic representation.

---

[4] http://alchemy.cs.washington.edu/

6.2   Evaluation Results

Results of Concept Extraction

The number of extracted academic concepts and the performance values of the different methods on four domains are shown in Table 1. The performance values of *Method* extraction, *Task* extraction, and the overall extraction are calculated separately. As C-Value is an algorithm for recognition of useful phrases, and it cannot distinguish the concept types, so only the overall extraction performance is measured. The average values of F-measure are calculated based on F-measures on four domains.

We can see that the overall performance values of our proposed joint model (MLN) are much better than the baselines. Compared to the baseline methods that can only extract concepts independently, the MLN model infers concepts and relations jointly, and it can take into consideration the joint information in a domain. In addition, the rules make it possible to supplement some missing concepts and remove some incorrect concepts. So the extraction results are more accurate and complete.

Table 1. Comparison Results of Concept Extraction

Test Domain		Task Concept			Method Concept			Task + Method			
		MLN	CRF	SVM	MLN	CRF	SVM	MLN	CRF	SVM	CValue
SMT	No.	48	48	84	275	283	293	323	331	377	324
	P	0.875	0.542	0.529	0.920	0.618	0.870	0.913	0.607	0.796	0.574
	R	0.824	0.510	0.870	0.907	0.626	0.916	0.894	0.609	0.909	0.564
	F	**0.849**	0.526	0.658	**0.913**	0.622	0.892	**0.904**	0.608	0.849	0.569
DS	No.	96	107	269	172	181	183	268	288	452	373
	P	0.958	0.701	0.390	0.936	0.586	0.929	0.944	0.628	0.608	0.413
	R	0.836	0.682	0.955	0.885	0.582	0.934	0.866	0.620	0.942	0.527
	F	**0.893**	0.691	0.554	0.910	0.584	**0.932**	**0.904**	0.624	0.739	0.463
SAOM	No.	185	200	534	323	353	351	508	553	885	455
	P	0.789	0.750	0.322	0.944	0.572	0.895	0.888	0.637	0.549	0.673
	R	0.741	0.761	0.873	0.892	0.591	0.918	0.837	0.653	0.902	0.568
	F	**0.764**	0.756	0.471	**0.917**	0.581	0.906	**0.862**	0.645	0.683	0.616
RR	No.	95	111	281	177	196	195	272	307	476	297
	P	0.958	0.748	0.320	0.972	0.607	0.903	0.967	0.658	0.559	0.731
	R	0.858	0.783	0.849	0.901	0.623	0.921	0.886	0.680	0.896	0.731
	F	**0.905**	0.765	0.465	**0.935**	0.615	0.912	**0.924**	0.669	0.688	0.731
Average	P	0.895	0.685	0.390	0.943	0.596	0.899	0.928	0.633	0.628	0.598
	R	0.815	0.684	0.887	0.896	0.606	0.922	0.871	0.641	0.912	0.597
	F	**0.853**	0.685	0.537	**0.919**	0.601	0.911	**0.899**	0.637	0.740	0.595

We can also find that the extraction of *Methods* is generally more effective than the extraction of *Tasks*. Considering the characteristics of academic literatures, they are usually discussing several methods for one task, so the number of *Methods* is certainly larger than that of *Tasks*, which can be confirmed from the numbers of extracted concepts in Table 1. With more ground atoms of *Method* in training data, the MLN model can learn a better model and performs more effectively for *Method* extraction.

Finally, we can conclude that our joint model helps to extract concepts in a more comprehensive way by taking into account global information and the relations between concepts. This advantage will be more notable in relation extraction.

Results of Relation Extraction
In our MLN model, concepts and relations are extracted together. While in the baseline, concepts are firstly extracted by the baseline model with the best performance (i.e. the SVM model), and then the SVM classifier is used for relation classification based on the extracted concept pairs. The results are shown in Table 2.

The performance of MLN is much better than SVM. Making full use of joint information, the joint model helps to improve the performance of knowledge extraction.

Although the SVM classifier returns more relation pairs, the recall value it gets is lower than that of MLN. This shows that the SVM classifier brings more incorrect results and its ability of discrimination between positive and negative cases in relation extraction is relatively poor.

Table 2. Results of Relation Extraction

	SMT				DS			
	No.	P	R	F	No.	P	R	F
MLN	469	0.85	0.82	**0.84**	277	0.90	0.74	**0.81**
SVM	668	0.52	0.71	0.60	368	0.67	0.73	0.70
	SAOM				RR			
	No.	P	R	F	No.	P	R	F
MLN	745	0.92	0.76	**0.83**	344	0.94	0.86	**0.90**
SVM	687	0.69	0.53	0.60	440	0.53	0.62	0.57
	Average performance in four domains							
MLN	P: 90.2%; R: 79.6%; F: 84.6%.							
SVM	P: 60.1%; R: 64.7%; F: 62.3%.							

7 Knowledge Graph Generation

7.1 Concept Importance Assessment

Before generating knowledge graphs, we assess the importance of different concepts in a specific domain, which is the basis of concept cloud graph generation. Importance assessment is based on the frequency a concept occurs in the literatures and the importance of other concepts that have relations with this concept. We have two assumptions:

- The more frequently a concept appears, the more important it is.
- The more important the other concepts related to the current concept are, the more important the current concept is.

In addition, the position where a concept occurs is also considered into importance assessment. For instance, a concept occurs in title will get a larger importance weight than the ones in other sections. The position-based weight of a concept is multiplied with the frequency it occurs in the corresponding section. In this study, each concept's prior importance score is given by its frequency, and the frequency of a concept occurs in title is multiplied by a weight (we assign 1.2 in experiments).

In the experiments, the personalized PageRank algorithm [12] is applied for concept importance assessment. The personalized PageRank algorithm can take into account both the prior scores and the concept relations, and the importance score of a concept is iteratively calculated until convergence.

7.2   Concept Cloud Graph Generation

In order to help researchers have a clearer and more comprehensive understanding of the concepts, concept cloud graph is generated according to the importance of concepts. Given the numerical importance scores of academic concepts, concept-clouds are produced by the "Wordle" platform[5].

As a case, the concept-cloud graph for the "DS" domain is shown in Figures 2. Due to the limited space, the graph only contains a portion of the concepts. After taking a look at the graph, we can know the most important concepts in the "DS" domain clearly and quickly. It is obvious that the most notable concept is "multi-document summarization", while "document summarization" and "single-document summarization" are relatively less notable. This implies that more researches focus on multi-document summarization nowadays. The concept "topic detection and tracking" is also notable, as it has many associations with "Document Summarization". In addition, concepts with larger size are mostly *Task* concepts, indicating that *Task* concepts occur more frequently in literatures.

Fig. 2. Concept Cloud Graph on "DS" Domain

---

[5] http://www.wordle.net/advanceds

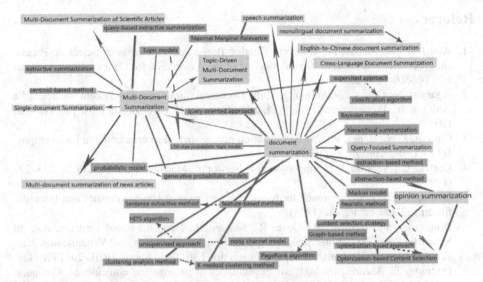

Fig. 3. Part of Relations Graph on "DS" Domain

(Yellow boxes denote *Tasks*, and green boxes represent *Method* concepts. Red arrows denote *Task-Method* relations, blue arrows denote *Task-Task* relations, and green arrows denote *Method-Method* relationships.)

7.3   Concept Relation Graph Generation

In order to present the extracted concept relations in academic literatures vividly, a concept relation graph is built on a research domain. Taking the "DS" domain as an example, the concept relation graph is shown in Figure 3. Again because of space limitation, only a portion of concept relations are shown. From the graph, we can learn that the "document summarization" *Task* evolves into a few *Tasks*, such as "single-document summarization", "multi-document summarization", "monolingual document summarization", "query-focused summarization" and so on. Many *Methods* applied to these *Tasks* are shown directly and vividly, and relations between *Methods* are also shown visually.

8   Conclusions and Future Work

In this paper, we propose a novel AKMiner system to extract academic concepts and relations between concepts from academic literatures in a research domain. The extracted results are visually presented to users via concept-cloud graph and concept relation graph. In future work, we will further classify relations into more specific categories, and even relations among three or more concepts will be investigated. We will also explore to recommend *Methods* for some *Tasks* through analysis of the network of concepts and their relations. This will bring much inspiration to scientific research work.

Acknowledgments. The work was supported by NSFC (61170166) and National High-Tech R&D Program (2012AA011101).

References

1. Abu-Jbara, A., Radev, D.: Coherent Citation-Based Summarization of Scientific Papers. In: Proceedings of the 49th Annual Meeting of the Association for Computational Linguistics, pp. 500–509 (2011)
2. Agarwal, N., Gvr, K.: SciSumm: A Multi-Document Summarization System for Scientific Articles. In: Proceedings of the ACL-HLT 2011 System Demonstrations, pp. 115–120 (2011)
3. Church, K.W., Hanks, P.: Word association norms, mutual information and lexicography. In: ACL 1989, pp. 76–83 (1989)
4. Cortes, C., Vapnik, V.: Support-vector Networks. Machine Learning 20(3), 273–297 (1995)
5. Daille, B.: Combined approach for terminology extraction: lexical statistics and linguistic filtering. Technical Report (1995)
6. Dunne, C., Shneiderman, B., Gove, R., Klavans, J., Dorr, B.: Rapid Understanding of Scientific Paper Collections: Integrating Statistics, Text Analytics, and Visualization. University of Maryland, Human-Computer Interaction Lab. Tech. Report HCIL-2011 (2011)
7. Dunning, T.: Accurate methods for the statistics of surprise and coincidence. Computational Linguistics 19(1), 61–74 (1994)
8. Earl, L.L.: Experiments in automatic extracting and indexing. Information Storage and Retrieval 6(X), 273–288 (1970)
9. EL-Arini, K., Guestrin, C.: Beyond Keyword Search: Discovering Relevant Scientific Literature. In: Proceedings of the 17th SIGKDD (2011)
10. Frantzi, K., Ananiadou, S., Mima, H.: Automatic recognition of multi-word terms: the C-value / NC-value method. International Journal of Digital Library 3, 115–130 (2000)
11. Han, H., Giles, C., Manavoglu, E., Zha, H., Zhang, Z., Fox, E.: Automatic Document Meta-data Extraction using Support Vector Machines. In: Proceedings of Joint Conference on Digital Libraries (2003)
12. Haveliwala, T.H.: Topic-sensitive pagerank. In: WWW 2002, pp. 517–526. ACM, New York (2002)
13. Isaac, Councill, G., Giles, C.L., Kan, M.Y.: ParsCit: An open-source CRF reference string parsing package. In: Proceedings of the Language Resources and Evaluation Conference (LREC 2008), Marrakesh, Morrocco (May 2008)
14. Jiang, X., Hu, Y., Li, H.: A ranking approach to keyphrase extraction. In Microsoft Research Technical Report (2009)
15. Justeson, J., Katz, S.: Technical terminology: some linguistic properties and an algorithm for identification in text. Natural Language Engineering 1, 9–27 (1995)
16. Lafferty, J., McCallum, A., Pereira, F.: Conditional random fields: Probabilistic models for segmenting and labeling sequence data. In: Proceedings of ICML, pp. 282–289 (2001)
17. Li, N., Zhu, L., Mitra, P., Mueller, K., Poweleit, E.: oreChem ChemXSeer: a semantic digital library for chemistry. In: JCDL (2010)
18. Mihalcea, R., Tarau, P.: TextRank: Bringing order into texts. In: Proceedings of EMNLP 2004 (2004)
19. Page, L., Brin, S., Motwani, R., Winograd, T.: The pagerank citation ranking: Bringing order to the web. Technical report, Stanford Digital Libraries (1998)
20. Poon, H., Domingos, P.: Joint inference in information extraction. In: Proceedings of AAAI 2007 (2007)

21. Poon, H., Vanderwende, L.: Joint inference for knowledge extraction from biomedical literature. In: Proceedings of the 2010 Annual Conference of the North American Chapter of the Association for Computational Linguistics (2010)
22. Poon, H., Domingos, P.: Joint unsupervised coreference resolution with Markov logic. In: Proceedings of the Conference on Empirical Methods in Natural Language Processing (2008)
23. Qazvinian, V., Radev, D.R.: Scientific Paper Summarization Using Citation Summary Networks. In: Proceedings of COLING 2008, vol. 1, pp. 689–696 (2008)
24. Richardson, M., Domingos, P.: Markov Logic Networks. Machine Learing 62(1-2), 107–136
25. Shahaf, D., Guestrin, C., Horvitz, E.: Metro Maps of Science. In: Proceedings of the 18th ACM SIGKDD (2012)
26. Singla, P., Domingos, P.: Entity resolution with markov logic. In: Proceedings of ICDM 2006 (2006)
27. Singla, P., Kautz, H., Luo, J.: Discovery of social relationships in consumer photo collections using Markov Logic. In: Workshops of CVPRW 2008 (2008)
28. S.K., Kan, M.: Scholarly paper recommendation via user's recent research interests. In: JCDL (2010)
29. Kondo, T., Nanba, H., Takezawa, T., Okumura, M.: Technical Trend Analysis by Analyzing Research Papers' Titles. In: Vetulani, Z. (ed.) LTC 2009. LNCS, vol. 6562, pp. 512–521. Springer, Heidelberg (2011)
30. Wan, X.J., Xiao, J.G.: Single document keyphrase extraction using neighborhood knowledge. In: Proceedings of AAAI 2008 (2008)
31. Yeloglu, O., Milios, E., Zincir-Heywood, N.: Multi-document Summarization of Scientific Corpora. In: SAC (2011)
32. Zhu, J., Nie, Z., Liu, X., Zhang, B.: StatSnowball: a statistical approach to extracting entity relationships. In: Proceedings of 18th WWW Conference (2009)

CDR Analysis Based Telco Churn Prediction and Customer Behavior Insights: A Case Study

Natwar Modani[1], Kuntal Dey[1], Ritesh Gupta[2,*], and Shantanu Godbole[1]

[1] IBM Research India, New Delhi and Bangalore, India
{namodani,kuntadey,shantanugodbole}@in.ibm.com
[2] Google, Bangalore, India
ritesh.iit@gmail.com

Abstract. Telecom churn has emerged as the single largest cause of revenue erosion for telecom operators. Predicting churners from the demographic and behavioral information of customers has been a topic of active research interest and industrial practice. In this case study paper, we present our experience of participating in a competitive evaluation for churn prediction and customer insights for a leading Asian telecom operator. We build a data mining model to predict churners using key performance indicators (KPI) based on customer Call Detail Records (CDR) and additional customer data available with the operator. Further, we analyze the social network formed between the (prepaid and postpaid) churners as well as the entire subscriber base. Our churn prediction method provided a lift of 8.4 over a nominal churn rate of 4.17% on 10% of the prepaid talking subscriber base on test data, and a lift of 7.62 on a nominal churn rate of 7.3% as reported in the customer evaluation on unseen data. This outperformed next best competitor in the study by more than twice. We also correlate social behavior patterns for churners and overall subscriber base. Our study indicates strong socially influenced churn among postpaid subscribers, in contrast with the prepaid subscribers. Our work provides guidelines and a template for conducting similar real-world studies for large telecom operators.

1 Introduction

The Telecom industry is booming in emerging markets like Asia and Africa, while witnessing consolidation in developed markets like America and Europe. Across the board, telecom companies are acquiring a laser focus on their customers and trying to enhance their products and services with an aim of maximizing satisfaction and life time value of users. The phenomenon of churn, where customers switch telecom operators due to financial, personal, or social reasons, is a major threat to the life time value of the subscriber. In such a scenario, churn management has emerged as a major problem area for the operators to focus on. Churn management involves not only predicting customers likely to churn based on demographic and behavioral data, it also involves preventive actions and running

---

* Ritesh Gupta contributed to this study when he was at IBM Research India.

X. Lin et al. (Eds.): WISE 2013, Part II, LNCS 8181, pp. 256–269, 2013.

campaigns to try and salvage predicted churners. Understanding behavior of customers involves several factors, like understanding calling patterns using social network call graph analysis techniques, modeling social influence, and finding interesting communities of users that are actionable for the operator.

In this case study paper, we present our experiences of participating in a competitive evaluation of churn prediction and customer understanding for a large Asian telecom operator. This Asian operator had a user base of about 12 million customers across prepaid and postpaid segments. We participated in the study along with two competitors and significantly out-performed them on churn prediction accuracy. Our customer behavior insights also received excellent response and evoked surprise from the operator.

While there are several methods in the literature for telecom churn prediction, we have not come across a method which attempts to combine social and non-social features, like usage, demographics, recharge history. Also, the churn prediction methods stop at providing only the list of predictions, and do not analyze the social network amongst the churners. In this article, we attempt to cover these aspects and also focus on helping the operator identify actionable customer insights beyond just churn prediction.

This study was performed on 3 months of Call Detail Record (CDR) data provided by the operator in raw format. We had 2 months to process all the data, build the KPIs, perform data mining, model tweaking, churn prediction, and come up with interesting actionable customer behavior insights. We performed CDR analysis and constructed a multitude of social interaction features which were utilized in building our churn prediction model. We ran several graph based community detection algorithms to find communities like cliques, stars, and dense subgraphs among customers to understand social calling behavior.

We divided the subscribers into four groups based upon two dimensions: (a) prepaid versus postpaid, and (b) talking versus silent (people who generated/did not generate CDRs in the previous month). Amongst the talking subscribers, our churn prediction process yielded a lift of 8.4 for postpaid and 4.4 for prepaid churners over the nominal churn rate for 10% prediction base. Nominal churn rate is defined as fraction of total number of churners to total subscriber population. Lift is ratio of precision and nominal churn rate.

Our analysis revealed several interesting insights. Among prepaid subscribers, the interaction between similar age subscribers is more frequent than between different ages with the same age pairs having a higher talk time (1.5 times more) as compared to different age pairs. In contrast, among postpaid subscribers, the interaction between different age subscribers is more frequent but a higher talk time per pair is observed for pairs having similar age (2 times more). Another interesting observation is that the ratio of friend pairs to acquaintance pairs is small (10 − 20%), but they contribute almost 80% of the talk time.

In the following, we recap related research work and industrial published efforts in this area. We then describe the problem setting and data set for our case study. We also recount challenges faced and our experiences as industrial researchers for this study. We then describe the methods and techniques we employed for churn prediction and understanding customer behavior. Subsequently, we present detailed results and their analysis, and finally conclude this article.

2   Related Work

With multiple telecom service providers penetrating almost all corners of the modern world, telecom churn has become extremely common. Analyzing and predicting telecom churn has emerged as a topic of research interest [1–6].

Churn has been studied in online social networks [7], service industries such as providing Internet service [8], banking [9] and Pay-TV [10]. In the services industry, churn has been observed to be the outcome of several factors such as pricing, convenience, service contract expiry, core service failures, customer service failures and dissatisfaction of provider ethics [11]. Subsequent studies such as [10] decompose such factors into features of customers and experiences to predict churn. Studies such as [12, 13] use features specific to churn prediction on online social networks, and model activity and role based churn.

Researchers have studied influences in multiple settings such as web search [14], [15], movies [16] and topics [17]. Diffusion-based spreading activation (SPA) has been studied in [18]. Subsequently, [5] observed social influence spreading to be a significant factor for telecom churn. They compare their results with the traditional machine learning based approach using social parameters based upon usage, connectivity and interconnectivity as feature set. [7] subsequently shows the applicability of spreading activation based models in online game settings.

Discovery of communities [19–21] have been active areas of research in social network analysis. [6] investigates the behavior of communities with respect to churn using a 'group first' approach for prediction. This study defines groups within the social networks, and finds node influences within groups. A machine learning model is built to classify the groups as churners. Although it appears to be interesting, this approach does not provide method for determining several parameters needed to apply the model to a new dataset. The incomplete specifications of the method makes it difficult to reproduce.

In contrast to this related work, we provide a method that combines social and non-social features, like usage, demographics, recharge history for churn prediction, and also helps identify actionable customer insights.

3   Study Setting

A leading Asian telecom operator with over 12 million subscribers was looking for a churn management and customer insights solution. They requested industrial analytics solution providers to participate in a competitive evaluation where they would give the same dataset to all participants and expect back results within 2 months. This competitive evaluation required addressing two aspects: the primary focus was churn prediction for their prepaid and postpaid subscriber bases, and the secondary focus was on customer behavior insights to help them understand non-obvious patterns of subscriber behavior.

We participated in this competitive evaluation using our research solution Telecom Social Network Analysis Churn Solution (T-SNACS) described in this paper to target the two parts of the evaluation. T-SNACS fits the bill perfectly in terms of satisfying the evaluation requirements, as it comprises of a churn prediction method and a suite of graph-based social network analysis algorithms.

3.1 Input Data and Basic Statistics

The data provided to the participants of the competitive evaluation was CDR dumps of voice calls and SMS for prepaid and postpaid subscribers of the telecom operator for the months of $M1$, $M2$, and $M3$. The data spans across the entire country, with no sampling. We were also separately provided demographic information about the subscribers.

The data provided by the telecom operator is also used to run their commercial operations, complying with government and legal requirements, including billing and record maintenance, and is clean and reliable. Table 1 shows the distribution of the prepaid and postpaid subscribers. The *Talking* customers were part of at least one phone call within the corresponding time period, and the *Silent* did not make or receive any call during this period.

We observe that the population is skewed. The number of prepaid subscribers are more than 3.2 times the number of postpaid subscribers. Further, the nominal churn rate among prepaid customers (5.777) is significantly higher compared to that of postpaid customers (3.509). The skew becomes even more stark in terms of the absolute number of churners, as we find 6 times prepaid customers to have churned, compared to postpaid customers. We further observe that the churn rate the among silent customers is about 5-6 times higher compared to the talking subscribers for both prepaid and postpaid subscribers.

The operator expected a list of customers likely to churn as the output, which they then validated internally against CDR data from $M4$ and beyond (evaluation participants did not have access to this information). Additionally, we

Table 1. Basic statistics about the social network and churn behavior of the subscribers (* denotes previous month talking behavior - useful for our predictions)

Prepaid	Num Active Subscribers			Num Churners			Nominal Churn Rate		
Month	Talking	Silent	Total	Talking*	Silent*	Total	Talking	Silent	Total
M1	7,813,510	1,083,731	8,897,241	n/a	n/a	621,940	n/a	n/a	6.9903
M2	8,138,041	936,987	9,075,028	329,368	291,048	620,416	4.2154	26.8561	6.8365
M3	8,008,762	1,121,261	9,130,023	339,403	188,078	527,481	4.1706	20.0726	5.7774
All Months			10,348,303			1,769,837			
Postpaid	Num Active Subscribers			Num Churners			Nominal Churn Rate		
Month	Talking	Silent	Total	Talking*	Silent*	Total	Talking	Silent	Total
M1	2,243,376	585,936	2,829,312	n/a	n/a	105,497	n/a	n/a	3.7287
M2	2,306,719	539,665	2,846,384	38,268	52,199	90,467	1.7058	8.9087	3.1783
M3	2,296,957	569,432	2,866,389	43,582	57,012	100,594	1.8894	10.5643	3.509
All Months			3,239,194			296,558			

provided insights about customer calling behaviors, subscriber communities, and potential acquisition targets for value added services for the operator's revenue enhancement. We now outline some of the interesting challenges and experiences.

3.2 Challenges: Data Scale and Domain Understanding

The total size of the data given to the evaluation participants was about 2.5 TB in compressed format. As the data was coming from a live system, the operator was not willing to do any preprocessing on the data, which could have reduced the size significantly. Such a large dataset required a few days for decompression, cleaning and all our pre-processing activities. No sample was provided before the actual data and the overall evaluation timeline was tight, it was imperative to make sure that we quickly understood the data and still, carried out the extraction process correctly at the first attempt. Since this was a competitive evaluation with multiple vendors, the operator was not willing to entertain too many questions about the data semantics. The data was provided on an "as-is" basis, and applying domain knowledge to understand the data was important.

4 T-SNACS

We now present the details of the methods and techniques comprising T-SNACS. First, we describe the data preparation and social network construction. Then, we briefly talk about the data mining methods we used for churn prediction task. Then we describe the KPIs that we derived from the data to build the data mining models. Finally, we discuss some of the techniques that were instrumental in deriving insights from the social network analysis. These insights are the result of domain specific understanding of trends and their appropriate interpretation – and as we will see, some very useful insights resulted.

4.1 Data Preparation

The first step was to decompress the data. The overall CDR dump contained many types of calls (some of which are intermediate format), so we filtered to data to retain only the MOC (mobile originating call, for an outgoing call), MTC (mobile terminating call, for an incoming call), SMS MO (mobile originating SMS, for an outgoing SMS) and SMS MT (mobile terminating SMS, for an incoming SMS) CDRs. Also, the CDRs contain many fields, several of which are not relevant from out point of view. We eliminated such fields.

To construct the social network for a given month, we combined all the calls between a pair of people to generate pairwise summary statistics. The dataset being large, the process required several days of computing to generate the summary statistics for the 3 months of data. This pairwise summary was used to construct an adjacency format representation of the social network graph. These pairwise summary statistics were further aggregated to generate last call date related statistics (described later), which was an important set of KPIs for our data mining models. Some statistics about the data is given in Table 1.

4.2   Data Mining Models

Feature selection is an important aspect of building data mining models. We determined the most significant features amongst the set of features we had derived based on mutual information measure (between the class variable and the feature in question). As the data had considerable skew (Table 1), we also needed to account for it in our approach. We used undersampling of non-churners to get a class balanced subset of data points. We built our data mining models with CHAID decision tree and Logistic Regression from a commercially available data mining package. We present our findings in the next section.

4.3   Features Used for Modeling

We constructed several features from the CDR data, which are described below. We divided the features in the following 6 categories.

Pure Social KPIs
1. OutDegree, InDegree, TotalDegree: Number of subscribers who make/receive calls/SMS to/from this person.
2. ChurnerOutDegree, ChurnerInDegree, ChurnerTotalDegree: Number of churners who make/receive calls/SMS to/from this person.
3. CompSubsOutDegree, CompSubsInDegree, CompSubsTotalDegree: Number of competition subscribers who make/receive calls/SMS to/from this person.

Usage KPIs
1. NumSMSOut, NumSMSIn: Number of SMS sent/received.
2. NumCallsOut, NumCallsIn: Number of calls made/received.
3. TotalCallDurationOut, TotalCallDurationIn: Total duration of calls made/received.
4. TotalCallAndSMSOutWeight, TotalCallAndSMSInWeight: Combined (SMS+Call) effective usage for outward/inward communications.

Social Usage and SNA KPIs
1. ChurnerNumSMSOut, ChurnerNumSMSIn, ChurnerNumCallsOut, Churner-NumCallsIn: Number of SMS/calls made/received to/from churners.
2. ChurnerOutCallDuration, ChurnerInCallDuration: Total duration of calls made/received to/from churners.
3. CompSubNumSMSOut, CompSubNumSMSIn, CompSubNumCallsOut, CompSubNumCallsIn: Number of SMS/calls made/received to/from competition subscribers.
4. CompSubOutCallDuration, CompSubInCallDuration: Total duration of calls made/received to/from competition subscribers.

Demographics KPIs
1. Age, Gender and Ethnicity of the subscriber.

Recharge KPIs for Prepaid
1. NumRecharges, TotalRecharge, LastRechargeAmt, LastRechargeDate.

Last Usage KPIs
1. LastCallMadeAsHomeDate, LastCallReceivedAsHomeDate,
LastCallAsHomeDate: Date of the last call made/received, or either, as home
subscriber.
2. LastSMSMadeAsHomeDate, LastSMSReceivedAsHomeDate,
LastSMSAsHomeDate: Date of the last SMS sent/received, or either, as home
subscriber.
3. LastCallDate, LastSMSDate: Date of the last call or SMS, made/received.
4. LastCallOrSMSAsHomeSentRecDate: Date of the last call/SMS made/
received as home subscriber.
5. LastCallOrSMSSentRecDate: Date of the last call/SMS made/received.

4.4   Graph-Based SNA Communities

We found the following types of communities on the churner induced subgraphs
as well as on the overall subscriber base, and derived further insights from these.

- Maximal Cliques: A clique is a set of vertices such that an edge exists between
  each pair of vertices within the set, i.e., the induced subgraph of the group
  vertices is complete. A maximal clique is a clique that is not contained in
  any other clique, and represents a strong social network community. We used
  [21] to find all maximal cliques in the churner-induced subgraphs.
- Stars: Stars are 'hub-and-spoke' structures. The hub is connected to a set of
  spokes. The spokes are largely disconnected. Stars can be of two types, global
  or local. In case of global stars, the spokes are (largely) not connected to any
  other nodes, whereas for local stars, the spokes are (largely) not connected
  to other spokes. Stars are on the other end of the spectrum compared to the
  dense maximal cliques, as these are inherently sparse structures.
- Densest Subgraph: Densest subgraph is the subgraph representing the social
  network, which has the highest value of the density measure for any subgraph
  in the graph. The popular density measures are minimum degree (lowest
  degree of a node in the given graph), average degree (number of edges divided
  by number of nodes) and density (number of edges divided by the possible
  number of edges). We used [22] to find the densest subgraphs in the churner-
  induced subgraphs.

5   Results and Analysis

In this section, we present results of using T-SNACS on the telecom operator
dataset. First, we discuss the comparative performance of various alternatives
and point out the best combination. Then, we present our churn prediction results obtained from the chosen approach in terms of lift curves. Then we share

the evaluation of our results received from the customer. This evaluation clearly shows that our technique significantly outperformed the competition methods. Then, we explore customer behavior and obtain insights around social interactions. Finally, we discuss some additional interesting observations.

At this point, we also define the measures used for evaluating the performance. Let the total number of subscribers be S, total number of predictions be d, number of total churners in the population be C, and the number of correct predictions be p. Then, accuracy is defined as ratio of correct predictions p in the given set of predictions d (which we evaluate at various levels) in percentage terms, i.e., $\%Accuracy = 100 * p/d$.
Coverage is defined as fraction of churners correctly predicted p to the total number of churners C in the subscriber base S in percentage terms, i.e., $\%Coverage = 100 * p/C$.
Lift is defined as the ratio of %Accuracy and the nominal churn rate (in % terms, which is the ratio of total number of churners to the total population), i.e., $Lift = (p/d)/(C/S)$.
We characterize the goodness of the methods using these three quality measures.

5.1 Choosing the Method

As mentioned earlier, we experimented with taking all the samples versus an undersampling to get class balance, taking all attribute versus selecting the most significant attributes, and CHAID decision tree versus Logistics Regression. Figure 1 shows the small prediction base part of results obtained using various options. Some important observations are:

- CHAID outperform Logistic Regression in all cases.
- Sampling and Attribute selection helps to improve the performance of both CHAID and Logistic Regression.
- Class balance is more important then feature selection.

Hence, we use undersampling to obtain class balance and feature selection with CHAID decision tree for the rest of the experiments. The following features were found to be more discriminative, and hence used in the decision trees learnt mentioned in Figure 1. Age, Gender, Ethnicity, TotalDegree, TotalCallAndSM-SOutWeight, compSubsOutDegree, numCallsIn, compSubNumSMSOut, compSubNumSMSIn, compSubNumCallsIn, compSubOutCallDuration, compSubInCallDuration, numRecharges, TotalRecharge, LastRechargeAmt, LastRechargeDate. It is interesting and encouraging to note that as expected, demographic features are relevant to distinguish between subscribers in general. In addition, features pertaining to phone usage behaviour we also predominant. Also, features about social calling behaviour were relevant. This gives us confidence that the variety of these features points to the fact that the classes of features we start out with is diverse and captures different aspects of the problem. Further, numerical results show the effectiveness of our approach.

5.2   Predicting Churners

We run our churn prediction algorithm of T-SNACS on the input dataset both for the prepaid and postpaid subscriber segments. We investigate the lift curves attained by each of the prediction processes, using the ground truth data and known nominal churn rates. Since churn predicted using $M3$ data as input will not be verifiable, as the actual churn will happen at a future point of time, we use the $M2$ data to predict churners.

Figure 2 shows the lift obtained, accuracy and coverage for the prepaid and postpaid churn predictions made on month $M2$. The predictions captured in the figure represent only the population having participated in at least one voice or SMS call. For the prepaid subscribers, comparing with ground truth data shows that, out of the 8.138 million subscribers, our model could attain a lift of 7.3 over the first 1 million predicted churners. For the 2.3 million postpaid subscribers, our model attained a lift of 4.67 over the first 200,000 predicted churners. Both these lifts attained are substantial, especially given the high nominal churn rates (Table 1), indicating the soundness of our model.

We run our churn prediction algorithm on the prepaid and postpaid subscribers found to be silent for our chosen silence period of 30 days during $M2$, and churned in $M3$. We use the $M1$ call data for these subscribers to model their social calling behavior. Our experimental results in Figure 3 indicate a high level of accuracy of our model. For the prepaid segment, we attained an accuracy of 70% with a lift of about 3.5 over the first 200,000 predicted churners, out of

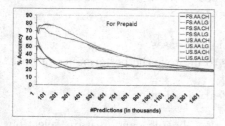
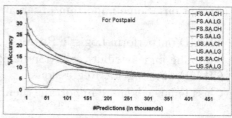

Fig. 1. Churn prediction lift curves for (a) prepaid and (b) postpaid, respectively. FS/US represents full/under sampling, AA/SA represents all and some attributes, and CH/LG represents CHAID and Logistic Regression models, respectively

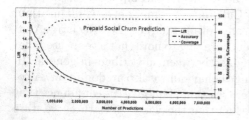
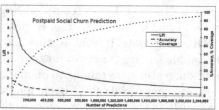

Fig. 2. Lift curves for social churn predictions for (a) prepaid and (b) postpaid

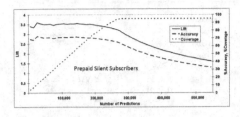 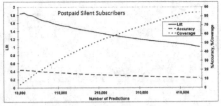

Fig. 3. Lift curves for predicting silent churners for (a) prepaid and (b) postpaid

about 0.93 million customers. For postpaid, the lift observed was about 1.6 over the first 100,000 predicted churners, from a population of 540,000.

5.3 Competitive Evaluation Results

We presented our results and insights to the executives of the large Asian telecom operator. We supplied the predicted churners for $M4$ and all our other results to the operator. The operator validated the predicted churners against the ground truth, to which only they had the access, by looking at churners of $M4$. This was a competitive evaluation. We present the validation results in Table 2. The results show that we outperformed the nearest competition by more than a factor of 2 for prepaid case and by 38% for the postpaid case.

5.4 Social Interactions amongst Churners

In order to understand and characterize the behavior of churners, we conducted an in-depth study of the social interaction patterns. We conducted several experiments that lead us to the following insights.

1. Overall, the socially contagious churn is not a very strong phenomenon for the data under consideration. Socially contagious churn is the case when a present churner is socially connected to some past churner.
2. However, we find that the social connections between the postpaid churners are much stronger compared to the social connections between the prepaid churners. This observation is based on the in and out degree distribution,

Table 2. Churn prediction results: (a) Prepaid: Subscriber base 8,504,682, number of predictions 850,540, nominal churn rate 7.3% (b) Postpaid: Subscriber base 1,260,422, number of predictions 63,400, nominal churn rate 1.14%

Partner	Prepaid			Postpaid		
	Num correct predictions	Percentage accuracy	Model Lift	Num correct predictions	Percentage accuracy	Model Lift
TSNACS	472,900	55.6	7.62	2,292	3.6	3.17
Partner B	206,598	24.3	3.33	-	-	-
Partner C	200,520	23.6	3.23	1,656	2.6	2.29

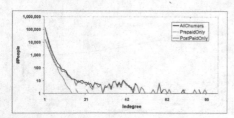

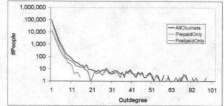

Fig. 4. (a) In and (b) Out Degree distribution for churner-churner induced subgraphs

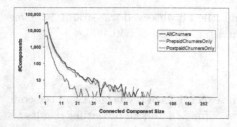

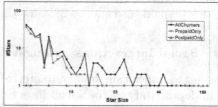

Fig. 5. (a) Connected component (CC) and (b) strongly connected components (SCC) size distribution for churner-churner induced subgraphs

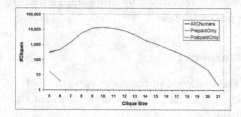

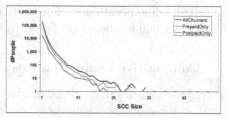

Fig. 6. (a) Clique and (b) Star size distributions in churner-churner induced subgraphs

shown in Figure 4. Most of the churners well-connected with their fellow-churners belong to the postpaid group both from in-degree and out-degree perspectives.

3. We find churn to be a socially localized phenomenon. This is indicated by the significant presence of connected components (ignoring edge directions) and strongly connected components (factoring in edge directions), for both prepaid and postpaid churner graphs. The graph exhibits a long-tail distribution, and is shown on a log-scale. There are large components, about 10,000 in size, present in both the graphs, and many smaller components. Figure 5 shows the distribution of connected and strongly connected components.

4. Although the number of churners in postpaid is much smaller compared to prepaid, we find that the postpaid churners form far more cliques amongst themselves. This further asserts our observation of existence of stronger social connections between the postpaid subscribers compared to prepaid. Figure 6(a), shown on a log-scale, indicates the much larger number (and sizes)

of cliques in postpaid data despite the relatively fewer postpaid churners. Thus, group churn is more prevalent in postpaid than prepaid.

5. Star structures, which are graphs with low (negative) assortativity, are more dominant in the prepaid chuners compared to postpaid churners. This is yet another indication of stronger social bonding of the postpaid churners. Figure 6(b) shows the star-size distribution in churner-churner interactions.

6. We find a large, densely connected group of postpaid churners, which is not the case with prepaid churners. Intuitively, dense groups are activity hotbeds, and here, for socially contagious churn. Table 3 shows the dense group characteristics of the churner graphs, where connection density is defined as the ratio of the number of connections to the number of people.

A detailed comparative study between the prepaid and postpaid dense groups reveals a set of stark behavioral contrasts. The postpaid dense group churners are found to be highly homogeneous in terms of behavior as opposed to the prepaid. All the postpaid dense group churners follow the same billing plan, while the prepaid dense group churners belong to 4 different billing plans. In the postpaid churner dense group, 79 out of 104 are of age 40, and 16 others are of age 25-26. In contrast, 9 out of 46 prepaid dense group churners are of age 40 and the rest are scattered between 29 to 42. The churn dates of 101 out of the 104 churners in postpaid were the same (a day in $M3$), while the churn dates for the prepaid were scattered over the first 28 days of $M3$. In the postpaid churner dense group, each churner knows between 44 to 186 people. This number varies between 8 and 90 for prepaid. In postpaid, 32 churners had *joined* the telecom operator on the same day and the rest had joined over a period of about 4 years. In the prepaid churner dense group, 11 had join on the same day while the rest had joined over a period of about 2 years.

5.5 Additional Observations

We investigate the behavior of the overall subscriber base, which helps us obtain deeper insights into the subscribers' social network. We attempt to discover the overall connection patterns and strengths by discovering graph properties for the overall graph as well as the overall prepaid and postpaid subscriber bases (not just the churners). We make the following observations by analyzing the prepaid and the postpaid graphs independently.

- Among the prepaid subscriber base, the interaction between people of similar ages (age difference $\leq$ 5 years) is more frequent, and also has about 1.5 times more talk time per pair.

Table 3. Dense group characteristics of the churner graphs (conn denotes connection)

Churners	Num people	Num conns	Conn density	Conn density in full graph	Conn density lift
All Churners	104	3,003	28.875	0.69	41.85
Prepaid	45	320	3.56	0.62	5.74
Postpaid	104	3,003	28.875	0.92	31.39

- However, among the postpaid subscriber base, interaction between people of different ages (age difference > 5 years) is more frequent. But the talk time per pair for similar age is 2 times higher in postpaid.
- We define connection strength as: total talk time in seconds on voice calls + 30 * total number of SMS sent or received (i.e., we treat an SMS as equivalent to a 30 second call based on domain knowledge: that the cost of a 30-second call was comparable to an SMS). We observe that the ratio of friend (those whose connection strength is ≥ 10,000) pairs to acquaintance (connection strength < 10,000) pairs is small (10-20%), but these friend pairs contribute almost 80% of the talk time. *This observation is surprising and unexpected, and can be thought of as a 80-20 rule of a novel kind.*
- The average number of postpaid males that a postpaid male talks to is higher than, male-to-male (prepaid) as well as female-to-female (both prepaid and postpaid). However, highest average talk time per pair occurs between female-to-female postpaid subscribers. Thus, we observe that prepaid males make the most calls and postpaid females make the longest calls.
- We observe the number of interactions to be most prevalent among people of the same gender both for prepaid and postpaid subscribers. The average talk time is also 30% higher for same-gender interactions for both the segments.

All the above observations suggest that a social networking behavior based profiling can provide significant insights, especially for prepaid customers where the profile data is frequently missing or is unreliable.

We observe significant social propensity in subscriber connection patterns, including connection patterns of churners. This makes a social networking based analytical study insightful and valuable from the perspective of understanding customer behavior and predicting events like churn on the operator's network.

6   Conclusion

In this study, we analyzed the data of a large Asian telecom operator to predict churners and gain customer behavior insights in a competitive evaluation. We used voice and SMS call CDRs, demographics, billing and customer churn history information. We used T-SNACS, a KPI-based model combined with graph-based analytical techniques and data mining methods, for churn prediction.

T-SNACS provided a lift of 8.4 times over the nominal churn rate of 4.17% on 10% of the prepaid talking subscriber base on test data and a lift of 7.62% on a nominal churn rate of 7.3% as reported in the customer evaluation on unseen data - more than twice as good as the next best competitor in the study.

We also investigated the social behavior of the churners, independently for the prepaid and postpaid subscriber base, as well as together. We observed higher social affinity among postpaid subscribers compared to prepaid. The overall graph, combined over voice and SMS calls, demonstrated significant social behavior.

Our study provides guidelines, and acts as a template, for KPI-based churn prediction model building. We also highlight the rich set of insights possible by social network analysis in real world telecom networks.

References

1. Huang, B., Buckley, B., Kechadi, T.M.: Multi-objective feature selection by using nsga-ii for customer churn prediction in telecommunications. Expert Syst. Appl. 37(5) (2010)
2. Hung, S.Y., Yen, D.C., Wang, H.Y.: Applying data mining to telecom churn management. Expert Syst. Appl. 31(3) (2006)
3. Tsai, C.F., Lu, Y.H.: Customer churn prediction by hybrid neural networks. NIPS 36(10) (2009)
4. Wei, C.P., Chiu, I.T.: Turning telecommunication call details to churn prediction: a data mining approach. Expert Syst. Appl. 23(2) (2002)
5. Dasgupta, K., Singh, R., Viswanathan, B., Chakraborty, D., Mukherjea, S., Nanavati, A.A., Joshi, A.: Social ties and their relevance to churn in mobile telecom networks. In: EDBT (2008)
6. Richter, Y., Yom-Tov, E., Slonim, N.: Predicting customer churn in mobile networks through analysis of social groups. In: SDM (2010)
7. Kawale, J., Pal, A., Srivastava, J.: Churn prediction in mmorpgs: A social influence based approach. In: CSE (2009)
8. Huang, B.Q., Kechadi, M.-T., Buckley, B.: Customer churn prediction for broadband internet services. In: Pedersen, T.B., Mohania, M.K., Tjoa, A.M. (eds.) DaWaK 2009. LNCS, vol. 5691, pp. 229–243. Springer, Heidelberg (2009)
9. den Poel, D.V., Lariviere, B.: Customer attrition analysis for financial services using proportional hazard models. Euro. J. of Operational Research 157(1) (2004)
10. Burez, J., den Poel, D.V.: Crm at a pay-tv company: Using analytical models to reduce customer attrition by targeted marketing for subscription services. Expert Syst. Appl. 32(2) (2007)
11. Gustafsson, A., Johnson, M., Roos, I.: The effects of customer satisfaction, relationship commitment dimensions, and triggers on customer retention. J. of Marketing 69(4) (2005)
12. Karnstedt, M., Rowe, M., Chan, J., Alani, H., Hayes, C.: The effect of user features on churn in social networks. In: Third ACM/ICA Web Science Conference (2011)
13. Karnstedt, M., Hennessy, T., Chan, J., Basuchowdhuri, P., Hayes, C., Strufe, T.: Churn in social networks. In: Handbook of Soc. Net. Tech. and App. Springer (2010)
14. Page, L., Brin, S., Motwani, R., Winograd, T.: The PageRank Citation Ranking: Bringing Order to the Web. Tech. rep., Stanford (1998)
15. Kleinberg, J.M.: Authoritative sources in a hyperlinked environment. J. ACM 46(5) (1999)
16. Goyal, A., Bonchi, F., Lakshmanan, L.V.: Discovering leaders from community actions. In: CIKM (2008)
17. Tang, J., Sun, J., Wang, C., Yang, Z.: Social influence analysis in large-scale networks. In: SIGKDD (2009)
18. Salton, G., Buckley, C.: On the use of spreading activation methods in automatic information. In: SIGIR (1988)
19. Clauset, A., Newman, M.E.J., Moore, C.: Finding community structure in very large networks. Phys. Rev. E 70(066111) (2004)
20. Gibson, D., Kleinberg, J., Raghavan, P.: Inferring web communities from link topology. In: HYPERTEXT (1998)
21. Modani, N., Dey, K.: Large maximal cliques enumeration in sparse graphs. In: CIKM (2008)
22. Pandit, V., Modani, N., Mukherjea, S., Nanavati, A., Roy, S., Agarwal, A.: Extracting dense communities from telecom call graphs. In: COMSWARE (2008)

Novel Client-Cloud Architecture
for Scalable Instance-Intensive Workflow Systems

Dahai Cao[1], Xiao Liu[2,*], and Yun Yang[1,*]

[1] Faculty of Information and Communication Technologies
Swinburne University of Technology Melbourne, Australia
{dcao,yyang}@swin.edu.au
[2] Software Engineering Institute East China Normal University Shanghai, China
xliu@sei.ecnu.edu.cn

Abstract. Though workflow technology is relatively mature and has been one of the most popular components of process aware systems over the last two decades, few workflow architectures can efficiently support a large number of concurrent workflow instances, i.e. instance-intensive workflows. The basic requirements include high throughput, elastic scalability, and cost-effectiveness. This paper proposes a novel client-cloud architecture which takes advantages of cloud computing to support instance-intensive workflows, presents an application level real-time resource utilization estimation model, and identifies two primary principles to ensure the sustainable scalability, namely: (1) the time for a load balancer checking must be less than the decaying time of a server instance when it is overloaded, (2) the sampling time for an alarming service plus the launching time of new server instance must be less than the decaying time of a server instance when it is overloaded. Based on the above, we design and implement the SwinFlow-Cloud prototype. Finally, we deploy and evaluate the prototype on Amazon Web Services cloud. The results show that the prototype is able to satisfy all the basic requirements for instance-intensive workflows.

Keywords: cloud computing, cloud workflow, instance-intensive workflows, client-cloud architecture.

1   Introduction

Workflow has been a core component in process aware systems since workflow reference model was proposed by workflow management coalition (WfMC) in 1995 [5, 14]. From the early centralised architecture to recent decentralized architecture such as Grid and Peer-to-Peer (P2P), the workflow system has been evolved significantly. However, there are still some challenges in front of today's workflow systems. One of the major challenges is that few of workflow architectures can elastically, efficiently, and economically handle a large number of concurrent workflow instances, i.e. instance-intensive workflows. For example, there are over 740 million mobile users for China Mobile Limited as of June 30, 2013

---

* Corresponding author.

X. Lin et al. (Eds.): WISE 2013, Part II, LNCS 8181, pp. 270–284, 2013.

(www.chinamobileltd.com/?lang=en). If each user sends 2 messages on average every day, there would be 1.36 billion messages to be charged on a daily basis. Therefore, its mobile short messaging service (SMS) charging workflow can be regarded as a typical instance-intensive workflow example which includes fee calculation task, billing task and others. With the massively increasing number of workflow instances, traditional workflow architectures cannot meet the requirements through expanding hardware or software investments. More importantly, the hardware investment may be insufficient or wasted due to the lack of elasticity if the resource demands change.

Recently, cloud workflow has been a new research focus in the workflow research community. As a SaaS (Software as a Service), cloud workflow is a workflow management system (WfMS) which is designed and deployed in the cloud, supporting dynamically-scalable cost-effective large-scale workflows. A cloud workflow system is able to automatically scale out by recruiting more virtual machine instances with workflow enactment services (workflow server instances) for handling large-scale workflows (during peak time) and scale in by releasing unnecessary workflow server instances for cost saving (during off-peak time). In theory, it can scale out without restrictions.

In this paper, we propose novel client-cloud architecture for scalable workflow systems to handle instance-intensive workflows. Here, client-cloud means that the client communicates with an elastic pool of workflow server instances on the cloud side, which is different from the traditional client-server model where the client communicates with one static or a cluster of physical workflow servers. The size of the pool in the cloud can elastically expand or shrink according to the dynamic resource demand. The client-cloud architecture is different from decentralized architectures such as Grid and P2P. The client-cloud architecture has a centralised cloud side to serve for all clients while decentralized architectures generally have no central server to control all the system components. With the novel client-cloud architecture, we design and implement a scalable SwinFlow-Cloud (*Swin*burne Work*flow* system on *Cloud*) prototype. Through our evaluation of the prototype on the AWS cloud, we demonstrate that our client-cloud based workflow system can meet all the basic requirements for handling instance-intensive workflows, as detailed in Section 2.

The main contributions of the paper include:

C1. Novel client-cloud architecture for scalable workflow system which can elastically and economically support instance-intensive workflows. The architecture promises a new solution for revamping traditional client-server model based workflow systems to handle instance-intensive workflows.

C2. An application level real-time resource utilization estimation model for scalable workflow system. It is a new approach to offer more sensitive scalability for a cloud workflow system.

C3. Two primary principles for sustainable scalability of the client-cloud based workflow system running on the cloud. The principles include: (1) the time for a load balancer checking must be less than the decaying time of a server instance when it is overloaded, (2) the sampling time for an alarming service plus the launching time of new server instance must be less than the decaying time of a server instance when it is overloaded.

This paper is organized as follows. Section 2 discusses further the motivation of the research and analyses the system requirements. Section 3 details the system architecture of the client-cloud based workflow system. Section 4 addresses the system implementation. Section 5 presents the system deployment. Section 6 describes the evaluation of the SwinFlow-Cloud prototype on the AWS cloud. Section 7 reviews the related work in workflow architecture design. Finally, Section 8 concludes this paper and outlines the future work.

2   Motivation and Requirements Analysis

Workflow facilitates business process automation through executing predefined process models with navigation rules. To improve the performance, workflow research communities focus on two types of workflow architectures: centralised workflow architecture and decentralized workflow architecture.

The centralised architecture is still the most popular paradigm in today's workflow community because it is simple and easy to rapidly implement a prototype or product to support workflows. The client-server model is a typical centralised architecture. Its advantage is that the server side is able to utilize more controls in workflow execution while the client side is able to concentrate on presenting workflow work list or interacting with workflow user interfaces. To handle instance-intensive workflows, the architecture needs to extend single workflow server to a cluster consisting of multiple logical or physical nodes. A cluster can improve the performance effectively. However, the central workflow server is still a traffic bottleneck for running workflows and bears the enormous risks that system performance may decrease sharply once overloaded. Furthermore, the scalability of architecture is still limited. The cost for upgrading the hardware or software of a centralised architecture is very expensive.

The decentralized architecture has been another area in workflow research for handling challenges of instance-intensive workflows. In this architecture, workflow processes or data are distributed to multiple enactment components that are physically or logically dispersed at separate locations or platforms. Its advantage is that the decentralization of workflow enactment components enhances the scalability and reduces the risk of bottleneck. However, with rapidly rising number of workflow instances, more enactment components are added, which brings with fast rising on the cost of time, coordination, and communication, especially when they are distributed in multiple physical locations.

Based on the analysis of the two architectures, for the instance-intensive workflows, we need to design a novel workflow system architecture which should be able to meet the following requirements:

R1. It is able to support high throughputs of workflow instances. Here throughput is measured by the sum of the launched workflow instances, i.e. input, and the completed workflow instances, i.e. output, in the workflow system within a set of time fragments (high throughput for short);

R2. It is able to elastically scale out to balance the requests from workflow clients and scale in for cost saving (elastic scalability for short);

R3. It is able to cost-effectively manage the execution of instance-intensive workflows (cost-effectiveness for short).

3   System Architecture

According to the requirements analysis above, we propose novel client-cloud architecture. It has a central group of sever instances in cloud to power large-scale applications but with very little risks of bottleneck. It enables provision of more virtual and cheaper workflow server instances with high performance on demand. The architecture consists of a client side and a cloud side, as shown in Fig. 1.

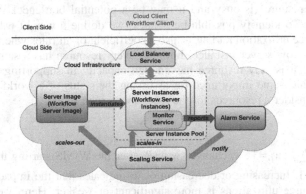

Fig. 1. Client-cloud architecture

The client side can communicate with the cloud side through various protocols. The cloud side is implemented in cloud infrastructures and includes load balancer service, workflow server image(s), server instance pool with many workflow server instances, alarm service, scaling service and so on. Each workflow server instance is accompanied by a monitor service, etc. Same as the traditional client-server model, the client side also sends requests to the cloud side for response. Once the cloud side receives the requests, the cloud side processes the requests through a sustainable coordination of the services on the cloud side and responds to the client side. We present these individual services and their coordination in detail next.

3.1   Client Side: Cloud Client

Cloud client on the client side is a set of Web-based or desktop-based applications that interact with the cloud side. The applications are to construct the requests to be sent to the cloud side and display the results received from the cloud side. In the client-cloud based workflow system, it has three functions: the first is to design and transfer a process model for handling on the cloud side; the second is to interact with the cloud side for completing the workflow; the third is to monitor and control the status of the system for optimizing the system performance.

3.2   Cloud Side: Monitor Service

The monitor service hosts each workflow server instance to watch and estimate the system status in real time. The service can watch various resource utilizations of the workflow server instance, which includes not only virtual machine level such as CPU utilization, memory utilization, I/O read/write utilization, etc., but also WfMS level, such as typical utilizations of thread pool in workflow enactment service, etc. In general, the overload of any one of these resources will result in performance decay of workflow server instance. Therefore, we must consider an overall metric to assess the status of workflow server instance.

Here, we formally define resource utilization rate as $R_i \in [0,1]$, $i=1..n$. The greater is R_i, the higher is the i resource utilization rate. The higher resource utilization rate means that the resource is busy and it may be a potential bottleneck of a workflow server instance. To identify possible bottleneck, we define a weight $w_i \in [0,1]$, $i=1..n$ for every resource utilization rate. w_i is also a coefficient of all rates when assessing the status of a workflow server instance. Initially, every resource has a same weight, denoted as W_0. That is, any resource has enough capacity in supporting the workflow enactment and has same importance for assessing the status of a workflow server instance. If we consider n resources for the assessment, then W_0 is:

$$w_0 = \frac{1}{n} \tag{1}$$

From Eq. (1), $\sum_{i=1}^{n} w_i = 1$. w_i changes with R_i after the WfMS starting up but the sum of all w_i is still 1. Increasing or decreasing of w_i indicates that the impact of R_i on the considered resource utilizations is more significant or weaker. Here, we define w_i as follows:

$$w_i = w_0 + \Delta w \tag{2}$$

$$\Delta w = R_i - \frac{\sum_{i=1}^{n} R_i}{n} \tag{3}$$

From Eq. (3), $\sum_{i=1}^{n} \Delta w = 0$. Thus, it ensures $\sum_{i=1}^{n} w_i = 1$. In Eq. (3), when the other resource utilizations keep stable, if the i resource utilization R_i is greater, then Δw is also greater. It indicates weight w_i of the i resource will be more significant for impacting on the overall system status. Otherwise, w_i is weaker for impacting. Based on the discussion above, we further define a metric, called composite index I, to assess the overall status of a workflow server instance below:

$$I = \sum_{i=1}^{n} w_i R_i \tag{4}$$

In Eq. (4), $R_i \in [0,1]$, $w_i \in [-1,1]$, then $I \in [0,1]$. Theoretically the minimum value of I means the system is idle due to any resource not being used, while the maximum value of I means that all resources of system are in full workload. But normally, I changes between the two extreme values. To estimate I, we use a threshold set $T = \{T_1, T_2,\ldots, T_n\}$, $T_k \in [0,1]$, $k = 1,..,n$. T divides the open interval of $[0,1]$ into $k+1$ parts. When $I \in [T_{k-1}, T_k]$, the workflow server instance will change from one status to another status, e.g., from the healthy to the unhealthy. The thresholds generally are not fixed values.

3.3   Cloud Side: Load Balancer Service

This service is a Web portal of the cloud side. All requests and responses between the client side and the cloud side are passed through this service. It has two primary functions: load balancing and dispatching. The former can communicate with the monitor service in each workflow server instance to obtain the system status periodically for assessing whether the server instance is overloaded or under loaded. The length of the period affects the coordination between the services on the cloud side. We denote the period as t_b. The latter parses, classifies, and finally dispatches a request to an appropriate server instance. When the service receives a workflow request, such as launching a workflow, submitting a workflow task and so on, it assesses the status of each workflow server instance firstly and then chooses an appropriate one to dispatch.

3.4   Cloud Side: Alarm Service

The service is aware of the statuses of all workflow server instances through timely communication with the monitor service in each workflow server instance and generates their statistics periodically. The length of the period also affects the coordination between the services on the cloud side. We denote the length of the period as t_a.

The alarm service uses the results of statistics to compare with a preconfigured threshold. If the results are less than the threshold, it may notify the scaling service to scale out by one or more server instances. Otherwise, it may notify the scaling service to scale in to save cost.

3.5   Cloud Side: Scaling Service

After receiving notifications from the alarm service, this service can automatically scale out by one or more server instances from workflow server image. Scaling-out action is to clone a new workflow server instance from the image and put it into the sever instance pool. Then the server instance will be launched and initiated to accept requests in a short period of time. Next, the server instance will be registered in the load balancer service. The launching time of the server instance, denoted as t_l, also influences the coordination of the services. Scaling-in action needs to firstly assess whether the server instances can be released to save cost.

3.6   Cloud Side: Server Instance Pool and Server Image

This server instance pool is a container that can contain many workflow server instances that are instantiated from the workflow server image by the scaling service. The workflow server instance in the client-cloud architecture is shown in Fig. 2. The workflow server instance innovates on the basis of the workflow reference model proposed by WfMC. We retain all the components and the interface definitions in the WfMC workflow reference model. For adapting to cloud computing environments, we add some cloud relevant components into the reference model:

- Cloud workflow accompaniment tools. As part of workflow client, the tools are able to define and configure various parameter and metrics such as weights of composite index, size of workflow process thread pool, size of workflow task thread pool, pricing policy and payment policy, tool agents, and parameters for invoking external applications.

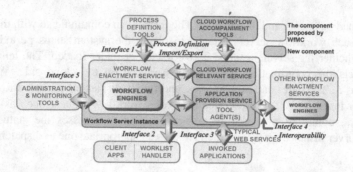

Fig. 2. Workflow server instance in client-cloud architecture

- Cloud workflow relevant service. As a bridge between workflow server instance and cloud infrastructure, the service is able to support workflow enactment service to communicate with cloud computing infrastructure, so that the cloud infrastructure can be aware of the status of the workflow server instances.
- Application provision service. The service reserves and enhances the tool agent component in the workflow reference model. It enables workflow system to schedule the tool agents to avoid crashing down the legacy applications under massive or frequent invocations.

3.7   Coordination of Services

As presented in Subsections 3.3, 3.4, 3.5 above, our system architecture applies a periodical sample statistic mechanism to supporting automatic scaling. All the services on the cloud side are coordinated: the load balancer service checks the statuses of all instances per t_b and evenly dispatches the requests to the server instances; the alarm service samples per t_a and generates results of statistics and notifies the scaling service when necessary; the scaling service scales out by more server instances with launching time t_l to balance the workload and scales in for cost saving.

Coordination keeps all workflow server instances in the pool to be recoverable and prevents them from crashing down due to overloading, or releases the idle to avoid waste. If some workflow server instances are unrecoverable, it means that they will crash down and the cloud side will lose the requests that are dispatched to them by the load balancer service. It will result in system errors or data inconsistency. If all instances are kept recoverable, the cloud side can bear more workload and every request should be handled correctly. Furthermore, the scaling service is able to scale out or in regularly under the coordination which can ensure a sustainable scalability. It meets requirement R2 presented in Section 2.

To achieve sustainable scalability, we propose two primary principles to facilitate coordination. The principles are discovered and abstracted based on our empirical analysis on the experimental results obtained after deployment of our system in cloud.

Some important time factors are denoted earlier, viz., t_b (the time period for the load balancer service to check the status of server instances), t_a (the sampling time period for alarm service to generate the statistics for the status of server instances), t_l (the launching time of a server instance), and we must be aware of the maximum time period when a server instance starts decaying and just before it cannot be recovered, we denote the period as t_p. If the server instance spends average 50 seconds to reach the threshold, then t_p is equal to 50 seconds. Here, we formulate two principles for configuring a system based on the client-cloud architecture.

Principle 1: When the load balancer service needs to assess the status of server instances before dispatching a request, if and only if $t_b<t_p$, then the server instances are able to recover from its decaying.

If $t_b\geq t_p$, that means the load balancer service is aware of the status of a server instance after it turns into the unrecoverable status. For example, t_b is 55 seconds, and t_p is 50 seconds, i.e., the load balancer service gets the status of the server instance after it has crashed down for 5 seconds.

Principle 2:   If and only if $t_a+t_l<t_p$, then the server instance is able to recover.

If $t_a+t_l\geq t_p$, it may result in that the decaying server instances have already crashed down but (1) new instances have not been launched yet; (2) alarm service is not aware of crashing down. For example, t_p is 50 seconds, t_a is 30 seconds, and t_l is 40 seconds. We assume that a workflow server instance crashed down at the 0th second. Then, at the 30th second, the alarm service is aware of crashing down and notifies the auto-scaling service to clone new workflow server instances for load balancing. A new workflow server instance needs 40 seconds to start up. However, at the 70th second, the decaying workflow server instance has already crashed down.

The accurate values of t_p, t_b, t_a, t_l are different from system to system. Especially, t_p depends heavily on, say, CPU, available memory size, and heap size (in Java virtual machine). It requires experiments with various workflows to get relatively accurate t_p before deploying and configuring a cloud-based workflow system.

The two principles aim at effective coordination of the services and correctly handling all client requests. Effective coordination is to guarantee sustainable scalability.

4   System Implementation

Based on the design above, we implement the SwinFlow-Cloud prototype in Java programming language. The implementation of the workflow server prototype is depicted in Fig. 3. The structure of the prototype has traditional functions of workflow system in WfMC. We further extend the traditional functions and implement the cloud accompaniment tools, cloud workflow relevant services, and application provision service. Process definition tools are able to graphically model a workflow and transfer it to workflow enactment service; administration tools are able to monitor and manage the running workflows; work list handler is able to fetch the work items for handling and submitting the results back to the workflow enactment service. We develop these tools on the basis of Eclipse Rich Client Platform and Eclipse Graphical Editing Framework (www.eclipse.org).

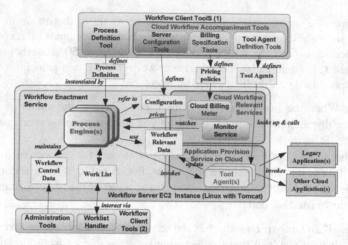

Fig. 3. SwinFlow-Cloud workflow server prototype

As a core component of the workflow system, the workflow enactment service can instantiate a workflow, allocate all essential resources, and create a process engine for execution. The thread will execute the ready tasks and activate their subsequent tasks, until all the tasks are completed. For invocation of the legacy applications or other cloud applications, the workflow enactment service may access tool agents hosted in the application provision service.

The implementation of the cloud workflow accompaniment tools and cloud workflow relevant services considers cloud features. The cloud workflow accompaniment tools are composed of:

- Server Configuration Tools, which are able to adjust basic configurations of workflow server instance, such as the size of the workflow process thread pools, the length of the waiting queues of the thread pools, and so on;
- Billing Specification Tools, which are able to specify pricing policies to charge cloud workflow systems on CPU usage, network usage, workflow execution time, size of workflow instance and so on;
- Tool Agent Definition Tools, which are able to create tool agents and the defined parameters for calling external applications.

As mentioned above, workflow relevant services are a bridge between cloud infrastructure and workflow server prototype. They utilize the APIs of specific cloud infrastructures. They are composed of:

- Monitor Service. As mentioned above, the service is to collect the considered resource utilization information and compute the composite index;
- Billing Meter. The service is based on the resource monitoring service provided by cloud infrastructure. It can conduct statistical analysis according to a pricing policy every short period of time and send results to the bill monitoring service in cloud infrastructure.

5   System Deployment

The previous two sections address the design and implementation of the client-cloud workflow system. This section will further discuss the deployment of the SwinFlow-Cloud prototype in cloud.

The deployment begins with the migration of the workflow server prototype onto the cloud infrastructure. The prototype is deployed on a virtualized machine in cloud similar to a traditional client-server model based workflow server. Then we create a workflow server image using the virtual machine with the prototype and deploy the load balancer service, the alarm service, and the auto-scaling service to support sustainable scalability. For the cloud computing environment, we choose the AWS cloud, a popular commercial public cloud service, as an example. As an IaaS (Infrastructure as a Service), the AWS cloud provides the essential services for load balancing, monitoring, and auto-scaling. We use Amazon Relational Database System (RDS, aws.amazon.com/rds) console tools to store the data of the prototype. To support invocation of various types of cloud applications, we use Amazon Simple Storage System (S3, aws.amazon.com/s3) to store API library files of external applications. We create an S3 bucket and folders for each tool agent to store the API library files. A workflow tool agent remotely accesses the external applications through the API library files in the bucket. It facilitates many legacy applications to migrate on the AWS cloud for workflow invocation.

During system deployment, launching time t_l and decaying time t_p need to be obtained. Furthermore, it is important to make sure that the relation between t_b and t_p obeys Principle 1, and the relation between t_a, t_l, and t_p obeys Principle 2 addressed in Subsection 3.7.

6   Evaluation

This section presents the evaluation of SwinFlow-Cloud using a simplified mobile charge workflow mentioned earlier.

6.1   Workflow Example

For evaluation of the SwinFlow-Cloud prototype, we select an instance-intensive workflow example: a simplified mobile charge workflow mentioned in Section 1 which consists of Start Task, Compute Bill Task, Charge Task, Count Task, and End Task. We use the process definition tools to model this process.

The Compute Bill Task calls an external application on the AWS cloud for computing the bill of one mobile short message. Then the Charge Task calls another external application on the AWS cloud for charging the bill. Because the application for charging the bill may have some delay, the Charge Task will check its state code to determine whether the charge is completed or not. If the charge is not completed, the Charge Task returns an unsuccessful message, the workflow will navigate to the Count Task; otherwise, the workflow will navigate to the End Task. The End task will complete this workflow instance. The SwinFlow-Cloud prototype creates a charge workflow instance for each mobile short message. The number of the workflow instances will increase accordingly with the growing number of mobile short messages.

6.2   Performance Experiment

We first deploy a workflow server onto an Amazon Linux server instance on the AWS cloud and configure the sizes of thread pools as 200 and the size of the waiting queue as unlimited.

For this experiment, we further design a client emulator to simulate massive requests to evaluate the SwinFlow-Cloud prototype on the AWS cloud. We asynchronously launch 12 emulators and each can send 200 requests within 2 to 8 seconds. That is, about 3,000 requests are sent to the workflow server instances on the cloud side every 5 seconds on average.

As in Eq. (4) discussed in Section 3 for the composite index, we consider the three typical resource utilization rates for calculating I: R_{cpu} (CPU resource utilization ratio), R_{proc} (the ratio of the number of workflow instances in workflow-instance-thread pool to the length of its waiting queue) and R_{task} (the ratio of the number of workflow tasks in workflow task-thread pool to the length of its waiting queue).

Then through experiments, we obtain that the launching time (i.e. t_l) of the workflow server instance is about 60 seconds.

Afterwards, we upload the process definition to the workflow server on Amazon Linux server instance and launch it to get the decaying period of a workflow server instance (i.e. t_p) of the workflow server instance. When the requests increase quickly, I begins to increase. The experimental result shows that the server will not be able to recover after I is greater than 0.8, i.e. T_{ol} is 0.8. It takes about 130 seconds for the Amazon Linux server instance to decay from 0.4 to 0.8. Thus t_p is about 130 seconds. The experimental data show that the larger the size of the thread pool, the slower it decays, i.e. the longer t_p is.

Initially, there is only one EC2 instance with a workflow server prototype for handling the requests from the client side. Our experiment consists of two separate parts. The first part is to demonstrate the throughput of the SwinFlow-Cloud prototype on the AWS cloud without implementing the two principles, while the second part is to demonstrate the throughput with the two principles implemented.

For the first part, we set t_b as 120 seconds ($t_b{\geq}t_p$) and set t_a as 150 seconds ($t_a{+}t_l{\geq}t_p$), which do not obey the two principles. The experimental results are shown in Fig. 4. The Y-axis is the sum of input (the started workflow instances) and output (the completed workflow instances) every 10 minutes. The X-axis is time. Its interval is 10 minutes. The line with dots demonstrates the changes of the input, while the line with crosses demonstrates the changes of the output.

In our experiments, the SwinFlow-Cloud prototype eventually created over 30 workflow server instances. The experimental results show that the pattern of changing throughput for every workflow server instance is very similar to each other (the reason is explained later) except that their start-up moments are different. Therefore, to clearly depict the typical pattern, we just demonstrate the throughput results of a representative workflow server instance in Fig. 4. The figure shows the record of the throughput for every 10 minutes. The throughput keeps increasing for the first 10 minutes and reaches to the maximum point of about 21,960. It then decreases sharply and approaches to zero after 1 hour.

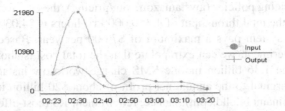

Fig. 4. Throughput of one workflow server instance without two principles implemented

The reason for such a sharp decrease is due to $t_b > t_p$, which means that the moment when the load balancer service finds out the workflow server instances have crashed down is later than the moment when the workflow server instances become unrecoverable. When the massive number of requests is sent to the workflow server instances on the cloud side, the workflow server instances crash down quickly and become unrecoverable before the timely detection by the load balancer service.

After the load balancer service has detected the unhealthy workflow server instances, it will stop dispatching new requests to them. In such a situation, the SwinFlow-Cloud prototype will be unable to handle the new requests from the client side. Furthermore, the alarm service samples the healthy status of the workflow server instances for every period of t_a. When the result shows that the workflow server instances are unhealthy, the auto-scaling service will create new workflow server instances to balance the incoming requests. However, the requests will make new workflow server instances become also unrecoverable due to $t_b{\geq}t_p$. The patterns of changing throughput of the new workflow server instance are thus similar to the old ones. Therefore, we just demonstrate the results of a representative workflow server instance in Fig. 4.

For the second part, with the same initial conditions, we set t_b of the load balancer service as 40 seconds according to Principle 1 and set t_l as 60 seconds according to Principle 2. The experimental results are shown in Fig. 5. The experiment lasts for 5 hours. We observe that the prototype eventually only scales out up to 11 workflow server instances to handle the requests sent from the client testers. After the client testers stop sending requests, the prototype gradually scales in by releasing the idle workflow server instances which have completed all the workflow instances running on them. Figure 5 shows that the maximum total input of 11 workflow server instances is more than 81,000 workflow instances every 10 minutes with an average of about 75,000. The maximum total output of 11 workflow server instances is about 80,000 workflow instances every 10 minutes with an average of about 70,000. Therefore, the total throughput of the SwinFlow-Cloud prototype is about 4,350,000 in 5 hours. The experimental results show that the throughput of the SwinFlow-Cloud prototype with the two principles implemented is much better than the throughput without the two principles implemented. Moreover, it clearly meets requirement R1 (high throughput of workflow instances) in Section 2. In this 5-hour experiment, the auto-scaling service scaled out by a total of additional 10 workflow server instances to meet the resource demand. When the instances were idle, the service automatically scaled in. It meets requirement R2 (elastically scale out or in) in Section 2. According to the Amazon pricing policies (aws.amazon.com/pricing), the cost for our experiment which simulated the total throughput of 4,350,000 in 5 hours is $4.03 maximum. That is, the workflow system pays a maximum of $7,000 per year. Based on the performance in our experiment, we can extrapolate that the total cost is about $500,000 per year to handle the 1.36 billion mobile SMS charge workflow instances on a daily basis which is marginal to the annual net profit of about $20 billion in China Mobile Limited (www.chinamobileltd.com). Hence, requirement R3 (cost-effectively manage the execution of the instance-intensive workflows) in Section 2 is also met.

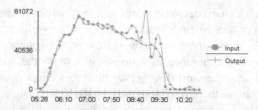

Fig. 5. Total throughput of SwinFlow-Cloud with two principles implemented

Furthermore, our client-cloud workflow system significantly improves the traditional workflow system in handling instance-intensive workflows (i.e., contribution C1).The resource utilisation estimation model can also be sensitively aware of the system status to support the auto-scaling mechanism (i.e., contribution C2). The experiments demonstrate the sustainable scalability of our client-cloud workflow system due to undertaking the two principles (i.e., contribution C3).

7   Related Work

Workflow research communities have put many efforts on the research of the workflow system architecture in the last two decades. As discussed in Section 2, the research of centralised architectures has gained significant achievements. Nowadays, the typical workflow systems based on the centralised structure include IBM FileNet, TIBCO iProcess Suite and so on [1]. Some researchers studied the decentralized architecture to improve the scalability since the last century, for example, Grid workflows and P2P workflows. Their scalabilities are better than centralised architecture. Grid workflow is a successful architecture for scientific workflow. The typical Grid workflow systems include Kepler, Pegasus, Taverna, ASKALON and so on [8]. So far, Grid workflow mainly focuses on data-intensive and/or computation-intensive scientific workflows.

Nowadays, workflow system vendors and research communities are investigating the cloud-based workflow systems, such as Apache Hadoop's MapReduce based CloudWF [4, 15], Gridbus based Cloudbus Workflow Engine [2, 3], Pegasus (the latest cloud version), SwinDeW-G (SwinDeW for Grid) based SwinDeW-C (SwinDeW for Cloud) [6-8, 12-14] and so on. These workflow systems either utilize the cloud features, or are extended on the basis of the Grid, P2P infrastructures to support specific workflow requirements. However, these workflow systems mainly focus on scientific workflows rather than instance-intensive workflows. Amazon Simple Workflow service (SWF, aws.amazon.com/swf) is an orchestration service for building distributed applications. However, it is a simple workflow service which cannot support high throughputs workflow applications. At present, there is very limited work specifically on scalability of cloud workflow. Some researchers propose the predication approaches based on the infrastructure-level metrics, e.g., deadline, response time, etc, to implement an auto-scaling model in cloud [9-11]. However, few of their approaches focus on the cloud application-level metrics of WfMS.

As discussed in Section 2, the basic requirements for instance-intensive workflows are high throughput, elastic scalability, and cost-effectiveness. Unfortunately, most existing workflow system architectures cannot meet these requirements satisfactorily where our work aims at sealing this gap.

8   Conclusion and Future Work

In this paper, we have proposed novel client-cloud architecture for scalable instance-intensive workflow systems. The client-cloud architecture has revamped the traditional client-server architecture for supporting instance-intensive workflows. Based on empirical analysis, we have also discovered and abstracted two principles to achieve a dynamic, elastic, and sustainable scalability. Given the three basic requirements (high throughput, elastic scalability, and cost-effectiveness) for instance-intensive workflows in the cloud, we have evaluated the SwinFlow-Cloud prototype deployed on the AWS cloud. The experimental results have shown that the prototype is able to meet all the requirements. Therefore, our work demonstrated in this paper can contribute to traditional client-server based workflow systems to handle instance-intensive workflows. In the future, further research on improving the performance of the Swin-Flow-Cloud prototype will be conducted. We will also deploy the prototype to other cloud computing environments.

Acknowledgment. The research work reported in this paper is partly supported by the Australian Research Council under Linkage Project LP0990393. Xiao Liu's contribution was primarily made when he was at Swinburne University of Technology.

References

1. Aalst, W.V.D., Hee, K.M.V.: Workflow Management: Models, Methods, and Systems, vol. 368. MIT Press (2004)
2. Buyya, R., Venugopal, S.: The Gridbus Toolkit for Service Oriented Grid and Utility Computing: An Overview and Status Report. In: Proceedings of the 1st IEEE International Workshop on Grid Economics and Business Models (GECON 2004), pp. 19–66. IEEE Computer Society, Seoul (2004)
3. The CLOUDS Lab: Cloudbus Workflow Engine, http://www.cloudbus.org/workflow/ (accessed on August 18, 2013)

4. Dean, J., Ghemawat, S.: MapReduce: Simplified Data Processing on Large Clusters. Communications of the ACM - 50th Anniversary Issue: 1958 - 2008 51(1), 107–113 (2008)

5. Hollingsworth, D.: Workflow Management Coalition: The Workflow Reference Model, Workflow Management Coalition, Winchester, Hampshire, UK. pp. 1–55 (1995)

6. Liu, X., Yang, Y., Jiang, Y., Chen, J.: Preventing Temporal Violations in Scientific Workflows: Where and How. IEEE Transactions on Software Engineering 37(6), 805–825 (2011)

7. Liu, X., Yuan, D., Zhang, G., Chen, J., Yang, Y.: SwinDeW-C: A Peer-to-Peer based Cloud Workflow System. In: Furht, B., Escalante, A. (eds.) Handbook of Cloud Computing, pp. 309–332. Springer US (2010)

8. Liu, X., Yuan, D., Zhang, G., Li, W., Cao, D., He, Q., Chen, J., Yang, Y.: The Design of Cloud Workflow Systems. Springer (2012)

9. Mao, M., Humphrey, M.: Auto-Scaling to Minimize Cost and Meet Application Deadlines in Cloud Workflows. In: International Conference for High Performance Computing, Networking, Storage and Analysis (SC), Seattle, USA, pp. 1–12 (2011)

10. Shen, Z., Subbiah, S., Gu, X., Wilkes, J.: CloudScale: Elastic Resource Scaling for Multi-Tenant Cloud Systems. In: The 2nd ACM Symposium on Cloud Computing (SoCC 2011), pp. 1–14. ACM, Cascais (2011)

11. Vaquero, L.M., Rodero-Merino, L., Buyya, R.: Dynamically Scaling Applications in the Cloud. ACM SIGCOMM Computer Communication Review 41(1), 45–52 (2011)

12. Yan, J., Yang, Y., Raikundalia, G.K.: A Decentralised Architecture for Workflow Support. In: Proceedings of the 7th International Symposium on Future Software Technology (ISFST 2002), pp. 23–25. Software Engineers Association, Wuhan (2002)

13. Yan, J., Yang, Y., Raikundalia, G.K.: SwinDeW—A p2p-Based Decentralized Workflow Management System. IEEE Transactions on Systems, Man, and Cybernrtics—Part A: Systems and Humans 36(5), 922–935 (2006)

14. Yang, Y., Liu, K., Chen, J., Lignier, J., Jin, H.: Peer-to-Peer Based Grid Workflow Runtime Environment of SwinDeW-G. In: Proceedings of the 3rd IEEE International Conference on e-Science and Grid Computing, pp. 51–58. IEEE Computer Society, Bangalore (2007)

15. Zhang, C., De Sterck, H.: CloudWF: A Computational Workflow System for Clouds Based on Hadoop. In: Jaatun, M.G., Zhao, G., Rong, C. (eds.) Cloud Computing. LNCS, vol. 5931, pp. 393–404. Springer, Heidelberg (2009)

A Study on the Evolution of Cooperation in Networks

Dayong Ye and Minjie Zhang

University of Wollongong, Wollongong, Australia
{dayong,minjie}@uow.edu.au

Abstract. This paper studies the phenomenon of the evolution of co-operation in networks, where each player in networks plays an iterated game against its neighbours. An iterated game in a network is a multiple round game, where, in each round, a player gains payoff by playing a game with its neighbours and updates its action by using the actions and/or payoffs of its neighbours. The interaction model between the players is usually represented as a two-player, two-action (i.e., cooperation and defection) Prisoner's Dilemma game. Currently, many researchers developed strategies for the evolution of cooperation in structured networks in order to enhance cooperation, i.e., to increase the proportion of cooperators. However, experimental results, reported in current literature, demonstrated that each of these strategies has advantages and disadvantages. In this paper, a self-organisation based strategy is proposed for the evolution of cooperation in networks, which can utilise the strengths of current strategies and avoid the limitations of current strategies. The proposed strategy is empirically evaluated and its good performance is exhibited. Moreover, we also theoretically find that, in static networks, the final proportion of cooperators evolved by any pure (or deterministic) strategies fluctuates cyclically irrespective of the initial proportion of cooperators.

1 Introduction

The evolution of cooperation among selfish individuals is one of the most fundamental issues in various disciplines, such as physics [4, 5], biology [1, 2], artificial intelligence [3], sociology [6, 7] and economics [8]. It is well known that in unstructured populations, natural selection favours defectors over cooperators [9]. However, in the real world, who-meets-whom is not random but determined by spatial relationships or social networks [10–13]. Recently, there is much interest in studying the evolution of cooperation in structured populations or on graphs [14–17]. In these studies, the Prisoner's Dilemma game is the leading metaphor for the evolution of cooperative behaviours in populations of selfish players, especially since the well known computer tournaments of Axelrod [18]. In each round, two players engaged in an iterated Prisoner's Dilemma game have to choose between cooperation and defection. In any given round, the two players receive a payoff R if both cooperate with each other, and a payoff P if both

X. Lin et al. (Eds.): WISE 2013, Part II, LNCS 8181, pp. 285–298, 2013.

defect against each other; but if one player chooses cooperation whereas another player uses defection, the defector gets a payoff T while the cooperator gains only a payoff S, where $T > R > P > S$ and $2R > T + S$. Thus, in a single round, it is always best to defect, but in an iterated Prisoner's Dilemma game, cooperation may be better [14].

Currently, several strategies have been proposed towards increasing the proportion of cooperators in the iterated Prisoner's Dilemma game. The most famous strategy is the Tit-For-Tat (TFT) [2], where each player cooperates in the first round and then adopts the behaviour used by its opponent in the former round. A TFT player concurs with cooperators and retaliates against defectors, but a TFT player also forgives those defectors that switch to be cooperators. Thus, although TFT is quite simple, it does encourage cooperation among players. However, TFT suffers from a stochastic perturbation, namely that occasional mistakes between two TFT players cause long runs of mutual retaliation. In order to overcome this drawback of TFT, two revised versions of TFT were developed: Tit-For-Two-Tats ($TF2T$) [19] and Generous TFT ($GTFT$) [20]. A $TF2T$ player allows two consecutive defections before retaliating. A $GTFT$ player chooses cooperation after an opponent's cooperation but still uses cooperation, with a certain probability, after an opponent's defection. Later, Nowak and Sigmund [21] claimed that their strategy, Win-Stay, Lose-Shift ($WSLS$), outperformed TFT in the Prisoner's Dilemma game. A $WSLS$ player maintains its action, i.e., cooperation or defection, only if its current payoff is at least as high as in the former round. In [14], Nowak and May presented a strategy, called Imitate-Best-Neighbour (IBN), where each player imitates the action of the player (including itself), which achieves the most payoff in the former round. Another strategy, devised by Tang et al. [22], is that in each round, a player selects a neighbour, based on the "degree" of each neighbour, and then probabilistically adopts the selected neighbour's action, based on the payoff difference between itself and the selected neighbour. Here, the term "degree" means the number of neighbours of a player. Santos et al.'s strategy [23] is similar to Tang et al's strategy. In Santos et al.'s strategy, named Stochastic Imitate-Random-Neighbour ($StIRN$), in each round, a player randomly selects a neighbour. The player imitates the selected neighbour's action, with a probability based on their payoff difference, only if in the former round, the selected neighbour's payoff is greater than the player's payoff. Recently, Hofmann et al. [3] developed a strategy, named Imitate-Best-Strategy (IBS), where each player sums the payoffs of all cooperating neighbours and the payoffs of all defecting neighbours (including itself) in the former round, and copies the action, which achieves the greater total payoff. Hofmann et al. also experimentally studied some of the current strategies, and analysed their advantages and limitations. Based on Hofmann et al.'s analysis, it can be found that the results of evolution of cooperation, i.e., the final proportions of cooperators, derived by different strategies depend heavily on the initial proportions of cooperators and the network types. For example, the strategy IBS can increase the proportion of cooperators only when the initial proportion of cooperators is greater than 0.6; the strategy $WSLS$ can improve the proportion of cooperators only when the initial proportion of cooperators is less than 0.5; Santos et al.'s strategy can advance the proportion of cooperators only in a scale-free network. Obviously, these dependence relationships will limit the applicability of these strategies. In order to overcome these dependence relationships of current strategies, in this paper, a self-organisation based strategy is proposed, which aims to provide a good result for the evolution of cooperation and can be independent of the initial proportion of cooperators and the network types. Additionally, Hofmann et al. [3] found an interesting phenomenon regarding $WSLS$, that is, in a given type of static network, $WSLS$ leads to the same final proportion of cooperators irrespective of the initial proportion of cooperators. Hofmann et al. left the theoretical investigation about that phenomenon as one of their future studies. Against this motivation, after study, we theoretically find a more general phenomenon that is, in static networks, the final proportion of cooperators evolved by any deterministic strategies, including $WSLS$, fluctuates cyclically irrespective of the initial proportion of cooperators. This theoratical finding can support Hofmann et al.'s empirical finding about $WSLS$, because a uniform final proportion of cooperators can be considered that the period of the cycle is 1. The proposed self-organisation based strategy and the theoretical finding are claimed as a two-fold contribution of this paper.

The rest of the paper is organised as follows. In the next section, the proposed self-organisation based strategy will be described in detail. In Section 3, the experimental results and analysis will be provided in both static and dynamic networks. In Section 4, the proof of the theoretical finding, regarding the proportion of cooperators evolved by deterministic strategies, will be given. Finally, the paper will be concluded and some future studies will be outlined in Section 5.

2   The Self-organisation Based Strategy

Before the self-organisation based strategy is described, we first depict the network model and the iterated Prisoner's Dilemma game in networks. The player interactions can be modeled as an undirected graph (or network) $G = (V, E)$, where $V = \{v_1, ..., v_n\}$ is a set of n nodes (or players) and $E \subseteq V \times V$ is a set of edges. Two players v_i and v_j are neighbours if $(v_i, v_j) \in E$. Neighbours can play with each other. $N_i = \{v_j | \langle v_i, v_j \rangle \in E\}$ indicates a player v_i's neighbour set, while $N_i^+ = N_i \cup \{v_i\}$ is the neighour set including v_i itself. A Prisoner's Dilemma game is a two-player game, where each player has two actions, i.e., cooperation and defection. The payoffs for two players are symmetric. If the two players cooperate with each other, both get a reward R; while mutual defection results in punishment P to both. If one player chooses cooperation while the other uses defection, the cooperator gains the lowest sucker's payoff S, while the defector gets the highest temptation payoff T. For a single Prisoner's Dilemma game, $T > R > P > S$ holds and for an iterated Prisoner's Dilemma game, additionally, $T + S < 2R$ holds. In an iterated Prisoner's Dilemma game in a network, there are n players, which form the nodes of the graph, and the game proceeds in rounds. In each round, there are two phases (for static networks) or three phases (for dynamic networks). First, in the game playing phase, the players play the game with all their neighbours and compute their total payoffs. Second, (for dynamic networks only), in the interaction adaptation phase, each player removes some unsatisfactory neighbours and form new interactions with other players. Finally, in the action update phase, each player updates its action according to a strategy.

The proposed self-oragnisation based strategy, used by each player for action update, is described as follows. The essence of the self-organisation based strategy is that it embodies existing strategies into each player's knowledge base. Then, in each round, each player autonomously selects one strategy to update its action and plays with its neighbours. The knowledge of a player j is represented as $S_j = \langle s_1, ..., s_m \rangle$, where each $s_i \in S_j$ is an existing strategy. S_j is also called the strategy set of a player j. One of the merits of this strategy is that, in the future, if a new strategy is developed, it can also be embodied into each player's knowledge base. Thus, the proposed self-organisation based strategy is extendable. We develop a Q-learning algorithm to realise the self-organisation based strategy, which consists of the following three steps.

(i) In each round, each player autonomously selects a strategy from its strategy set based on the probability distribution over strategies. Then, each player uses its selected strategy to update its action to play in the current round.

(ii) When the current round is finished, each player adjusts the probability distribution for strategies based on the payoffs that it receives and its neighbours receive in the current round. The adjustment of the probability on each strategy is executed by using the following equations in sequence (taking a player j for example).

$$Q(s_i) \leftarrow (1 - \alpha) \cdot Q(s_i) + \alpha \cdot p_j \qquad (1)$$

$$\pi(s_i) \leftarrow \pi(s_i) + \beta \cdot (Q(s_i) - \gamma \cdot \bar{p}_j) \qquad (2)$$

$$\Pi \leftarrow Normalise(\Pi) \qquad (3)$$

In Equation 1, $Q(s_i)$ is the Q-value of a strategy s_i, which is a real number and is used to calculate the probability of selecting s_i (see Equation 2); p_j is the player j's payoff received in the current round; and α is the learning rate, which is in the range of $(0, 1)$. In Equation 2, $\pi(s_i)$ is the probability of selecting a strategy s_i; $\bar{p}_j$ is the average payoff of a player j's neighbours (excluding j itself), namely $\bar{p}_j = \frac{\sum_{i \in N_j} p_i}{|N_j|}$, where $|N_j|$ means the number of elements in the set N_j; β is the probability (policy) adaptation rate, which is in the range of $(0, 1)$; γ indicates how important a player's neighbours' average payoff is for that player's probability adaptation, and γ is in the range of $[0, 1]$. In Equation 3, $\Pi = \langle \pi(s_1), ..., \pi(s_m) \rangle$ demonstrates a probability distribution for strategies, and the function $Normalise()$ is used to constrain Π to a legal probability distribution, i.e., $\sum_{1 \le i \le m} \pi(s_i) = 1$, and the result of $Normalise(\Pi)$ has the minimum Euclidean distance to Π.

Initially, the probability for each strategy is equal, namely that each strategy has an equal opportunity to be selected, and all the Q-values are set to 0.

(iii) The above two steps are iterated untill the game ends.

This learning algorithm is simple but powerful enough for our problem. A more efficient learning algorithm might enable the proposed strategy to achieve better results, and we leave this as one of our future studies. Readers might suggest that in each round, each player just simply chooses the strategy with the highest Q-value from its strategy set. Nevertheless, Kaelbling et al. [24] stated that players, who always choose the highest Q-values actions, (called strategies in this paper), can lead a system to a suboptimal status.

3   Experiment and Analysis

After the proposed strategy is described, the experimental results are presented. In this experiment, to implement the proposed self-organisation based strategy, three strategies are selected to be embodied into each player's knowledge base. The three selected strategies are IBN, $WSLS$ and IBS, which have been depicted in the previous section. IBN and $WSLS$ are classical and very famous, while IBS is newly developed. Certainly, other strategies, even those strategies developed in the future, could also be embodied into each player's knowledge base, and additionally, different players could even have different knowledge, i.e., different strategy sets. In this paper, since the aim of this experiment is to provide readers with a general insight into the operation and performance of the proposed self-organisation based strategy, the strategy set, i.e., knowledge, of players is simplified, which contains only three existing strategies, namely IBN, $WSLS$ and IBS, and each player's strategy set is set to be identical. The performance of a strategy, s_1, is considered to be better than another strategy, s_2, if s_1 derives a higher final proportion of cooperators than s_2 does.

The experiment was conducted in two different networks: a scale-free network [25] and a small-world network [26] with both static and dynamic versions. In a scale-free network, the distribution of node degree follows a power law, $n_d \propto d^{-\tau}$, where n_d is the number of players of degree d and $\tau > 0$ is a constant (typically $\tau \in [2,3]$). τ is set to 2.5, which makes an average degree of the network approximately 4. A small-world network shows a high clustering coefficient and a short average path length. First, a ring is built and each player is connected to the two neighbouring players on each side. Then, links are randomly released and reconnected to other players. The rewiring probability is set to 0.2, which leads to an average degree of roughly 4.

A static network means that the topology of the network is fixed, and a dynamic network means that the topology of the network could be changed, namely that players could dynamically break interactions with their neighbours and form new interactions with other players. The parameters, which are used in the experiment, are set as follows, $T = 5$, $R = 4$, $P = 0$, $S = -1$, $\alpha = 0.2$, $\beta = 0.3$ and $\gamma = 0.85$. The values of the four payoffs T, R, P, S can be set arbitrarily, which need only to obey the conditions $T > R > P > S$ and

$2R > T + S$. Different combinations of the values of these four payoffs affect all of the strategies in a same extent, so they do not affect the comparison of these strategies. For the learning parameters α, β, γ, different combinations of their values were attempted and the best combination was selected, which could make the proposed strategy achieve the highest final proportion of cooperators. In addition, each network consists of 1000 players and the average degree of each player is set to 4. For each strategy, in both static and dynamic networks, one running is composed of 500 rounds.

3.1   Experimental Results: Static Networks

The experimental results regarding static networks are displayed in Figure 1, where in all the four sub-figures, the x-axis indicates rounds of the iterated game, and the y-axis demonstrates the proportion of cooperators. Figures 1(a) and 1(b) exhibit the evolution of cooperation in a static scale-free network with the initial proportions of cooperators 0.2 and 0.7, respectively. Figures 1(c) and 1(d) show the evolution of cooperation in a static small-world network with the initial proportions of cooperators 0.2 and 0.7, separately. According to Figure 1, it can be seen that the proposed self-organisation based strategy outperforms other three strategies under different conditions. Specifically, in Figures 1(a) and 1(c), the proportion of cooperators cannot be improved by using IBN and IBS. This phenomenon can be explained as follows. If a player, especially a high degree player, defects, it can exploit all its neighbouring cooperators and gain a high payoff. Then the strategies, i.e., imitating the best neighbour (IBN) or the best strategy (IBS), encourage that defecting player's neighbours to switch to defection. In Figures 1(a) and 1(c), the initial proportion of cooperators is set very low (0.2), so it is not likely that defectors can find a wealthy neighbouring cooperator and then imitate the cooperator's action, because cooperators surrounded by many defectors cannot obtain high payoffs. This situation is improved when the initial proportion of cooperators is increased. This is because that when the initial proportion of cooperators is very high (0.7), wealthy cooperators are quite likely to exist and then the few defecting neighbours will

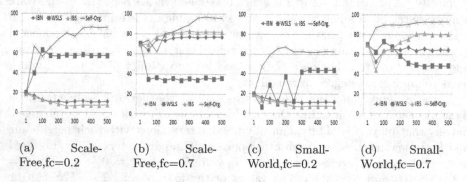

| (a)   Scale-Free,fc=0.2 | (b)   Scale-Free,fc=0.7 | (c)   Small-World,fc=0.2 | (d)   Small-World,fc=0.7 |

Fig. 1. The evolution of cooperation by using different strategies in static networks

switch to cooperation, i.e., imitate the wealthy cooperators' action. However, under $WSLS$, the situation is reversed. In Figures 1(b) and 1(d), when the initial proportion of cooperators is very high (0.7), the proportion of cooperators cannot be increased by using $WSLS$. This is due to the fact that the update of a player's action is based only on its own payoff over time irrespective of its neighbours' actions. Thus, even if a cooperating player has many cooperating neighbours, once one of the cooperating neighours defects, the cooperating player will switch to defection in the next round. Hence, under $WSLS$, a high initial proportion of cooperators cannot guarantee the promotion of proportion of cooperators. However, when the initial proportion of cooperators is very low (0.2), a defector can turn to be a cooperator, once any one of its defecting neighbours switches to cooperation. Thereby, a low initial proportion of cooperators is somewhat helpful for the evolution of cooperation. These results demonstrate that each strategy, i.e., IBN, $WSLS$ and IBS, has limitations. However, in all these situations, the proposed self-organisation based strategy can promote the proportion of cooperators. This can be explained by the fact that players, using the proposed strategy, can benefit from the strengths while avoid the weaknesses of each existing strategy. When a strategy brings a player very low payoff, this strategy will correspondingly be assigned a low Q-value by the player (Equation 1) and the probability of selecting this strategy in the next round will also be reduced, especially when the average payoff of the player's neighbours is high (Equation 2). When the average payoff of a player's neighbours is high, the player is very likely to be exploited by its defecting neighbours. Thus, the strategy used by this player is not good in such a situation and has to be assigned a low probability for selection in the next round.

3.2 Experimental Results: Dynamic Networks

In the above subsection, the performance of the proposed self-organisation based strategy in static networks has been exhibited. However, in many social systems, individuals are continuously creating or removing interactions according to the benefits of interactions. One example occurs in scientific collaboration networks

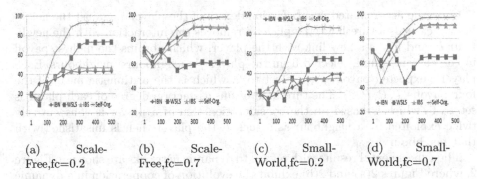

(a)   Scale-Free,fc=0.2

(b)   Scale-Free,fc=0.7

(c)   Small-World,fc=0.2

(d)   Small-World,fc=0.7

Fig. 2. The evolution of cooperation by using different strategies in dynamic networks

[27], where scientists usually work in small groups of collaborations, and the relationships evolve under performance and self-interest of the individual members. Therefore, it is necessary to evaluate the proposed strategy in dynamic networks, where players can remove existing interactions with neighbours and establish new links with other players. Currently, the study on the evolution of cooperation in dynamic networks received less attention than that in static networks [28, 29]. Zimmermann and Eguiluz [29] studied the evolution of cooperation in the iterated Prisoner's Dilemma game with adaptive local interactions, and showed that interaction adaptation could boost the evolution of cooperation in a society. Their interaction adaptation approach allows only unsatisfied defecting players to change their interactions with their defecting neighbours, and then randomly select new players in the whole network to form links. However, allowing only unsatisfied defecting players to change interactions is unfair, because cooperating players may be also unsatisfied with several of their neighbours. Moreover, as mentioned in Section 1 that in the real world, who-meets-whom is not random but determined by spatial relationships or social networks, therefore, it is somewhat impractical to randomly select new players in the whole network to create interactions. Thus, the interaction adaptation approach presented in this paper is that in each round, after playing with neighbours, any players can remove interactions with unsatisfactory neighbours and build new interactions with other players from their neighbours' neighours. This consideration, selecting players from neighbours' neighbours, is based on Kautz et al.'s [30] discussion that shorter interaction chains are more likely to be fruitful and accurate. Here, the unsatisfaction degree of a player against a neighbour depends on the difference between that player's expected payoff and the payoff it actually receives from the neighbour, and the unsatisfaction degree is accumulated along with the game over time. Specifically, the unsatisfaction degree of a player j against a neighbour k can be calculated by using Equation 4, where r is the number of rounds till now, p_{exp} is player j's expected payoff, and $p^i_{j \leftarrow k}$ is player j's actual payoff received from the neighbour k in round i.

$$unsat_{j \rightarrow k} = \sum_{1 \leq i \leq r} p_{exp} - p^i_{j \leftarrow k} \tag{4}$$

Once the accumulated unsatisfaction degree of the player j against the neighbour k exceeds a threshold, the player j removes the interaction with the neighbour k, and forms a new link with the player, which obtains the highest payoff in the current round, selected from the player j's neighbours' neighbours. Each player's expected payoff, p_{exp}, is set to 2, which is the arithmetic mean of the four payoffs T, R, P, S). The threshold of unsatisfaction degree of each player is set to 6, which is based on the consideration that if a player is consecutively twice exploited by a neighbour as a sucker, the player then is unsatisfied with that neighbour.

The experimental results with regard to dynamic networks are shown in Figure 2, where Figures 2(a) and 2(b) exhibit the evolution of cooperation in a dynamic scale-free network with the initial proportions of cooperators 0.2 and 0.7, respectively, and Figures 2(c) and 2(d) demonstrate the evolution of cooperation in a dynamic small-world network with the initial proportions of cooperators 0.2 and 0.7, separately. It can be seen that the final proportion of cooperators derived by each strategy in dynamic networks is improved more or less compared with that in static networks. This is ascribed to the fact that each player is allowed to dismiss the interactions with their neighbours, with whom the player is quite unsatisfied, and create new interactions with other wealthy players, who have the potential to satisfy the player. An interesting finding is that in Figures 2(a) and 2(c), where the initial proportion of cooperators is only 0.2, the final proportion of cooperators is improved by using IBN and IBS, whereas in static networks, when the initial proportion of cooperators is 0.2, the final proportion of cooperators cannot be increased by using IBN and IBS. This phenomenon can be explained as follows. In dynamic networks, when a player attempts to form a new link, it will search its neighbours' neighbours and find the wealthiest player. The wealthiest player is very likely a cooperator with several cooperating neighbours. Certainly, a defecting player can also achieve a high payoff by exploiting its cooperating neighbours, but this situation will immediately be changed in the next round, as its neighbours will imitate the wealthy defecting player's action. Therefore, when the game progresses over time, a wealthy player is more likely a cooperator instead of a defector. Obviously, when a player establishes an interaction with a wealthy cooperator, this player is easy to be assimilated as a cooperator by using IBN and IBS. Hence, the final proportion of cooperators can be improved. Again, the proposed self-oragnisation based strategy achieves the best results.

Overall, it can be seen that the self-organisation based strategy works very well in any situations, while each of the three strategies embodied into the self-organisation based strategy has limitations. Thus, it is a promising idea to combine current strategies into a single strategy set and let each player autonomously select a strategy to update its action in each round, because in this manner, the limitations of each strategy can be avoided and the strengths of each strategy can be utilised comprehensively.

4   A Theoretical Finding

As described in Section 1, motivated by Hofmann et al.'s empirical finding about $WSLS$ [3], we made further investigation and found a general phenomenon that in static networks, the final proportion of cooperators evolved by any deterministic strategies, including $WSLS$, fluctuates cyclically irrespective of the initial proportion of cooperators.

In an iterated Prisoner's Dilemma game, in a single round, a player v_i has only two actions, i.e., cooperate (represented as 1) or defect (represented as 0). Thereby, the action of a player v_i, recorded as act_i, is in $\{0, 1\}$. Let us format the payoff parameters, i.e., T, R, P, S, to be a matrix as follows.

$$\begin{bmatrix} P & T \\ S & R \end{bmatrix} \tag{5}$$

For convenience, we use the following matrix in place of the above matrix,

$$M = \begin{bmatrix} m_{00} & m_{01} \\ m_{10} & m_{11} \end{bmatrix} \tag{6}$$

where $m_{00} = P$, $m_{01} = T$, $m_{10} = S$ and $m_{11} = R$. Then, if player v_i plays against v_j, then v_i's payoff is $p_{ij} = m_{act_i, act_j}$ and v_j's payoff is $p_{ji} = m_{act_j, act_i}$. Hence, in a single round, the total payoff of v_i is $p_i = \sum_{v_j \in N_i} p_{ij}$. Obviously, given a network $G = (V, E)$ and a payoff matrix M, the payoff vector of a network $\mathbf{p} = \langle p_1, ..., p_n \rangle$ is determined only by the action vector of a network $\mathbf{act} = \langle act_1, ..., act_n \rangle$. (Actually, the phrase "determined by" means that there exists a mapping from the action vector $\mathbf{act}$ to the payoff vector $\mathbf{p}$, but for the readability purpose, we use natural language to make the description.) It should be noted that an action vector indicates a proportion of cooperators in a network. The components of an action vector are either 0 (defect) or 1 (cooperate), so by counting the number of "1" in an action vector and then divided by n, the proportion of cooperators is obtained.

Similarly, in an iterated game, in each round t, the payoff vector of a network $\mathbf{p}(t) = \langle p_1(t), ..., p_n(t) \rangle$ is also determined only by the action vector of a network $\mathbf{act}(t) = \langle act_1(t), ..., act_n(t) \rangle$, if the network G and the payoff matrix M are fixed, and the strategy, which each player uses to update its action, is fixed as well.

Definition 1. (*A k-order Strategy*) For a strategy, if there exists a minimum positive integer k, such that the action vector of a network in round t, i.e., $\mathbf{act}(t)$, where players use this strategy, is determined by the action vectors $\mathbf{act}(t-1)$, ..., $\mathbf{act}(t-k)$ or the payoff vectors $\mathbf{p}(t-1)$, ..., $\mathbf{p}(t-k)$ of the network, then this strategy is called a *k-order* strategy.

For example, the strategy $WSLS$ is a *2-order* strategy, because a $WSLS$ player updates its action based on its current and former payoffs. The strategies IBN and IBS are *1-order* strategies, as players, using these two strategies, update their actions based only on the current payoff. The definition of "*k-order*" is somewhat like the term "memory length", which was often mentioned in current literature, e.g., [31]. However, there is a significant difference between "*k-order*" and "memory length". "Memory length" describes only how many previous rounds are required to determine an action vector in the current round, while the definition of "*k-order*" not only specifies how many previous rounds are required but also dictates that the action vector in current round is determined only by the previous action vectors and payoff vectors. For example, the strategy TFT is a *1-order* strategy, and its memory length is 1. Nonetheless, the memory length of the strategy $GTFT$ is 1, but it is not a *1-order* strategy, because the current action vector is based not only on the former action vector but also on a certain probability, which contradicts with **Definition 1**. It can also be deduced that if a strategy is *k-order*, it must be a deterministic strategy.

Lemma 1. *If a strategy is k-order, the action vector of a network in round t, where players use this strategy, is determined only by the action vectors $\mathbf{act}(t-1)$, ..., $\mathbf{act}(t-k)$.*

Proof. The proof of this lemma is straightforward. As described above, in each round t, the payoff vector of a network $\mathbf{p}(t)$ is determined only by the action vector of a network $\mathbf{act}(t)$. Therefore, each payoff vector $\mathbf{p}(t-i)$, where $1 \leq i \leq k$, is determined only by the action vector $\mathbf{act}(t-i)$, where $1 \leq i \leq k$, respectively. Thus, according to **Definition 1**, the action vector of a network in round t, where players use a k-*order* strategy, is determined only by the action vectors $\mathbf{act}(t-1)$, ..., $\mathbf{act}(t-k)$. $\square$

Definition 2. (*A Cyclic Game*) For an iterated game in a network, if there exists a minimum positive integer r, such that after a round t_0, the equation $\mathbf{act}(t) = \mathbf{act}(t+r)$ always holds, where $t \geq t_0$, then this iterated game is called a cyclic game, and the integer r is called the cyclic period of the game.

Theorem 1. *Given a network G and a payoff matrix M, if a strategy is k-order, then players in the network, using this strategy, will lead an iterated game to be a cyclic game, and the upper bound of the cyclic period of the game is 2^{kn}, i.e., $r \leq 2^{kn}$, where n is the number of players in the network.*

Proof. In a network with n players, there are 2^n different action vectors. If k action vectors are selected from the 2^n action vectors, the selection has 2^{kn} different compositions. Let the selected k action vectors be $\mathbf{act}(2^{kn}+1)$, ..., $\mathbf{act}(2^{kn}+k)$. Thus, the k action vectors, $\mathbf{act}(2^{kn}+1)$, ..., $\mathbf{act}(2^{kn}+k)$, are equal to at least one set of k sequential action vectors taken from $\mathbf{act}(1)$, ..., $\mathbf{act}(2^{kn})$. Suppose that the equal set of k sequential action vectors, taken from $\mathbf{act}(1)$, ..., $\mathbf{act}(2^{kn})$, is $\mathbf{act}(t_0+1)$, ..., $\mathbf{act}(t_0+k)$. According to Lemma 1, in round t, $\mathbf{act}(t)$ is determined only by $\mathbf{act}(t-1)$, ..., $\mathbf{act}(t-k)$. Hence, $\mathbf{act}(t_0+k+1)$ is determined only by $\mathbf{act}(t_0+k)$, ..., $\mathbf{act}(t_0+1)$, and likewise, $\mathbf{act}(2^{kn}+k+1)$ is determined only by $\mathbf{act}(2^{kn}+k)$, ..., $\mathbf{act}(2^{kn}+1)$. Since $\mathbf{act}(t_0+k) = \mathbf{act}(2^{kn}+k)$, ..., $\mathbf{act}(t_0+1) = \mathbf{act}(2^{kn}+1)$ (the above assumption), it can be obtained that $\mathbf{act}(t_0+k+1) = \mathbf{act}(2^{kn}+k+1)$, and similarly, it can also be obtained that $\mathbf{act}(t_0+k+2) = \mathbf{act}(2^{kn}+k+2)$, $\mathbf{act}(t_0+k+3) = \mathbf{act}(2^{kn}+k+3)$, ... Let $r' = 2^{kn} - t_0$, when $t > t_0$, the equation $\mathbf{act}(t) = \mathbf{act}(t+r')$ always holds[1]. Therefore, the game is cyclic and the upper bound of its cyclic period is 2^{kn}, as $r \leq r' \leq 2^{kn}$. $\square$

Based on the above description, the following two deductions can be obtained.

(i) The conclusion of Theorem 1 is that any deterministic strategy will lead an iterated game to be a cyclic game. This conclusion is independent of the initial proportion of cooperators in a network, the topology of a network and a specific game, because in the proof of Theorem 1, neither initial proportion of cooperators is set nor specific network structures or orders of the parameters, T, R, P, S, are imposed. Thus, this conclusion still holds in other games, e.g., Snowdrift game, where $T > R > S > P$, and Stag Hunt game, where $R > T > P > S$

---

[1] Suppose that $t = t_0 + 1$, then $\mathbf{act}(t)$ becomes $\mathbf{act}(t_0+1)$ and $\mathbf{act}(t+r')$ becomes $\mathbf{act}(2^{kn}+1)$. As described above, $\mathbf{act}(t_0+1) = \mathbf{act}(2^{kn}+1)$ and this equation also holds for $t = t_0 + 2$, $t = t_0 + 3$, ..., so $\mathbf{act}(t) = \mathbf{act}(t+r')$ always holds when $t > t_0$.

[23]. However, the exact cyclic period of an iterated game depends on several factors, such as topology of the network, the specific strategy used by players in the network, the specific game and the initial proportion of cooperators. There may be a function or an approximate function to compute the cyclic period of a game based on the listed factors. We leave investigating such a function as one of our future studies.

(ii) A corollary can be derived from Theorem 1.

Corollary 1. *In a network G, even if different players use different deterministic strategies, the iterated game is still a cyclic game and the upper bound of its cyclic period is $2^{k'n}$, where $k' = MAX(k_1, ..., k_m)$, m is the number of different strategies used by players, and each k_i demonstrates that the strategy s_i is a k_i-order strategy.*

Proof. The proof of this corollary is similar to the proof of Theorem 1 by altering k to k'. If different players use different strategies, the action vector in round t is determined by $\mathbf{act}(t-1), ..., \mathbf{act}(t-k')$. For example, in a network, a group of players use a 2-*order* strategy (recorded as a 2-*order* group) and another group of players use a 3-*order* strategy (recorded as a 3-*order* group). The action vector of the 2-*order* group in round t is based on the action vectors of the 2-*order* group in $t-1$ and $t-2$ rounds, while the action vector of the 3-*order* group in round t is based on the action vectors of the 3-*order* group in $t-1$, $t-2$ and $t-3$ rounds. It is noted that the action vector of the 2-*order* group in round $t-1$ is based on the action vectors of the 2-*order* group in $t-2$ and $t-3$ rounds, so the action vector of the 2-*order* group in round t can also be considered to be based on the action vectors of the 2-*order* group in $t-1$, $t-2$ and $t-3$. Hence, the situation has been the same as that in Theorem 1.   □

5   Conclusion and Future Work

In this paper, a self-organisation based strategy for evolution of cooperation was proposed, which embodies current stretegies as players' knowledge and enables each player to autonomously select a strategy to update its action to play the game. Thereby, the proposed strategy can avoid the flaws of each strategy and can harness the advantages of each strategy. Furthermore, the proposed strategy is extendable, as in the future, a newly developed strategy can also be included into the proposed strategy, which embodies the new strategy as a new piece of knowledge into each player. The experimental results demonstrated the good performance of the proposed strategy in different situations. In addition, the theoretical description of the phenomenon of the evolution of cooperation in static networks was also given, which can support Hofmann et al.'s experimental finding in [3].

In the future, as mentioned in Section 2, we attempt to develop a more complex and efficient learning algorithm to realise the self-organisation based strategy. Furthermore, we intend to test the self-organisation based strategy in different games, such as the aforementioned Snowdrift and Stag Hunt games [23]. We also want to test the proposed strategy in asynchronous games, because in biology, political science and sociology, players do not always decide simultaneously but rather make their decisions in turns [32]. In addition, it is also interesting to test the proposed strategy in an open network, where existing players may leave the network and new players may join the network. On the other hand, for the theoretical study, at the current stage, we determined only the upper bound cyclic period of a game, so in the future, we plan to determine more precisely the cyclic period of a game, such as the precise spectrum of the cyclic period of a game, and to investigate when the game gets into a cycle. It is also interesting to do further theoretical investigation in asynchronous games to see whether there are similar conclusions to those observed in synchronous games. Moreover, for application domains, the proposed evolution of cooperation strategy could be employed in wireless sensor networks for sleep/wake-up scheduling. In wireless sensor networks, the energy of each node is usually limited, so it is necessary to keep the nodes in sleep state as long as possible to reduce the energy consumption and then to increase the lifetime of sensor networks. A simple approach for this issue is that all nodes keep in sleep state, unless data have to be transferred. However, this approach will incur a heavy data transmission delay, which should also be avoided. Then, the scheduling problem arises. Here, nodes can be considered as players, and sleep and wake-up can be considered as the actions in an iterated game, i.e., cooperate and defect. One point needs to be paid attention to is that instead of a single payoff matrix, two payoff matrices are probably required, because if two neighbouring nodes do not have data to transfer, the best choice for both of them is sleep, so in this situation, sleep should be considered as cooperation; but if any one of the two neighbouring nodes has data to transfer, then wake-up becomes the best choice, so in this case, reversely, wake-up should be considered as cooperation. This is only a preliminary idea, and many efforts have to be done on it. We will keep track of these future studies.

References

1. Hamilton, W.D.: The genetical evolution of social behaviour. J. Theor. Biol. 7, 1–16 (1964)
2. Axelrod, R., Hamilton, W.D.: The evolution of cooperation. Science 211, 1390–1396 (1981)
3. Hofmann, L.M., Chakraborty, N., Sycara, K.: The evolution of cooperation in self-interested agent societies: a critical study. In: Proc. of AAMAS 2011, Taipei, Taiwan, pp. 685–692 (May 2011)
4. Guan, J.Y., Wu, Z.X., Wang, Y.H.: Effects of inhomogeneous activity of players and noise on cooperation in spatial public goods games. Phys. Rev. E 76, 056101 (2007)
5. Szolnoki, A., Perc, M.: Group-size effects on the evolution of cooperation in the spatial public goods game. Phys. Rev. E 84, 047102 (2011)
6. Wedekind, C., Milinski, M.: Cooperation through image scoring in humans. Science 288, 850–852 (2000)
7. Fehr, E., Fischbacher, U.: The nature of human altruism. Nature 425, 785–791 (2003)

8. Kagel, J.H., Roth, A.E.: The handbook of experimental economics. Princeton University Press (1995)
9. Dawes, R.M.: Social dilemmas. Ann. Rev. Psychol. 31, 169–193 (1980)
10. Hassell, M.P., Comins, H.N., May, R.M.: Species coexistence and self-organizing spatial dynamics. Nature 370, 290–292 (1994)
11. Abramson, G., Kuperman, M.: Social games in a social network. Phys. Rev. E 63, 030901 (2001)
12. Szabo, G., Vukov, J.: Cooperation for volunteering and partially random partnership. Phys. Rev. E 69, 036107 (2004)
13. Lieberman, E., Hauert, C., Nowak, M.A.: Evolutionary dynamics on graphs. Nature 433, 312–316 (2005)
14. Nowak, M.A., May, R.M.: Evolutionary games and spatial chaos. Nature 359, 826–829 (1992)
15. Nakamaru, M., Matsuda, H., Iwasa, Y.: The evolution of cooperation in a lattice-structured population. J. Theor. Biol. 184, 65–81 (1997)
16. Ifti, M., Killingback, T., Doebeli, M.: Effects of neighbourhood size and connectivity on the spatial continuous prisoner's dilemma. J. Theor. Biol. 231, 97–106 (2004)
17. Santos, F.C., Pacheco, J.M.: Scale-free networks provide a unifying framework for the emergence of cooperation. Phys. Rev. Lett. 95, 098104 (2005)
18. Axelrod, R.: The evolution of cooperation. New York (1984)
19. Axelrod, R., Dion, D.: The further evolution of cooperation. Science 242, 1385–1390 (1988)
20. Nowak, M., Sigmund, K.: Tit for tat in heterogeneous populations. Nature 355, 250–253 (1992)
21. Nowak, M., Sigmund, K.: A strategy of win-stay, lose-shift that outperforms tit-for-tat in the prisoner's dilemma game. Nature 364, 56–58 (1993)
22. Tang, C., Wang, W., Wu, X., Wang, B.: Effects of average degree on cooperation in networked evolutionary game. Eur. Phys. J. B 53, 411–415 (2006)
23. Santos, F.C., Pacheco, J.M., Lenaerts, T.: Evolutionary dynamics of social dilemmas in structured heterogeneous populations. Proc. Natl. Acad. Sci. 103, 3490–3494 (2006)
24. Kaelbling, L.P., Littman, M.L., Moore, A.W.: Reinforcement learning: A survey. Journal of AI Research 4, 237–285 (1996)
25. Barabasi, A.L., Albert, R.: Emergence of scaling in random networks. Science 286(5439), 509–512 (1999)
26. Watts, D.J., Strogatz, S.H.: Collective dynamics of 'small-world' networks. Nature 393, 440–442 (1998)
27. Newman, M.E.J.: The structure of scientific collaboration networks. Proc. Natl. Acad. Sci. 98, 404–409 (2001)
28. Ebel, H., Bornholdt, S.: Coevolutionary games on networks. Phys. Rev. E 66, 056118 (2002)
29. Zimmermann, M.G., Eguiluz, V.M.: Cooperation, social networks, and the emergence of leardership in a prisoner's dilemma with adaptive local interactions. Phys. Rev. E 72, 056118 (2005)
30. Kautz, H., Selman, B., Milewski, A.: Agent amplified communication. In: AAAI 1996, Portland, pp. 3–9 (1996)
31. Szabo, G., Fathb, G.: Evolutionary games on graphs. Phys. Rep. 446, 97–216 (2007)
32. Hauert, C., Schuster, H.G.: Extending the iterated prisoner's dilemma without synchrony. J. Theor. Biol. 192, 155–166 (1998)

Supporting Adaptation Patterns in the Event-Driven Business Process Modeling Paradigm

Malinda Kapuruge, Jun Han, and Alan Colman

Faculty of Information and Communication Technologies
Swinburne University of Technology, Melbourne, Australia
{mkapuruge,jhan,acolman}@swin.edu.au

Abstract. The event-driven business process modeling has been emerged as an alternative paradigm to the traditional workflow-based business process-modeling paradigm. One of the influencing factors for using this new paradigm is the flexibility and the agility it provides in supporting business process change. Weber et al. have proposed 14 high-level adaptation patterns that need to be supported by a flexible process-aware information system irrespective of the chosen paradigm. In this paper we investigate whether the event-driven paradigm can satisfactorily support these adaptation patterns; if so, how an event-driven approach should support these adaptation patterns. We also point to the future research directions that this work can lead to.

Keywords: Adaptation Patterns, Event-driven, BPM.

1 Introduction

Business Process Management (BPM) has been widely adopted by business organizations as part of their core business strategy [1]. However, real-world business processes are subject to change, demanding changes to their IT counterparts, in particular, the control-flow structures. For example, activities of a business process can be added, removed or reordered. The flexibility provided by the process modeling languages plays a major role in supporting changes in process types (evolutionary changes) and the enacted instances (ad-hoc changes) [2, 3].

Weber et al. [4] identified 14 *adaptation patterns* that need to be supported by a process-aware information system (PAIS), to support both evolutionary and ad-hoc adaptations. Instead of using low-level change primitives (e.g., add a node, delete a control flow edge), these patterns allow users to structurally modify a process type or a process instance using high-level change operations (e.g., insert a process fragment). Their work focused on the **workflow-based process modeling** paradigm, and used BPMN [5] notation to show the changes of business processes.

In the recent past, the **event-driven process modeling** has emerged as an alternative process modeling paradigm [6-8]. Instead of using nodes and edges organized as a graph to describe a process flow (as is the case with the workflow paradigm), the event-driven paradigm uses *event-action* rules to describe a process

X. Lin et al. (Eds.): WISE 2013, Part II, LNCS 8181, pp. 299–308, 2013.
© Springer-Verlag Berlin Heidelberg 2013

flow. As such, the complex event patterns are used as pre-conditions to trigger business activities and also as post-conditions upon the completion of the business activities [9, 10]. Although the event-driven processes provide a highly decoupled structure, it is challenging for a process designer/engineer to realize the high-level change requirements by modifying the event-driven dependencies captured by event-action rules.

In this context, the adaptation patterns defined by Weber et al. [4] provide suitable abstractions that represent generic adaptation requirements of real-world business processes. It is therefore, important that such patterns should be supported in the event-driven paradigm as well. This paper shows how those adaptation patterns can be mapped to the event-driven paradigm. This is important primarily due to the differences in change primitives of the two paradigms. The change primitives presented in [4] such as *addEdge*, *removeEdge* are directly applicable for workflow-based process modeling, e.g., remove/add an edge in BPMN. However, in the event-driven paradigm, there are no explicit edges to be removed/added. Instead, the control-flow is modified by altering the pre- and post-conditions of activities.

In this paper, we do not limit the discussion to a specific event-driven language/approach, as there are many [8-13]. We have examined these event-driven approaches and suitably generalized them to produce a simple yet sufficient meta-model that captures the essence of an event-driven process modeling language. Using this meta-model, we explore the possibility of supporting each adaptation pattern.

The contribution of this work is to show how all 14 adaptation patterns can be mapped to the generic event-driven meta-model. This work can also be used as a reference to support/implement adaptations in event-driven business processes.

The rest of the paper is organized as follows. In Section 2, we introduce event-driven business process modeling and example approaches from the research literature. In the same section, we define the abstract event-driven meta-model and a supporting language that will be used for the explanation. How the adaptation patterns are supported is explained in Section 3. A detailed analysis and sighting of several research directions that this work can lead to are presented in Section 4, before we conclude the paper in Section 5.

2   Event-Driven Business Process Modeling

The event-driven process modeling has emerged as an alternative paradigm for the traditional workflow-based business process modeling. In the event-driven paradigm, the event, which describes a *situation of interest* [14], is used as an *explicit entity*. Usually, an event or a complex composition of multiple events specifies the pre- and post-conditions of business activities defining a control-flow structure.

Krishnamurthy et al. [9] describe *"an event-action system as a software system in which, events occurring in the environment of the system trigger actions in response to the events. The triggered actions may generate other events, which trigger other actions, and so on"*. Adopting above style into business process modeling, the events can be used to define the business activity dependencies. The triggering of events initiates business activities. Further events could be triggered as a result of activity execution. In this sense, the events are employed to decouple the activities [15]. This style of process modeling is evident in a number of approaches proposed in the past.

For example, Alexopoulou et al. [10] uses an event-action rule-based approach in defining business processes to improve the agility of changes. Events are causally related and lead to action execution. YEAST [9] is a general purpose event-action system that also could be used for flexible process automation in a distributed environment. Geppert et al. [11] propose an event-condition-action rule-based process engine (EVE). EVE uses a layered event-based approach to describe the behaviors of processing entities in a heterogeneous environment.

Dayal et al.[14] explore the use of event-triggering and transactions to organize long running activities. The control-flow is described in an event-driven manner to achieve the flexibility and modularity. ARIS toolset is a pioneer of using Event-driven Process Chain (EPC) to model business processes [13]. EPC graphs use events and their combinations to visualize the relationships between multiple functions/activities. In other works, we also exploit the benefits of the event-driven business process modeling to provide the loose coupling between activities and support variability in multi-tenanted service compositions [12]. In common, all the above approaches use events to define the relationship between activities. The flexibility and agility of performing changes has been the influencing benefits.

To explore how the adaptation patterns are supported in the event-driven paradigm, we need an abstract event-driven language. Such a language should, *(a)* consist of minimal concepts to support process modeling constructs used in [16] (e.g., activity, process fragments and definition), so that the discussion is consistent yet not overly complicated; *(b)* Be independent of the existing event-driven approaches, yet provide an abstract representation of them so that this work could be generally applicable.

EventLang: To fulfill the above requirements, we define an abstract language called EventLang. In EventLang, there are four main elements, i.e., *Activity, Process Fragment, Process Definition* and *Event* as shown in the meta-model in Fig. 1.

Fig. 1. The EventLang meta-model

Here, an *Activity* represents an atomic or a single step of a business process, e.g., *TreatPatient*. A *Process Fragment* groups these activities, could be a sub-process or even reduced to a single activity. A *Process Definition* is a special type of a process fragment with its own start- and end-conditions that define when a process instance should be initiated or terminated. These three elements, i.e., *Activity, Process Fragment, Process Definition* correspond to the terms of the original article and are sufficient to show how the adaptation patterns are supported in the event-driven paradigm.

Whilst the terminology is similar to [16], there are differences too due to the event-driven nature of the EventLang. Unlike the workflow-based process models used by the original article, the activities are not related with edges to define the control-flow. Instead, the activities define their pre- and post-conditions in terms of event patterns.

Here, an *Event* is a passive element that defines a situation [14]. For example, AllergyTestDone is an event that defines the situation *allergy test is completed* in a business process. Such atomic events can be combined/correlated with operators (AND, OR, XOR) to construct complex pre- and post-conditions, as represented by the pre-event-pattern (pre_ep) and post-event-pattern (post_ep) attributes of an activity. For example, the activity *TreatPatient* will be carried out when the pre_ep="DoctorReferralReceived AND AllergyTestDone" is evaluated to be true. Upon the completion of this activity, the events could be triggered according to the post_ep="PatientTreated AND (NextAppointmentReqd XOR NoAppointmentReqd)". Fig. 2 describes the above example using EventLang and illustrates using an EPC notation [13] for clarity.

The attribute evaluation_rules represents the business rules that are fired during the execution of the activity. This is important to support several adaptation patterns that require evaluating a condition in the control-flow, e.g., AP1c. The rule could be implemented using a language such as Drools [17] or JESS [18].

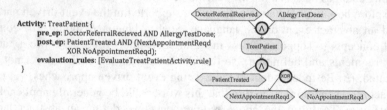

Fig. 2. The activity described in EventLang and illustrated using EPC notation

A *Process Fragment* may include logically related activities. For example, the activities related to treating patients could be included in a single process fragment whilst the activities related to retrieving patient data in another.

A *Process Definition* is a special type of a process fragment, which explicitly defines its start and end conditions. A process instance needs to be instantiated and terminated as specified by its definition. The attributes condition-of-start (CoS) and condition-of-termination (CoT) of a process definition serve this purpose. Similar to pre_ep and post_ep, the CoS and CoT are also specified in terms of event patterns.

An example process definition, its fragments and activities are illustrated in Fig. 3. As shown, the process defines CoS, which instantiate an instance of it. One or more process fragments may exist within the process definition. Once a process is instantiated, the activities are executed depending on the triggered events. An event triggered by one activity may be used to instantiate one or more activities. For example, the event PatinetTreated establishes the dependency between the two activities TreatPatient and ClosePatientFile. Once the CoT is met, the process is terminated.

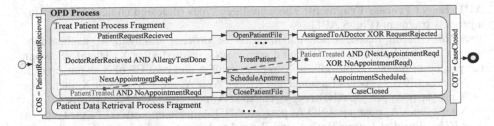

Fig. 3. A sample process definition, process fragment and activities

3   Supporting Adaptation Patterns

In this section, we examine how each of the adaptation patterns (AP) proposed by Weber et al. [4] can be mapped to the event-driven process modeling paradigm.

In order to be consistent with the original article [4] we use the same figures as shown in Fig. 5 (AP1-AP14). In the discussion, firstly, we analyze how the adaptation is supported from the event-driven point of view, in a way the same order of activity execution that of the original article, could be achieved. Secondly, we show the modification steps (alongside APs) using several *change primitives* introduced below.

Change Primitives. We use following change primitives to add, delete activities and to modify their properties (pre_ep, post_ep).

- *add A* : add an activity A.
- *delete A* : delete an activity A.
- *assign A.prop1=B.prop2* : the value of a property *prop1* of an activity A is assigned as the value of a property *prop2* of an activity B.

Apart from the above change primitives, we also use following additional change primitives to add/delete or modify a business rule R1.

- *addRule A.R1* : add rule R1 to be evaluated upon activity A.
- *deleteRule A.R1* : remove rule R1 scheduled to be evaluated upon activity A.
- *modifyRule A.R1* : modify rule R1 scheduled to be evaluated upon activity A.

Example: Consider a process of a restaurant where activities readMenu, orderMeal, orderDrinks, orderDessert and payBill ordered sequentially as shown in the box to the left of Fig. 4. By applying AP9 (explained later) the activities orderFood, orderDrinks and orderDessert are parallelized via the above change primitives (specified along the arrow). The modified process is available in the box to the right.

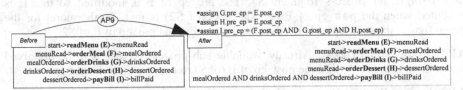

Fig. 4. Applying AP9 upon event-driven process description

Let us now examine how each adaptation pattern is supported. The change operations are listed alongside the illustrations for each adaptation pattern in Fig. 5.

AP1: Insert Process Fragment: There are three types of design choices for inserting a process fragment (Fig. 5.AP1).

AP1a - Serial Insert: Upon the insertion of process fragment X, it should be always executed after A, but B should be always executed after X. Therefore assign the pre_ep of the newly introduced X to post_ep of A and post EP of X to pre_ep of B.

Note: Let X is a process fragment and A_1, A_2, A_3... A_n are activities of X, then,

X.pre_ep = A_1.pre_ep AND A_2.pre_ep AND A_3.pre_ep ... AND A_n.pre_ep.

X.post_ep = A_1.post_ep AND A_2.post_ep AND A_3.post_ep ... AND A_n.post_ep.

AP1b - Parallel Insert: Both X and B need to be executed after A. C needs to be executed after both X and B. So, assign pre_ep of X to post_ep of A. Then modify pre_ep of C to combine post_eps of B and X.

AP1c - Conditional Insert: Add rule R1, which will be evaluated upon activity A. Then the post_ep of A needs to be modified to trigger either B.pre_ep or X.pre_ep. The rule R1 contains the condition, which will determine the triggering of A.post_ep.

AP2: Delete Process Fragment: The activities D and E need to be sequentially executed when the activity C is deleted. Therefore, the post_ep of B is now assigned with pre_ep of D. In addition, the post_ep of D is assigned as pre_ep of E. Then delete C.

AP3: Move Process Fragment: There are three design choices.

AP3a - Serial Move: Here, an activity A will be executed after B but before D. First, assign the pre_ep of A into the pre_ep of B to ensure B continues the execution. Then to ensure A get executed in place of D, assign D's pre_ep to A's pre_ep. To ensure D get executed after A, replace D's pre_ep with A's post_ep.

AP3b - Parallel Move: Here, A will be executed in parallel to C. First, assign the pre_ep of A into pre_ep of B. To make sure A and C both are executed in parallel, assign C's pre_ep to A's pre_ep. Then update E's pre_ep to ensure that E is executed after the execution of either (A and C) or D.

AP3c - Conditional Move: Here, the succeeding activities after B need to have different pre_eps to support the conditional branching. Therefore A needs a new event pattern (newEP) for its pre_ep different from C's and D's.

AP4: Replace Process Fragment: B is replaced by X. To support this pattern, simply assign the pre_ and post_eps of B for those of the new X. Then delete B.

AP5: Swap Process Fragment: Swap the pre_ and post_eps of the swapping fragments. To facilitate this, a temporary·variable *var1* is used to back up the values.

AP6: Extract Sub Process: Group common tasks {F, G, H} of processes S1and S2 into a *single* sub-process fragment S3. The pre_ep of F is modified so that E is executed when the post_ep of E (of S1) or C (of S2) is met. Furthermore, for the given figure, the post_ep of H needs to be assigned to the pre_ep of E of S2.

AP7: Inline Sub Process: Already available sub-process S1 needs to be dissolved into a main process. Assign the post_ep of E (of S) to the pre_ep of F (of S1).

AP8: Embed Process Fragment in Loop: A loop needs to have an exit condition. If the exit condition is not met, the loop continues. This pattern needs to be supported with help of a rule R1, which specifies the exit condition and the repetition condition. The rule R1 will be evaluated upon the completion of E, and triggers B.pre if a repetition (looping) is required. Otherwise, trigger F.pre, to exit the loop.

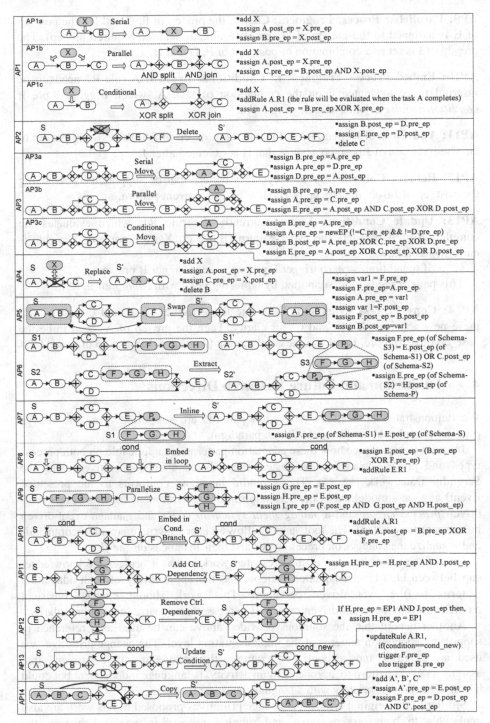

Fig. 5. Adaptation Patterns (AP1 – AP14)

AP9: Parallelize Process Fragment: Change the pre_ep of F, G and H. The post_ep of E is assigned to the pre_ep of G and H (F already has the required pre_ep). The pre_ep of I is set as a combination of post_eps of F, G and H.

AP10: Embed Process Fragment in Conditional Branch: A rule R1 needs to be inserted to be evaluated upon the completion of task A, which triggers B.pre if *condition==false* or otherwise triggers F.pre.

AP11: Add Control Dependency: Include the post_ep of J to the existing pre_ep of H. The inclusion of J.post_ep ensures that H is executed after J.

AP12: Remove Control Dependency: This pattern can be supported by removing the post_ep of J from the pre_ep of H. Here EP1 is in event pattern.

AP13: Update Condition: Assume currently the rule R1, which evaluates the completion of A, triggers the event pattern (B.pre_ep XOR F.pre_ep). The rule is currently as follows,

      if (condition==*cond*) trigger F.pre_ep;   else trigger B.pre_ep

This pattern needs to be supported by replacing *cond* with *cond_new* in R1.

AP14: Copy Process Fragment: Duplicate the activities A, B and C. Let us call these new activities A', B' and C' for clarity. Then the pre_ep of A' need to be the post_ep of E. The pre_Ep of F is the combination (AND) of the post_ep of D and C'.

4   Discussion and Future Research Directions

We demonstrated how all 14 adaptation patterns are supported using the change primitives applicable to the event-driven paradigm.

In order to experience how these adaptation patterns are applicable on both process types and process instances, we tested above listed change primitives using the Serendip framework [12, 15]. The Serendip language conforms to the presented EventLang meta-model. Also, the Serendip runtime environment consists of an event-driven *process engine* and a *management system to* enact and manage event-driven business processes [15]. The tool support allows a user to write adaptation commands and visualize the changes on process types and their instances *on-the-fly*.

One of the major benefits of the presented work is that it bridges the knowledge-gap between high-level adaptation patterns and the applicability of such adaptation patterns to the event-driven paradigm. This work should enable consistent modifications to the event-driven processes expressed at a coarse level of granularity.

Another benefit that the solutions provided above can be used as a reference in supporting adaptation patterns at the implementation level. The change primitives we used are language independent. For example, the change primitive *"assign"* that assigns an event pattern to a property of an activity, is language independent but could be substituted appropriately with a language-specific manner.

Similar to the original article, the change primitives proposed in this work are also applicable to both process types and running process instances. However, the adaptations on running process instances demand a mechanism to ensure the adaptations are state and transaction safe.

The presented EventLang is a highly simplified language. The way events are evaluated may vary between existing languages thereby changing the mapping from the meta-model. For example, an event-condition-action rule (e.g., [11]) may specify the pre-event-pattern as part of the condition, whereas an event-action (e.g., [10]) rule may embed the pre-event-pattern as a composite event. Therefore, the change primitives used in this article need to be consistently mapped in a language-specific manner considering these subtle differences among event-driven approaches.

Modern Complex Event Processing (CEP) literature supports more expressive event correlation operations, e.g., temporal-intervals and event windows [17, 19]. However, we did not consider all the event correlation operators in our simplified language, but focused on using a sufficient sub-set of operators (i.e., AND, OR, XOR), because our intention is to show that event-driven paradigm can support the adaptations proposed in the workflow-based process paradigm. Similarly, we also avoid considering the concepts such as event-sources, event-sinks and event channels to avoid the complexity. We refer to [7] for a comprehensive glossary for CEP.

While this work shows the support for adaptation patterns in the event-driven paradigm, it does not compare the comprehensiveness of applying those patterns to that of the workflow-paradigm. Intuitively, performing adaptations in the event-driven paradigm looks comprehensive due to the loosely-coupled [15] nature of activities compared to rigidity in workflows [20, 21]. However, such claim needs to be experimentally validated, e.g., based on the number of steps/time taken to perform a particular change. We leave such evaluations for the future work.

Another research question is how to automatically translate workflows into event-driven processes to reap the benefits of both paradigms. The workflows, e.g., BPMN [5], are easy to understand for a business user. On the other hand the activity decoupling [15] in the event-driven paradigm enhances the flexibility in adapting enacted business processes. An automated translation would allow the change operations performed by a business user upon (high-level, easy-to-understand) workflow model, e.g., BPMN, are systematically applied on enacted (low-level) event-driven processes.

5   Conclusion

In this paper, we demonstrated the applicability of the adaptation patterns proposed by Weber et al. [4] to change the control-flow structures in the event-driven process modeling paradigm. We showed how all 14 adaptation patterns can be supported by a generic event-driven meta-model. Subject to its translation to the implementation of specific event-driven process languages, such a mapping should facilitate the reliable adaptation of event-driven process descriptions.

Acknowledgments. This research was partly supported by the Smart Services Cooperative Research Centre, through the Australian Government's CRC Programme.

References

1. Weske, M.: Business Process Management: Concepts, Languages, Architectures. Springer (2010)
2. Schonenberg, H., Mans, R., Russell, N., Mulyar, N., van der Aalst, W.M.P.: Process Flexibility: A Survey of Contemporary Approaches. In: Advances in Enterprise Engineering, pp. 16–30 (2008)
3. Nurcan, S.: A Survey on the Flexibility Requirements Related to Business Processes and Modeling Artifacts. In: Hawaii International Conference on System Sciences, pp. 378–388. IEEE Computer Society (2008)
4. Weber, B., Reichert, M., Rinderle-Ma, S.: Change patterns and change support features – Enhancing flexibility in process-aware information systems. Data & Knowledge Engineering 66, 438–466 (2008)
5. OMG, Business Process Modeling Notation (BPMN) Specification 1.1, http://www.omg.org/spec/BPMN/1.1
6. Leavitt, N.: Complex-Event Processing Poised for Growth. Computer 42, 17–20 (2009)
7. Luckham, D.C.: The Power of Events. An Introduction to Complex Event Processing in Distributed Enterprise Systems. Addison-Wesley Longman Publishing Co., Inc. (2001)
8. Suntinger, M., Obweger, H., Schiefer, J., Gröller, M.E.: Event Tunnel: Exploring Event-Driven Business Processes. IEEE Comput. Graph. Appl. 28, 46–55 (2008)
9. Krishnamurthy, B., Rosenblum, D.S.: Yeast: A General Purpose Event-Action System. IEEE Transactions on Software Engineering 21 (1995)
10. Alexopoulou, N., Nikolaidou, M., Chamodrakas, Y., Martakos, D.: Enabling On-the-Fly Business Process Composition through an Event-Based Approach. In: Hawaii International Conference on System Sciences, pp. 379–389. IEEE Computer Society Press (2008)
11. Geppert, A., Tombros, D.: Event-based distributed workflow execution with EVE. In: Proceedings of the IFIP International Conference on Distributed Systems Platforms and Open Distributed Processing, pp. 427–442. Springer (1998)
12. Kapuruge, M., Colman, A., Han, J.: Achieving Multi-tenanted Business Processes in SaaS Applications. In: Bouguettaya, A., Hauswirth, M., Liu, L. (eds.) WISE 2011. LNCS, vol. 6997, pp. 143–157. Springer, Heidelberg (2011)
13. Scheer, A.W.: Aris: Business Process Modeling. Springer (2000)
14. Dayal, U., Hsu, M., Ladin, R.: Organizing long-running activities with triggers and transactions. SIGMOD Rec. 19, 204–214 (1990)
15. Kapuruge, M., Han, J., Colman, A., Kumara, I.: Enabling Ad-hoc Business Process Adaptations through Event-Driven Task Decoupling. In: Salinesi, C., Norrie, M.C., Pastor, Ó. (eds.) CAiSE 2013. LNCS, vol. 7908, pp. 384–399. Springer, Heidelberg (2013)
16. Weber, B.: Beyond rigidity - Dynamic process lifecycle support: A Survey on dynamic changes in process-aware information systems. Computer Science 23, 47–65 (2009)
17. Amador, L.: Drools Developer's Cookbook. Packt Publishing (2012)
18. JESS: The Java Expert System Shell, http://www.jessrules.com/
19. Paschke, A., Kozlenkov, A.: Rule-Based Event Processing and Reaction Rules. In: Governatori, G., Hall, J., Paschke, A. (eds.) RuleML 2009. LNCS, vol. 5858, pp. 53–66. Springer, Heidelberg (2009)
20. van der Aalst, W.M.P., Jablonski, S.: Dealing with workflow change: identification of issues and solutions. International Journal of Computer Systems Science and Engineering 15, 267–276 (2000)
21. van der Aalst, W.M.P., Weske, M.: Case Handling: A New Paradigm for Business Process Support. Data and Knowledge Engineering 53, 129–162 (2005)

An Improved Genetic Algorithm for Service Selection under Temporal Constraints in Cloud Computing

Helan Liang, Yanhua Du, and Sujian Li

School of Mechanical Engineering, University of Science and Technology Beijing,
Beijing, 100083, China
lianghelan@gmail.com

Abstract. To guarantee the successful execution of service based processes in cloud computing, one important requirement is the QoS-driven selection of candidate services under temporal constraints. In this paper, a new approach based on improved genetic algorithm (HPGA) is proposed where the hamming similarity degree is used to avoid inbreeding and the pheromone strategy is designed with considering not only the individual fitness but also the global information of best chromosomes. Compared with the existing works, this approach is more precise and especially suitable for the service selection of large-scale and complex processes with vast amounts of candidate services.

Keywords: Temporal constraint, Service selection, Petri net, Genetic algorithm, Hamming similarity degree, Pheromone strategy.

1 Introduction

Recently, cloud computing has become popular because it can bring many cost and efficiency benefits to enterprises when they build the service based processes derived from the requirements of users [1-2]. Usually, a large number of Web services (abbreviated to services) can provide similar functions in the cloud computing paradigm, so that it gives the challenge to QoS-driven service selection.

As one kind of important QoS requirements, temporal constraints [2-3] are usually set explicitly by process designers or implicitly enforced by laws, regulations or business rules. The violation of temporal constraints may affect the execution of service based processes and even lead to the termination of processes [2]. Therefore, it is vital to check if temporal constraints are satisfied when selecting candidate services.

Currently, there exist some works related to the problem of service selection, and they can be classified into two categories: *analyzing methods considering time factors* [4-6], and *validating methods for temporal constraints* [7-14]. The former only regards time factors as a kind of QoS criteria which are used for the service evaluation, and they can not be applied directly to the problems we focus on. The latter considers temporal constraints according to the requirements of users, but they are low efficient and imprecise when encountering complex and large-scale problems.

In this paper, an improved genetic algorithm based on the hamming similarity degree and pheromone strategy (HPGA) is proposed for the service selection under

X. Lin et al. (Eds.): WISE 2013, Part II, LNCS 8181, pp. 309–318, 2013.

temporal constraints. Firstly, a service selection model considering both the QoS objective and temporal constraints is defined. Secondly, the solving procedure is presented explicitly. In the crossover operator of this approach, the hamming similarity degree is designed to prevent premature convergence and the pheromone strategy is proposed to increase the searching efficiency. Compared with the existing works, our approach is more precise and especially suitable for large-scale and complex processes with vast amounts of candidate services.

2 Preliminary Formalism

2.1 Abstract Processes and Temporal Constraints

In recent years, cloud computing has offered a new way of implementing enterprise business processes [15]. In order to realize higher-order business transactions, at the beginning, the abstract processes should be defined. In this paper, the WorkFlow net (*WF-net*) [16] is used to describe the abstract processes.

Usually, the five generic QoS criteria for candidate services including *time*, *cost*, *reputation*, *success rate* and *availability* are taken into account. The QoS aggregation models [7, 14] are used for calculating the five QoS criteria for the four basic patterns of a *WF-net* including sequence, choice, parallel and iterative structures.

According to [8,14], we formulize each temporal constraint as $D_R(t_i, t_j) \leq s$ which denotes that a task t_j should end no later than s time units after t_i starts.

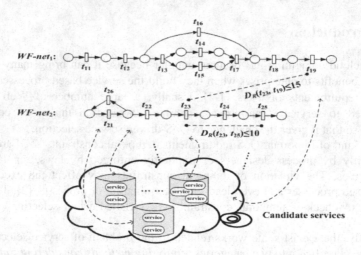

Fig. 1. An example of *WF-net* with candidate services and temporal constraints

An example of *WF-net* based abstract processes is shown in Fig.1. It is composed of two abstract processes *WF-net₁* and *WF-net₂*, and two temporal constraints are used to describe the temporal requirements in and between the processes. The detailed description is presented in Section 3.5.

2.2　Formal Problem Statement

Due to a large amount of candidate services which have the same functionality and different QoS values for each task, it is necessary to present a more precise and practical method for the selection of candidate services under temporal constraints.

Before presenting the formal statement of the problem, the following parameters are summarized:

(1) $T = \{t_1, t_2, \cdots, t_n\}$ is the set of tasks of the WF-nets.

(2) $C_i = \{c_{i1}, c_{i2}, \cdots, c_{im}\}$ is the set of candidate services for each task t_i. As to each candidate service c_{ij}, $Q_{ij} = \{q_{ij1}, q_{ij2}, q_{ij3}, q_{ij4}, q_{ij5}\}$ is used to describe its QoS level in aspects of time, cost, reputation, success rate and availability, respectively.

(3) $W = \{w_1, w_2, w_3, w_4, w_5\}$ defines the weights for the five QoS criteria including time, cost, reputation, success rate and availability respectively, and $\sum_{p=1}^{5} w_p = 1$.

(4) $\psi = \{ \cdots, D_R(t_i, t_j) \le d_u, \cdots, \}$ is the set of temporal constraints where d_u is the uth temporal requirement.

The problem we focus on is to find an execution plan $X = (x_1, x_2, \cdots, x_n)$ where service x_i belongs to C_i to maximize the QoS of the abstract processes. In sum, the formal statement of the problem which is addressed is presented as Equation (1) ~ (3). $Q_p(X)$ is the QoS level of the execution plan X; Q_p^{max} and Q_p^{min} are the maximum and minimum QoS values of the pth criterion, respectively; s_i and p_i denote the starting time and execution duration of task t_i; U is the number of temporal constraints.

$$F(X) = \sum_{p=1}^{2} \left(\frac{Q_p^{max} - Q_p(X)}{Q_p^{max} - Q_p^{min}} * w_p \right) + \sum_{p=3}^{5} \left(\frac{Q_p(X) - Q_p^{min}}{Q_p^{max} - Q_p^{min}} * w_p \right) \tag{1}$$

St.
$$s_j \ge (s_i + p_i), \quad \forall t_i \rightarrow t_j, \quad t_i, t_j \in T \tag{2}$$

$$\psi = \{..., D_R(t_i, t_j) \le d_u, ...\}, \quad t_i, t_j \in T, \quad u = 1...U \tag{3}$$

For simplification, we assume that the processes all begin at the same time and the iterative structure in the processes is pre-definied as some certain times. If one process begins later, we can always insert a dummy transition with exactly the same interval before the initial place of this process [9].

3　Our Approach Based on HPGA

3.1　Chromosome Encoding and Fitness Function

First of all, an integer string $X = \{x_1, x_2, \cdots x_n\}$ is used for the chromosome definition where n is the number of tasks involved in the abstract processes, and x_i ($1 \le i \le n$) in the ith gene position is from 1 to the number of candidate services for task t_i. For example, the tasks of the abstract processes in Fig.1 is sequenced as $[t_{11}, t_{12}, t_{14}, t_{15}, t_{16}, t_{18}, t_{19}, t_{21}, t_{22}, t_{23}, t_{24}, t_{25}, t_{26}]$, so a chromosome $\{2,1,1,1,3,1,1,2,3,1,4,1,2\}$ represents the second candidate service in group one is selected for task t_{11}, and the first candidate service in group two for task t_{12} and so on.

Then, a novel penalty and price fitness function is developed as follow:

$$Fitness(X) = F(X) + (1 - \frac{V(X)}{V_{num}}) + P \qquad (4)$$

Where $Fitness(X)$ is the combination of penalty and price of X and its range is [0, 3]; $F(X)$ is the objective function defined in Section 2.2; $V(X)$ is the total number of temporal violations of X; V_{num} is the total number of temporal constraints; P is the price of X which equals to 1 for a feasible individual and 0 for an infeasible one.

The above fitness function grantees: 1) the more temporal violations of an infeasible individual the more penalties it obtains; 2) any infeasible individuals will have less fitness than feasible ones. Therefore, it is helpful for feasible individuals with higher fitness to pass on to the next generation.

3.2   Hamming Similarity Degree

The hamming similarity degree [17] between each two chromosomes is calculated by $H_{ij}=h_{ij}/n$. Herein, H_{ij} is the hamming similarity degree between the ith and jth chromosome and its range is [0, 1]; h_{ij} is the number of services which both ith and jth chromosome select; n is the total number of tasks according to a given abstract processes.

The parent selection strategy based on the hamming similarity degree is designed as follows: if some pairs of chromosomes whose hamming similarity degree is less than the threshold P_h, one pair of them would be selected by the roulette wheel selection rule [18] where the chromosome is selected with probability consistent to its fitness; otherwise, the parent chromosomes will be randomly selected from the chromosomes group by the roulette wheel selection rule.

The hamming similarity degree can force chromosomes from different regions of the solution space to breed. Therefore, it is very helpful to keep the diversity and prevent premature converge which may be caused by inbreeding.

3.3   Pheromone Strategy

After the parent chromosomes are selected, the offspring is generated by Equation (5), where the crossover operator will choose the better parent with probability p_{p1}, potential "fittest" candidate service which has high visibility and pheromone with probability $p_{p2}-p_{p1}$, and with probability $(1-p_{p2})$ randomly.

$$N^k(g) = \begin{cases} \arg Max\{\eta_m, \eta_f\}, & if\ r \le p_{p1}; \\ \arg\underset{j \in C_k}{Max}\{\tau_j(g)*[\eta_j]^\alpha\}, & if\ p_{p1} < r \le p_{p2}; \\ R, & otherwise \end{cases} \qquad (5)$$

Where $N^k(g)$ is the service selected for the kth task in the gth iteration; m and f are the number of parent chromosomes; $\tau_j(g)$ is the value of pheromone of the jth service in the gth iteration; η_j is the visibility of the jth service and it is defined as Equation (6); α represents the relative importance of the pheromone and visibility; r is a decimal chosen randomly with uniform probability in $[0, 1]$; p_{p1} and p_{p2} are the thresholds; R is a random variable defined as Equation (7).

η_j is the visibility which is used to make the service with high qualities have a higher probability to be selected, and it is defined by Equation (6).

$$\eta_j = \sum_{p=1}^{2} \frac{q_{kp}^{max} - q_{kjp}}{q_{kp}^{max} - q_{kp}^{min}} * w_p + \sum_{p=3}^{5} \frac{q_{kjp} - q_{kp}^{min}}{q_{kp}^{max} - q_{kp}^{min}} * w_p, \quad if \ j \in C_k \tag{6}$$

Herein, q_{kp}^{min} and q_{kp}^{max} represent the maximum and minimum QoS values of the pth criterion of candidate services for the kth task, respectively.

R is a random variable selected according to the distribution [19] defined as follow:

$$p_j^k(g) = \frac{\tau_j(g)*[\eta_j]^{\alpha}}{\sum_{l \in C_k} \tau_l(g)*[\eta_l]^{\alpha}}, \quad if \ j \in C_k \tag{7}$$

Herein, $p_j^k(g)$ is the probability to select the jth service for the kth task in the gth iteration.

At the end of each iteration, the pheromone of each service will be updated according to Equation (8) ~ (9).

$$\tau_i(g+1) = \rho\tau_i(g) + \Delta\tau_i(g) \tag{8}$$

$$\Delta\tau_i(g) = \begin{cases} Fitness(FG(g)), & if \ i \in FG(g), \\ 0, & otherwise \end{cases} \tag{9}$$

Where $\tau_i(g)$ is the pheromone of the ith service in the gth iteration, $\tau_i(0)=1/(1-\rho)$; ρ is the evaporation coefficient, $\rho \in [0,1]$; $\Delta\tau_i(g)$ is the pheromone new added to the ith service; $FG(g)$ is the best chromosome before the gth iteration.

The pheromone strategy can enhance the possibility of finding better solutions, because it uses not only the local information inherited from parents, but also the local QoS information recorded by parameter "visibility" and the global information inherited from the previous good solutions which are recorded by parameter "pheromone" to guide the offspring generation. Besides, the random selection is helpful to increase the variety of chromosomes, so as to prevent premature converge.

3.4   The Procedure of Our Approach

Based on the above discussion, the pseudo code of our approach is shown as Fig.2.

3.5   A Case Study of Our Approach

In this section, a real-life example in [14] describing the production processes of a new cell phone is used to illustrate the proposed approach. As shown in Fig.1, two processes include: WF-net₁ (production of electronic main board) and WF-net₂ (production of peripheral parts). In order to shorten product's release time into market and obtain a bigger market share, two temporal constraints are specified on them. Because of the limited scope of paper, the detailed descriptions of activities and the QoS attributes about the candidate services can be found in literature [14].

Algorithm: Improved GA for Service Selection under Temporal Constraints
Input: Parameters for the algorithm such as pop_size, P_h ...etc.;
    Abstract processes with temporal constraints $PN= (P, T, F)$, $\psi =\{ ...,D_R(t_i,t_j)\leq d_k ... \}$;
    Candidate services $C=\{c_{11}, c_{12}, \cdots, c_{nm}\}$.
Output: An execution plan $X=\{x_1, x_2, \cdots, x_n\}$.

L1:   Begin
L2:   Create pop_size initial chromosomes as G_0 randomly, and $g=0$;
L3:   While $g\leq L$ do
L4: $g=g+1$, $G_g= G_{g-1}$ and $TempG= G_{g-1}$;
L5:   Calculate the Hamming similarity degree H_{ij} of each two chromosomes in G_g;
L6: $Candi=$ a set of chromosomes whose H_{ij} is no more than P_h;
L7:   For $d=1$ to $pop\_size/2$ do
L8:   If ($Candi$ is not $null$)
L9:   Select parent chromosomes (S_f, S_m) from $Candi$ according to the roulette strategy;
L10:   Else
L11:   Select parent chromosomes (S_f, S_m) from $TempG$ according to the roulette strategy;
L12:   Initially create a new offspring chromosome S_d randomly from S_f and S_m;
L13:   Update S_d base on the pheromone strategy with probability P_c;
L14:   Update S_d base on the improved mutation strategy [14] with probability P_m;
L15:   If (S_d is not the same as individuals in G_g and it is better than at least one individual in G_g)
L16: S_d enters G_g and delete the worst one in G_g;
L17:   End for;
L18:   If (the best fitness in G_g is better than that of the previous iterations)
L19:   Update the best solution and its fitness;
L20:   Update the pheromone base on the pheromone updating strategy;
L21:   End while;
L22:   End;

Fig. 2. The pseudo code of our approach

According to the proposed algorithm, parameters are set as follows: $pop\_size=30$, $L=30$, $p_h=0.5$, $p_c=0.8$, $p_m=0.2$, $\alpha=3$, $p_{p1}=0.3$, $p_{p2}=0.9$, $\rho=0.95$. The encoding sequence of tasks is t_{11}, t_{12}, t_{14}, t_{15}, t_{16}, t_{18}, t_{19}, t_{21}, t_{22}, t_{23}, t_{24}, t_{25}, t_{26}. At the end of each run, we can get the best solution as $\{1,2,1,1,3,1,1,2,3,1,4,1,2\}$ whose QoS value 0.8766. Therefore, the customer can select the specific service for each task, so as to fulfill the processes correctly and efficiently.

Noting that the solution obtained by the proposed approach is much better than the best one $\{1,2,1,2,3,1,1,2,3,1,4,1,2\}$ whose QoS is 0.8730 in [14].

4   Validation of Our Approach by Test Cases

In order to validate the precision of our approach, we compare our approach with the traditional GA method [20] and existing improved GA method [14] in the following three aspects: the number of tasks involved in the problem, the number of candidate services for each task, and the number of temporal constraints.

All above methods are implemented in C# under the Visual Studio.net environment, and run on a 2.2 GHz Intel Core 2 Duo CPU and a 2 GB RAM. For all three methods, the population sizes are 100, the probabilities for crossover and mutation are 0.8 and 0.2, respectively, and the termination conditions are stopping after 1500 iterations. Besides, more parameters necessary in our method are set as follows: p_h=0.5, α=3, p_{p1}=0.3, p_{p2}=0.9, ρ=0.95.

(1) Test Cases with Different Number of Tasks
In this experiment, the number of tasks varies from 5 to 50 with an increase of 5, and the number of candidate services for each task is fixed as 10 and the number of temporal constraints as 5. The results are shown in Fig.3. We can see the improvement we obtain over the improved GA is enlarging from 3.57% to 25.86% as the number of tasks increase. (Improvement %= (avg.value of our GA- avg. value of improved GA)/ avg. value of improved GA*100)

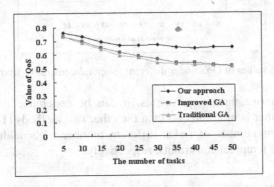

Fig. 3. The values of QoS under different tasks from 5 to 50

(2) Test Cases with Different Number of Candidate Services
In this experiment, the number of candidate services for each task varies from 10 to 100 with an increment of 10, and the number of tasks is fixed at 30 and the number of temporal constraints at 10, respectively. Fig.4 shows the experimental results. The improvement we obtain over the improved GA is enlarging from 19.15% to 27.22% as the number of candidate services increase.

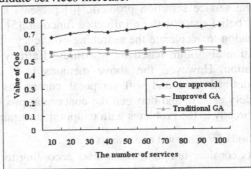

Fig. 4. The values of QoS under different candidate services from 10 to 100

(3) Test Cases with Different Number of Temporal Constraints
In this experiment, the number of temporal constraints varies from 5 to 50, and the number of tasks is fixed as 30 and the number of candidate services for each task as 10. The experimental results are shown in Fig.5. We can see the improvement we obtain over the improved GA is enlarging from 17.55% to 35.23% as the number of temporal constraints increase.

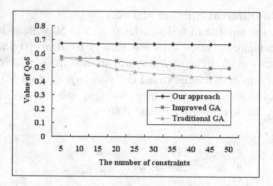

Fig. 5. The values of QoS under different temporal constraints from 5 to 50

Based on the above three experiments, it can be concluded that the approach proposed in this paper is more precise than the other two methods [14, 20] in various cases with different number of tasks, different number of candidate services and different number of temporal constraints, respectively.

5   Related Work

As it is mentioned that there exist other researches [4-14] related to the global optimization for QoS-aware service selection, and they can be classified into two categories: *analyzing methods considering time factors* [4-6], and *validating methods for temporal constraints* [7-14].

(1) Analyzing methods considering time factors
In order to solve the service selection problem, [4] presents a Cultural Algorithm which uses a global belief space with an influence function. [5] applies a GA with a seeded initial population to accelerate the searching efficiency. [6] proposes a mixed intelligent optimization algorithm based on Maximum Entropy Method and Social Cognitive Optimization. However, the above methods [4-6] do not support the definition of user-defined hard and soft temporal constraints. Because temporal constraints may widely exist in and between the concurrent processes, such methods can not be applied directly to the problems with temporal constraints.

(2) Validating methods for temporal constraints
Such kinds of works consider temporal constraints according to the requirements of users when selecting services. Recently, two kinds of methods are mainly used for the problem solving: exhaustive methods [7-9] and approximate methods [10-14].

An Integer Linear Programming based method [7] is proposed to solve the problem of service selection with global QoS constraints. In order to solve the service selection including local temporal constraints, the Mixed Integer Linear Programming [8, 9] methods are proposed. The exhaustive methods [7-9] are very effective when the size of the problem is small. However, these methods suffer from poor scalability.

Compared with exhaustive methods, the approximate methods such as meta-heuristic methods are more efficient thanks to their less time consuming. In order to solve the problem with global duration constraints, an improved PSO algorithm [10] and an adaptive GA [11] are proposed. [12] proposes a mutation operator for the service selection problem considering both the duration and atomic services constraints. [13] uses the GA for service composition and resources scheduling with local temporal constraints. An improved GA [14] is also proposed based on uniform crossover and improved mutation for the services selection with local temporal constraints. Although several improved meta-heuristic algorithms are proposed, most of them concentrate on the problems with just global duration or atomic services constraints which do not consider the local temporal constraints in or between processes. Besides, many of them only use the random evolutionary information to guide the evolution, and they are easy to fall into premature convergence.

6   Conclusion

In this paper, a novel approach to selecting the appropriate services for service based processes under temporal constraints is proposed. The hamming similarity degree is used to avoid inbreeding, and the specific pheromone is also designed to increase the searching efficiency. Compared with the existing works by various test cases, our approach is more precise and is particularly applicable for large-scale and complex processes.

In the future, we would like to extend researches on the following aspects: 1) How to consider the multi-objective optimization of service selection, and 2) How to consider the soft temporal constraints under uncertain environment.

Acknowledgments. This work was supported by the National Natural Science Foundation of China under Grant No. 61004109, the Beijing Natural Science Foundation (4133087) and the Fundamental Research Funds for the Central Universities (FRF-TP-12-047A, FRF-TP-12-052A).

References

1. Marston, S., Li, Z., Bandyopadhyay, S.: Cloud Computing-The Business Perspective. Decision Support Systems 51, 176–189 (2011)
2. Du, Y., Xiong, P., Fan, Y., Li, X.: Dynamic Checking and Solution to Temporal Violations in Concurrent Workflow Processes. IEEE Transactions on Systems, Man, and Cybernetics, Part A: Systems and Humans 41, 1166–1181 (2011)
3. Li, H., Yang, Y.: Dynamic checking of temporal constraints for concurrent workflows. Electronic Commerce Research and Applications 4, 124–142 (2005)

4. Liu, Y., Ngu, A., Zeng, L.: QoS computation and policing in dynamic Web service selection. In: Proceedings of the13th International Conference on World Wide Web, New York, pp. 66–73 (2004)
5. Kobti, Z., Zhiyang, W.: An Adaptive Approach for QoS-Aware Web Service Composition Using Cultural Algorithms. In: Orgun, M.A., Thornton, J. (eds.) AI 2007. LNCS (LNAI), vol. 4830, pp. 140–149. Springer, Heidelberg (2007)
6. Chen, Y., Zhang, J., Sun, J., Zheng, Q., et al.: A Service Selection Model Using Mixed Intel ligent Optimi-zation. Chinese Journal of Computers 33, 2116–2125 (2010)
7. Zeng, L., Benatallah, B., Ngu, A.: QoS-aware middleware for Web services composition. IEEE Transactions on Software Engineering 30, 311–327 (2004)
8. Ardagna, D., Pernici, B.: Adaptive Service Composition in Flexible Processes. IEEE Transactions on Software Engineering 33, 369–384 (2007)
9. Du, Y., Li, H., Ai, L.: Dynamic Configuration of Service based Processes in Cloud Computing using Linear Programming. In: 9th World Congress on Intelligent Control and Automation (WCICA 2012), pp. 3509–3514 (2012)
10. Wang, W., Sun, Q., Yang, F., Zhao, X.: An Improved Particle Swarm Optimization Algorithm for QoS-Aware Web Service Selection in Service Oriented Communication. International Journal of Computa-tional Intelligence Systems 4, 18–30 (2010)
11. Zhang, C.: Adaptive Genetic Algorithm for QoS-aware Service Selection. In: 2011 Workshops of International Conference on Advanced Information Networking and Applications, pp. 273–278 (2011)
12. Rosenberg, F., Muller, M., Leitner, P., et al.: Metaheuristic Optimization of Large-Scale QoS-aware Service Compositions. In: 2010 IEEE International Conference on Services Computing, pp. 97–104. IEEE, Los Alamitos (2010)
13. Ye, Z., Zhou, X., Bouguettaya, A.: Genetic Algorithm Based QoS-Aware Service Compositions in Cloud Computing. In: Yu, J.X., Kim, M.H., Unland, R. (eds.) DASFAA 2011, Part II. LNCS, vol. 6588, pp. 321–334. Springer, Heidelberg (2011)
14. Du, Y., Wang, X., Ai, L.: Dynamic Selection of Services under Temporal Constraints in Cloud Computing. In: 2012 International Conference on e-Business Engineering (ICEBE 2012), pp. 252–259. IEEE Press, China (2012)
15. Phan, T., Han, J., Schneider, J.-G., Wilson, K.: Quality-Driven Business Policy Specification and Refinement for Service-Oriented Systems. In: Bouguettaya, A., Krueger, I., Margaria, T. (eds.) ICSOC 2008. LNCS, vol. 5364, pp. 5–21. Springer, Heidelberg (2008)
16. Du, Y., Li, X., Xiong, P.: A Petri Net Approach to Mediation-aided Composition of Web Services. IEEE Transactions on Automation Science and Engineering 9, 429–435 (2012)
17. Wang, J., Ma, Y., Wang, F.: Study of improved Genetic Algorithm based on dual mutation and its simulation. Computer Engineering and Applications 44, 57–61 (2008)
18. Freisleben, B., Merz, P.: A genetic local search algorithm for solving symmetric and asymmetric traveling salesman problems. In: Proceedings of IEEE International Conference on Evolutionary Computation, pp. 616–621 (1996)
19. Zhao, F., Zhao, F., Li, T.: A new Pheromone Trail-based Genetic Algorithm for Comparative Genome Assembly. Nucleic Acids Research 36, 3455–3462 (2008)
20. Zhang, T., Gruver, W., Smith, M.: Team scheduling by genetic search. In: Proceedings of the Second International Conference on Intelligent Processing and Manufacturing of Materials, pp. 839–844 (1999)

GEAM: A General and Event-Related Aspects Model for Twitter Event Detection

Yue You[1,2], Guangyan Huang[2], Jian Cao[1,*], Enhong Chen[3], Jing He[2], Yanchun Zhang[2], and Liang Hu[1]

[1] Department of Computer Science and Engineering,
Shanghai Jiao Tong University, China
[2] Centre for Applied Informatics,
Victoria University, Australia
[3] School of Computer Science and Technology
University of Science and Technology of China, China
{yy71107103,cao-jian,lianghu}@sjtu.edu.cn,
{guangyan.huang,jing.he,yanchun.zhang}@vu.edu.au, cheneh@ustc.edu.cn

Abstract. Event detection on Twitter has become a promising research direction due to Twitter's popularity, up-to-date feature, free writing style and so on. Unfortunately, it's a challenge to analyze Twitter dataset for event detection, since the informal expressions of short messages comprise many abbreviations, Internet buzzwords, spelling mistakes, meaningless contents etc. Previous techniques proposed for Twitter event detection mainly focus on clustering bursty words related to the events, while ignoring that these words may not be easily interpreted to clear event descriptions. In this paper, we propose a General and Event-related Aspects Model (GEAM), a new topic model for event detection from Twitter that associates General topics and Event-related Aspects with events. We then introduce a collapsed Gibbs sampling algorithm to estimate the word distributions of General topics and Event-related Aspects in GEAM. Our experiments based on over 7 million tweets demonstrate that GEAM outperforms the state-of-the-art topic model in terms of both Precision and DERate (measuring Duplicated Events Rate detected). Particularly, GEAM can get better event representation by providing a 4-tuple (*Time, Locations, Entities, Keywords*) structure of the detected events. We show that GEAM not only can be used to effectively detect events but also can be used to analyze event trends.

Keywords: Event detection, Twitter, General and Event-Related Aspects Model (GEAM), Topic model, Gibbs sampling.

1 Introduction

Social network services such as Twitter[1], Facebook[2] have experienced a rapid growth in recent years. Millions of people turn from traditional news websites to

---

* Corresponding author.
[1] https://twitter.com
[2] http://www.facebook.com/

X. Lin et al. (Eds.): WISE 2013, Part II, LNCS 8181, pp. 319–332, 2013.

these services to gather real-time news, share opinions, or read hot comments. As a real-time information network to share *"the latest stories, ideas, opinions and news*[3]*"*, Twitter allows users to express everything by writing a *"tweet"* up to 140 characters. According to the Paris-based analyst group Semiocast, as of July 2012, Twitter has more than 500 million users all over the world[4]. Twitter has some unique characteristics that make it a better source for event detection than traditional news articles, blogs, etc. Firstly, Twitter is the most up-to-date information channel of real events. Because of the length limit and the popularity of Twitter mobile applications, users can update information instantly at a pretty low cost, which makes Twitter a fresher resource than others. Secondly, Twitter covers a wider range of information than other sources. Millions of users in Twitter are not only information consumers but also news publishers, they can post tweets describing everything in their life. So, Twitter almost covers every aspect of the society, from breaking news to personal opinions. Thirdly, we can also analyze the opinions or sentiments related to the detected events in Twitter, which can help the organizations respond to the upcoming event quickly [1]. Thus, event detection on Twitter has attracted more and more interest recently.

Although event detection has been well studied in formal text such as news articles, blogs [2–4], Twitter dataset brings several challenges to us because it contains many abbreviations, Internet buzzwords, spelling mistakes, meaningless contents etc. In existing work on formal text, the underlying assumption is that all the documents in the corpus is related to some undiscovered events, but it is not reasonable in Twitter. According to Pear Analytics study[5], about 40% of tweets are *"pointless babbles"* like "I'm eating a sandwich". Such tweets are limited to users' personal feelings, and should not draw attention from the majority audience. Unlike well-written text, Twitter dataset also contains a lot of Internet buzzwords and misspelling due to the length limit and the free writing style[5]. All these unique characteristics make most existing techniques in formal text unavailable on Twitter dataset. Besides, most recently proposed methods for Twitter event detection are either limited to certain types of tweets (e.g. related to earthquakes or crimes and disasters) [6–8] or based on clustering bursty words [9–11], and provide results in the form of single terms, hashtags or *n*-grams. The outcome of these approaches may be uninformative or meaningless, and can't help the users obtain more fine-grained insights of the whole event.

To tackle the above challenges, we propose GEAM, a General and Event-Related Aspects Model for event detection in Twitter. GEAM is an unsupervised generative topic model that differentiates General words (describing general opinions) from Event-related Aspects words (describing different aspects of an event) in an event-related tweet. In this paper, we focus on detecting realistic events, which are discussed by a large number of users. We define a 4-tuple representation of a realistic event, e.g. *(Time, Locations, Entities, Keywords)*, which are corresponding to

---

[3] https://twitter.com/about

[4] http://techcrunch.com/2012/07/30/analyst-twitter-passed-500m-users-in-june-2012-140m-of-them-in-us-jakarta-biggest-tweeting-city/

[5] http://news.bbc.co.uk/2/hi/technology/8204842.stm

4 Event-related Aspects in GEAM respectively. Here, Time and Locations represent when and where the event happened; Entities represent main subjects (persons, organizations, movies, etc.) in the event; Keywords describes the meaning of the event. Unlike formal news articles, users in Twitter also use General words in tweets. For example, during Obama's victory speech, there are 327k tweets per second posted/reposted to discuss Obama's re-election[6]. One of the typical tweets like "Thank The Lord for that!! Well done Obama in Chicago, Illinois!! That was a Great victory speech!". In this tweet: Time is *"November 7, 2012"*, *"Chicago"* and *"Illinois"* represent Locations, *"Obama"* represents Entity, Keywords are *"Great"*, *"victory"*, and *"speech"*, and the rest are General words. GEAM can provide a clearer description of events by separating General words from Event-related Aspects words.

The most related work is the model proposed by Lau et al. [12]. They introduce an online processing variant of topic model Latent Dirichlet Allocation (LDA) [13, 14] to analyze tweets trend. Their graphic structure is the same as LDA, which models each tweet as a multinomial mixture of all topics or events. However, this assumption is obviously unreasonable due to Twitter's short text length, most users only discuss one event in a tweet. In this paper, we improve LDA for Twitter event detection by assigning only one event to an event-related tweet and differentiating General words from Event-related Aspects words.

To the best of our knowledge, we are the first to extend LDA to detect event in Twitter by simulating the generative process of the General words and Event-related Aspects words in a tweet. The advantages of our GEAM include: (1) GEAM outperforms the state-of-the-art topic models LDA in terms of both Precision and DERate (measuring Duplicated Events Rate detected); (2) the result is more informative than previous clustering algorithms which based on unigram text model, since we treat the words in a tweet differently based on different information corresponding words express (General words and Event-related Aspects words); (3) the event detection results of GEAM can also be easily utilized to perform event trend analysis.

The rest of the paper is organized as follows. Section 2 presents related work. Section 3 describes GEAM and its components in detail, and introduces the approximate inference algorithm for GEAM based on Gibbs sampling. In Section 4, we present the experimental results. Finally, we conclude the paper in Section 5.

2   Related Work

We present a survey of the state-of-the-art Twitter event detection methods and topic models in this section.

---

[6] http://www.kansascity.com/2012/11/07/3903945/
   social-media-pickup-lines-illegal.html

2.1   Twitter Event Detection

Although event detection from formal texts [2–4] has been studied for over a decade, event detection from Twitter is a relatively new topic which receives more interests in recent years. Most existing approaches for event detection from Twitter are limited to detect only certain types of events. For example, Sakaki et al. [6] devise a classifier to recognize tweets describing earthquakes in Japan, then they introduce a probabilistic spatiotemporal model to find the center and the trajectory of the earthquake. Rui et al. [7] focus on detecting Crime and Disaster related Events (CDEs), they also use a classifier to determine whether a crawled tweet is related to CDEs or not. Then, they utilize the author's network information to predict the location of a tweet. Benson et al. [8] utilize a graphic model to identify artists and venues mentioned with tweets posted by users in New York. Very recently, some open domain Twitter event detection approaches are proposed. Ritter et al. [9] extract an open-domain calendar of significant events from Twitter, they utilize a named entity tagger and sequence-labeling technology to extract event-related words. Chenliang et al. [10] propose a segment-based algorithm to detect events. They utilize Microsoft Web N-Gram service and Wikipedia to segment the tweets, detect the bursty event segments, and cluster the event segments using k-Nearest Neighbor Graph.

So, existing Twitter event detection approaches are either limited to certain types of tweets (e.g. related to earthquakes or CDEs, containing a predefined location) or based on detecting bursty words (single terms, hashtags or n-grams) which may be uninformative or meaningless.

2.2   Topic Models and Variants

Topic models (LDA) [13, 14] proposed by Blei et al. is popular for modeling latent topics in a corpus. Many variants of LDA has been proposed for different applications since it has been proven to be effective. Particularly, in social network or social media, Jonathan et al. [15] analyze documents on Wikipedia and infer descriptions of its entities and of relationships between those entities by proposing a probabilistic topic model. Hong et al. [16] study how the standard topic models can be trained on the microblogging dataset, they apply several schemes to train the model and compare their quality and effectiveness. Hu et al. [17] propose a Bayesian model called Event and Tweets LDA (ET-LDA) that performs topic modeling and event segmentation in one unified framework. Given transcripts of a known event from both New York Times and Twitter dataset, their work focus on how to jointly extract the topics covered by the two different datasets and segment the event.

The most related work is the model proposed by Lau et al. [12] to track emerging events in Twitter. They describe a topic model that processes documents in an online fashion. The model can update automatically based on time slices and can cope with dynamic vocabulary. However, they model each tweet as a multinomial mixture of all topics or events, which is obviously unreasonable due to short lengths of tweets. In this paper, we assign only one event to an event-related tweet.

3   General and Event-Related Aspects Model

We present an overview of our proposed system in Section 3.1. In Section 3.2, we develop the General and Event-Related Aspects Model (GEAM), and explain how to inference the model via collapsed Gibbs sampling.

3.1   Overview of the System

We aim at detecting open domain realistic events that a large number of people talking about in Twitter. Figure 1 plots the main components and procedures in our system for Twitter event detection. Given a raw stream of tweets, we remove "*pointless babbles*", which may not attract the majority users' attention based on the Named Entity Tagger information. Then, we send tweets (labeled with named entity and time information) to the General and Event-Related Aspects Model (GEAM) to estimate the General topic or Event-related Aspect word distributions. Finally, we rank the detected events based on the number of tweets assigned to each event.

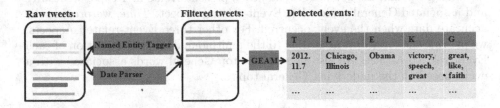

Fig. 1. Overview of our system

We manually investigate a dataset of one day's tweets on June 25, 2009. We observe that almost all the tweets that related to an realistic event contain named entities or hashtags. Table 1 shows 3 typical events on June 25, 2009, along with the number of whole tweets and tweets that contain named entities or hashtags corresponding to each event. From the numbers in Table 1, we can see that over 95% of tweets related to one realistic event contain named entities or hashtags. Based on this observation, we apply a named entity tagger for Twitter provided by Ritter et al. [5] to segment and label tweets. If no named entity or hashtag exists in a tweet, we ignore it. In other words, if a tweet contains some named entities or hashtags, it has great possibility related to a realistic event, and we will process it further. We also analyze the time information in tweets. There are several different ways to refer the same date in Twitter, for example "last night" and "the next Friday" may represent the same day depending on when the tweet was posted. So, we use natty[7] to extract the time information referred in a tweet. Natty is a natural language date parser that can recognize dates described in many ways, like "*the next Friday*".

---

[7] http://natty.joestelmach.com/

Table 1. Number of whole tweets and tweets which contain named entities or hashtags corresponding to 3 typical events on June 25, 2009

Event	Iran election	"Transformer 2" released	Mark Sanford scandal
all tweets	24612	5644	3397
tweets contain named entities or hashtags	23583	5368	3234
proportion	95.8%	95.1%	95.3%

After preprocessing and filtering, event-related tweets, which are labeled with named entities and time information, are sent to the General and Event-Related Aspects Model (GEAM), the kernel component of our system. GEAM is a probabilistic graphical model that simulates the generative process of a tweet related to a realistic event, and identifies General words and Event-related Aspects words in the tweet. A collapsed Gibbs sampling algorithm is designed to estimate word distributions of each event.

Finally, we rank the detected events, then present events in 4-tuple structures and associated General topics. For Event-related Aspect, *Time*, we provide the accurate date when the event occurred. For other three Event-related Aspects, we provide the top words according to the multinomial word distribution of each Event-related Aspect. We also provide the top General words associated with a event based on the underlying General topics.

3.2   General and Event-Related Aspects Model (GEAM)

We present how to model the realistic event and the corresponding general words associated with it by General and Event-related Aspects Model (GEAM), a hierarchical Bayesian model based on Latent Dirichlet Allocation (LDA). The inference of GEAM via collapsed Gibbs sampling will be introduced under the definition of the model.

Model Description. We use a generative process to model the tweets that describe a realistic event. GEAM models each tweet as a mixture of Event-related Aspects and General topics, then generates each word from the Event-related Aspects or General topics word distribution. We define a 4-tuple to represent an event in Twitter, i.e. *(Time, Locations, Entities, Keywords)*, each tuple reflects an aspect of the event. Due to informal expression of Twitter, people always use several words that doesn't belong to any Event-related Aspects mentioned before. We regard these part of words as General words in GEAM. For each tweet in the corpus, we assign a latent variable *event* to it, the *event* variable in GEAM is similar to a *topic* in LDA, and each *event* is characterized by 4 multinomial distributions over words, according to 4 Event-related Aspects. We assign an Event-related Aspect (i.e. *Time, Locations, Entities* or *Keywords*) to each named entity or hashtag word in a tweet, according to the named entity tagger results and hashtags. For other words, we use a switch variable to indicate whether the word comes from General topic word distribution or Event-related Aspect word distribution. If the word is chosen from the General topic word distribution, then a *General topic* based on the whole corpus will be assigned to this word. Otherwise it will be assigned to the *Keyword Aspect* of an event. The graphical structure is shown in Figure 2.

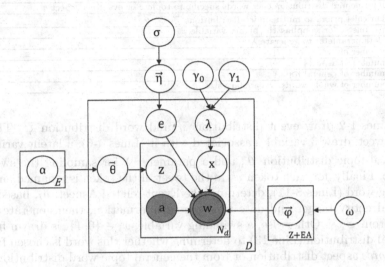

Fig. 2. Graphical representation of GEAM

Let us briefly introduce notations in Figure 2. e is an latent variable assigned to each tweet specifying which event the tweets are describing, for example, a typhoon or a sports event. Each token in a tweet is associated with 4 variables (only 3 of them are meaningful at one time): word, w, switch variable, $x \in \{0, 1\}$, to indicate whether this word is an Event-related Aspect word or a General word, Event-related Aspect, a, to reflect which aspect (i.e. *Time, Locations, Entities* or *Keywords*) this word belongs to (in this case $x = 1$), and General topic, z, to specify which general topic the word corresponds to (in this case $x = 0$). Here we assume that there are E events and Z general topics in the whole corpus, each event has A Event-related Aspects, and the size of vocabulary is V. $\vec{\eta}$ is the whole corpus' multinomial distribution over events, which is E dimensional and drawn from a Dirichlet distribution parameterized by σ. $\vec{\theta}$ is a multinomial distribution over General topics for each tweet, which is Z dimensional and drawn from a Dirichlet distribution parameterized by α_e. $\vec{\varphi}$ is a multinomial distribution over words specific to topic z or event e, which is V dimensional and drawn from a Dirichlet distribution parameterized by ω. λ is the parameter for sampling the binary variable, x, and both γ_0 and γ_1 are beta parameters to generate λ. We summarizes the notations used in the GEAM in Table 2. Formally, the generative process is described in Algorithm 1: For the whole

Table 2. Notations in the model

Symbol	Description
w	a word in a tweet
x	indicate if word w is from a general topic z or an event-aspect pair ea
e	the event described by the tweet
z	the topic assigned to w, $x = 0$
a	the event aspect assigned to w, $x = 1$
$\vec{\eta}$	multinomial distribution over events
$\vec{\theta}$	multinomial distribution over topics
$\vec{\varphi}_z, \vec{\varphi}_{ea}$	multinomial distribution over words specific to topic z or event e's aspect a
σ, ω, α_e	Dirichlet priors to multinomial distributions $\vec{\eta}, \vec{\varphi}, \vec{\theta}$
λ	parameter for sampling the binary variable x
γ_0, γ_1	Beta parameter to generate λ
E	number of events
A	number of aspects
Z	number of general topics
V	number of whole words, vocabulary size

corpus, Lines 1-2 draw event distribution $\vec{\eta}$ and word distribution $\vec{\varphi}$. Then, for each tweet, draw 3 variables associated with it (Lines 4-6): a latent variable e, a general topic distribution $\vec{\theta}$, and a parameter λ for sampling the switch variable x. Finally, for each token w_i in the tweet: if the token is a named entity or hashtag word (Lines 8-11), determine the Event-related Aspect, a_i, based on the named entity tagger and date time parser information, then generate the word w_i from $\vec{\varphi}_{ea_i}$. Otherwise, a switching variable, $x_i \in \{0,1\}$, is drawn from a Binomial distribution (Line 13) to determine whether this word is chosen from the *Keywords* aspect distribution or from the general topic word distribution. If $x_i = 0$, choose a non-event (general) topic, z_i, and draw the word w_i from $\vec{\varphi}_{z_i}$ (Lines 14-17). If $x_i = 1$, draw the word w_i from event-relates *Keywords* aspect $\vec{\varphi}_{e3}$ (Lines 18-20).

Model Inference. We use Gibbs sampling to estimate unknown parameters $\vec{\eta}, \vec{\theta}, \vec{\varphi}$ and λ in GEAM. Gibbs sampling allows the learning of a model by iteratively updating each latent variable given the remaining variables. In particular, we follow the idea of collapsed Gibbs sampling to approximate the posterior distribution of e, x, a, and z. We alternately sample the document-level variable e and the token-level variables x, a, and z. Then, given the sampling results of e, x, a, and z, we can easily infer $\vec{\eta}, \vec{\theta}, \vec{\varphi}$ and λ.

Firstly, **sampling tweet level event** e according to:

$$
p(e_d | \mathbf{e_{-d}}, \mathbf{w}, \mathbf{x}, \mathbf{z}, \mathbf{a}) \propto \frac{n_{e_d, -\mathbf{d}} + \sigma_{e_d}}{\sum_{e=1}^{E} n_{e, -\mathbf{d}} + \sigma_e} \times \prod_{k=1}^{K} \frac{\prod_{t=1}^{V} \prod_{i=0}^{n_{k,\mathbf{d}}^t - 1} (n_{k,-\mathbf{d}}^t + \omega^t + i)}{\prod_{i=0}^{n_{k,\mathbf{d}}^* - 1} (n_{k,-\mathbf{d}}^* + \omega^* + i)}
$$
$$
\times \prod_{a=1}^{A} \frac{\prod_{t=1}^{V} \prod_{i=0}^{n_{e_d a, \mathbf{d}}^t - 1} (n_{e_d a, -\mathbf{d}}^t + \omega^t + i)}{\prod_{i=0}^{n_{e_d a, \mathbf{d}}^* - 1} (n_{e_d a, -\mathbf{d}}^* + \omega^* + i)} \tag{1}
$$

where $\mathbf{e_{-d}}$ is the *event* vector associated with each tweet in corpus D excluding tweet d, $n_{e, -\mathbf{d}}$ is the number of tweets in the whole corpus assigned to event e

```
 1  Draw an event distribution $\vec{\eta} \sim Dir(\sigma)$ ;
 2  Draw multinomial word distribution $\vec{\varphi} \sim Dir(\omega)$ for General topics (Z) and Event-related
    Aspects (EA);
 3  foreach tweet $d \in [1, D]$ do
 4  │   Draw an event $e \sim Mult(\vec{\eta})$ ;
 5  │   Draw a multinomial topic distribution $\vec{\theta} \sim Dir(\alpha_e)$ ;
 6  │   Draw a switching distribution $\lambda \sim Beta(\gamma_0, \gamma_1)$ ;
 7  │   foreach word $w_i \in N_d$ do
 8  │   │   if $w_i \in \{named\ entities\ or\ hashtag\ words\}$ then
 9  │   │   │   Determine aspect $a_i \in \{0, 1, 2, 3\}$ (Time, Locations, Entities, Keywords
    │   │   │   aspect);
10  │   │   │   Draw word $w_i \sim Mult(\vec{\varphi}_{ea_i})$;
11  │   │   end
12  │   │   else
13  │   │   │   Draw $x_i \in \{0, 1\} \sim Bi(\lambda)$ ;
14  │   │   │   if $x_i = 0$ then
15  │   │   │   │   Draw topic $z_i \sim Mult(\vec{\theta})$ ;
16  │   │   │   │   Draw word $w_i \sim Mult(\vec{\varphi}_{z_i})$ ;
17  │   │   │   end
18  │   │   │   else
19  │   │   │   │   Draw word $w_i \sim Mult(\vec{\varphi}_{e3})$ (Keywords aspect);
20  │   │   │   end
21  │   │   end
22  │   end
23  end
```

Algorithm 1. Probabilistic generative process in GEAM

expect tweet d, $n_{k,\mathbf{d}}^t$ is the number of word t assigned to general topic k ($x = 0$) in tweet d and $n_{k,\mathbf{d}}^* = \sum_{t=1}^V n_{k,\mathbf{d}}^t$, $n_{e_d a,\mathbf{d}}^t$ is the number of word t assigned to event-related aspect $e_d a$ ($x = 1$) in tweet d and $n_{e_d a,\mathbf{d}}^* = \sum_{t=1}^V n_{e_d a,\mathbf{d}}^t$.

Secondly, for each token in a tweet, if the token is recognized as a named entity or a hashtag word, then the aspect is observed based on the preprocessing result. Otherwise, we jointly **sample the token-level variables** $x, a,$ and z. For $x_i = 0$:

$$p(x_i = 0, z_i = k | e_d, \mathbf{w}, \mathbf{x}_{-\mathbf{i}}, \mathbf{z}_{-\mathbf{i}}, \mathbf{a}) \propto \frac{N_{d,0,-\mathbf{i}} + \gamma_0}{N_{d,-\mathbf{i}} + \gamma_0 + \gamma_1}$$

$$\times \frac{N_{d,-\mathbf{i}}^k + \alpha_{e_d}^k}{\sum_{k=1}^Z N_{d,-\mathbf{i}}^k + \alpha_{e_d}^k} \times \frac{N_{k,-\mathbf{i}}^t + \omega^t}{\sum_{t=1}^V N_{k,-\mathbf{i}}^t + \omega^t} \qquad (2)$$

where $N_{d,0,-\mathbf{i}}$ is the number of words in tweet d associated with $x = 0$ excepting word i, $N_{d,-\mathbf{i}}^k$ is the number of words in tweet d associated with general topic k excepting the ith token, $N_{k,-\mathbf{i}}^t$ reflects the number of term t associated with general topic k excluding the ith token. For $x_i = 1$, the derivation formula is very similar to Eq.(2) ($x_i = 0$). Due to the length limit, thus, we omit the detail equation here. Finally, we can easily obtain the multinomial or binomial parameters $\vec{\eta}, \vec{\theta}, \vec{\varphi}$ and λ based on the previous sampling results.

4   Experiments

We demonstrate GEAM's performance from two aspects: effectiveness of event detection and capability for event trend analysis. For event detection effectiveness, the experiments show that GEAM outperforms the state-of-the-art topic models [13, 14] with better Precision and DERate (measuring Duplicated Events Rate detected). Also, 4-tuple structure *(Time, Locations, Entities, Keywords)* makes the detected events much easier to be understood by users. And majority people's reaction to the detected events can also be provided by the General topic words in GEAM. For event trend analysis, we demonstrate 3 typical events' trend based on detected results, and find different time patterns of these events.

4.1   Dataset and Experimental Setting

Twitter Dataset. We use one week data (from June 25, 2009 to July 1, 2009) from the tweets published by Stanford[8] to evaluate our system. There are a total of 7,088,229 tweets in the dataset. A number of realistic events happened in this period, such as "Iran Election 2009", "Micheal Jackson died", etc. Figure 3 shows the average of tweets published within each hour of a day. From Figure 3, we can see that most tweets were posted during midnight and afternoon. The average *"pointless babbles"* (tweets that contain neither named entities nor hashtags) that we filtered away in the dataset is about 47.32%.

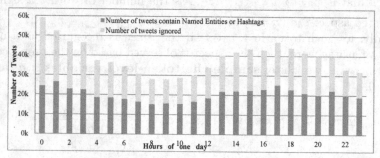

Fig. 3. Tweet volume against hour of day

Parameter Setting. There are several parameters in GEAM, and we use a validation set to find the optimal parameters. More specifically, GEAM archives good performance, when we set five parameters as follows: (1) $\gamma_0 = 0.8$ and $\gamma_1 = 0.2$ (to indicate that a word is more likely to be a General word), (2) $\alpha = 0.01$, $\omega = 0.01$, and $\sigma = 0.01$ (for hyperparameters of Dirichlet distribution). We can vary the number of events, E, and the number of general topics, Z, according to the data time-window (*e.g.*, one day or one hour). Note that, we fix time-window to one hour in this experiment, and GEAM is flexible enough to set the time-window to any time unit.

---
[8] http://snap.stanford.edu/data/twitter7.html

Evaluation Metric. A common problem existing in Twitter event detection is that we can't evaluate the quality of detected events by labeled ground truth, since it is infeasible to label over 7 million tweets manually. Alternatively, we manually search related real world facts to evaluate the detected events in terms of **Precision** and **DERate**. The **Precision** is defined as the fraction of detected events that are related to a realistic event. Note that, if there are several detected events related to the same realistic event, then all of them are regarded correct in **Precision**. Actually, the same realistic event may not be provided several times in the output. So, we use the metric **DERate** proposed in [10] to evaluate our GEAM, which is defined as the fraction of duplicate detected events among all all detected realistic events.

4.2   Event Detection Effectiveness

Our GEAM is different from LDA [13, 14]. LDA [13, 14] is a widely used statistical topic modeling technique, which aims at discovering the latent "topics" in a document corpus. For LDA, we set the number of topics K equals to the number of events E in GEAM; and we set the Dirichlet hyperparameters $\alpha = 0.01$, $\beta = 0.01$ similarly as α, ω and σ in GEAM.

Event Detection Performance. We compare GEAM and LDA in terms of Precision and DERate in Figure 4. We manually evaluate the top 5 words outputted by each model, if all the words are related to some realistic event, we regard it as true positive. We range the number of events (E) in GEAM and the number of topics (T) in LDA from 5 to 30. From the figure, we can see that GEAM outperforms LDA in terms of both Precision and DERate. Figure 4(a) demonstrates that based on one hour time-window dataset,the Precision of GEAM first increases with the increase of event number E, and archives highest of 80.6% at $E = 20$, then decreases when E becomes larger. Similar behaviors are also observed in LDA, who receives highest Precision at $T = 25$. It is reasonable that the ptimal setting of E based on one hour time-window Twitter dataset is about 20. When we set E smaller, the GEAM may combine two similar realistic events into one detected event. This phenomenon is more obvious for LDA,

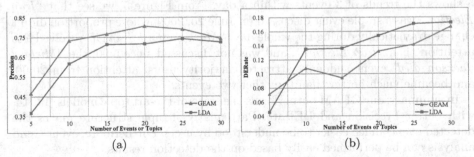

(a)   (b)

Fig. 4. Comparison of GEAM and LDA. (a) shows the Precision of two models on different event numbers. (b) shows the DERate of two models on different event numbers.

which models each tweet as a mixture of K latent topics. It shows in Figure 4(b) that generally the DERate of GEAM is smaller than LDA, which means that there are fewer duplicate events detected in GEAM. For each model, DERate increases with the increase of event number E and topic number K. The reason for this phenomenon is that, when event number E becomes larger, GEAM is more likely to assign different emphasis of one realistic event to different detected events. For example, when we set $E = 30$, the 2009 Iran election protest event will be divided into two events, one is for the video of Neda's death, the other is for an online activity to "Show support for democracy in Iran add green overlay to your Twitter avatar".

Event Interpretation. Five typical examples of the event detected by our GEAM in June 25, 2009 from 00:00 to 01:00 are demonstrated in Table 3, where T for *Time*, L for *Locations*, E for *Entities*, K for *Keywords*, and G for *General* words. For each event, we display top 10 Event-related Aspects words and top 3 General words. Due to the page limit, other events are not demonstrated. We illustrate that the output of GEAM is more informative, and can be easily understood with little background knowledge. From Table 3, we can see that the 4-tuple structure makes the detected events much clearer, and thus, users can easily access the different aspects of the event. GEAM can also provide reactions of majority people to the event by listing the General words. For example, e_3 expresses that the major audience enjoy the movie "Transformers 2". Among the 5 typical events listed in Table 3, we see that GEAM is an open-domain event detection model. It covers a wide range of events, from political events (Iran election), to entertainment events (Transformers 2), to sports (FIFA Confederations Cup), and to online hotspot (#lolquiz application).

4.3 Event Trend Analysis

By now, we evaluate GEAM on the basis of one hour time-window. It is more interseting to explore what can be found if we combine the detected events of continuous hours. As stated in Section 3.2, $\overrightarrow{\eta}$ in GEAM represents the multinomial distribution over events and $\overrightarrow{\eta}_e$ represents the fraction of tweets that are assigned to $event_e$. So, we can track the event in terms of hours. Figure 5 shows the trends of 3 events within a day. From Figure 5, we see that "*Iran Election*" was a stable event discussed on June 25, 2009, whose proportion didn't change dramatically during the whole day; "*Mark Sanford Scandal*" received less attention than "*Iran Election*", and was not discussed since 5:00; the "*Michael Jackson died*" event suddenly attracted majority users' attention from 20:00, and became much hotter than the other two events.

In summary, our GEAM outperforms the state-of-the-art topic models [13, 14] with better Precision and DERate. Particularly, 4-tuple structure makes the detected events much easier to be understood by users. Furthermore, event trend analysis can be performed easily based on the detection results.

Table 3. Examples of detected events

Event		GEAM output	Description
e_1	T:	2009-06-25	Footage of the death of Neda drew international attention after she was shot dead during the 2009 Iranian election protests.
	L:	Iran, Tehran	
	E:		
	K:	#Iranelection, #Neda, #Iran, Mousavi, support, democracy, green, avatar	
	G:	video, crime, freedom	
e_2	T:	2009-06-24	The U.S. National Team won World No. 1 Spain, 2-0, in the semifinals of the FIFA Confederations Cup on June 24, advancing to next final against the winner of Brazil and South Africa.
	L:		
	E:	U.S., Spain, USA, Brazil, South Africa	
	K:	win, soccer, beat, team, school, 2-0	
	G:	great, watching, shocked	
e_3	T:	2009-06-25	"Transformers 2" released on June 24, 2009 in North America. Many people tweet about watching the film.
	L:	America	
	E:	Transformers 2, Transformers	
	K:	good, movie, watch, tonight, wait, wish, lines	
	G:	tonight, join, happy	
e_4	T:	2009-06-25	South Carolina Governor Mark Sanford's disappearance and extramarital affair in June, 2009 was reported. He was in Argentina with his mistress for six days.
	L:	Argentina, South Carolina	
	E:	Make Sanford, Sanford	
	K:	governor, mistress, e-mails, affair, exposed, bizarre	
	G:	shamed, news, dumbass	
e_5	T:	2009-06-25	A web entertainment application called #lolquiz gets popular. Many users try it to know which star available for them.
	L:		
	E:	twilight, Knotb Song, Jonas Brothers Song, Mcfly	
	K:	try, quiz, #lolquiz, song, fan, what	
	G:	took, lover, try	

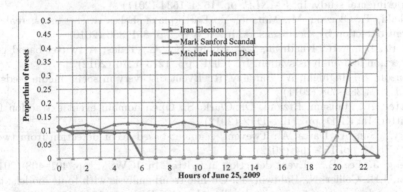

Fig. 5. Different proportion of tweets related to 3 events

5   Conclusions

In this work, we introduce a General and Event-Related Aspects Model (GEAM) for detect realistic event in Twitter. We design a collapsed Gibbs sampling algorithm to estimate the word distributions of an event. We also divide the words in an event-related tweet into General words or Event-related Aspect words, which matches the Twitter characteristic better than any unigram model. Our experiments demonstrate that GEAM outperforms the state-of-the-art topic model LDA in both Precision and DERate. Particularly, GEAM can get better event representation by providing a 4-tuple ($Time, Locations, Entities, Keywords$) of the detected events and the associated General topics. Moreover, GEAM can be used to analyze event trends in continuous hours.

Acknowledgments. This work is partially supported by China National Science Foundation (Granted Number 61073021, 61272438), Research Funds of Science and Technology Commission of Shanghai Municipality (Granted Number 11511500102, 12511502704), Cross Research Fund of Biomedical Engineering of Shanghai Jiaotong University (YG2011MS38).

References

1. Liu, K.L., Li, W.J., Guo, M.: Emoticon smoothed language models for twitter sentiment analysis. In: AAAI, pp. 1678–1684 (2012)
2. Fung, G.P.C., Yu, J.X., Liu, H., Yu, P.S.: Time-dependent event hierarchy construction. In: KDD, pp. 300–309 (2007)
3. Gabrilovich, E., Dumais, S.T., Horvitz, E.: Newsjunkie: providing personalized newsfeeds via analysis of information novelty. In: WWW, pp. 482–490 (2004)
4. He, Q., Chang, K., Lim, E.P.: Analyzing feature trajectories for event detection. In: SIGIR, pp. 207–214 (2007)
5. Ritter, A., Sam, Clark, M., Etzioni, O.: Named entity recognition in tweets: an experimental study. In: EMNLP, pp. 1524–1534 (2011)
6. Sakaki, T., Okazaki, M., Matsuo, Y.: Earthquake shakes twitter users: real-time event detection by social sensors. In: WWW, pp. 851–860 (2010)
7. Li, R., Lei, K.H., Khadiwala, R., Chang, K.C.C.: Tedas: A twitter-based event detection and analysis system. In: ICDE, pp. 1273–1276 (2012)
8. Benson, E., Haghighi, A., Barzilay, R.: Event discovery in social media feeds. In: ACL, pp. 389–398 (2011)
9. Ritter, A., Mausam, Etzioni, O., Clark, S.: Open domain event extraction from twitter. In: KDD, pp. 1104–1112 (2012)
10. Li, C., Sun, A., Datta, A.: Twevent: segment-based event detection from tweets. In: CIKM, pp. 155–164 (2012)
11. Weng, J., Lee, B.S.: Event detection in twitter. In: ICWSM, pp. 401–408 (2011)
12. Lau, J.H., Collier, N., Baldwin, T.: On-line trend analysis with topic models: twitter trends detection topic model online. In: COLING, pp. 1519–1534 (2012)
13. Blei, D.M., Ng, A.Y., Jordan, M.I.: Latent dirichlet allocation. In: NIPS, pp. 601–608 (2001)
14. Griffiths, T.L., Steyvers, M.: Finding scientific topics. Proceedings of the National Academy of Science, 5228–5235 (2004)
15. Chang, J., Boyd-Graber, J.L., Blei, D.M.: Connections between the lines: augmenting social networks with text. In: KDD, pp. 169–178 (2009)
16. Hong, L., Davison, B.D.: Empirical study of topic modeling in twitter. In: SOMA, pp. 80–88 (2010)
17. Hu, Y., John, A., Wang, F., Kambhampati, S.: Et-lda: Joint topic modeling for aligning events and their twitter feedback. In: AAAI, pp. 59–65 (2012)

Bitemporal Complex Event Processing
of Web Event Advertisements*

Tim Furche[1], Giovanni Grasso[1], Michael Huemer[2],
Christian Schallhart[1], and Michael Schrefl[2]

[1] Department of Computer Science, Oxford University,
Wolfson Building, Parks Road, Oxford OX1 3QD
firstname.lastname@cs.ox.ac.uk

[2] Department of Business Informatics – Data & Knowledge Engineering,
Johannes Kepler University, Altenberger Str. 69, Linz, Austria
lastname@dke.uni-linz.ac.at

Abstract. The web is the largest bulletin board of the world. Events of all types, from flight arrivals to business meetings, are announced on this board. Tracking and reacting to such event announcements, however, is a tedious manual task, only slightly alleviated by email or similar notifications. Announcements are published with human readers in mind, and updates or delayed announcements are frequent. These characteristics have hampered attempts at automatic tracking.

PEACE provides the first integrated framework for event processing on top of web event ads. Given a schema of events to be tracked, the framework populates this schema through compact wrappers for event announcement sources. These wrappers produce events including updates and retractions. PEACE then queries these events to detect complex events, often combining announcements from multiple sources. To deal with updates and delayed announcements, PEACE's schemas are bitemporal so as to distinguish between occurrence and detection time. This allows complex event specifications to track updates and to react to differences in occurrence and detection time. Our evaluation shows that extracting the event from an announcement dominates the processing of PEACE and that the complex event processor deals with several event announcement sources even with moderate resources. We further show, that simple restrictions on the complex event specifications suffice to guarantee that PEACE only requires a constant buffer to process arbitrarily many event announcements.

1 Introduction

Most events are announced first and often only on the web these days. This trend is even more pronounced for time critical events, as the web is a ubiquitous and prompt information source. While the immediate availability of up-to-date information is a blessing

* The research leading to these results has received funding from the European Research Council under the European Community's Seventh Framework Programme (FP7/2007–2013) / ERC grant agreement DIADEM, no. 246858. Michael Huemer has been supported by a Marietta Blau Scholarship granted by the Austrian Federal Ministry of Science and Research (BMWF) for a research stay at Oxford University's Department of Computer Science.

X. Lin et al. (Eds.): WISE 2013, Part II, LNCS 8181, pp. 333–346, 2013.

in enabling much more complex, rapid interactions, it also imposes a challenge: The immediacy of web-published events allows for frequently and quickly distributed updates, leading to fragmentary and preliminary but inaccurate advertisements which are fixed later. Therefore, modern coordination tasks often boil down to continuously checking all relevant event advertisement sources for changes, ready to react on violations of our constraints. For example, awaiting a person on a flight with a stopover, we need to check that both flights are in time. If the second flight is delayed, the person will arrive late the same day; in contrast, if the first flight is late, the person might not even arrive on the same day, depending on the timeliness of the second flight and other available connections. But these distinctions do not suffice, as incomplete, incorrect, and late announcements complicate the situation even more. For example, not only may scheduled flights arrive late, but also the event announcements for such events may be advertised late themselves. This requires us to consider the *bitemporality* of events: Each event has an *occurrence time,* when it supposedly takes place, and a *detection time,* when its advertisement is detected on the web.

Checking all these cases is a boring, stressful, and repetitive task, ideally suited for automation – at least at first glance. Except for specific domains, however, such systems are rare as the available technologies do not allow for a rapid development of event announcement detection integrated with complex event processing: The employed complex event processor must be able to deal with event announcements that are unreliable, volatile, and out of time, as published on the web, i.e., the event processing must be fully bitemporal. While complex events have been studied extensively, see [8] for a survey, existing systems deal only with some aspects of the bitemporality. For example, [2] considers events with two time dimensions, but does not consider mutable events or delayed events; [17] only allows to deal with mutable events in terms of new and unconsolidated update events. Most systems [7,9,10,18,21] assume a single time dimension, i.e., events are detected instantaneously when they occur. To extract event announcements from the web, however, an easy to use, scalable extraction system is needed that is able to produce a continuous stream of extracted event announcements.

The PEACE *(Processing Event Ads into Complex Events)* framework introduced in this paper addresses these challenges by an integrated framework for extracting event announcements and detecting complex events over such announcements through a bitemporal complex event processor. PEACE is driven by event models that specify the attributes of events in a domain and are used both in the specification of wrappers (for extracting event announcements) and in the specification of complex event queries. For the former, we use an extension of OXPATH [11], a highly efficient web automation language, slightly adapted to the needs of event announcement extraction. OXPATH is particularly well suited for this task as it is able to extract data from even heavily scripted modern web sites such as Ebay. For the latter, we introduce a novel bitemporal complex event processing language BICEPL *(Bitemporal Complex Event Processing Language)* that can be evaluated directly on top of the event announcements extracted by OXPATH. As required, BICEPL distinguishes between detection and occurrence time of events. This bitemporal event handling allows for enacting different actions, depending on the events' properties and timing information, e.g., considering delays between event detection and occurrence time. For a long-running complex event

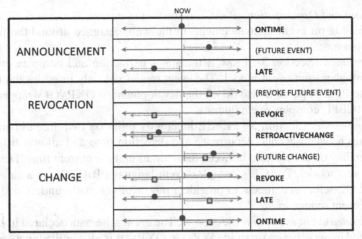

Fig. 1. Complex Event Publishing Cases

processing system it is essential that the memory use does not increase over time but remains constant (after a certain warm-up). This is guaranteed in BICEPL through the use of sliding windows, and, uniquely, in OXPATH through a novel buffer manager that guarantees constant memory regardless of the number of extracted events or web pages visited. This is key to an efficient implementation of BICEPL, since web event processing is dominated by loading and rendering web pages, as shown in Section 6.

PEACE is designed to support developers in adding new event announcement sources or new complex event queries. Both parts are driven by the event model and are minimal extensions of established query languages: OXPATH extends XPATH for web extraction. Most wrappers in OXPATH are a series of XPATH expressions, interspersed with action (such as clicking a button) and extraction instructions, that specify the attribute of the event model to populate with the selected web data. The XPATH portions can be created by a myriad of mature developer tools in Browsers such as Firebug, and then only (simple) actions and extraction instructions must be added to complete the wrapper. OXPATH also provides a visual interface that further simplifies this task [15].

BICEPL extends SQL-select statements to define *complex events* described with SQL-select statements that are extended, first with temporal comparisons between events, and second, with a definition of the occurrence time of the complex event in relation to the timing information of the involved events. The resulting events are continuously updated in their attributes, such as location or ticket price, and in their timing information. This leads to 10 different cases when complex events should be published, updated, or retracted, see Figure 1 and Section 5. Notice, Allan [3] defined possible relations between immutable, *unitemporal* intervals; in contrast Figure 1 displays possible cases in a life cycle of a single mutable, *bitemporal* event.

Contributions and Organization. PEACE is the first integrated framework for complex event processing on event announcements in the web, designed around the following four contributions:

(1) *Integration* (Section 3). PEACE integrates extraction and complex event processing through a joint event model. The event model not only provides the interface between the PEACE parts, it also drives the development of OXPATH wrappers just as much as BICEPL complex event queries.

(2) *Bitemporality* (Section 5). PEACE inherently relies on two time dimensions: It distinguishes between events' occurrence and detection time and allows different actions, depending on the relation between those times and the current time. Despite this powerful event model, PEACE's complex event language BICEPL is a small extension atop SQL-select statements to provide a powerful yet easily understood way for describing event schemata.

(3) *Integrated event extraction* (Section 4). The event schemata declared in BICEPL are populated by wrappers written in OXPATH. OXPATH is also built atop a commonly known declarative language, namely XPATH.

(4) *Lightweight, memory efficient implementation* (Section 6). The prototype implementation of PEACE is highly efficient and lightweight, requiring only constant memory regardless of the number of events or sources to extract from.

1.1 Running Example

We illustrate PEACE through a simple scenario taken from the daily live of a business man, who travels regularly by plane to business meetings (we discuss the details of the code shown in Sections 4 and 5). Flights are often delayed and hence, he must update his business partners frequently about delays. This is not only costly in time and effort but sometimes impossible, e.g., when he is on a flight without access to communication services. However, this task could be delegated to PEACE, such that it informs his colleagues whenever he would get late to a meeting. To simplify the example for presentation the business man only takes direct flights. To detect potential delays, the web sites of airlines and airports are observed continuously.

In this example we identify the following event classes: `FlightArrival` signifies the landing of a plane with the attributes `flightDay`, `flightNo`, `fromLoc`, and `toLoc`, the latter describing the departure and destination locations. Instances of `FlightArrival` are extracted from the web by wrappers covering airline and airport sites. `OneDayToArrival` is a complex event that occurs one day before a flight arrival associated with a business meeting, with attributes `flightDay`, `flightno`, `meetingId`, and `toLoc`. It announces relevant flight arrivals one day before the actual expected arrival takes place and is kept up-to-date on changes of the estimated arrival time. All classes have an additional implicit attribute occ for the occurrence time of the event and det for the detection time. Finally, the `BusinessMeeting` is provided through some database for this example, though they could also be extracted from a web calendar or some other meeting planning system.

To detect subscribed events on the web, PEACE integrates OXPATH wrappers that are executed on (possibly several) target pages. A wrapper for a certain event class produces data compliant with the schema of the event it observes. For this example, the wrapper in Figure 2 detects `FlightArrival` events from `flightarrivals.com`. This

```
1 doc("http://www.flightarrivals.com")
2   //a#panel0/{click /}//form#qbaForm/descendant::field()[1]/{$airport }
3   /following::field()[3]//option{select }/following::field()[1]/{click /}
4   /(/descendant::a[string(.)='Next >'][1]/{click /})*
5     //table#flifo//tr[position()>1]/self():<FlightArrival>
6       [./td[1]:<fromLoc=string(.)>]          [./td[2]:<flightNo=string(.)>]
7       [./td[3]/div:<flightDay=string(.)>]  [.:<toLoc=$airport>]
8       [./td[3]/text()[1]:<occTime=toUnixTime(.)>]
```

Fig. 2. OXPATH Wrapper

```
1 CREATE MUTABLE SUBSCRIBED EVENT CLASS FlightArrival(flightDay TEXT, flightNo
     TEXT, fromLoc TEXT, toLoc TEXT)
2   ID (flightDay, flightNo)
3   LIFESPAN (2d);
4
5 CREATE COMPLEX EVENT CLASS OneDayToArrival(flightDay TEXT, flightNo TEXT,
     toLoc TEXT, meetingId TEXT)
6   ID (flightDay, flightNo, meetingId)
7   AS SELECT fa.flightDay, fa.flightNo, fa.toLoc, bm.meetingId
8   FROM FlightArrival fa, BusinessMeeting bm
9   WHERE fa.flightNo = bm.flightNo AND fa.flightDay = bm.flightDay
10    OCCURRING AT fa - 1d
11  PUBLISH OneDayToArrival_OnTime
12    CASE LATE(0s,1m) OneDayToArrival_OnTime
13    CASE LATE(1m,1h) OneDayToArrival_Late
14    CASE RETROACTIVECHANGE OneDayToArrival_RAChanged
15    CASE REVOKE OneDayToArrival_Revoked;
```

Fig. 3. Event Definition for Flight Arrivals

wrapper fetches the target page (Line 1), then selects and submits the form for flights arriving at the parameterised arrival airport at any of the possible time windows in the day (Lines 2–3). The wrapper deals with paginated results by repeatedly clicking on "next" (Line 4). On each result page, FlightArrival objects are extracted along with their attributes (Lines 5–7) for every flight arrival entry listed, resulting in an event tuple as shown below:

flightDay	flightNo	fromLoc	toLoc	occ
▮▮▮	BA 112	New York JFK	London LHR	1369116000

Figure 3 shows the BICEPL specification for importing the subscribed event class FlightArrival and defining the complex event class OneDayToArrival. The first statement (Lines 1-3) declares the subscribed event class FlightArrival with its explicit attributes flightDay, flightNo, fromLoc and toLoc, with flightDay and flightNo as key (Line 2). FlightArrival is defined mutable (Line 1), since estimated flight arrival times change over time, and with a lifespan of 2d, meaning 2 days (Line 3).

The second statement (Lines 5-15) defines the complex event type OneDayToArrival based on the constituent event classes FlightArrival and BusinessMeeting. It features as attributes flightDay, flightNo, toLoc, and meetingId, with flightDay, flightNo, and meetingId as key (Line 6). A OneDayToArrival event occurs one day before the flight arrives, as defined with the **OCCURRING AT** clause (Line 10). When the complex event is detected on time, or with up to one minute delay (0s,1m), the complex event publishes OneDayToArrival_OnTime (Lines 11-12). If the event is detected within a delay of more than one minute (1m) up to one hour(1h), OneDayToArrival_Late is published (Line 13).

If an already published event must be revised, e.g.. the flight arrives later than assumed beforehand, OneDayToArrival_RAChanged is published (Line 14). If an already published event is revoked, e.g., if the flight is canceled less than a day before it is planned to arrive, OneDayToArrival_Revoked is published (Line 15). All these following the schema of the associated complex event OneDayToArrival.

2   Related Work

To the best of our knowledge, PEACE is the first system that addresses the complex event processing for event announcements from the Web. This differs from mining events from Twitter or other sources [5,13] but is related to typical complex event processing where many event sources are integrated to detect complex events and react to these. Therefore, we focus on the difference of PEACE and its complex event processor and language BICEPL with existing complex event processing systems.

Event processing approaches can be classified into three different approaches: Active Database Management Systems (ADMS) [1,12,20], Data Stream Management Systems (DSMS) [4,6,16] and Complex Event Processing Systems (CEP) [2,7,9,10,18,21]. Unfortunately, all ADMS and DSMS approaches do not fit PEACE for not supporting bitemporality, as these systems make the perfect technology assumption [23], i.e., they assume an event is known instantaneously after it occurred. Even most CEPs make this assumption: The occurrence time of an event is given to it when entering the system. This restrictive time model disables reasoning over events in an imperfect world, as for this task the occurrence time and detection time are essential. To the best of our knowledge AMIT [2] is the only CEP system thinking about two time dimensions. Yet, these time dimensions are not supported to be used for temporal comparisons in complex event processing. Situations, as they call complex events, are defined by their highly expressive, imperative complex event language which is conceptually similar to the ones used in active database systems, e.g., in [20].

Mapping our complex event declarations to existing complex event processing approaches is only partly possible: The *on time* case, defining the reaction if there are no delays, changes, or errors in the event planning and announcement, is the standard scenario assumed by all approaches. This is the only case that is directly supported. The *late* case, is not supported at all by existing approaches due to a missing second time dimension. Notice, the occurrence time is event inherent (implicit) and may not be compared with user defined (explicit) attributes. Mapping the *retroactive change* case as well as the *revoke* case to existing languages is possible, though a tenuous task and requires the subscription to multiple event types.

Beside event processing, also Event Calculus (EC) [14,19,22] in knowledge representation deal with events, representing knowledge about events for reasoning purposes. Originally EC is unitemporal, though, there exist extensions [19,22] to implement bitemporal deductive database systems. EC rules are expressive enough to model complex events, but would require a number of low-level rules to represent a single complex event declaration of BICEPL: EC provides a general event model while BICEPL is a succinct yet expressive language tailored for event management.

3   PEACE Approach and Event Model

PEACE is designed for quickly instantiating a complex event detection system on top of event announcements from new sources. The setup of PEACE application requires the provision of the following: **(1)** *Subscribed event classes,* specifying the schema of their events. **(2)** OXPATH *wrappers* for those subscribed event classes which feed from web sources, matching the schema of their subscribed event classes. **(3)** *Wrappers* for other sources, if any, matching again the schema of their subscribed event classes; examples include database triggers, e.g., to retrieve background information on a business meeting. **(4)** *Complex event classes* to aggregate the subscribed events into complex events for capturing the conditions and information required by the application and for driving the publication of events to be delivered to the client of PEACE.

In our running example, for **(1)**, we show the subscribed event class FlightArrival in Lines 1-3 of Figure 3, omitting the BusinessMeeting which is similar. For **(2)**, we show the OXPATH wrapper in Figure 2 while, for **(3)**, BusinessMeeting is filled via a database trigger. At last, Lines 5-15 of Figure 3, show the definition of a complex event. In total, the entire example takes 19 lines of BICEPL and 13 lines of OXPATH plus a database tigger to implement.

Event Model. We identify such a BICEPL program with the event classes $\mathbb{E}$ it declares. Every event e processed by $\mathbb{E}$ belongs to one such class $e.\text{class} = E \in \mathbb{E}$. Depending on the concrete class E, the event features certain attributes $e.attr$ for the attributes $attr \in E.\text{schema}$, as specified in the schema $E.\text{schema}$ of E. Each event schema contains a set of key attributes $E.\text{key} \subseteq E.\text{schema}$ and we denote with $e.\text{key}$ the values of the key attributes in event $e \in E$. Further, the occurrence time and detection time of each event e is accessible via the implicit attribute $e.\text{occ}$ and $e.\text{det}$, respectively. These attributes are implicit, since we want to control the way they are computed and accessed, providing occurring-at and checking-at clauses. Note that the key of an event identifies the event but not the announcement, i.e., there may coexist several announcements e and e' referring to the same event $e.\text{key} = e'.\text{key}$ detected at different times, i.e., with $e.\text{det} \neq e'.\text{det}$.

Event classes in $\mathbb{E}$ are partitioned into subscribed and complex events $\mathbb{S}$ and $\mathbb{C}$. In addition to the schema, subscribed event classes $S \in \mathbb{S}$ also have a lifespan $S.\text{lifespan}$, determining how long an event is retained before being purged, and an associated wrapper $S.\text{wrap}$ or trigger expression. In contrast, complex event classes $C \in \mathbb{C}$ have a set of publication statements $C.\text{pub}$ and a query function $C.\text{query}$. The publication statements in $C.\text{pub}$ are subdivided into $C.\text{pub}[O]$, $C.\text{pub}[L]$, $C.\text{pub}[C]$, $C.\text{pub}[R]$ for

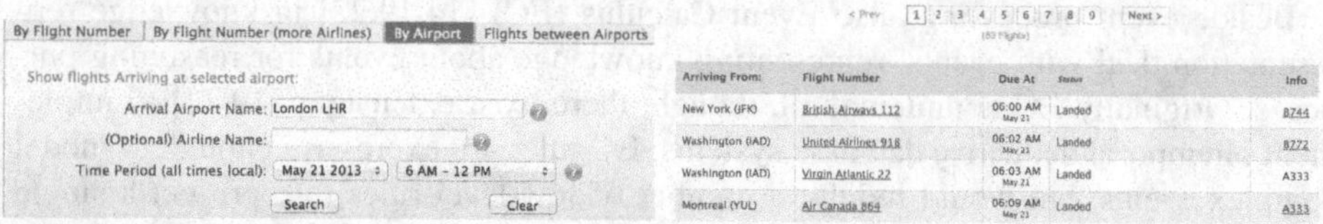

Fig. 4. Form and Result Page on www.flightarrivals.com

on-time, late, change, and revoke publication events. If the query C.query depends on observed subscribed event classes or other complex event classes (derived from those observed events), then we call these classes *constituent event classes* of C. The constituent classes do not only include direct dependencies but also indirect dependencies via other constituent complex classes. If $E \in \mathbb{E}$ is a constituent event class of $C \in \mathbb{C}$, we write $E \subset C$; hence $\subset$ is transitive by definition. Additionally, we require $\subset$ to be irreflexive and asymmetric, i.e., we allow *no cyclic dependencies* in complex event queries. Likewise, we write $e \subset c$ for concrete event instances e and c, if e is a constituent event of c. For the set of C-events obtained by evaluating C.query, we write C.query(O, D, t), where O is the set of so far observed subscribed events, D is the set of derived constituent complex classes, and t is the wall clock time.

4   Extracting Event Announcements

For detecting events announcements on the web, PEACE integrates OXPATH, a recent state-of-the-art tool for highly efficient data extraction [11]. OXPATH's main strengths lie in its ability to deal with modern scripted websites necessitating complex interaction (e.g., Ajax forms, auto-completion fields), and the capability to scale well in time and memory, handling even millions of pages and extracted results at ease. Indeed, OX-PATH's output handling requires no buffer as the extracted data is streamed out once matched on the page, see [11]. This behaviour fits perfectly with the needs of PEACE. A full description of OXPATH is out of this paper's scope, and can be found in [11]. In short, OXPATH is an extension of XPATH with four features: **(1)** *actions*, such as mouse events, form filling, for simulating user interactions with web pages; **(2)** iteration via *Kleene stars*, e.g., to deal with sites that present their results across several pages or use any pagination techniques; **(3)** more expressive node selection through the *style axis*, querying the actual visual attributes as rendered by the browser, to select e.g. all elements coloured green; and, **(4)** specification of data to be extracted in terms of (nested) records and attributes, via *extraction markers*. To illustrate OXPATH's capabilities, we discuss how to derive a wrapper from a PEACE event schema $\mathbb{E}$ in OXPATH along the running example from Figure 2.

There are two steps in the derivation for a specific event class $E \in \mathbb{E}$: First we define the navigation on the event source to the event announcements. Second, we specify how to map each of the event model's attributes to fragments of the event announcements. In our example, Lines 1–4 perform the navigation and Lines 5–8 the attribute mapping. Figure 4 shows the form on the left hand that the navigation part first fills by selecting the "By airport" tab and then filling the airport into the arrival airport field. It then iterates over all options of the second select in the time period and submits the form once for each of those options. On the result page, it iterates through the next links connecting the paginated results using a *Kleene star* expression. This entire expression can be easily obtained using standard web developer tools present in most browsers or in Firebug, though we here use some of OXPath's extensions to obtain a more readable expression. We also provide a visual tool that allows the recording of interaction sequences and automatically generates the corresponding navigation sequence so that the user only has to change the parameterized values or adjust the iteration parts.

For the attribute mapping, one can again use standard web developer tools to obtain XPath expressions for identifying which peace of a page maps to which attribute of E. This is shown in Lines 5–8. Each result page reached contains a table with flight arrival entries in its rows. We skip the first row as it only contains the column headers (Line 5), and for all the remaining we use OXPATH's *extraction markers* to shape the extracted data in compliance with the schema of the corresponding event class FlightArrival defined in Figure 3: One event :<FlightArrival> is created for each table row, along with attributes such as <fromLoc=string(.)> for the departure airport in the first column ./td[1]; similarly for flight number, flight day, and occurrence time in adjacent columns. Notice, that all key attributes (E.key) must be present explicitly on the web site to allow for tracking changes to the event (here flight day and number).

The resulting wrapper is used in PEACE to poll the website repeatedly and produce the input events for the complex event processor. The polling frequency and behaviour can be adjusted by the user, though PEACE provides a simple, but effective change detection based on a sample of the result pages by default. If within E.lifespan the same event is detected multiple times (in different pollings), it is only reported if its attributes have changed.

5 Event Processing with BICEPL

We designed BICEPL to handle event advertisements as they occur on the web. In a perfect world, each event, as identified by its key attributes, would be announced once and would have a single event occurrence time. Web announcements are certainly subject to frequent updates, hence BICEPL features events which are updated independently of the event's actual occurrence time, e.g., allowing for retro-actively changes or revocations. Moving into such an imperfect world, we still assume that events occur only once, but the attributes and occurrence times for past and future events may change. If there are different announcements e and e' for the same event e.key $= e'$.key, we assume that those announcements have different detection times e.det $\neq e'$.det. But given det and key, only a single event announcement e may exist with e.det $=$ det and e.key $=$ key.

In BICEPL, all events belong to a *event class*, defining a schema for its events' explicit attributes. *Subscribed events* are produced by OXPATH wrappers or other sources, hence BICEPL only define their lifespan. *Complex events* are derived from constituent events which are either subscribed or other complex. The definition of a complex event class comes in two parts, **(1)** as an *event selector* based on an SQL select statement which aggregates constituent events into complex events. The event selector describes the attributes of the corresponding complex events in a perfect world, i.e., it disregards event updates. (2) For each complex event, we define publication statements in reaction to certain update types. We consider situations when events (a) are announced on time, (b) have been detected late, after having already occurred, (c) are retro-actively changed, or (d) are retro-actively revoked.

Syntax of BICEPL BICEPL's syntax in Figure 5 defines a program ⟨*program*⟩ as sequence of event class declarations, each describing either a simple or complex event class. We declare a subscribed event class $S \in \mathbb{S}$ via ⟨*sclass*⟩ as either mutable or immutable, with an event schema S.schema, a key S.key, and a lifespan S.lifespan, given as ⟨*time_literal*⟩ which consists of a positive integer with s, m, h, or d for seconds, minutes, hours, or days. A complex event class $C \in \mathbb{C}$ is declared via ⟨*cclass*⟩, consisting of a schema C.schema, an SQL select statement C.query, and event publication statements C.pub. The SQL select statement may refer to subscribed event classes in $\mathbb{S}$ and to other complex event classes in $\mathbb{C}$, as long as no circular dependency arises. The schema ⟨*schema*⟩ describes with ⟨*attributes*⟩ the typed attributes E.schema and with ⟨*key*⟩ the attributes forming the key E.key of event class E. The select statements are extended with *occurring-at* and *checking-at* clauses. An occurring-at clause describes a complex event's occurrence time, while a checking-at clause describes when the event is checked for, parameterized with start and end times of those checks and the interval between two consecutive checks. Both clauses are based on time expressions ⟨*time*⟩ which refer to occurrence times of constituent events. The occurrence time of a constituent event is accessed via ⟨*table_ref*⟩, e.g., in a statement SELECT FROM FlightArrival fa..., we use fa to refer to the occurrence time of the selected flight arrival event. In occurring-at clauses only, time expressions may also contain 'NOW' to refer to the current system time. This implies that the boundaries of checking-at clause are determined by occurrence times of constituent events. Based on these basic expressions, time expressions involve recursively min/max and increment/decrement computations. The where-clause of a select statement may also involve comparisons of ⟨*time*⟩ expressions (not shown in the grammar). Next to the select statement, complex events also describe publication events C.pub in ⟨*publication*⟩. We distinguish publication events for on-time, late, retro-actively changed, and revoked event announcements, referred to as C.pub$[O]$, C.pub$[L]$, C.pub$[C]$, and C.pub$[R]$, and declared with ⟨*ontime*⟩, ⟨*late*⟩, ⟨*change*⟩, and ⟨*revoke*⟩. In all four cases, BICEPL requires the name of the publication event to generate. Late events are optionally parameterized with an interval restricting the considered delay.

Semantics of BICEPL We define the semantics of BICEPL in two variants – first as *idealized semantics* without ever purging observed events, and second as *sliding window semantics* by considering the lifespan of the subscribed events and purging them when they turn stale. As a technical prerequisite, we start with rewriting complex event queries into standard SQL.

Query rewriting: Given an expanded SQL select statement, we turn C.query into standard SQL by performing three rewriting steps: (1) We expand all table references so as to access the implicit occurrence time attribute, e.g., in our running example we rewrite OCCURRING AT fa - 1d into OCCURRING AT fa.occ - 84600, as 1 day equals 84600 seconds. (2) We rewrite occurring-at clauses into a definition of the implicit occurrence time attribute occ, e.g., OCCURRING AT fa.occ - 84600 yields SELECT fa.occ - 84600 as

$\langle program \rangle$ ::= { $\langle sclass \rangle$ | $\langle cclass \rangle$ }

$\langle sclass \rangle$ ::= 'CREATE' ('MUTABLE' | 'IMMUTABLE') 'SUBSCRIBED EVENT CLASS'
                 $\langle schema \rangle$ 'LIFESPAN' $\langle time\_literal \rangle$ ';'

$\langle cclass \rangle$ ::= 'CREATE' 'COMPLEX EVENT CLASS'
                 $\langle schema \rangle$ 'AS' $\langle selection \rangle$ ['PUBLISH' $\langle publication \rangle$] ';'

$\langle schema \rangle$ ::= $\langle name \rangle$ '(' $\langle attributes \rangle$ ')' 'ID' '(' $\langle key \rangle$ ')'

$\langle attributes \rangle$ ::= $\langle name \rangle$ $\langle type \rangle$ { ',' $\langle name \rangle$ $\langle type \rangle$ }

$\langle key \rangle$ ::= $\langle name \rangle$ { ',' $\langle name \rangle$ }

$\langle selection \rangle$ ::= 'SELECT' $\langle select\_clauses \rangle$ 'OCCURRING AT' $\langle time \rangle$
                 ['CHECKING AT (' $\langle time \rangle$ ',' $\langle time \rangle$ ',' $\langle time\_literal \rangle$ ')']

$\langle time \rangle$ ::= $\langle table\_ref \rangle$ | 'NOW' | $\langle time \rangle$ [('+' | '-') $\langle time\_literal \rangle$]
                 | ('MAX' | 'MIN') '(' $\langle time \rangle$ { ',' $\langle time \rangle$ } ')'

$\langle publication \rangle$ ::= $\langle ontime \rangle$ { $\langle late \rangle$ } { $\langle change \rangle$ } { $\langle revoke \rangle$ }

$\langle ontime \rangle$ ::= $\langle name \rangle$

$\langle late \rangle$ ::= 'CASE LATE' ['(' $\langle min\_delay \rangle$ ',' $\langle max\_delay \rangle$ ')'] $\langle name \rangle$

$\langle change \rangle$ ::= 'CASE RETROACTIVECHANGE' $\langle name \rangle$

$\langle revoke \rangle$ ::= 'CASE REVOKE' $\langle name \rangle$

(with $\langle select\_clauses \rangle$ and $\langle table\_ref \rangle$ taken from a SQL grammar)

Fig. 5. BICEPL Syntax

occ $\ldots$, defining the occurrence time as first attribute in the newly created event. **(3)** We turn checking-at clauses into additional where-clauses.

Idealized Semantics: A BICEPL program, identified by its event classes $\mathbb{E} = \mathbb{S} \cup \mathbb{C}$, observes a sequence of pairs O_i, t_i, where O_i is the set of subscribed events detected up to time stamp t_i. We set t_0 to the system start-up time, and require $t_i > t_{i-1}$ for all $i > 0$. We start with $O_0 = \emptyset$, and for $i > 0$, we set $O_i = \{e \in O_{i-1} \mid \nexists e' \in \Delta_i$ and $e.\text{key} = e'.\text{key}\} \cup \Delta_i$ where Δ_i contains the subscribed events observed between t_{i-1} and t_i. Depending on the differences between O_i and O_{i-1}, program $\mathbb{E}$ publishes a set of publication events, as specified in $C.\text{pub}$ for $C \in \mathbb{C}$. Hence we define the semantics $[\![\mathbb{E}]\!] (O_i, t_i, O_{i-1}, t_{i-1})$ of program $\mathbb{E}$ over two pairs O_i, t_i and O_{i-1}, t_{i-1} to be compared. The distinction between mutable and immutable events has no semantic effect but enables an optimized treatment of immutable events.

The semantics

$$[\![\mathbb{E}]\!] (O_i, t_i, O_{i-1}, t_{i-1}) = \text{compare}(\text{derive}(O_i, t_i), \text{derive}(O_{i-1}, t_{i-1}))$$

is computed in two steps: **(A)** We *derive* the complex events $D_i = \text{derive}(O_i, t_i)$ and $D_{i-1} = \text{derive}(O_{i-1}, t_{i-1})$ with $\text{derive}(O, t) = D^l$ for complex classes $\mathbb{C} = \{C^1 \ldots C^l\}$. Herein, we set $D^0 = \emptyset$ and $D^j = D^{j-1} \cup C^j.\text{query}(O, D^{j-1}, t)$. We assume $C^j \subset C^k$ for all $j < k$, as $\subset$ is irreflexive and asymmetric. **(B)** We determine the resulting publication events by *comparing* D_i and D_{i-1} and adding to $\text{compare}(D_i, D_{i-1})$ the publication events arising in the following cases (see Figure 1):

(1) *Announcement* – there exists $e \in D_i$ but no $e' \in D_{i-1}$ with $e.\text{key} = e'.\text{key}$.

(a) For $e.\text{occ} = t_i$ we add $p \in e.\text{class.pub}[O]$ (on-time).

(b) For $e.\text{occ} < t_i$ we add $p \in e.\text{class.pub}[L]$ if $p.\text{min} < t_i - e.\text{occ} \le p.\text{max}$ (late).

(c) For $e.\text{occ} > t_i$ we add nothing (future event).

(2) *Revocation* – there exists $e \in D_{i-1}$ but no $e' \in D_i$ with $e.\text{key} = e'.\text{key}$.
(a) For $e.\text{occ} < t_i$ we publish $p \in e.\text{class.pub}[R]$ (revoke).
(b) For $e.\text{occ} \geq t_i$ we add nothing (future revoke).
(3) *Change* – there exists $e \in D_{i-1}$ and a $e' \in D_i$ with $e.\text{key} = e'.\text{key}$ but $e \neq e'$.
(a) For $e.\text{occ} \leq t_i$ and $e'.\text{occ} \leq t_i$ we add $p \in e.\text{class.pub}[C]$ (retroactive change)
(b) For $e.\text{occ} > t_i$ and $e'.\text{occ} > t_i$ we add nothing (future change).
(c) For $e.\text{occ} > t_i$ and $e'.\text{occ} \leq t_i$ we add $p \in e'.\text{class.pub}[R]$ (revoke).
(d) For $e.\text{occ} < t_i$ and $e'.\text{occ} > t_i$ we add $p \in e.\text{class.pub}[L]$
　　　with $p.\text{min} < t_i - e.\text{occ} \leq p.\text{max}$ holds (late).
(e) For $e.\text{occ} = t_i$ and $e'.\text{occ} > t_i$ we add $p \in e.\text{class.pub}[O]$ (on-time).

Sliding Window Semantics: Ideally, we would never drop observed, un-revoked subscribed events, but due to limited resources, we have to drop events eventually. To avoid altering the semantics, we need not only consider the lifespan of each subscribed event e but also the lifespan of all subscribed events e' such that e and e' are involved into some complex event c. We set $c.\text{expiration} = \max\{e.\text{occ} + e.\text{class.lifespan} \mid e \subset c\}$ and $e.\text{expiration} = \max\{c.\text{expiration} \mid e \subset c\}$. Then we purge events with $\text{purge}(O,t) = \{e \in O \mid e.\text{expiration} \geq t\}$, keeping only unexpired events. Finally, we need apply the same time stamp t_{i-1} in purging O_i and O_{i-1} (instead of t_i and t_{i-1}), leading to

$$[\![\mathbb{E}]\!]_{\text{purge}}(O_i, t_i, O_{i-1}, t_{i-1}) = [\![\mathbb{E}]\!](\text{purge}(O_i, t_{i-1}), t_i, \text{purge}(O_{i-1}, t_{i-1}), t_{i-1}) .$$

This windowing and idealized semantics behave identically if $\mathbb{E}$ uses only **(A1)** *monotone queries*, i.e., $C.\text{query}(O,D,t) \subseteq C.\text{query}(O',D',t)$ for all $O \subseteq O'$, $D \subseteq D'$, and $C \in \mathbb{C}$, and is **(A2)** *key subsuming*, i.e., $E.\text{key} \subseteq C.\text{key}$ for all $E \subset C$; and if the subscribed events $O_0, O_1, \ldots$ fed to $\mathbb{E}$ have **(B1)** *timely updates*, never updating an event beyond its lifespan (for $e' \in O_{i-1}$ and $e \in O_i$ with $e'.\text{key} = e.\text{key}$, we have $e'.\text{occ} + e.\text{class.lifespan} \geq e.\text{det}$), and **(B2)** *cohesive updates*, i.e., the constituent events of a complex event must share an overlapping lifespan as matching constituent events.

Theorem 1. *If a program $\mathbb{E}$ satisfies (A1-2) and if the events $O_0, O_1, \ldots$ satisfy (B1-2), then $[\![\mathbb{E}]\!]_{\text{purge}}(O_i, t_i, O_{i-1}, t_{i-1}) = [\![\mathbb{E}]\!](O_i, t_i, O_{i-1}, t_{i-1})$ holds.*

6   Implementation and Evaluation

PEACE should not only support server systems but also applications on small mobile devices. Therefore, our current implementation has been designed to be lightweight and portable, implemented in Java and SQLite in-memory database, and tested on Android and Ubuntu. We evaluate PEACE in three scenarios: The first two scenarios both extract and process flight arrivals from the web site of Heathrow Airport but differ in scale. The third scenario involves a stress test on PEACE's complex event processor to demonstrate its scalability.

In Scenario 1 we extract all flights from Frankfurt Airport to London Heathrow Airport, which were 36 flights on the day we performed the test. Requiring to load multiple pages, this test took 17 seconds on average. In Fig. 6a we show the time spent in different of PEACE modules; the initialization of the event detector (EDT), performing the OXPATH query, forwarding the events to the buffer (Event Buffer Maintenance),

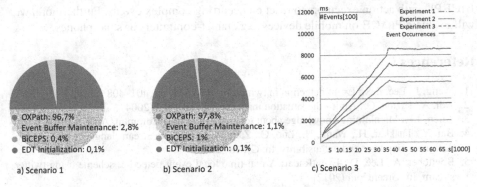

a) Scenario 1   b) Scenario 2   c) Scenario 3

Fig. 6. Profiling PEACE components

and processing the BiCEPL programs. OXPATH dominates the complex event processor with about 96,7% of the time needed to perform one processing step. As OXPATH spends 98% of its time in browser overhead [11], PEACE is dominated by browser overhead as well.

In Scenario 2 we increase the number of extracted and processed events to estimate the behaviour of the system when scaling up. We extracted all 1680 flight arrivals at London Heathrow Airport. As shown in Fig. 6b OXPATH dominates PEACE with 97,8% even stronger, as initialization overhead becomes less important.

In Scenario 3 we performed a stress test on BiCEPL's complex event processor to show the scaling behaviour of the complex event processor. To significantly increase the number of events, we would have to deploy a multitude of OXPATH wrappers. Therefore, we perform three experiments: Experiment 1 involves 5 subscribed event classes and 3 complex classes, each producing 10.000 events per hour. In Experiment 2, we add 2 more complex classes, and in Experiment 3 we use 2 complex classes and another complex class atop the former two. Fig. 6c shows the performance curve in milliseconds per processing step, and the number of subscribed events present in the repository (in hundreds, the same in all three experiments). All complex events include on-time, late, retroactive change, and revoke event publication statements. As Fig. 6c shows, the runtime of the complex event processor increases with the number of events in the repository. Keeping the number of event occurrences in the repository constant, also the time needed for processing steps remains constant. Due to lifespan specifications of subscribed events and the subsequent purging of old events the processing time can be capped together with the number of events in the repository. In the three experiments the database size was bounded by 165 MB, 221 MB, and 158 MB, respectively.

7   Conclusion

We have presented PEACE as an integrated framework to extract and process event announcements on the web. PEACE relies on a bitemporal event model which supports BiCEPL, a compact language for complex event processing, and an efficiently implementable semantics, requiring limited buffering only. We are currently expanding

BiCEPL with action executors to react on occurring complex events. Furthermore, we will evaluate PEACE on mobile devices, e.g., tablet-computers or smartphones.

References

1. Weng, J., Lee, B.S.: Event detection in twitter. In: ICWSM, pp. 401–408 (2011)
2. Adi, A., Etzion, O.: Amit - the situation manager. VLDB J. 13 (2004)
3. Allen, J.F.: Maintaining knowledge about temporal intervals. Commun. ACM 26 (1983)
4. Bai, Y., Thakkar, H., Wang, H., Luo, C., Zaniolo, C.: A data stream language and system designed for power and extensibility. In: CIKM (2006)
5. Boettcher, A., Lee, D.: EventRadar: A real-time local event detection scheme using twitter stream. In: GreenCom (2012)
6. Chen, J., DeWitt, D.J., Tian, F., Wang, Y.: NiagaraCQ: A scalable continuous query system for internet databases. In: SIGMOD Conference (2000)
7. Cugola, G., Margara, A.: TESLA: a formally defined event specification language. In: DEBS (2010)
8. Cugola, G., Margara, A.: Processing flows of information: From data stream to complex event processing. ACM Comput. Surv. 44 (2012)
9. Demers, A.J., Gehrke, J., Panda, B., Riedewald, M., Sharma, V., White, W.M.: Cayuga: A general purpose event monitoring system. In: CIDR (2007)
10. Eckert, M., Bry, F.: Rule-based composite event queries: the language XChange^{EQ} and its semantics. Knowl. Inf. Syst. 25 (2010)
11. Furche, T., Gottlob, G., Grasso, G., Schallhart, C., Sellers, A.: OXPath: A Language for Scalable Data Extraction, Automation, and Crawling on the Deep Web. VLDB Journal (2013)
12. Gehani, N.H., Jagadish, H.V.: Ode as an active database: Constraints and triggers. In: VLDB (1991)
13. Ilina, E., Hauff, C., Celik, I., Abel, F., Houben, G.-J.: Social event detection on twitter. In: Brambilla, M., Tokuda, T., Tolksdorf, R. (eds.) ICWE 2012. LNCS, vol. 7387, pp. 169–176. Springer, Heidelberg (2012)
14. Kowalski, R.A., Sergot, M.J.: A logic-based calculus of events. New Generation Comput. 4 (1986)
15. Kranzdorf, J., Sellers, A., Grasso, G., Schallhart, C., Furche, T.: In: Proc. of WWW
16. Liu, L., Pu, C., Táng, W.: Continual queries for internet scale event-driven information delivery. IEEE Trans. Knowl. Data Eng. 11 (1999)
17. Luckham, D.: Event Processing for Business. John Wiley & Sons, Inc., Hoboken (2012)
18. Luckham, D.C.: Rapide: A language and toolset for causal event modeling of distributed system architectures. In: Masunaga, Y., Tsukamoto, M. (eds.) WWCA 1998. LNCS, vol. 1368, pp. 88–96. Springer, Heidelberg (1998)
19. Mareco, C.A., Bertossi, L.E.: Specification and implementation of temporal databases in a bitemporal event calculus. In: Kouloumdjian, J., Roddick, J., Chen, P.P., Embley, D.W., Liddle, S.W. (eds.) ER Workshops 1999. LNCS, vol. 1727, pp. 74–85. Springer, Heidelberg (1999)
20. McCarthy, D.R., Dayal, U.: The architecture of an active data base management system. In: SIGMOD Conference (1989)
21. Schultz-Møller, N.P., Migliavacca, M., Pietzuch, P.R.: Distributed complex event processing with query rewriting. In: DEBS (2009)
22. Sripada, S.M.: A logical framework for temporal deductive databases. In: VLDB (1988)
23. Wieringa, R.: Design methods for reactive systems - Yourdon, Statemate, and the UML. Morgan Kaufmann (2003)

An Approach for Bursty and Self-similar Workload Generation

Xingjian Lu, Jianwei Yin, Hanwei Chen, and Xinkui Zhao

College of Computer Science and Technology,
Zhejiang University. Hangzhou, China
{zjulxj,zjuyjw,chw,zhaoxinkui}@zju.edu.cn

Abstract. As two of the most important characteristics of Web systems' workloads, burstiness and self-similarity are gaining more and more attentions. And synthetically generating bursty and self-similar workloads is a key technique for Web system performance analysis. In this paper, a configurable synthetic approach for bursty and self-similar workload generation has been proposed based on a superposition of 2-state Markovian arrival processes (MAP2). This method can generate workload with both specified intension of burstiness and self-similarity. The detailed evaluation show the accuracy and robustness of our method.

Keywords: Workload Generation, Markovian, Burstiness, Self-similarity.

1 Introduction

In recent years, more and more Web-based systems have been moved to cloud computing platforms such as Amazon EC2 and Google App Engine, which can promise of on-demand resource provisioning based on virtualization techniques. And some characteristics such as burstiness of workloads can have critical impact on resource provisioning strategies and performance of cloud platforms. For example, flash-crowd service requests can cause resource allocation problems and seriously degrade system performance [1]; Simultaneously launching jobs for different cloud applications, which are no longer single-program-single-execution applications, during a short time period can immediately aggravate resource competitions and load unbalancing among computing sites [16]. So synthetically generating bursty workloads is an important technique for Web system performance analysis, especially in the context of cloud computing.

Burstiness, which means highly variable request arrival rate or service time, has been observed in Ethernet LAN, Web applications, storage systems [15] and grid systems [10]. Many mathematical methods, including peakedness, peak-to-mean ratio, coefficient of variation, and indices of dispersion for count (IDC) have been proposed to characterize the intension of burstiness. The Markovian Arrival Process (MAP) [5,13], which is a generalization of Markov Modulated Poisson Process (MMPP), is usually leveraged to model bursty request arrivals. And some workload generators such as SWAT [8] and Geist [7] can also support the bursty workload generation.

X. Lin et al. (Eds.): WISE 2013, Part II, LNCS 8181, pp. 347–360, 2013.
© Springer-Verlag Berlin Heidelberg 2013

However, these methods only focus on the burstiness at some specific time-scale, while self-similarity, which has been also observed in a variety of working communication networks and computing systems, presents a process displaying similar-looking workload burstiness over all or a wide range of time-scales. The intension of self-similarity is often characterized by the Hurst parameter. And some new models, such as chaotic maps [11], fractional brownian motion (FBM) [17] and fractional autoregressive integrated moving average (FARIMA) model [9], have been proposed to describe self-similar behavior in a relatively simple manner. Also, a number of self-similarity models have been developed based on traditional traffic models. Similarly, these methods merely focus on modeling and fitting self-similarity, none of them can synthetically generate workloads that exhibit both specified burstiness and self-similarity degree.

In order to deal with these deficiencies, a markovian approach for bursty and self-similar workload generation has been proposed in this paper based on a superposition of 2-state Markovian arrival processes (MAP2). Our approach can leverage some simple traffic parameters, which can be straightforwardly derived from real system logs or provided by performance analysts, to compose a MAP with both required intension of bursiness and self-similarity. The detailed analysis and experiment results show the accuracy and robustness of our method.

The remainder of this paper is organized as follows. Section 2 introduces the motivation. Section 3 describes the Markovian approach for bursty and self-similar workload generation. Section 4 evaluates our workload generation method by conducting a detailed accuracy and robustness analysis. Finally, section 5 concludes and describes the future work.

2   Motivation

For workload analysis, the IDC has been widely used to characterize the bursti-ness of arrival. This is a standard burstiness index first used in networking, and then applied to model workload burstiness in Multi-Tier applications [12]. The IDC at time t is the variance of the number of requests arrived in an interval of length t divided by the mean number of requests arrived in this interval:

$$I_t = \frac{Var(N_t)}{E(N_t)} \tag{1}$$

where N_t represents the number of arrivals in the continuous interval of $(0, t)$.

Traditional workload generators such as Surge [4] and Httperf [14], can not support burstiness generation. Then, SWAT [8], Geist [7] and the method pro-posed in [13] were developed to provide mechanisms for burstiness injection. Although these methods can support injecting burstiness into workloads, the resulting models, based on burstiness characterizations using IDC, are adequate only over a limited range of time-scales. No one can synthetically generate work-load with specified long range bursty behavior across large time-scales, i.e. the self-similarity.

Let the discrete-time stochastic process $X = \{X_i, i = 0, 1, ...\}$ is used to describe the number of arrivals in the i-th interval (length is Δ). And the aggregated process of X is defined as follows:

$$X^{(m)} = \{X_i^{(m)}\} = \{\frac{X_1 + ... + X_m}{m}, ..., \frac{X_{mk+1} + ... + X_{(m+1)k}}{m}, ...\}$$

Then X is called exactly second-order self-similar with the Hurst parameter $H = 1 - \beta/2$ if

$$Var(X^{(m)}) = \sigma^2 m^{-\beta}. \tag{2}$$

where σ^2 is the variance of X, m is the aggregate level. There are some other equivalent definitions of self-similarity, we mainly consider the one relates to IDC in this paper. That is if X satisfies the following formula, X is self-similar.

$$I_m = I_{(t=m\Delta)} = \frac{Var(N_{(m\Delta)})}{E(N_{(m\Delta)})} = I_1 m^{2H-1}. \tag{3}$$

where I_1 denotes the IDC of the arrival process at the unit interval Δ.

Some new models such as chaotic maps, FBM and FAIMA have been developed to describe and model the self-similar behavior, and the corresponding approaches are also developed to generate self-similar traffic or workload based on these models. Because the queueing theoretical techniques are hardly to be used for these new models, a number of self-similarity models have been developed based on traditional traffic models too. For instance, MMPP as a superposition of 2-state Markov processes, is used to emulate self-similarity over a certain range of time scales in [2,3,18]. In [6], markovian arrival process as a superposition of a phase type renewal process and an interrupted Poisson Process (IPP), is proposed to approximate real traffic behavior. However, all these methods proposed to generate self-similar workload are only dedicated to the long range bursty behavior across large time-scales, they can't be used to generate self-similar workloads with specified burstiness on certain time-scales.

Motivated by the fact that current workload generation methods only focus on either burstiness or self-similarity, we claim a complete and practical workload generator should support workload generation with specified intension of burstiness and self-similarity. Then two questions may come into being:

First, why should we consider both the burstiness and self-similarity during workload generation. Here we use a real case to describe the reasons in the following. Three workloads with identical burstiness profiles ($IDC = 400$) and different intensions of self-similarity ($H = 0.62, 0.76, 0.9$ separately) are described in Fig. 1. For each plot of this figure, there are 10^5 inter-arrival time samples, whose mean value is 0.001 seconds. If single burstiness is enough to describe the bursty characteristic of the workload, the performance of these three workloads should be identical or nearly identical when they are imposed to the same system with identical environment. In order to verify this claim, we show the queueing performance of these workloads with $\cdot/D/1$ queueing network simulation in Fig. 2. The constant service time for each request is set to 0.001 seconds.

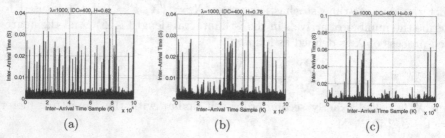

Fig. 1. Three workloads with identical burstiness ($IDC = 400$) but different self-similarity ($H = 0.62, 0.76, 0.9$) separately

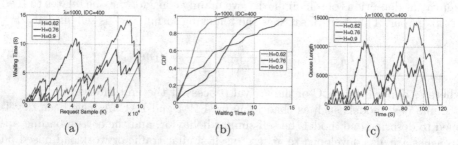

Fig. 2. Performance of the workloads depicted in Fig. 1 with $\cdot/D/1$ queue

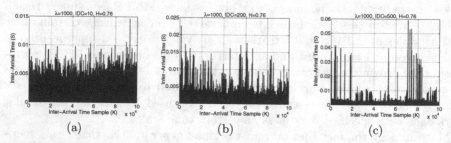

Fig. 3. Three workloads with identical self-similarity $H = 0.76$ but different burstiness ($IDC = 10, 200, 400$) separately

As shown in Fig. 2(a), the waiting time of each request when $H = 0.62$ is much smaller than $H = 0.76$ and $H = 0.9$. Though the waiting time when $H = 0.76$ is larger than $H = 0.9$ for some requests, the average and maximum value when $H = 0.9$ is larger than $H = 0.76$. This observation is validated by Fig. 2(b) and Fig. 2(c). From Fig. 2(b), we can see the waiting time for 80% of the requests is 2.72, 7.26 and 10.05 separately for the corresponding workload ($H = 0.62, 0.76, 0.9$). The queue length curve plotted in Fig. 2(c) also show evident performance differentiation when H is assigned to different values, even the IDC is identical. So we can see it's apparently inaccurate to describe the bursty characteristic of the workload merely by the burstiness parameter IDC.

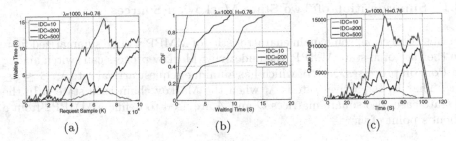

Fig. 4. Performance of the workloads depicted in Fig. 3 with $\cdot/D/1$ queue

Similarly, Fig. 3 shows the inter-arrival time samples for another three workloads with identical self-similarity ($H = 0.76$) but different burstiness ($IDC = 10, 200, 500$ separately), while Fig. 4 shows the corresponding performance for these workloads with the same $\cdot/D/1$ queueing network. As shown in Fig. 4, even with identical self-similarity intension, the performance is still significantly different for different value of IDC. And the higher the IDC the worse the performance. So single self-similarity is also not adequate to model the bursty behavior of practical workload. That means combining the burstiness and self-similarity is more completed and practical to generate workloads.

And the second question is how can we combine the burstiness and self-similarity. From previous description we know MMPP2 is often used to model workload burstiness, and some superpositions of Markov processes can be used to generate self-similar workload. So the natural way we may consider is to develop a method to generate the bursty and self-similar workload based on Markovian models, and with which the queueing theoretic techniques developed in the past can be used to guide the performance evaluation.

In a word, motivated by current workload generation methods usually focus on single burstiness or self-similarity, we aim to seek for a completed and practical approach for bursty and self-similar workload generation. Considering the computational tractability and the convenience for performance evaluation based on queuing theoretic techniques, the proposed approach for workload generation is based on Markovian models.

3 Markovian Modeling for Bursty and Self-similar Workload Generation

The proposed Markovian approach for bursty and self-similar workload generation is based on the model by Anderson et al. [2,3], where workload is modeled by the superposition of several 2-state MMPPs. The benefits of using a Markov model is that it is possible to re-use the well-known queuing theoretical techniques developed before and a whole array of tools for calculating performance measures is already available.

3.1   Superposition of Two State Markovian Sources

In this subsection, some main characteristics of MMPP will be summarized first. In the case of m-state MMPP, the underlying Markov process can switch among m Poisson processes, each of which has a unique request arrival rate λ_i, $(1 \leq i \leq m)$. That is, the arrival rate is λ_i when the Markov chain is in state i. In the 2-state case, two square matrices Q and Λ are used to define a MMPP2 from a client's point of view.

$$\mathbf{Q} = \begin{bmatrix} -r_1 & r_1 \\ r_2 & -r_2 \end{bmatrix}. \qquad \mathbf{\Lambda} = \begin{bmatrix} \lambda_1 & 0 \\ 0 & \lambda_2 \end{bmatrix},$$

For the case of MMPP2, the mean value of N_t is given by

$$E(N_t) = \frac{r_2 \lambda_1 + r_1 \lambda_2}{r_1 + r_2} t. \tag{4}$$

And the variance of N_t is can be calculated as follows:

$$Var(N_t) = \frac{r_2 \lambda_1 + r_1 \lambda_2}{r_1 + r_2} t + 2A_1 t - \frac{2A_1}{r_1 + r_2}(1 - e^{-(r_1+r_2)t})$$

where $A_1 = \frac{r_1 r_2 (\lambda_1 - \lambda_2)^2}{(r_1+r_2)^3}$.

Since any MMPP obtained by superposing several MMPP2s can be described by a superposition of several interrupted Poisson processes (IPP) and one Poisson process. We consider the required MMPP is composed of $d(> 1)$ IPPs and one Poisson process. ith IPP can be give by

$$\mathbf{Q}_i = \begin{bmatrix} -r_{1i} & r_{1i} \\ r_{2i} & -r_{2i} \end{bmatrix}, \qquad \mathbf{\Lambda}_i = \begin{bmatrix} r_i & 0 \\ 0 & 0 \end{bmatrix}.$$

The superposition can be described as follows

$$\mathbf{Q} = \mathbf{Q}_1 \bigoplus \mathbf{Q}_2 \bigoplus \cdots \bigoplus \mathbf{Q}_d$$
$$\mathbf{\Lambda} = \mathbf{\Lambda}_1 \bigoplus \mathbf{\Lambda}_2 \bigoplus \cdots \bigoplus \mathbf{\Lambda}_d \bigoplus \lambda_p,$$

where $\bigoplus$ is the Kronecker's sum and λ_p means the arrival rate of the Poisson process. The whole arrival rate of the superposition process λ can be given by

$$\lambda = \lambda_p + \sum_{i=1}^{d} \frac{r_{2i} \lambda_i}{r_{1i} + r_{2i}} \tag{5}$$

In the next subsection, we show how to determine the parameters of the IPPs and the Poisson process.

Table 1. Preliminarily Required Parameters

Parameter	Meaning
λ	Average arrival rate of the whole process.
m_{min}, m_{max}	Minimum and Maximum of the time-scales over which self-similarity is taken into consideration.
$I_{m_{min}}$	The IDC value at the minimum time-scale.
H	Hurst parameter.
d	Number of IPPs.

3.2 Applied Parameterizing Algorithm

In this subsection, a procedure is given to determine the parameters of the IPPs and the Poisson procss to construct a MMPP such that the properties of the workload generated by our approach match predefined values. Table 1 shows the preliminary required parameters for our generation model.

Let $N_{t|i}$ and $N_{t|p}$ be the number of arrivals during the t-th time slot in the i-th IPP and the Poisson process separately, and let $N_{t|i}^m$ and $N_{t|p}^m$ be the corresponding aggregated processes of them. Considering the computational tractability, we assume the r_1 and r_2 satisfy the following relation, for each IPP.

$$f = \frac{r_{2i}}{r_{1i} + r_{2i}}, \qquad (1 \le i \le d) \tag{6}$$

Then using (5), we have

$$\lambda = \lambda_p + \sum_{i=1}^{d} f\lambda_i \tag{7}$$

and using (4), we obtain the variance of the i-th IPP as

$$Var(N_{t|i}) = f\lambda_i t + \frac{2(1-f)^2 \lambda^2 t}{r_{1i} f} - \frac{2(1-f)^3 \lambda_i^2}{f r_{1i}^2}(1 - e^{-\frac{r_{1i}}{1-f}t}) \tag{8}$$

The variance of aggregated arrival process $N_{t|i}^{(m)}$ can be expressed as

$$Var(N_{t|i}^{(m)}) = \frac{Var(N_{(m\Delta)|i})}{(m\Delta)^2} \tag{9}$$

where Δ is previous mentioned sampling resolution. Here we consider Δ one time unit, using (8) and (9), we can get

$$Var(N_{t|i}^{(m)}) = \frac{f\lambda_i}{m} + 2f(1-f)^2 \eta_i \lambda_i^2 \tag{10}$$

where

$$\eta_i = \frac{1}{mr_{1i}} - \frac{1-f}{m^2 r_{1i}^2}(1 - e^{\frac{1}{1-f}mr_{1i}}) \tag{11}$$

The corresponding variance of the Poisson process is λ_p/m. For independent subprocesses, the variance of the superposition equals the sum of individual variances, so the variance of the whole process is given by

$$Var(X_t^{(m)}) = \frac{\lambda_p}{m} + \Sigma_{i=1}^d Var(N_{t|i}^m) = \frac{\lambda}{m} + 2f(1-f)^2 \sum_{i=1}^d \eta_i \lambda_{1i}^2 \qquad (12)$$

where we used (7). Then using (12) and (1), we can get

$$I_m = \frac{Var(N_{(m\Delta)})}{E(N_{(m\Delta)})} = \frac{m^2 Var(X_t^{(m)})}{m\lambda} = 2f + \frac{2mf(1-f)^2}{\lambda} \sum_{i=1}^d \eta_i \lambda_{1i}^2 \qquad (13)$$

Since the superposition of d IPPs and a Poisson process is expected to show self-similarity over d different time-scaless, and the sojourn time of each IPP is in accordance with the different time-scales, so there are d different points $m_i(1 \le i \le d)$. According to the range of time-scales specified by the input parameters, we have $m_{min} \le m_i \le m_{max}$, let

$$m_i = m_{min} a^{i-1} \qquad (14)$$

where

$$a = (\frac{m_{max}}{m_{min}})^{\frac{1}{d-1}}, \quad d > 1. \qquad (15)$$

In order to reduce the number of parameters which have to be determined, we also assume $m_i r_{1i} = 1$, i.e.

$$r_{1i} = \frac{1}{m_i}, \quad (1 \le i \le d). \qquad (16)$$

Then using (6), (14)-(16), we can obtain r_{2i} for each IPP. Now the parameters we need to obtain are only f and λ_i, since λ_p can be derived from (7) if λ_i is determined. Based on the above analysis, the applied parameterizing algorithm is in the following:

– **SETP1. Determine λ_i as the function of f.** From (4) and (14), we have

$$I_1 \begin{bmatrix} m_1^{(2H-1)} \\ m_2^{(2H-1)} \\ \vdots \\ m_d^{(2H-1)} \end{bmatrix} = 1 + \mathbf{B} \begin{bmatrix} \lambda_1^2 \\ \lambda_2^2 \\ \vdots \\ \lambda_d^2 \end{bmatrix} \qquad (17)$$

where $\mathbf{B}$ is the $d \times d$ matrix whose (i,j) element is

$$\mathbf{B}_{ij} = \frac{2f(1-f)^2}{r_{1i}\lambda} - \frac{2f(1-f)^3}{m_i r_{1i}^2 \lambda}(1 - e^{\frac{m_i r_{1i}}{f-1}}) \qquad (18)$$

Solving this, we can determine λ_i as the function of f.

– **STEP2. Find the value of f heuristically.** First, find the range of f heuristically, and set an initial value for f and the largest number of iterations. Then, use **STEP1** to obtain λ_i and further other needed parameters to determine the MMPP. Next, use this model to generate specified number of the inter-arrival time sample. Then we calculate the value of the average arrival rate, IDC and H from the generated sample data, to obtain the combined error. The value of f that minimizes the combined error is selected as the final value of f. Then other parameters can also be determined to generate the required workload.

To conclude, we compare our method with that of [2] and [18] in Table 2. Here, we call the procedure of [2] covariance method, the procedure of [18] variance method, and ours IDC method. The generation procedure of our method and the variance method are exactly constructed while that of [2] contains some approximations. Furthermore, the variance method does not hold when $Var(N_t^{(m)}) \leq \lambda/m$ or $Var(N_t^{(m)}) \geq \lambda/m + \lambda^2$, while our method does not have this constraint. This is significant to workload generator, which not only needs to fit the original trace, but also need to allow the generation of workload with desired characteristics which may cover a large different range.

Table 2. Comparison between IDC, Variance, and Covariance Methods

	IDC	Variance	Covariance
Required Parameters	$\lambda, H, I_1, d,$ Time scale	$\lambda, H, d, \sigma^2,$ Time scale	$\lambda, H, d, r(1),$ Time scale
Type of Component MMPPs	IPP	IPP	SPP
Parameter Fitting	Exact	Exact	Approximation
Constraint	None	$\frac{\lambda}{m} < Var(N_t^{(m)}) < \frac{\lambda}{m} + \lambda^2$	None

4   Evaluation

Accurately and robustly generating required workloads is the most important criteria to evaluate a workload generator. Thus in this section, we mainly evaluate the accuracy and robustness of our bursty and self-similar workload generation approach with the notion of average deviation, which is the relative error between the derived indicator parameter values with the expected ones. And it can be calculated as follows:

$$Avg\_Dev = \frac{1}{n} * \sum_{i=1}^{n} \frac{Dev(X_i) - Expec(X_i)}{Expec(X_i)} \tag{19}$$

where $Dev(X_i)$ and $Expec(X_i)$ denotes the derived and expected value of λ, IDC or H during i-th execution. For each indicator parameter, we execute the generation approach $n = 100$ times to derive the average value of deviations.

4.1 Accuracy Analysis

In this subsection, we evaluate the accuracy of generating workloads with specified intension of IDC and H. During these experiments, we experimentally set the expected average arrival rate $\lambda = 1000$, the number of IPPs $d = 4$, the burstiness $IDC = 10, 50, 100, 200, 400, 500$ (in practice, the maximum value 500 is enough to present the typical large value of IDC when $\lambda = 1000$), the self-similarity $H = 0.55, 0.62, 0.69, 0.76, 0.83, 0.9, 0.97$, and the minimum and maximum time-scale is 1 and 10^4 separately. By changing the value of IDC and H, we can derive the generating accuracy of our approach under different intension of burstiness and self-similarity. For giving an intuitive presentation of the generated workload by our approach and describing the motivation of this paper, we plot one set of the inter-arrival time samples with identical burstiness and different self-similarity in Fig.1, while inter-arrival time samples with identical self-similarity but different burstiness in Fig.3. And the corresponding queueing performance of these samples is depicted in Fig.2 and Fig.4 separately.

In table 3, we describe the average deviation of λ for each composition of IDC and H. From this table, we can see the deviation of λ is low, even when $IDC = 500$ and $H = 0.97$ the value is only 6.63%. And the tendency is evident that the deviation of λ increases with IDC (or H) when the value of H (or IDC) is identical. That is the higher the intension of burstiness or self-similarity the lower the accuracy of our method. The main reason for this behavior is that higher intension of burstiness or self-similarity means more variability of the inter-arrival times, which often brings more difficulty in accurately estimating the mean value, so the resulted average arrival rate may have a larger deviation.

The average deviation of IDC is described in table 4, generally the value of these deviations is larger than the ones of λ, since the calculation of IDC is more complex and inaccurate than the mean value of average arrival rate. It is also obvious that the deviation of IDC increases with the expected value of IDC and H. For instance, the deviation of IDC is only 0.75% when $IDC = 10$ and $H = 0.55$, while the value of deviation reaches 10.06% when $IDC = 500$ and $H = 0.97$. The reason is similar to the one of the deviation of λ.

However, the average deviation of H, as shown in table 5, shows a different tendency compared with the one of λ and IDC. First, the deviation of H doesn't show a complete increasing or decreasing tendency during the entire range of H. It decreases initially and then increases with the value of H. The main reason can be explained as follows: Our approach is developed based on the assumption that the required workload is self-similar, it doesn't work well under the case of no or low intension of self-similarity. Thus the lower the value of H the higher the inaccuracy to generate self-similar inter-arrival time samples, and further the higher the inaccuracy to derive the expected value of H. Furthermore, when the value of H closes to the maximum value 1, the relative error to get the required parameters of the MMPP model is larger than the low or moderate H, so the deviation of H begins to increase again after the minimum value. Second, although the deviation of H show an increasing tendency with the value of IDC, the increasing rate is not identical for different intensions of self-similarity. From table 5, we can see the increasing rate when H closes to the extreme value (1/2 or 1) is much larger than the one when H is moderate. That means the value of IDC plays less influences on the deviation of H when the self-similarity is moderate. The reason is also due to the extreme values of H make the fitting method more inaccurate during the workload generation process.

From above analysis, we can see our bursty and self-similar workload generation approach can ensure the accuracy within a reasonable range ($< 10\%$) for a wide range of specified intension of IDC and H.

Table 3. Average deviation of average arrival rate λ (%)

IDC	H						
	0.55	0.62	0.69	0.76	0.83	0.90	0.97
10	0.11	0.13	0.18	0.23	0.44	0.51	0.91
50	0.20	0.32	0.53	0.51	0.93	1.01	2.12
100	0.35	0.51	0.62	0.79	0.80	1.61	3.07
200	0.49	0.54	1.10	1.23	1.73	2.34	4.82
400	0.85	0.92	1.56	2.37	3.15	4.26	5.89
500	1.54	2.03	2.98	3.52	4.36	5.49	6.63

Table 4. Average deviation of burstiness intension IDC (%)

IDC	H						
	0.55	0.62	0.69	0.76	0.83	0.90	0.97
10	0.75	0.80	0.94	1.25	1.97	3.51	5.81
50	0.67	0.71	1.06	1.37	2.33	3.63	6.32
100	0.59	0.71	1.11	1.45	2.23	3.58	6.85
200	0.84	0.95	1.47	1.55	2.56	3.79	7.94
400	1.11	1.56	2.16	2.79	3.25	4.56	8.60
500	1.94	2.23	2.94	3.44	4.32	5.07	10.06

4.2 Robustness Analysis

For a robust workload generation approach, it is not only to ensure the accuracy for different input parameters of the generation model, but also required to make sure the number of samples won't influence the generation accuracy. In order to evaluate the impact of the number of samples on the accuracy, we test the deviation of λ, IDC and H with different number of generation samples. During these experiments, we change the number of samples from 10^5 to 10^6 with the interval of 10^5, and for each of which, we generate the specified number of inter-arrival samples 100 times for different compositions of the value of IDC and H. And other parameters are also set $\lambda = 1000$, $d = 4$, the minimum and maximum time-scale 1 and 10^4 separately.

Table 5. Average deviation of self-similarity intension H (%)

IDC	H						
	0.55	0.62	0.69	0.76	0.83	0.90	0.97
10	2.36	0.95	0.81	0.90	0.87	0.91	0.96
50	2.71	1.05	0.85	0.87	0.84	0.89	1.04
100	2.83	1.38	0.93	0.92	0.86	0.91	1.16
200	3.09	1.17	0.84	0.95	1.08	0.115	1.32
400	3.42	1.32	0.89	0.93	1.32	1.28	1.67
500	3.61	1.43	0.95	0.97	1.41	1.35	1.95

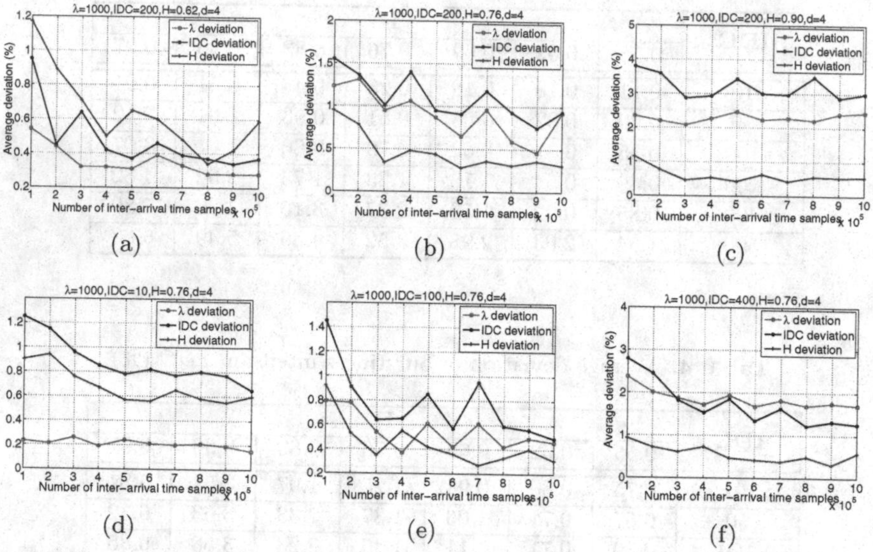

Fig. 5. Accuracy analysis with different number of samples for different compositions of the value of IDC and H

During these experiments, the deviation of λ, IDC and H show little fluctuation with the number of samples. The deviations roughly stay around a constant value when the number of samples exceed 2×10^5 or 3×10^5. We plot the deviations in Fig.5 with six typical compositions of IDC and H. The first three with identical value of $IDC = 200$ and different value of H, while the last three with identical value of $H = 0.76$ but different value of IDC. As shown by these plots, the deviations generally vary a little even though the higher the value of H or IDC, the larger the average value of these deviations. Furthermore, the deviations show significant improvement when the number of samples start to increase initially, while then the improvement begins to ease up until reaching around a constant value. The main reason for this kind of behavior lies in that when the number of samples is very small (e.g. 10^5), there maybe no enough data samples to fitting the expected value of λ, IDC and H. Thus the deviations decrease greatly when the number of samples start to increase initially. However, once the number of samples is large enough to fit the required parameters, the deviations begin to stay around a constant value, even though the number of samples is still keep increasing.

The results in above experiments show strong robustness of our approach. Even with extremely large number of samples to be generated, our approach can still ensure the accuracy of the required parameters within a reasonable range. This property is meaningful to the practical Web system workload generation, especially in the context of cloud computing, in which the large scale of system architecture and users often require a large number of workload samples to evaluate system performance or do optimal resource provision.

5   Conclusion and Future Work

Synthetical workloads modeling emerging or future applications is extremely important in the design of efficient system architecture. However, current approaches for workload generation only focus on either burstiness or self-similarity. With accurate characterization of the two key properties of the workloads by IDC and Hurst parameter separately, we developed a markovian approach for bursty and self-similar workload generation by fitting a MMPP model as a superposition of several IPPs and one Poisson process. The main contribution of the proposed approach lies in workload generation with specified intension of both burstiness and self-similarity, the simultaneous occurrence of which is the real case for cloud applications in the production. And the experiments and evaluation show the accuracy and robustness of our approach. After focusing on bursty and self-similar workload generation in this paper, our future work on this subject is mainly to evaluate the system performance under such kind of workloads, and find approaches for performance optimalization and resource efficient utilization to reduce the negative impacts of burstiness and self-similarity.

Acknowledgments. This work was supported by National Science and Technology Supporting Program of China (No. 2012BAH06F02), National Natural Science Foundation of China under Grant (No. 61272129), Research Foundation for the Doctoral Program by Ministry of Education of China (No. 20110101110066), New-Century Excellent Talents Program by Ministry of Education of China (No. NCET-12-0491), and Zhejiang Science Fund for Distinguished Young Scholars (R13F020004).

References

1. Amini, L., Jain, N., Sehgal, A., Silber, J., Verscheure, O.: Adaptive control of extreme-scale stream processing systems. In: Proceedings of the 26th IEEE International Conference on Distributed Computing Systems, p. 71. IEEE Computer Society, Washington, DC (2006)
2. Andersen, A., Nielsen, B.: An application of superpositions of two state markovian source to the modelling of self-similar behaviour. In: Proceedings IEEE INFOCOM 1997, vol. 1, pp. 196–204 (1997)

3. Andersen, A., Nielsen, B.: A markovian approach for modeling packet traffic with long-range dependence. IEEE Journal on Selected Areas in Communications 16(5), 719–732 (1998)

4. Barford, P., Crovella, M.: Generating representative web workloads for network and server performance evaluation. SIGMETRICS Perform. Eval. Rev. 26(1), 151–160 (1998)

5. Bolch, G., Greiner, S., Meer, H.D., Trivedi, K.S.: Queueing Networks and Markov Chains. Wiley-Interscience (2005)

6. Horváth, A., Rózsa, G.I., Telek, M.: A map fitting method to approximate real traffic behaviour. In: 8th IFIP Workshop on Performance Modelling and Evaluation of ATM & IP Networks, p. 32 (2000)

7. Kant, K., Tewari, V., Iyer, R.K.: Geist: A web traffic generation tool. In: Field, T., Harrison, P.G., Bradley, J., Harder, U. (eds.) TOOLS 2002. LNCS, vol. 2324, pp. 227–232. Springer, Heidelberg (2002)

8. Krishnamurthy, D., Rolia, J.A., Majumdar, S.: A synthetic workload generation technique for stress testing session-based systems. IEEE Trans. Softw. Eng. 32(11), 868–882 (2006)

9. Lakehal, M.R., Ferdi, Y., Taleb-Ahmed, A.: Generation of farima $(0, \alpha, 0)$ sequences by recursive filtering: Testing for self-similarity. In: Electronics, Circuits, and Systems, ICECS 2009, pp. 563–566. IEEE (2009)

10. Li, H., Muskulus, M.: Analysis and modeling of job arrivals in a production grid. SIGMETRICS Perform. Eval. Rev. 34(4), 59–70 (2007)

11. Lo, S.C., Cho, H.J.: Chaos and control of discrete dynamic traffic model. Journal of the Franklin Institute 342(7), 839–851 (2005)

12. Mi, N., Casale, G., Cherkasova, L., Smirni, E.: Burstiness in multi-tier applications: Symptoms, causes, and new models. In: Issarny, V., Schantz, R. (eds.) Middleware 2008. LNCS, vol. 5346, pp. 265–286. Springer, Heidelberg (2008)

13. Mi, N., Casale, G., Cherkasova, L., Smirni, E.: Injecting realistic burstiness to a traditional client-server benchmark. In: Proceedings of the 6th International Conference on Autonomic Computing, pp. 149–158. ACM, New York (2009)

14. Mosberger, D., Jin, T.: httperf: a tool for measuring web server performance. SIGMETRICS Perform. Eval. Rev. 26(3), 31–37 (1998)

15. Riska, A., Riedel, E.: Disk drive level workload characterization. In: Proceedings of the Annual Conference on USENIX 2006 Annual Technical Conference, p. 9. SENIX Association, Berkeley (2006)

16. Tai, J., Zhang, J., Li, J., Meleis, W., Mi, N.: Ara: Adaptive resource allocation for cloud computing environments under bursty workloads. In: 30th IEEE International Performance Computing and Communications Conference (IPCCC), pp. 1–8 (2011)

17. Tan, X., Huang, Y., Jin, W.: Modeling and performance analysis of self-similar traffic based on fbm. In: IFIP International Conference on Network and Parallel Computing Workshops, NPC Workshops, pp. 543–548. IEEE (2007)

18. Yoshihara, T., Kasahara, S., Takahashi, Y.: Practical time-scale fitting of self-similar traffic with markov-modulated poisson process. Telecommunication Systems 17, 185–211 (2001)

Spatio-temporal Event Modeling and Ranking

Xuefei Li, Hongyun Cai, Zi Huang, Yang Yang, and Xiaofang Zhou

The University of Queensland, QLD 4072 Australia
{x.li14,h.cai2}@uq.edu.au
{huang,yang.yang,zxf}@itee.uq.edu.au

Abstract. Effective event modeling allows accurate event identification and monitoring to enable timely response to emergencies occurring in various applications. Although event identification has been extensively studied in the last decade, the triggering relationship among initial and subsequent events has not been well studied, which limits the understanding of event evolvements from both spatial and temporal dimensions. Furthermore, it is also useful to measure the impact of events to the public so that the important events can be first seen. In this paper, we propose to systematically study event modeling and ranking in a novel framework. A new method is introduced to effectively identify events by considering the spreading effect of event in the spatio-temporal space. To capture the triggering relationships among events, we adapt the self-exciting point process model by jointly considering event spatial, temporal and content similarities. As a step further, we define the event impact and rank them at different time stamps. Extensive experimental results on real-life datasets demonstrate promising performance of our proposal in identifying, monitoring and ranking events.

Keywords: Spatio-temporal, Event Identification, Modeling, Ranking.

1 Introduction

Event identification, monitoring and ranking play a critical role in many applications such as health monitoring and environmental management. For example, early warnings of impending natural disasters or disease are critical for the safety and security of populations within the affected areas. Timely access to detailed event information provides an aggregated source of information on events of significance to the public, enables rapid response to public emergencies, and facilitates the monitoring of crowd sentiment among affected people. Thus, under a variety of situations, it is in high demand to derive an effective approach to identify events, track their evolvements and report their impact to the public.

In recent years, event identification and tracking in social network has attracted a lot of attention from different research communities. With the rapid development of Web 2.0, social media has become a common platform for communication and useful resource for facilitate various database and multimedia applications [19]. The unprecedented public access to large streams of real-time human communication presents a prime opportunity for automated analysis of important events, their evolving trends, and the corresponding public sentiment [12,16]. Thus, it is promising to detect events from social media data by utilizing its associated rich information.

X. Lin et al. (Eds.): WISE 2013, Part II, LNCS 8181, pp. 361–374, 2013.

However, the complexities of event detection pose an array of research challenges, including how to incorporate spatio-temporal context and content; how to distinguish different events and model their relationships; how to measure the event impact to the public, and so on. Some existing work incorporates temporal or spatial dynamics into event detection [9,12,20], while others combine spatio-temporal and content information into one similarity metric [2,14]. They either do not well address the problem of modeling events from a comprehensive spatio-temporal viewpoint or spend too much time tuning coefficient of each feature. Furthermore, event relationships and impact to the public have not been systematically studied.

In this paper, we propose a framework for systematic study of event modeling and ranking. Our principle contributions are summarized as follows:

- We propose a new method to effectively identify events from the social media data by jointly considering the spatio-temporal context and textual content. Particularly, spatial and temporal expansions are employed to reflect the natural spreading of events in the spatio-temporal space.
- We model the triggering relationships among events by a triggering probability graph. We adapt the self-exciting point process model to capture the spatio-temporal and content triggering relationships among events. Conducting the graph visualization along the time line, the evolvements of events can be monitored.
- We further define and compute the event impact from the constructed triggering probability graph. Events can then be ranked and recommended to users according to their impact to the public.

We organize the rest of the paper as follows. Section 2 provides the problem definition and introduces the overall framework. We describe event modeling with event identification and triggering analysis in Section 3. Event ranking is discussed in Section 4. The results are shown in Section 5. The related work is summarized in Section 6, followed by conclusion in Section 7.

2   Definitions and Framework

Definition 1 (Incident). *An incident is defined as a spatio-temporal object. It is denoted as* $d_i = \{t_i, l_i, c_i\}$, *where* t_i *and* l_i *are the time and location spot where the incident* d_i *takes place respectively, and* c_i *is the content describing the incident.*

Definition 2 (Incidents Similarity). *Given a set of incidents* $\mathbb{D} = \{d_1, d_2, \cdots, d_n\}$, $\forall d_i, d_j \in \mathbb{D}$, *we define that two incidents* d_i, d_j *are similar and denoted as* $d_i \leftrightarrow d_j$ *iff the difference of* d_i *and* d_j *in temporal, spatial and content properties are respectively not greater than three given thresholds* θ_t, θ_l *and* θ_c, *i.e.* $|t_i - t_j| \leq \theta_t$, $|l_i - l_j| \leq \theta_l$ *and* $|c_i - c_j| \leq \theta_c$.

Definition 3 (Event). *An event* E_i *can be characterized as* $E_i = \{T_i, L_i, C_i, N_i\}$, *where* T_i *and* L_i *are the time and location information of all the incidents in* E_i, C_i *represents the combination of the content from all the incidents in* E_i, *and* N_i *stands for the total number of incidents in* E_i.

For simplicity, an event's time is represented as the earliest time of all the incidents, its location is represented as the centroid of all the incidents' location and its content is represented by combining all the texts into a single bag-of-word vector.

Definition 4 (Event Relationship). *The relationship among events is illustrated by conditional probability matrix P, which indicates the triggering relationship between any pair of events. Given a set of N events, $P_{i,i}$ indicates how possible E_i is an initiative event and $P_{j,i}$ stands for the probability of E_i being triggered by E_j.*

Apparently, E_j must occur before E_i to achieve a non-negative $P_{j,i}$. Otherwise, $P_{j,i} = 0$. After events are detected and their triggering relationship is obtained, we define event impact based characteristics of events. Ranking result is obtained from impact. As illustrated in Figure 1, we aim to identify social events from the Flickr photos and further analyse their relationships to calculate event impact and there are four major tasks in our work, including event identification (step 1), event triggering relationship analysis (step 2), event impact calculation (step 3) and event ranking at a certain time point (step 4).

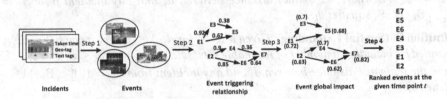

Fig. 1. An overview of the framework

3 Event Modeling

In this section, a novel model is derived to represent events, reveal the triggering relationships among events and give a ranking result based on event impact.

3.1 Event Identification

According to the Definition 1, spatial, temporal and content are three major characteristics of an incident. However, grouping incidents into events considering all aspects is not easy work. Our goal of event identification is to partition all incidents into representative sets(events) so that incidents in one set strictly satisfied constrains from three features, maximum number of incidents are covered by one set and minimum number of sets is discovered. We would also like to extend the event boundary in spatio-temporal space so that relatively far away relevant incidents can be grouped together.

Incident Set Initialization and Expansion. With the definition of incidents similarity, for each incident $d_i \in \mathbb{D}$, we collect all similar incidents to d_i into a set, which can be represented as $\mathbb{D}_i = \{d_l | d_l \leftrightarrow d_i, d_l \in \mathbb{D}\}$. Hence, the set $\mathbb{D}_i$ covers d_i and all incidents similar to d_i. Our general idea is to find events to cover all incidents $\mathbb{D}$. It can also be treated as to find minimum sets to cover all incidents. Furthermore, by considering the overlapping between two sets, we define the set connectivity and reachability below.

Definition 5 (Set Connectivity). *Given two sets $\mathbb{S}_i$ and $\mathbb{S}_j$, $1 \leq i, j \leq n$, they are connected if there exists a bridging $d_k \in \mathbb{D}$ such that $d_k \in \mathbb{S}_i$ and $d_k \in \mathbb{S}_j$.*

Definition 6 (Set Reachability). *Given set $\mathbb{S}_i$ and set $\mathbb{S}_j$, they are reachable if there exists $\{\mathbb{S}_{k_0}, ..., \mathbb{S}_{k_r}\}$, $\mathbb{S}_{k_0} = \mathbb{S}_i$ and $\mathbb{S}_{k_r} = \mathbb{S}_j$, such that $\mathbb{S}_{k_{l-1}}$ and $\mathbb{S}_{k_l}$ are connected for all $1 \leq l \leq r$.*

For spatial and temporal flexibility, we use set reachability to define spatial and temporal expanding as follows, based on which the expanded incident sets are generated. A strict constraint is applied in the spatial/temporal expansion that content dissimilarity for any two incidents to be grouped together must not be greater than θ_c. It is because that from the perspective of semantics, compared with the spatial and temporal information, the content information has greater potential to distinguish one incident from another in event identification.

Definition 7 (Temporal Expanding). *Given a set $\mathbb{S}_i$ for d_i, $1 \leq i \leq n$, it can be temporally expanded by including a number of reachable sets $\mathbb{S}_{k_1}, ..., \mathbb{S}_{k_Z}$, $1 \leq k_z \leq n$, $z = 1 ... Z$ where the spatial distance between d_i and any incident from $\mathbb{S}' = (\bigcup_{z=1}^{Z} \mathbb{S}_{k_z}) \cup \mathbb{S}_i$ satisfies θ_l.*

Definition 8 (Spatial Expanding). *Given a set $\mathbb{S}_i$ for d_i, $1 \leq i \leq n$, it can be spatially expanded by including a number of reachable sets $\mathbb{S}_{k_1}, ..., \mathbb{S}_{k_Z}$, $1 \leq k_z \leq n$, $z = 1 ... Z$ where the temporal distance between d_i and any incident from $\mathbb{S}'' = (\bigcup_{z=1}^{Z} \mathbb{S}_{k_z}) \cup \mathbb{S}_i$ satisfies θ_t.*

For each incident d_i, its initial set $\mathbb{S}_i$ can be expanded in the spatio-temporal space to include more relevant incidents via transitive connectivity. With spatial or temporal expanding, incidents across different locations or time stamps can still be grouped together, which avoids splitting a complete event into several pieces due to the location or time threshold.

Event Generation. Given the expanded incident sets $\{\mathbb{S}_i\}$, $i = 1 ... n$, we aim to find the minimal number of sets to cover all incidents, where each set is considered as an event. This objective can be transferred into a minimum set cover problem. Thus, the basic idea of our algorithm is to repeatedly select the current largest set to generate a new event and remove the incidents belonging to this set from any other sets until no set is left. The proposed algorithm is greedy in nature, and it has approximation ratio of $(1 + \ln r)$, where r is the cardinality of the largest set.

3.2 Event Triggering Analysis

Events can be divided into different types. Some events arise independently while others are triggered by previous events or influence their subsequent events. One typical example is the earthquake model [21] where the outbreak of an fore-shock earthquake is inevitable to increase the likelihood of the main shock and other aftershocks nearby, and another example is the crime model in criminology [13] where burglars may repeatedly attack nearby targets. Inspired from the above event models, we believe that social events happen in similar ways with more complex mutual relationship. This provides an intuition for analysing the triggering relationship among events.

Basic Self-exciting Point Process. A point process $N(t)$ is a type of random process which counts the number of events and the time that these events occur in a given time interval t. It is generally characterized via its conditional intensity $\lambda(t)$. The self-exciting point process model [13] is a more appropriate way to model the event evolvement, because it also considers the emergence of the history independent events. Every event is possible to enhance the emergence chance of its following events and the kernel function $\phi(\Delta t)$ models the enhanced chance which spreads in time. The intensity function of self-exciting point process model is defined as follows [13]:

Definition 9 (Self-exciting Point Process).

$$\lambda(t) = \eta(t) + \sum_{i,t_i < t} \phi(t - t_i) \tag{1}$$

where t_i is the occurrence time of the i-th event, $\eta(t)$ is a Poisson background rate of the event occurs at t, ϕ is the memory kernel function indicating the increase rate in the intensity triggered by previous events at t_i ($t_i < t$), and $\lambda(t)$ is the conditional intensity of a point process.

Adapted Self-exciting Model for Social Events. Naturally, the self-exciting point process can also be used to model social events' evolvements. Here we adapt the self-exciting point process to utilize the spatio-temporal context and content for social event relationship analysis.

For each event $E_i=\{T_i, L_i, C_i, N_i\}$, we define its intensity λ as:

$$\lambda(E_i) = \eta_t(T_i)\eta_l(L_i)\eta_c(C_i) + \sum_j \phi(\Delta T_j, \Delta L_j, \Delta C_j) \tag{2}$$

where $\eta_t(\cdot), \eta_l(\cdot)$, and $\eta_c(\cdot)$ are the Poisson background rates of the event E_i occurring at time T_i, in location L_i, and with content C_i respectively. ϕ is the memory kernel function indicating the increase rate in the intensity triggered by previous events before T_i. $\Delta T_i, \Delta L_i$ and ΔC_i are the distances between E_i and its previous event E_j along the temporal, spatial, and content dimensions respectively.

In the above adapted model, the first component $\eta_t(\cdot)\eta_l(\cdot)\eta_c(\cdot)$ can be understood as the opportunity for the event to self-occur, and the second component ϕ indicates the probability of the event being triggered by previous events.

When calculate the intensity of each event, we aim to model the relationships between events by the event relationship matrix P, as defined in Definition 4. For a specific event E_i, it has a probability $P_{i,i}$ of being a history independent event and $P_{j,i}$ indicating the probability of E_i being trigged by E_j. For event E_i, its total sum of $P_{i,i}$ and $P_{j,i}$ should add up to 1. For example, if $P_{i,i} = 1$, then E_i is self-triggered and it is an isolated event or initiative event.

Based on the adapted self-exciting point process model, $P_{i,i}$ and $P_{j,i}$ are calculated as:

$$P_{i,i} = \frac{\eta_t(T_i)\eta_l(L_i)\eta_c(C_i)}{\lambda(E_i)} \tag{3}$$

and

$$P_{j,i} = \frac{\phi(\Delta T_j, \Delta L_j, \Delta C_j)}{\lambda(E_i)} \qquad (4)$$

To compute P based on the adapted self-exciting point process model, the events need to be separated into two groups: background (or history independent) events $\mathbb{E}_b$ corresponding to the first component, and triggered events $\mathbb{E} - \mathbb{E}_b$ corresponding to the second component in the model. Here we apply the stochastic declustering method [21] to separate events, estimate η_t, η_l, η_c, and ϕ, and update P iteratively until convergence is achieved. The scheme is sketched in Algorithm 1.

We first provide an initial guess for P (line 1). With this initial guess, we use an expectation maximization (EM) algorithm to finalize P. While P is changed from last iteration, the following E-step and M-step are repeated (lines 2-8). In the E-step, a threshold on $P_{i,i}$ is firstly generated using Monte Carlo method, and the events whose $P_{i,i}$ values in P are greater than the threshold will be added into $\mathbb{E}_b$ and the rest belong to $\mathbb{E} - \mathbb{E}_b$ (line 4). Based on $\mathbb{E}_b$ and $\mathbb{E} - \mathbb{E}_b$, η_t, η_l, η_c and ϕ for each event are then estimated using the Variable Bandwidth Kernel Density Estimation (VBKDE) (line 5). In the M-step, P is updated from η_t, η_l, η_c, and ϕ using Equations 4 and 5. These steps are repeated until P remains unchanged.

Algorithm 1. P Estimation

Input : $\mathbb{E}$
Output: P

1 *Initialize P;*

2 **while** *P is changed from last iteration* **do**

3 | **E-step:**

4 | *Generate $\mathbb{E}_b$ and $\mathbb{E} - \mathbb{E}_b$ using Monte Carlo method and $P_{i,i}$;*

5 | *Estimate η_t, η_l, η_c, and ϕ from P, $\mathbb{E}_b$ and $\mathbb{E} - \mathbb{E}_b$ using VBKDE (Variable Bandwidth Kernel Density Estimation);*

6 | **M-step:**

7 | *Update P from η_t, η_l, η_c, and ϕ using Equations 4 and 5 ;*

8 **end**

We use Kernel Density Estimation (KDE) to estimate the values of both Poisson background rates and the increase rate in the intensity triggered by previous events. Kernel density estimation is a non-parametric estimation way to estimate the probability distribution of a random variable [13]. The quality of KDE depends mainly on its bandwidth. Small values of bandwidth lead to very spiky estimates with small bias and large variance while large values of bandwidth lead to small variance and large bias which will cause over-smoothing. Therefore choosing appropriate bandwidth is very important. Since the memberships of background events $\mathbb{E}_b$ and triggered events $\mathbb{E} - \mathbb{E}_b$ change in each iteration, a fixed bandwidth will thus lead to the problem of over-smoothing or under-smoothing at different iterations. To address this problem, we also choose Variable Bandwidth KDE as in [13] which is outlined in Algorithm 2.

Algorithm 2. VBKDE

Input : P, $\mathbb{E}_b$ and $\mathbb{E} - \mathbb{E}_b$
Output: Estimated η_t, η_l, η_c, and ϕ for each Event

1 Scale inter-event distance to have unit variance and zero mean along temporal, spatial, and content dimensions respectively;
2 Compute bandwidths based on scaled data in three dimensions;
3 Transfer scaled data back to the original scales;
4 Estimate η_t, η_l, η_c, and ϕ for each event using Gaussian Kernel.

In the above VBKDE, the first three steps are used to generate the bandwidths for three dimensions. The inter-event distances on different dimensions could be computed differently since different applications may have different types of data to represent temporal, spatial and content information. We will present the distance functions for the tested datasets in the experiment section. For each event, the distance of its k-th nearest neighbor is regarded as the bandwidth for that event. In the fourth step, the Gaussian kernel is used to estimate the values of η_t, η_l, η_c, and ϕ.

The time complexity of computing P based on the adapted self-exciting point process model is $O(iter \times N^2)$, where $iter$ is the number of iterations. Based on the fact that people will only care about the top 10 or top 200 events in a ranking, events with small impact are ignored and less than 10% of all events remain, which leads to a rather small N.

From the generated matrix P, a triggering probability graph can be constructed, where each node represents an event, and the directed edge from one event E_j to another event E_i represents the probability of E_i being triggered by E_j. Recall the three types of events. Subgraphs with single nodes represent isolated events. Nodes with outgoing links but without incoming links are initiative events, and nodes with incoming links are subsequent events. By visualizing the graph, event relationships and their evolvements along time line can be easily monitored. With such a graph, events can also be ranked according to their impact to the public so that the most important events can be identified and recommended earlier.

4 Event Ranking

In this section, we rank events from the triggering probability graph represented by P.

4.1 Event Impact

Numerous new events occur continuously in the world. Some of them have significant impact to the society while some others fade away quickly with not much effect. We consider several factors in determining the event impact. Firstly, it is observed that if an event has major impact to the society, it usually draws high attention from lots of people, reflected by the number of photos uploaded for the event and the number of users who upload the photos. The more photos related to the event and the more unique users participating in the event, the more impact an event can gain. Secondly, it is highly expected that events spreading over larger regions and lasting longer can affect the public in greater spatial and temporal scales. Therefore, here we use four factors to define the event impact as below.

Definition 10 (Event Impact). *Given an event E_i, its impact denoted as $I(E_i)$ is computed as:*

$$I(E_i) = \alpha_1 logN_i + \alpha_2 N_i^u + \alpha_3 \rho_i^l + \alpha_4 \rho_i^t \tag{5}$$

where N_i is the number of incidents in the event, N_i^u is the number of unique users participating in the event, ρ_i^l and ρ_i^t are the area and the period covered by all the incidents in the event respectively. $\alpha_1 + \alpha_2 + \alpha_3 + \alpha_4 = 1$. They are the coefficients of the linear combination. Logarithmic function is used on the number of incidents to avoid the effect of large number of photos uploaded by a single user.

Obviously, the above four factors are not directly comparable. We have to normalize all of them before they can be combined. They are normalized by dividing by the maximum values among all the events on each individual factors respectively. The weights of different factors will be empirically tuned in the experiments.

4.2 Event Ranking

The global event impact reflects the importance of an event in the whole graph. However, the impact of an event at a particular time point could be different. It is usual that the impact of an event decays as time goes by. Given an event, to calculate its current impact to the society is critical. By taking into account the time decay effect on event impact, we derive the following formula to calculate the local impact of event E_i at a given time point t.

$$\hat{I}_t(E_i) = \begin{cases} I_k(E_i) \times (1 - \text{Sigmoid}(t - t_i)), & t \geq t_i \\ 0, & t < t_i \end{cases} \tag{6}$$

Where k is the last number of iteration and the sigmoid function is $\text{Sigmoid}(t) = \frac{1}{1+e^{-t}}$. $1 - \text{Sigmoid}(t - t_i)$ is adopted to model the time decay. Given a time point t, the events identified from the social media data will be ranked according to their $\hat{I}_t$ values.

5 Experiments

5.1 Set Up

Data Sets. Three real-life datasets are used in our experiments. All these datasets are extracted from Flickr by applying Flickr search API. **(a)London dataset.** We extract 29,682 photos from Flickr, which were taken between July 1, 2012 to September 15, 2012, within the region bounding box from (longitude:-0.557, latitude:51.283) to (longitude:0.327, latitude:51.686). It covers greater London area. **(b)Brisbane dataset.** A total number of 11,330 photos taken between January 1, 2011 to May 1, 2013 are extracted from the online group *Brisbanites: Brisbane Photos*. **(c)New York dataset.** By setting the search bounding box as (longitude: -74.5, latitude: 40) to (longitude: -73.5, latitude: 41.2) and the time interval as 1 Jan, 2011 to 1 May, 2013, we collect 372,357 photos taken in New York City and its neighboring areas.

All extracted Flickr photos are associated with taken time, location and text tags and the text tags are pre-processed by removing stop words, splitting compound words, stemming, etc. The proposed algorithms are applied to the above three datasets to identify the events and also calculate their impacts for event ranking.

Distance Functions. To measure the distance between two events, in this paper we use a relatively simple event representation. The earliest of all the incidents' time stamps in an event is taken as the event's time, and the centroid of all the incidents' locations in an event is taken as the event location. An event content is constructed by combining all the tags from the member incidents and the content distance is measured by the cosine distance function in the vector space model.

Ground Truth Generation. Two groups of ground truth are generated for event identification and event ranking separately. We manually identify all events for London, Brisbane and New York data sets. A web system is developed to collect the crowd intelligence for event ranking ground truth generation. Each event is represented by its most frequent tags and four photos which are randomly selected to depict the event. Users are required to give an impact rate to each event from five levels from 5 to 1 indicating the decreasing importance of the event. The final impact of each event is calculated by averaging the rates from 30 users.

Performance Indicators. Two measures are used for performance indication.

- **Event coverage**. This indicator is used to measure the effectiveness of the event identification algorithm. For each ground truth event and its best matched event returned from the algorithm, its coverage is defined as the ratio of *the size of the overlapped incidents* to *the size of the union of incidents* in two events. The averaged coverage over all the ground truth events is used as the final indicator to indicate the effectiveness of the algorithm. Note that every event identified by the algorithm can only be matched at most once with the ground truth events. Obviously, the bigger the value of event coverage, the more effective the algorithm is to identify events and assign incidents to correct events.
- **Normalized Discounted Cumulative Gain (nDCG)** [10]. nDCG is used to evaluate the event ranking performance. It uses a graded relevance scale of events (5 levels in our experiments) in the result list, and measures the gain of an event based on its position in the result list. The gain is accumulated from the top of the result list to the bottom with the gain of each result discounted at lower ranks.

Compared Methods. Two different comparison strategies are used in our experiment for event identification and event ranking respectively.

Event identification. We compare our event identification work with three existing work [2], [4] and [14]. In [2], a supervised learning method is introduced to learn similarity metrics for time, location and text features in event identification. [4] analyse the temporal and spatial distributions of tags by means of discrete wavelet transform and identify events from related tag clusters. In [14], a content-similarity graph is first constructed by pair-wisely comparing images' content, followed by applying SCAN approach [18] to do graph clustering. Each generated subgraph is regarded as an event.

In this paper, we use *incremental clustering*, *wavelet-based clustering* and *graph clustering* to refer to the method in [2], [4] and [18] respectively.

Event Ranking. Due to the lack of existing work on event ranking, we evaluate the proposed framework in the way of replacing our proposal in each step by one or several baseline algorithms and study their differences on performance. The effect of event triggering relationship analysis, the effect of event impact initialization for random walk and the effect of random walk itself are quantitatively analyzed. All the experiments are implemented on a PC with Windows 7, Intel(R) Core(TM) i7-2600 CPU, and 8GB of RAM.

5.2 Results on Event Identification

In our event identification algorithm, we have three parameters θ_t, θ_l and θ_c to tune. Our preliminary results on textual content only show that a content distance smaller than 0.8 can find high quality events from the perspective of semantics. Therefore, we fix $\theta_c = 0.2$ and tune θ_t and θ_l in this experiment. Different users may have interest in different event types. Those who focus on large global events may provide larger thresholds while others who care about local events may provide relatively small thresholds. Here we aim to tune the thresholds such that events can be identified as many as possible, based on the given ground truth.

Figure 2 (a)-(c) shows the effects of θ_t and θ_l on event coverage without spatial or temporal expansion, and Figure 2 (d)-(f) shows the results with spatial and temporal expansions. It is noticed that events in London and New York typically span shorter time period and smaller location range than events in Brisbane. One reason is that there are less photos in Brisbane dataset which also have a longer time range than the other two datasets. Thus, there are less events which potentially last longer in larger area in Brisbane dataset. Comparing (a)-(c) with (d)-(f) in Figure 2, it is clear that the identification algorithm with spatial and temporal expansions improves the base algorithm significantly on all three datasets. This confirms that proper expansions can better capture the event in both spatial and temporal dimensions. Based on the results in Figure 2 (d)-(f), for London dataset, we set θ_t=2 hours and θ_l=1 km. For New York dataset, we set θ_t=1 hour and θ_l=2 km. As for Brisbane dataset, we set θ_t=4 hours and θ_l=4 km.

The comparison of average coverage among different methods is illustrated in Figure 3. We select top 10, 30, 50 largest events from ground truth and compare them with generated events from three methods. As can be seen, our method outperforms other three methods consistently. Incremental clustering in [2] applies supervised learning which relies on the prior knowledge of ground truth. However, it is still outperformed by our method. One main reason we believe is that the enforced similarity combination from three different features using a voting scheme may not work very well. Three features have completely different properties. It is not intuitive to combine them in a sensible way. However, our method deals with different features individually by providing different thresholds to them, according to their own properties. This shows that it might not be necessary to combine the evidences from different features as a single indicator, when they are not comparable at all. Wavelet-based clustering in [4] analyse spatial and temporal distribution of tag usage to identify event. Since the periodic events do not match the definition of events in our method, it has relatively lower coverage value.

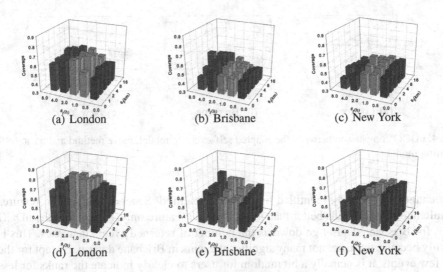

(a) London   (b) Brisbane   (c) New York

(d) London   (e) Brisbane   (f) New York

Fig. 2. Event coverage under different thresholds without expansion ((a)-(c)) v.s. with expansion ((d)-(f))

Graph clustering in [14] performs worst since the information on location and time is not fully utilized in recognizing events.

5.3   Results on Event Ranking

Effect of Event Relationship Generation. To evaluate the effectiveness of the adapted self-exciting model, we design a naive algorithm which does not consider the probability that an event occurs independently. We also apply the basic self-exciting model to each individual feature respectively to generate relationship for comparison. The global ranking results of the adapted self-exciting model, the naive algorithm and basic self-exciting model are compared in Figure 4, where the x-axis indicates the number of top ranked results and y-axis indicates the nDCG value. We can see that the adapted self-exciting model outperforms the naive method and basic self-exciting models significantly, which implies that first, there are many isolated events and initiative events

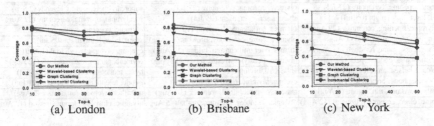

(a) London   (b) Brisbane   (c) New York

Fig. 3. Average coverage comparison among different event identification methods

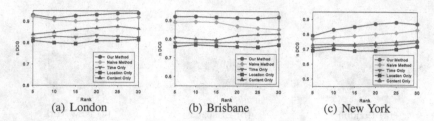

(a) London   (b) Brisbane   (c) New York

Fig. 4. nDCG comparison between the adapted self-exciting model, naive method and basic self-exciting model

that cannot be properly identified by the naive method. Second, using three features simultaneously performs better than using a single feature only. Notice that the nDCG lines for Brisbane dataset go down as the number of returned events increases. This is mainly because there are not many significant events in Brisbane dataset. Except for the top few events, it is actually a bit random for users to clearly indicate the ranks for less important events.

Effect of Time Decay. In the last experiment, we also look at the effect of time decay in ranking events. As it is really difficult to obtain ground truth for event ranking at different time stamps, we use a real example generated by our method to show the clear differences for event ranking with time decay in Table 1. Each event is represented by a collection of most representative tags and occurring date. Each column of the table contains a list of top ten ranked events at different time points. The events in the first column are ranked by their global impact, while the events in the other two columns are ranked by their local impact on 13 Aug and 11 Sep respectively. As we can see, Olympic ceremony is ranked the first based on the global impact (the first column). However, when September arrives, it is ranked the third (the third column), due to the time decay. This makes sense since people start to look at other important events such as supercar race and the Paralympic Games opening ceremony after Olympic Games all finish.

Table 1. Time decay illustration of London dataset

Ranked by Global Impact	Ranked by Local Impact on 13/08/12	Ranked by Local Impact on 11/09/12
olympic, opening, ceremony (27 Jul)	olympic, opening, ceremony (27 Jul)	supercar, chelsea, auto, legends (2 Sep)
shard, laser, light, night (05 Jul)	shard, laser, light, night (05 Jul)	paralympic, opening, ceremony (29 Aug)
paralympic, opening, ceremony (29 Aug)	olympic, park, game, handball (12 Aug)	olympic, opening, ceremony (27 Jul)
cycling, olympic, road, trial (1 Aug)	olympic, close, ceremony, firework(12 Aug)	shard, laser, light, night (05 Jul)
olympic, park, game, handball (12 Aug)	thames, river, millennium, bridge (12 Aug)	cycling, olympic, road, trial (1 Aug)
hyde, madonna, concert, park (27 Jul)	olympic, freestyle, wrestle, medal (12 Aug)	olympic, park, game, handball (12 Aug)
supercar, chelsea, auto, legends (2 Sep)	volleyball, olympic, beach (12 Aug)	hyde, madonna, concert, park (27 Jul)
public, telephone, britain, icon (21 Jun)	cycling, olympic, road, trial (1 Aug)	public, telephone, britain, icon (21 Jun)
marathon, women, run, square (5 Aug)	mascot, olympic, stroll, wenlock (10 Aug)	marathon, women, run, square (5 Aug)
olympic, close, ceremony, firework(12 Aug)	olympic, wembley, stadium, football (5 Aug)	olympic, close, ceremony, firework(12 Aug)

6   Related Work

Event detection has been widely studied in rencently. In early time, events are extracted from news documents through textual similarity, where text is the major feature used in the detection. As one of the earliest work on news event detection, [1] deals with a stream of news stories to detect new event. As event detection is considered as part of topic detection and tracking (TDT) problem, many topic models such as latent Dirichlet allocation (LDA) [3] and probabilistic latent semantic analysis (pLSA) [8] are naturally applied to discover the latent topic structures embedded in the document collection, based on which a number of events related to different topics are identified. By monitoring the topic changes over time, the event evolvement is assumed to be captured [7,9,12].

With the extensive use of social media data recently, non-textual features such as geo-tags, time, visual content has been involved in event detection. In [20], a location-driven model and text-driven model are derived to detect location-based activities from Flickr images. [2] studies different similarity metrics for both textual and non-textual features contained by the social media data to facilitate effective event clustering. A variety of techniques for learning multi-feature similarity metrics for social media document are explored. Utilizing the tag's usage distribution to detect events is studied in [15,4]. Both of them deal with Flickr photos. [4] analyzes the temporal and locational distributions of tag usage through wavelet-based spatial analysis. Tags with similar spatio-temporal usage distributions are clustered and assumed to represent an event. [15] uses GPS data from Flickr images to extract event and place semantics. [17] puts forward a model-based framework GeoFolk which combines textual feature with geographical attributes, to improve content classification and clustering. [20] introduces a text-location joint model called latent geographical topic analysis by combining a probabilistic topic model with a Gaussian mixture model to detect geographical topics as well as to estimate distribution of locations for topic comparison.

Event relationship analysis has been widely studied in IR community. In [5], dynamic connections among entities are discovered from an event which can be further consolidated from the discovered entity dynamic relationship. In [6], it is proposed to trace paths of diffusion and influence through networks and then infer the networks over which contagions propagate. In this way the optimal network that best explains the observed infection can be identified. In [11], a framework is developed for tracking short, distinctive phrases that travel relatively intact through on-line text, coupled with a scalable algorithm to identify and cluster textual variants of such phrases. All these works focus on relationship analysis among different entities to consolidate or trace event. However, our work considers the triggering and self-triggering relationships among events simultaneously and utilizes such relationships to reinforce each other's impact in event ranking.

7   Conclusions

In this paper, we propose a novel framework for event ranking. Most of the current work focus on event detection and do not consider the event relationships and the impact of events to the society. Our approach tackles these issues by effectively analyzing the event relationships with an adapted self-exciting point process model and ranking events with random walk. Our experimental results verify the effectiveness of our method. In future, we plan to study sophisticated event representation models and introduce scalable algorithms to detect events from large-scale datasets.

Acknowledgement. This research is partially supported by National 863 High-tech Program (Grant No. 2012AA011001) and the Australian Research Council (DP120102829 and DP110103871).

References

1. Allan, J., Papka, R., Lavrenko, V.: On-line new event detection and tracking. In: SIGIR, pp. 37–45 (1998)
2. Becker, H., Naaman, M., Gravano, L.: Learning similarity metrics for event identification in social media. In: WSDM, pp. 291–300 (2010)
3. Blei, D.M., Ng, A.Y., Jordan, M.I.: Latent dirichlet allocation. JMLR 3, 993–1022 (2003)
4. Chen, L., Roy, A.: Event detection from flickr data through wavelet-based spatial analysis. In: CIKM, pp. 523–532 (2009)
5. Sarma, A.D., Jain, A., Yu, C.: Dynamic relationship and event discovery. In: WSDM, pp. 207–216 (2011)
6. Rodriguez, M.G., Leskovec, J., Krause, A.: Inferring networks of diffusion and influence. In: KDD, pp. 1019–1028 (2010)
7. Ha-Thuc, V., Mejova, Y., Harris, C., Srinivasan, P.: A relevance-based topic model for news event tracking. In: SIGIR, pp. 764–765 (2009)
8. Hofmann, T.: Probabilistic latent semantic indexing. In: SIGIR, pp. 50–57 (1999)
9. Hong, L., Yin, D., Guo, J., Davison, B.D.: Tracking trends: incorporating term volume into temporal topic models. In: KDD, pp. 484–492 (2011)
10. Järvelin, K., Kekäläinen, J.: Cumulated gain-based evaluation of ir techniques. TOIS 20(4), 422–446 (2002)
11. Leskovec, J., Backstrom, L., Kleinberg, J.: Meme-tracking and the dynamics of the news cycle. In: KDD, pp. 497–506 (2009)
12. Lin, C.X., Zhao, B., Mei, Q., Han, J.: Pet: a statistical model for popular events tracking in social communities. In: KDD, pp. 929–938 (2010)
13. Mohler, G.O., Short, M.B., Brantingham, P.J., Schoenberg, F.P., Tita, G.E.: Self-exciting point process modeling of crime. JASA 106(493), 100–108 (2011)
14. Papadopoulos, S., Zigkolis, C., Kompatsiaris, Y., Vakali, A.: Cluster-Based Landmark and Event Detection for Tagged Photo Collections. IEEE MultiMedia 18, 52–63 (2011)
15. Rattenbury, T., Good, N., Naaman, M.: Towards automatic extraction of event and place semantics from flickr tags. In: SIGIR, pp. 103–110 (2007)
16. Sakaki, T., Okazaki, M., Matsuo, Y.: Earthquake shakes twitter users: real-time event detection by social sensors. In: WWW, pp. 851–860 (2010)
17. Sizov, S.: Geofolk: latent spatial semantics in web 2.0 social media. In: WSDM, pp. 281–290 (2010)
18. Xu, X., Yuruk, N., Feng, Z., Schweiger, T.A.J.: Scan: a structural clustering algorithm for networks. In: KDD, pp. 824–833. ACM (2007)
19. Yang, Y., Yang, Y., Shen, H.T.: Effective transfer tagging from image to video. TOMCCAP 9(2), 14 (2013)
20. Yin, Z., Cao, L., Han, J., Zhai, C., Huang, T.S.: Geographical topic discovery and comparison. In: WWW, pp. 247–256 (2011)
21. Zhuang, J., Ogata, Y., Vere-Jones, D.: Stochastic declustering of space-time earthquake occurrences. JASA 97(458), 369–380 (2002)

Processing Ubiquitous Personal Event Streams to Provide User-Controlled Support*

Jeremy Debattista, Simon Scerri, Ismael Rivera, and Siegfried Handschuh

Digital Enterprise Research Institute, National University of Ireland, Galway
firstname.lastname@deri.org

Abstract. The increase in use of smart devices nowadays provides us with a lot of personal data and context information. In this paper we describe an approach which allows users to define and register rules based on their personal data activities in an event processor, which continuously listens to perceived context data and triggers any satisfied rules. We describe the Rule Management Ontology (DRMO) as a means to define rules using a standard format, whilst providing a scalable solution in the form of a Rule Network Event Processor which detects and analyses events, triggering rules which are satisfied. Following an evaluation of the network v.s. a simplistic sequential approach, we justify a trade-off between initialisation time and processing time.

1 Introduction

The increase in the use of smart mobile devices provides us with ample data regarding the users' surroundings, activities and information. This data can be collected from various applications, embedded physical sensors (such as GPS), and users' online presence. As this information is heterogeneous in nature, it cannot be readily unified under a common domain model. This limits the potential of having smart devices operating on this combined user data, in order to provide added value. If this limitation is addressed, smart devices can become increasingly aware of a user's activities, situations and habits. As a result, daily repetitive tasks (e.g. changing the mobile mode to silent when arriving at work) can be automated or suggested to the user.

The di.me[1] project addresses the above limitation by unifying the user's personal information across various heterogeneous sources, including social networks and personal devices, into one standardised data representation format. di.me restricts the working Knowledge Base (KB) to cover a personal closed-world environment; introducing and extending ontologies modeling the user's Personal Information Model (PIM). Thus we represent both information *about me*[2] (such as current location, nearby persons, live

---

* This work is supported in part by the European Commission under the Seventh Framework Program FP7/2007-2013 (*digital.me* – ICT-257787) and in part by Science Foundation Ireland under Grant No. SFI/08/CE/I1380 (*Líon-2*).

[1] http://www.dime-project.eu

[2] Term coined by David Karger in the blog: http://groups.csail.mit.edu/haystack/blog/2012/02/17/personal-information-management-is-not-personal-information-management/

X. Lin et al. (Eds.): WISE 2013, Part II, LNCS 8181, pp. 375–384, 2013.
© Springer-Verlag Berlin Heidelberg 2013

posts, etc) and *for me*[2] (such as calendar, emails, contacts, etc), rather than just the latter, as covered by the Social Semantic Desktop [8]. The availability of the di.me PIM enables us to tap this personal data and allow users to define declarative rules.

In di.me, we develop a scalable solution that assists the user with daily repetitive tasks within a personal information sphere. This paper's objectives are to:

1. Enable the declarative representation of event patterns and associated actions (i.e. rules), so as to enable both human users and their machines to make sense of them;
2. Create an *Event Processor* that interprets declarative rules, detects events and triggers the desired actions;
3. Evaluate the performance and scalability of the *Event Processor* implemented.

The rest of this paper is organised as follows: Section 2 compares related work; Section 3 and 4 focus on the DRMO ontology and Event Processor respectively; whereas Section 5 presents the process and results of our evaluation. Future work will be discussed in the concluding remarks in Section 6.

2 Related Work

Every solution which implements event processing requires an underlying rule-language. Systems such as [2] propose rule languages which are specific to their particular framework. One major issue in such approaches is that the modelled rules cannot be easily reused on other frameworks. RuleML[3] is an XML markup language which allows rules to be defined using a formal notation. Since our user-defined rules are dependant on PIM data, the use of the Resource Description Framework (RDF) to model rules was a natural choice. In contrast to XML-based frameworks, RDF helps us achieve semantic interoperability more easy [4]. Since in di.me we strive to have one unified model to represent the user's KB, XML based rules would need to be transformed into RDF prior to being stored in the user's PIM. Another advantage of RDF over XML is that known knowledge can be reused in rule instances having the same semantics, thus for example saved rule conditions can be reused in other rule instances.

Complex Event Processing (CEP) and Rule-based systems are commonly used to allow rules to define how perceived data (or events) is processed. Whilst CEP (using the Event-Condition-Action pattern) rules are only triggered by specific events and do not need advanced matching techniques such as RETE, production rules are constantly being monitored for activation using such algorithms. The DRMO vocabulary allows users to express rules whose conditions can be expressed in a sequential fashion using the *succeeded by* and *preceded by* operators, though rules are not necessarily activated via specific events. Therefore DRMO rules can be classified as production rules with the added value of temporal constraints. Our approach in creating an event processor is to exploit CEP properties that enable the use of temporal constraints in rules and perceive data from multiple sources, whilst having a rule processing algorithm to filter and trigger relevant rules. Traditionally, the Rete algorithm [5] is used in rule-based systems to match facts with rule patterns. This algorithm was extended by various efforts

---

[3] http://ruleml.org

including in [9], where the authors tackle the missing temporal aspect to support complex event detection. One main problem of the original Rete Algorithm is that values in the network should be explicit, where non-constants and variables requiring further querying in a KB are not allowed. In this paper we create a *Rule Network* based on the idea of the Rete network, supporting the two mentioned shortcomings; temporal aspect, and allowing implicit values in the tree. Unlike Rete, the proposed *Rule Network* will be used to efficiently filter rules, rather than to match conditions with nodes.

3   di.me Rule Management Ontology (DRMO)

The Rule Management Ontology (Figure 1) concepts are inspired by the *Event-Condition-Action* (ECA) pattern. The ECA pattern is used in event-driven architectures, where the event specifies what triggers the rule, what conditions to specify, and what actions are executed, unlike in our case where DRMO rules do not expect any transaction event signals (such as "on update" or "on delete") to invoke the rule. The DRMO is based on the following rule pattern:

$$if\ R \implies \{a_1, .., a_n\}; R = \{c_1, ..., c_m\}; C_m = \{c_{m_1}, ..., c_{m_n}\} \tag{1}$$

where R represents a *rule* that consists of a combination of *conditions* c_m, triggering one or more resulting *actions* a_n. c_m consists of a number of constraints c_{m_n}, possibly recursive. In theory, constraints can be applied recursively, thus allowing an infinite number of embedded conditions. For practical reasons, we expect applications of DRMO to set a limit on the amount of embedded conditions (more on this later). We refer the reader to [3] or the schema online[4] for a full description of the ontology.

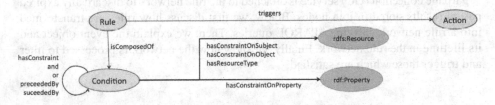

Fig. 1. An extract of the Rule Management Ontology[5]

3.1   Defining Rules

Figure 2 demonstrates how a user can define a rule using the intelligent di.me User Interface(UI) [7]. The defined rule consists of three adjoined conditions (a specific user-created Situation, a Contact and an Image) and an action (share image). The latter two conditions have been constrained. The full rule can be described as: "When in Situation 'going out' and in the vicinity of any Contact(s) belonging to a Group 'Friends' and a new Image is created then: Share the Image with that Contact(s)". This functionality

---

[4] http://www.semanticdesktop.org/ontologies/drmo/
[5] The full visualisation and description of the DRMO ontology can be found online:
  http://www.semanticdesktop.org/ontologies/drmo/

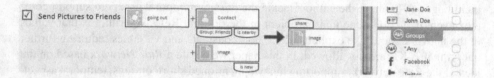

Fig. 2. Creating rules using the Rule Manager in di.me userware

is similar to that used in IFTTT[6] services, where users can define rules in terms of condition blocks and actions. JSON[7] is used as a communication interface between the DRMO instances stored in the PIM and the UI.

4 Rule Network Event Processor

For the Rule Network Event Processor, a network is generated based on the idea of Forgy's Rete network [5], where DRMO rule instances are registered in the event processor, forming a network structure. Rule instances are also transformed into SPARQL queries, each of which is stored in a level above the network's terminal node. An algorithm based on the K-Shortest Path [10] is proposed to filter and trigger candidate rules from the network. The system architecture is illustrated in Figure 3. Different sources send data about the user's activities, surroundings and information as registered in the PIM. A broadcaster broadcasts the new events to the registered processes (the rule network), which is then processed to trigger any rules relevant to the newly perceived event. A garbage collection (GC) service is attached to the rule network to discard any expired partial results stored in join nodes. Below, we first discuss how rules are transformed into a rule network carrying SPARQL queries. Then, we explain the event object and its lifetime in the rule network. Finally we show how the network is processed to filter and trigger those which are satisfied.

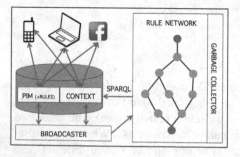

Fig. 3. General System Architecture

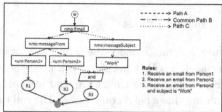

Fig. 4. A Rule Network Example

---

[6] http://ifttt.com
[7] http://www.json.org

4.1   Building the Network

In our rule network we define six different nodes. The *Root* node is the one with no incoming vertices and the *Terminal* node is the one with no outgoing vertices. Figure 4 is an example of a rule network.

Each rule instance registered to the event processor is fed through the root node of the network to start the process of adding rules to the network and transforming them into SPARQL queries. A rule instance is first broken down into a number of conditions defined by the property *drmo:isComposedOf*. These conditions and their constraints are attached to the network's root as nodes, starting with the *Object-Type* node (in Figure 4, *nmo:Email*) and finishing with the *Rule* node (represented as circle nodes : R1, R2, R3), storing the transformed SPARQL query. Then, for each of condition's constraints, we check the *drmo:hasConstraintOnProperty* and add the node to the network (attached to the previously added *Object-Type* node), followed by the associated constraint value (subject/object) node. These last two mentioned nodes are *Triple-type* nodes (shown as rectangles in Figure 4) and store the triple patterns required to create the final query.

Two conditions (in a rule) can be joined together using the *Join* node, apart from those linked together with a *drmo:or* operator. In the latter case, the event processor handles the *drmo:or* joined conditions as two distinct rules triggering the same action since these are independent from each other (see all supported operators in [3]). An advantage of this decision is that for such queries the response time is decreased, as we are adding less triple patterns to the queries. A *Join* node refers to two ordered inputs. For each input we store the intermediate query and intermediate result. The advantage of storing intermediate results is that during processing they enable us to check if a rule can be triggered based on the intermediate result. Intermediate queries are used in order to avoid using SPARQL FILTERS to compare the event's occurrence time for rule conditions joined by the *drmo:succeededBy* and *drmo:precededBy* operators. Another advantage of storing intermediate results is that unlike the sequential processor defined in [3], there is no need for an *Event Log* to keep track of the perceived events. Similar to [9] the ordered inputs are used to check the temporal constraints of a rule. If the join node is a *succeededBy* node, then the left input must be satisfied first in order to trigger the right input (and vice-versa for a *preceededBy* node, where the right input is the first input to be satisfied). There is no particular order for *and* join nodes.

The rule transformation ends with the *Rule* node, in which the transformed SPARQL query [3] and the action instances are stored. SPARQL allows the fast querying of the perceived events stored in the user's KB. These queries also help in filling the "blank" in a rule which does not have an explicit value, for example a rule "Receive an email from <urn:Person1>" would require a further query to the PIM to get all the email addresses linked to <urn:Person1>, since the email event metadata would contain a raw email address (e.g. "person1@email.com") rather than the URI defined in the user's PIM.

4.2   Events and Validity

Events perceived by the sensors and stored in the user's KB (PIM), can trigger any user-defined rule. When events are perceived (this could be a created, modified or deleted resource in the PIM), semantic lifting is performed on relevant graphs in order to store the metadata for the event perceived. The event's resource type together with a times-tamp, a pointer to the graph this resource is stored in, and the event operation (e.g. Resource Modified), is then broadcasted to the event processor.

The lifetime of an event is difficult to predict since different event types might have different time-spans when correlated to other events in rules. In order for the rule "If I receive an email succeededby a new document created" to trigger, two events (receive an email) and (new document created) need to occur after each other. This could lead the rule to trigger at an indefinite time if the first event does not expire after a certain time. On the other hand, premature event expiry might also lead to relevant rules being missed. In our system we employ the *consumption mode* technique, *maximum event lifetime* technique, and the *time-based windows* technique, similar to how these are described in [9]. The first technique is used to keep the most recent resource URI in a join node. On the other hand the *maximum event lifetime* expires resource URIs in join nodes after an X amount of time, irrelevant of the resource type, whilst for the last technique, these expire according to a time period assigned to the respective resource type. Events defining a context state [1] (e.g. current availability) are automatically invalidated and removed from the join nodes according to the changed state.

4.3 Processing the Network Algorithm

Algorithm 1. Processing Rule Network to trigger rule(s) from detected events

Data: Perceived Event E; Rule Network N; ResultSet S
get resource type T for event E ;
$P_L \leftarrow$ ksp(N,T) ;
from P_L find common subpaths C_p ;
while $\exists$ *path P in P_L* **do**
  if *path P contains join node J* **then**
    inputPath $\leftarrow$ get left or right input of P in J ;
    if J = *AndNode* **then**
      inputPath.result $\leftarrow$ execute(inputPath.query) ;
    checkEventJoinNode(N, inputPath, J) ;
    continue iteration ;
  else
    if *path execute(P.rule)* $\neq$ *null* **then**
      remove all paths in P_L where P is not a subpath (C_p) ;
      S.add(P.rule)
    remove all paths in P_L where P is a subpath (C_p) ;

Procedure checkEventJoinNode(N,P,J)

Input: Rule Network N; InputPath P; JoinNode J
if J = *SucceededByNode* **then**
  if *inputPath is the left input* **then**
    inputPath.result $\leftarrow$ execute(inputPath.query) ;
  else
    if *left input result is not empty* **then**
      if *execute(inputPath.query)* $\neq$ *null* **then**
        S.add(P.rule) ;

if J = *PrecededByNode* **then**
  if *inputPath is the right input* **then**
    inputPath.result $\leftarrow$ execute(inputPath.query) ;
  else
    if *right input result is not empty* **then**
      if *execute(inputPath.query)* $\neq$ *null* **then**
        S.add(P.rule) ;

In contrast to the Rete Algorithm [5], the improved processor we introduce does not process the network by matching patterns, but aims to efficiently filter and find candidate rules. This is done by forming a subgraph, with the perceived resource type as the root node. Algorithm 1 shows how the processing of the network is done. Candidate rules are filtered (P_L) using Yen's K-Shortest Path algorithm [10]. Common subpaths (C_p) are also discovered during the execution of the k-shortest path algorithm. This has the benefit of reducing the number of queries performed to check for potential rules to be triggered. Once a set of paths is ordered, these are iterated and the query in the *rule* node is executed. Rules whose queries return a result (indicating a rule triggered) are stored in a set which is then passed to the action executor. The rule matcher runs its matching process until all paths (in P_L) are checked. At worst, the matcher iterates on all candidate paths found by the KSP algorithm. When a rule is matched, all other paths which do not have the same condition as the rule in question, are removed and not checked by the matcher. This approach is due to the non-repetitive nature of the network, where rules with the same condition are represented by the same branch in the network.

5 Evaluation

Our evaluation serves two purposes: to compare the efficiency of the new network-based approach to the sequential approach proposed earlier [3], and to investigate its scalability. We base our investigation on a study on the performance of event processing systems [6], which focus on the bottlenecks and the degrading of performance as the load is increased. The authors suggest benchmarking various factors, including the selection and pattern detection. The key performance aspects to prove the scalability of the proposed event processor are:

- The initialisation of the Rule Network;
- The selection of rules on event detection;
- The maximum load of events the event processor can handle.

For these tests, we require a knowledge base (KB), DRMO rule instances and event data. As a KB, we use a sample di.me PIM containing data based on various OSCAF[8] ontologies. Test rules were manually created. In the context of di.me, we limit the number of recursive conditions (defined in Section 3 - equation (1)) to 2. For these evaluation tests no repeated rules were used so that the Rule Network event processor is not given an advantage over the Sequential. We also propose the following limits: 50 different rules each having 5 conditions, based on the sample PIM and following a small user study. For the generation of the event data we developed an event generator that generates realistic[9] daily events based on the given PIM and scenarios. The evaluation tests have been carried out on a Macbook Pro, Intel Core i5 with 2.4Ghz processor speed and 4GB of RAM. Data is stored in-memory using Sesame[10].

---

[8] http://www.oscaf.org
[9] All data and evaluation results can be found:
   http://smile.deri.ie/Rule-Network-Evaluation
[10] http://www.openrdf.org

5.1 Initialisation Test

The aim of this test is to measure the time taken to register rule instances in both event processors (sequential vs rule network). We initialise our event processors ten times each, starting with five rules and progressively incrementing the number of loaded rules by five. Figure 5 shows the time taken (in milliseconds), against the number of rules loaded in the event processors with each initialisation. The graph shows that the behaviour of the time consumed in relation to the number of rules loaded in the system is linear. The rule network event processor consistently took more time to initialise than the sequential event processor, due to the extra overhead needed to update the network.

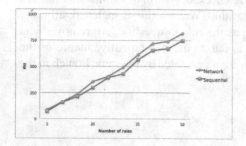
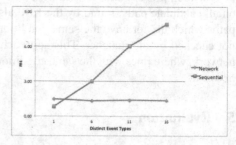

Fig. 5. Time taken to initialise Event Processors

Fig. 6. No. of distinct events perceived over time

5.2 Filtering Test

In this test we measure the time taken (in ms) by the processors to select candidate rules with each event perceived. We use data generated over 24-hours, thus assuming that events in the event log (required only for the sequential event processor) remain valid throughout this period. After loading all 50 rules in both processors, the test was repeated four times, each time producing a number of events until the desired number of distinct types were available on the event log. For the experiment we require that the filtering algorithm returns candidate rules satisfying either of the following criteria:

1. A rule with one condition where the associated resource type (*drmo:hasResourceType*) is the same as the perceived event's type;
2. A multiple-condition rule such that one of it's conditions is satisfied by the type of the new event.

The graph in Figure 6 shows how the sequential process grows in time as the number of distinct event types increase in the event log, whilst the network filtering process remains constant. This result was expected, since the sequential's filtering process needs to go through all the rules to check their resource types against the distinct event types on the event log, in order to satisfy either of the mentioned criteria. On the other hand the network is built such that the associated rules can be automatically found by forming a sub-graph, using the resource type as the start vertex. The network filtering process takes time in order to sort the possible rules by shortest path first, allowing simple rules to be evaluated before the more complex ones.

5.3 Load Testing

The aim of this experiment is to understand to which extent the event processors can work sufficiently and in an acceptable manner. For this experiment we created a producer/consumer service, where the producer sends a stream of generated events to the consumer (the event processor), which filters and triggers rules simultaneously. After both processors were loaded with the 50 rules, we ran this test five times, progressively increasing the event load up to a maximum 16000 events. We established that an acceptable time-frame for the processing of consumed events should be less than 20 seconds. Since the sequential process failed this test outright, with 100 events being processed in 92.48 seconds, it was subsequently eliminated from this experiment.

Figure 7 shows time time taken (in seconds) to perform all consumed events for the rule network event processor. In particular, we observe that 10000 events were consumed and processed in around 2 seconds. In the di.me userware, we do not foresee a user to have anything near 10000 events being perceived at the same time. More realistically, we observe that the rule network event processor can process up to 2000 events in a reasonable time (≈ 0.4 sec), with 100 events being processed in less than 0.1 sec.

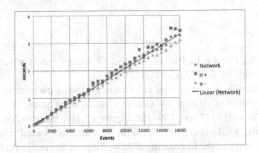

Fig. 7. Testing time taken to process events

From these results we conclude that although during the initialisation stage the Rule Network event processor takes longer than the Sequential event processor, the runtime process for filtering and triggering takes considerably less time, regardless of the load of events being consumed by the processor. The trade-off between the time taken to initialise the rule, and to process events, is justified for the following reasons:

1. The initialisation process is done only once;
2. The difference in milliseconds between both event processors is almost insignificant (to initialise 50 rules, the network takes < 100ms more);
3. The filtering of events takes considerably less time in the network event processor;
4. Even with a load of 10000 events, the rule network performs within a reasonable time-frame.

6 Conclusion

In this paper we describe an ontology-driven event processor which operates on personal activities and context; as perceived by various devices and online sources. The event processor compares event streams to the antecedent of declarative rules, defined by a user through an intelligent UI. Prospectively, a JSON-LD[11] serialisation would enable us to interchange DRMO instances, ensuring that the full semantics are retained event at the UI level. Rules consist of various conditions corresponding to items in a unified PIM, and one or more resultant actions. Due to the nature of the rule conditions and their constraints, we propose to represent them by a rule management ontology. This also means that in theory, the personalised user-defined rules can be processed by various platforms. In the future, we also intend to investigate the possibility of enabling the automatic rule learning and discovery based on the availability of a user's context history.

Defined rules are then transformed and registered in a Rule Network, which is then operated upon by the event processor. After initialising the rule network, the system is ready to start perceiving events from multiple datasources. With each new event, the rule network filters candidate rules and triggers any which are satisfied. The proposed network-based event processor is compared to an earlier sequential approach. Evaluation results show that although the rule network takes more time at the initialisation stage, it performs considerably better than the alternative, during both the filtering and triggering processes. In particular, our event processor can process 100 events in less than 0.1 seconds.

References

1. Attard, J., Scerri, S., Rivera, I., Handschuh, S.: Ontology-based situation recognition for context-aware systems. In: I-SEMANTICS (2013)
2. Beltran, V., Arabshian, K., Schulzrinne, H.: Ontology-based user-defined rules and context-aware service composition system. In: García-Castro, R., Fensel, D., Antoniou, G. (eds.) ESWC 2011. LNCS, vol. 7117, pp. 139–155. Springer, Heidelberg (2012)
3. Debattista, J., Scerri, S., Rivera, I., Handschuh, S.: Ontology-based rules for recommender systems. In: Proceedings of the International Workshop on Semantic Technologies meet Recommender Systems & Big Data (2012)
4. Decker, S., Melnik, S., van Harmelen, F., Fensel, D., Klein, M., Broekstra, J., Erdmann, M., Horrocks, I.: The semantic web: the roles of xml and rdf. IEEE Internet Computing 4(5), 63–73 (2000)
5. Forgy, C.L.: Rete: A fast algorithm for the many pattern/many object pattern match problem. Artificial Intelligence 19(1), 17–37 (1982)
6. Mendes, M.R.N., Bizarro, P., Marques, P.: A performance study of event processing systems. In: Nambiar, R., Poess, M. (eds.) TPCTC 2009. LNCS, vol. 5895, pp. 221–236. Springer, Heidelberg (2009)
7. Scerri, S., Schuller, A., Rivera, I., Attard, J., Debattista, J., Valla, M., Hermann, F., Handschuh, S.: Interacting with a context-aware personal information sharing system. In: Kurosu, M. (ed.) HCII/HCI 2013, Part V. LNCS, vol. 8008, pp. 122–131. Springer, Heidelberg (2013)
8. Sintek, M., Handschuh, S., Scerri, S., van Elst, L.: Technologies for the social semantic desktop. In: Tessaris, S., Franconi, E., Eiter, T., Gutierrez, C., Handschuh, S., Rousset, M.-C., Schmidt, R.A. (eds.) Reasoning Web. LNCS, vol. 5689, pp. 222–254. Springer, Heidelberg (2009)
9. Walzer, K., Breddin, T., Groch, M.: Relative temporal constraints in the rete algorithm for complex event detection. In: DEBS, pp. 147–155 (2008)
10. Yen, J.Y.: Finding the K Shortest Loopless Paths in a Network. Management Science (1971)

---

[11] http://json-ld.org

On Co-occurrence Pattern Discovery from Spatio-temporal Event Stream

Jiangtao Huo, Jinzeng Zhang, and Xiaofeng Meng

School of Information, Renmin University of China, Beijing, China
{huojiangtao,zajize,xfmeng}@ruc.edu.cn

Abstract. The proliferation of location-acquisition technologies and online social networks such as twitter, Foursquare, Meetup lead to huge volumes of spatio-temporal events in the form of event stream. In this study, we investigate the problem of discovering spatio-temporal co-occurrence patterns from spatio-temporal event stream (CoPES). We propose an effective sliding-window based dynamic incremental and decayed (abbreviated as DIAD) algorithm for discovering CoPES. DIAD algorithm proposes a novel decay mechanism to calculate the prevalence of CoPES and a sliding-window to process the event stream time slot by time slot to discover CoPES. The algorithm utilizes a hash tree to store the closet COPES. Then the decay mechanism and the sliding-window exploit the superimposed spatio-temporal neighbor relationships between time slots to get the accurate prevalence from event stream and discover CoPES efficiently. The experimental results on real dataset show that our proposed algorithm has superior quality and excellent expansibility.

Keywords: Co-occurrence Patterns, Event Stream, Spatio-temporal Support-prevalence, Decay Mechanism.

1 Introduction

The advances in location-acquisition devices and social networks have generated huge amount of social events data: (1) Twitter has started supporting location service in message. Based on geo-twitter message, the social events are generated in the manner of event stream. (2) Location-based online social networks such as Foursquare, Gowalla enable mobile users check-in POIs (Point of Interests) in real-time. (3) Newly emerged event-based online social networks, such as Meetup and Eventbrite have provided convenient online platforms for people to create and organize social events. Therefore, the social events generated can be regards as the event stream.

CoPES discovery gives us important insights for many application domains, including ecology, public health, Earth science, electronic business and so forth. For example, spatio-temporal co-occurence patten can calibrate the urban planning of a city and contribute to the future planning. Besides, spatio-temporal co-occurence patten can also benefit hotspots choosing for a business and advertisement in differnt time domain.

X. Lin et al. (Eds.): WISE 2013, Part II, LNCS 8181, pp. 385–395, 2013.

A close work of CoPES discovery is conducted by Lee Chang-Hung[4]. He proposed a SWF algorithm [4] which uses a sliding-window to find frequent itemsets in the fixed number of recent data records. The method maintains candidate 2-itemsets of all transactions in the window separately. With the window being advanced, the oldest partition is disregarded and a new partition containing newly generated data records is appended to the window. Although this approach is available for mining frequent itemsets, but it is incompetent to find CoPES because it can't process the complex relationship between the new data records and the old data records.

In order to discover the spatio-temporal co-occurrence patterns, we propose a dynamic incremental and decayed algorithm (abbreviated as DIAD). Different from the SWF algorithm [4], we first distribute the event stream into grid partitions, then we will calculate the original spatio-temporal co-occurrence patterns. When new data records arrived, we distribute them into related grid partition, and refresh the gird record. Then we adopt a decaying technique to capture the dynamic changes of an event stream. The main contributions of this work are summarized as follows.

— We define the problem of discovering CoPES and propose a new method to measure the spatio-temporal co-occurrence patterns.
— We propose a novel algorithm DIAD to discover CoPES. The algorithm can efficiently process the relationship between the new data records and the old ones.
— Extensive experiments evaluate the efficiency of our algorithm and the results show that the proposed algorithm outperforms the baseline approach.

The remainder of the paper is organized as follows. Section 2 formally presents the related works of spatio-temporal co-location pattern mining. Section 3 gives the basic concepts and defines the problem of CoPES discovery. Section 4 describes the processing procedure of DIAD in detail. The experimental evaluation is shown in section 5. Finally, we conclude the paper in section 6.

2   Related Works

The research on spatio-temporal co-occurrence patterns can be divided into two areas, namely co-location pattern discovery and spatio-temporal co-occurrence pattern discovery.

In the spatial association mining literature, [2,3,5,6,7,11,13] proposed different approaches for mining co-location patterns. Morimoto et al. [3] first defined the problem of finding frequent neighboring co-locations in spatial databases. Huang et al. [6] proposed a general framework for apriori-gen [1] based co-location pattern mining. Based on the general approach, Yoo et al. [7,10] proposed methods that materialize neighborhoods of objects.

In the spatio-temporal co-occurrence pattern discovery area, many researchers conduct their work on spatio-temporal data with time labels [8,9,12,14]. They first discover the co-location patterns from a certain time slot. After all time slots having been processed. they calculate the temporal property, only the patterns which satisfy the temporal prevalence can be identified as a spatio-temporal co-occurrence   pattern.

3   Basic Concepts and Problem Statement

3.1   Basic Concepts

Formally, let E be a set of spatial event types, I be a set of their instances, R_S be a spatial neighbor relationship where the distance between two event instances in I is smaller than threshold θ_S.

Definition 3.1 (co-occurrence pattern). Given an event stream and a set of time slots, the **co-occurrence pattern** is a subset of spatio-temporal events whose instances form a clique using the spatio-temporal neighbor relationship.

Here, we regard the time as a dynamic parameter, we propose a novel method to measure the CoPES. The measure method is defined as follows.

Definition 3.2 (sliding-window). Given an event stream ES and a set of time slots, the **sliding-window** $SW^{i,j}$ represents the sliding-window processing the event stream arriving from time t_i to time t_j.

Definition 3.3 (original spatial support-prevalence ratio). Given an event stream and a $SW^{i,j}$, Let $CO=\{e_1,e_2,...,e_k\}$ be a co-occurrence pattern in $SW^{i,j}$ at time t_k which does not exist at time t_{k-1} (abbreviated as original-CoPES), D_i denotes the number of event instances of e_i in $SW^{i,j}$, ICO_i denotes the number of distinct event instances of e_i participating in CO. Then the **original spatial support-prevalence ratio** of event e_i in CO is defined as follows.

$$SPR_o(CO, e_i, t_k)=ICO_i/D_i \tag{1}$$

Definition 3.4 (spatio-temporal support-prevalence ratio). Given an event stream and a $SW^{i,j}$, let $CO=\{e_1,e_2,...,e_k\}$ be a co-occurrence pattern at time t_k, PD_i denotes the integrated number of event instances of e_i at time t_{k-1}, $PICO_i$ denotes the integrated number of event instances of e_i participating in CO at time t_{k-1}, CD_i denotes the number of new arrival event instances of e_i at time t_k. Let $CICO_i$ denotes the number of event instances of e_i participating in CO at time t_k. The **spatio-temporal support-prevalence ratio** of event e_i in CO at time t_k is measured as follows.

$$SPR_u(CO,e_i, t_k)=(\ \lambda PICO_i+CICO_i)\ /\ (\lambda PD_i+CD_i)\ \{0<\lambda<1\} \tag{2}$$

Specially, for the co-occurrence in definition 3.4,

$$SPR_u\ (CO,e_i,t_k)= SPR_o\ (CO,\ e_i,t_k) \tag{3}$$

Definition 3.5 (Spatio-temporal support-prevalence). Given an event stream and a set of time slots $(t_0,t_1,...)$, let $CO=\{e_1,e_2,...,e_k\}$ be a co-occurrence pattern at time t_k, the **spatio-temporal support-prevalence** of CO at t_k is the minimum of spatio-temporal support-prevalece ratio values of events in CO.

$$STSP(CO, t_k)=mim\{\ STSPR(CO,e_i, t_k)\} \tag{4}$$

Definition 3.6 (CoPES). Given an event stream and a set of time slots$(t_0, t_1, \ldots t_k \ldots)$, let μ be the spatio-temporal support-prevalence threshold, then the **CoPES** at time t_k is the co-occurrence patterns which satisfy $STSP(CO, t_k) > \mu$.

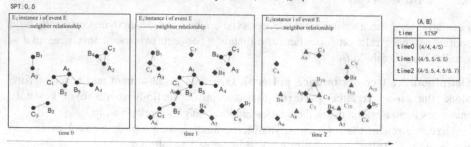

Fig. 1. An example of CoPES discovery from event stream

3.2   Problem Statement

Given:
1) A spatial framework SF
2) A time framework TF=$t_0 U \ldots .. U t_n$.
3) A set of spatio-temporal events E and a set of their instances I.
4) A spatio-temporal support-prevalence (abbreviated as STSP).
5) A spatio-temporal support-prevalence threshold SPT (defaulted as 0.5).
6) A temporal decayed factor λ over time sequence.

Develop: An algorithm to find CoPES whose STSP satisfy a given threshold.

Objective: Find a complete and correct set of CoPES on event stream.

Example: The spatio-temporal event stream given in Fig. 1 contains 3 event types, A,B and C for 3 time slots. Let the λ be 0.5. In this example, {A,B},{A,C}, {B,C} and {A,B,C} form candidates of CoPES. At time t_0, the SPR_u of {A,B} is (4/4,4/5), when it comes to time t_1,the SPR_u of {A,B} becomes (4/5,5/5.5) which integrated the previous result of time t_0. Therefore, {A,B} is a CoPES at time t_0 and t_1.

4   DIAD: Sliding-Window Based Dynamic Incremental and Decayed Mining Algorithm

Nowadays, few researchers concentrate attentions on spatio-temporal co-occurrence patterns discovery from event streams (CoPES). A straightforward method is to exploit the existing method (e.g. a density based approach in [15]) in each time window to identify the co-location patterns. This method first distribute all data records into grid partitions, then it identifies the size-2 co-occurrence patterns by processing these grids, after that, it joins the size-k co-occurrence patterns to generate the size-k+1 co-occurrence candidates and identifies them by processing the grids. However, this method has several mortal drawbacks. First, it can't differentiate the information between different time slots. Second, it can't dynamically process the new coming event stream. Therefore, in this paper, we propose a DIAD algorithm to discover CoPES. In section 4.1 we shall give a introduction about the decay mechanism of DIAD. Then, we give the main processing of algorithm DIAD in section 4.2.

4.1   Decay Mechanism for Calculating STSP

Let basic-CoPES represents all the new generated instances of CoPES at time t_k. Suppose that we have a CoPES $CO=(A,B)$ at time t_k, then the number of distinct event instances participating in basic-CoPES of CO at time t_k is $SP_k(A,B)= (Xk_{AB}, Yk_{AB})$, and the number of new arrival event instances of each event type at time t_k is $TN_k(A,B)=(Nk_A,Nk_B)$. The integrated number of event instances participating in CO at time t_k is $ISP_k(A,B)= (IXk_{AB}, IYk_{AB})$, the integrated number of event instances of each event type at time t_k is $ITN_k(A,B)=(INk_A, INk_B)$.

Assume that we have processed 11 time slots $t_0, t_1,...,t_{10}$ and (A,B) is always a CoPES at each time slot, then the basic-CoPES state of $CO=(A,B)$ in each time slot is

t_0: (AB)-$(X0_{AB}/ N0_A, Y0_{AB}/ N0_B)$

t_1: (AB)-$(X1_{AB}/ N1_A, Y1_{AB}/ N1_B)$

......

t_{10}: (AB)-$(X10_{AB}/ N10_A, Y10_{AB}/ N10_B)$

Then the CoPES of CO at time t_{10} will be:

T_{10}: (A,B)-$(IX10_{AB}/ IN10_A, IY10_{AB}/ IN10_B)$

Then we have the following relationship.

$IX10_{AB}=X10_{AB}+\lambda X9_{AB}+...+\lambda^{10}X0_{AB}$ $IY10_{AB}=Y10_{AB}+\lambda Y9_{AB}+...+\lambda^{10}Y0_{AB}$

$IN10_A= N10_A+\lambda N9_A+...+\lambda^{10}N0_A$ $IN10_B=$
$N10_B+\lambda N9_B+\lambda^2 N8_B+...+\lambda^{10}N0_B$

Therefore, the STSP of CO at time t_{10} is

$$STSP(CO, t_{10})=min(IX10_{AB} /IN10_A, IY10_{AB} /IN10_B) \qquad (5)$$

Here, the above method needs us to store all previous basic-CoPES for processing, and this is highly cost both on time and memory. Through our research work, we find that the above procedure can be simplified as one step.

Lemma 1. Given a spatio-temporal co-occurrence pattern $CO=(A,B)$ at time t_k, then the *STSP* of CO at time t_k is

$$STSP(CO, t_k)=min((Xk_{AB}+\lambda IXk\text{-}1_{AB})/(Nk_A+\lambda INk\text{-}1_A),$$

$$(Yk_{AB}+\lambda IYk\text{-}1_{AB})/(Nk_B+ \lambda INk\text{-}1_B)) \qquad (6)$$

Proof.

Let k represents the number of time slot.

(1) If k=0, then

t_0: (A,B)-$(X0_{AB}/N0_A, Y0_{AB}/ N0_B)$, $T_0(A,B)$-$(IX0_{AB}/IN0_A, IY0_{AB}/IN0_B)$

we have, $(X0_{AB}, Y0_{AB})= (IX0_{AB}, IY0_{AB})$, $(N0_A, N0_B)= (IN0_A, IN0_B)$

(2) If k>1, then

t_k: (A,B)-(Xk_{AB}/Nk_A, Yk_{AB}/Nk_B)

T_k:(A,B)-(IXk_{AB}/INk_A, IYk_{AB}/INk_B)

We have

$IXk_{AB}/ INk_A = (Xk_{AB}+\lambda Xk\text{-}1_{AB}+...+\lambda^k X0_{AB})/(Nk_A+\lambda Nk\text{-}1_A+...+\lambda^k N0_A)$

$IYk_{AB}/ INk_B= (Yk_{AB}+\lambda Yk\text{-}1_{AB}+...+\lambda^k Y0_{AB})/(Nk_B+\lambda Nk\text{-}1_B+...+\lambda^k N0_B)$

Then,

t_{k+1}: (A,B)-($Xk+1_{AB}/Nk+1_A$,$Yk+1_{AB}/Nk+1_B$)

T_{k+1}(A,B)- ($IXk+1_{AB}/Nk+1_A$,$IYk+1_{AB}/Nk+1_B$)

Then,

$IXk+1_{AB}/ INk+1_A =(Xk+1_{AB}+\lambda Xk_{AB}+...+\lambda^{k+1} X0_{AB})/(Nk+1_A+\lambda Nk_A+...+\lambda^{k+1} N0_A)$
$$=(Xk+1_{AB}+\lambda IXk_{AB})/(Nk+1_A+\lambda INk_A)$$

Similarly, $IYk+1_{AB}/ INk+1_B =(Yk+1_{AB}+\lambda IYk_{AB})/(Nk+1_B+\lambda INk_B)$

Therefore, $STSP(CO, t_{k+1})=\min((Xk+1_{AB}+\lambda IXk_{AB})/(Nk+1_A+\lambda INk_A),$
$$(Yk+1_{AB}+\lambda IYk_{AB})/(Nk+1_B+ \lambda INk_B))$$

According to **Lemma 1**, we can iteratively calculate the STSP at time t_k just by utilizing the STSP at time t_{k-1} and this method can drastically improve the efficiency.

4.2   The Processing Procedure for DIAD

Here, we first explain the meanings of symbols we used in our research in Table 1.

Table 1. The meanings of symbols

Symbols	meaning
P-CoPES	Previous integrated CoPES
E-CoPES	The basic-CoPES which is exist in P-CoPES
O-CoPES	The original-CoPES which is not exist in P-CoPES
$CoPES_k$	A set of size-k spatio-temporal co-occurrence patterns
AGP	Grid partitions which have received new data records
COC_k	A set of size-k candidates

The DIAD algorithm adopts different strategy to handle the event stream arriving at different time slot. For simplicity, we set the number of sliding-window to be two. The CoPES discovery can be decomposed into two steps:

- Preprocessing procedure: It deals with mining on the data sets of first time slot.
- Incremental mining procedure: It deals with the update of the CoPES for an ongoing time-variant event streams.

4.2.1   Preprocessing Procedure

The preprocessing procedure is only utilized for the initial mining of CoPES in the first time slot.

Algorithm 1. Preprocessing procedure of Algorithm DIAD
1.   n is the number of grid partitions, k is the spatio-temporal co-occurrence size
2.   while data records at time t_0 not end
3.   distribute the data records into grid partitions
4. $CoPES_1 = E, k = 1, P\text{-}CoPES \neq \varphi$
5.   while $CoPES_k \neq \varphi$ do
6.   generate size-k+1 candidates COC_{k+1}
7.   identify size-k+1 $CoPES_{k+1}$ in each AGP
8.   add the $CoPES_{k+1}$ into P-CoPES
9.   k=k+1
10. return CoPES;

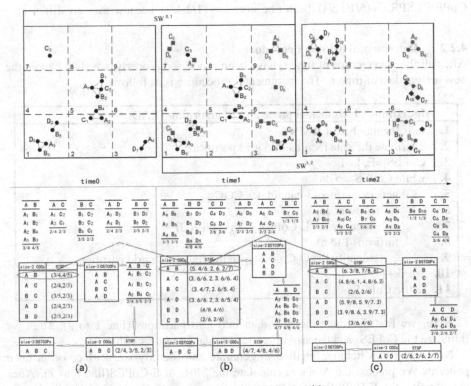

Fig. 2. An example of a DIAD algorithm

In **Algorithm 1**, we first hash the data records arriving at time t_0 into grid partitions (line 1-3). Then, we initialize the $CoPES_1$, k, P-CoPES (line 4).

After that, we identify all CoPES at time t_0 (line 5-10). We iteratively use the $CoPES_k$ to generate COC_{k+1}(line 6), and then we process all the AGPs to identify the STSP of $CoPES_{k+1}$ (line 7). Besides, we add the identified CoPES into P-CoPES for the incremental mining procedure (line 8).

Example: In Fig. 2(a), we first hash the data records arriving at time t_0 into grid partitions. After distribution, the grids-{1,2,3,5,7} have received data records. Then what we need to do is processing these girds to identify the size-2 co-occurrence candidates—{(A,B),(A,C),(A,D),(B,C),(B,D),(C,D)}. We can see that {(A,B),(A,C), (B, C),(A,D)} have enough co-occurrence instances, and the STSP of them satisfy the SPT(0.5), therefore, they are identified as CoPES$_2$.

After that, we will use the size-2 CoPES to generate the size-3 candidates. As only (A,B,C) has all size-2 subset {(A,B),(A,C), (B,C)}, then only (A,B,C) can be generated as a size-3 candidate. Here, the STSP(A,B,C,t_0) is min(2/4,3/5,2/5) =2/4, which satisfy the SPT(0.5). Then (A,B,C) shall be identified as CoPES$_3$.

After we identify all the CoPES, we add them into P-CoPES for future work. It is noted that we not only store the CoPES, but also store their SPR$_u$, it means that if a CoPES's SPR$_u$ is (5/10,2/4) it can't be stored as (1/2,1/2), because they are different.

4.2.2 Incremental Mining Procedure

After all the data records of time t_0 have been processed properly, we shall handle the new arrival data of time t_1. The incremental procedure is as follows.

Algorithm 2. Incremental procedure of algorithm DIAD
1.   m is the number of AGPs
2.   distribute the data records into grid partitions
3.   CoPES$_1$=E, k=1
4.   while CoPES$_k$≠φ do
5.   generate size-k+1 candidates COC$_{k+1}$
6.   for i=1 to m do
7.   identify the E-CoPES$_{k+1}$ in each AGP
8.   for(i=m+1 to n)
9.   identify the O-CoPES$_{k+1}$ in the remaining grid partition
10.   k=k+1
11. return CoPES;

Here, we first distribute the new data into grid partitions (line 1 to 2). Then we initialize the CoPES$_1$ and the size of co-occurrence pattern as 1 (line 3).

Next, we use a iterative method to generate and identify size-k+1 co-occurrence patterns. We process the AGPs to calculate the SPR$_u$ of E-CoPES(line 6 to 7). After we identify the E-CoPES, we process the remaining grid partitions to identify O-CoPES (line 8 to 9).

Example: In Fig. 2(b), we first distribute the event stream into grid partitions {1,3,6,7,9}.Then we use CoPES$_1$ to generate the size-2 co-occurrence candidates. We use two different strategy to process the different type of size-k(k≥2) candidates. If the size-k(k≥2) candidates happened in P-CoPES, we process the AGPs to identify whether they are CoPES or not, like {(A,B),(A,C),(B,C),(A,D)}. If the size-k(k≥2) candidates didn't happen in P-CoPES, we have to process all grid partitions to measure them, like {(B,D),(C,D)}.

After all $CoPES_2$ are identified, we shall identify $CoPES_3$. In **Fig. 2(b)**, the only COC_3 is (A,B,D), as (B,D) is a new identified size-2 CoPES, then (A,B,D) can't be a CoPES in the previous time slot. Then we will process all the grids to identify whether (A,B,D) is a pattern or not. After all CoPES identified, we add them into P-CoPES.

5   Experimental Evaluation

In this section, we present our experiment to compare the performance of DIAD algorithm with a baseline method proposed by X. Y. Xiao in [15]. Our experiments used a real data set—geo-twitter dataset (GTT). The GTT dataset contains profile information of 225098 users and more than 22 million check-ins from Jan. 2011 to Sep 2011 in the united States. After filtering, the dataset contains 300 event type and 1500000 event instances in total. The minimum instance number is 1856, the maximum instance number is 6843. The average instance number is 5000.

In order to simulate real data stream scenario, the r%(e.g. r=10) of check-in records in GTT can be regards as the original dataset. In each time slot, we select s% (e.g. s=5) check-in records of the rest of   dataset that are added into the original dataset in the form of data stream. Experiments were conducted on an intel Core™-2 quad 2.66 GHz computer with 2GB of RAM.

5.1   Effect of the Number of Time Slots Per Sliding-Window

In first experiment, we evaluated the effect of the number of time slots in a sliding-window on the execution time of both algorithms.

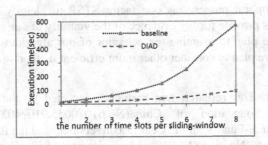

Fig. 3. Effect of the number of timeslots per sliding-window

In Fig. 3, we can see that, the DIAD algorithm is computationally more efficient than the baseline approach. It is vividly that the DIAD algorithm gets a linear growth with the number of timeslots increasing while the baseline approach is exponential growth.

5.2   Effect of the Decay Factor λ and SPT

We study the effect of the decay factor λ and the SPT on the efficiency of our algorithm. The default number of time slots per sliding window is 5.

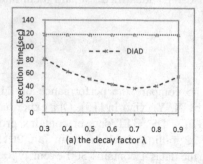 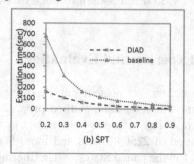

Fig. 4. (a) Effect of the decay factor λ   (b) Effect of the SPT on the execution time

In Fig. 4(a), we can see that with the decay factor increasing, the execution time of DIAD does not change linearly. The execution time first diminished quickly, however, when λ is bigger than 0.7, it bounced. In Fig. 4(b), with the SPT diminishing, the execution time of baseline approach is increasing drastically, and at the meantime, the DIAD algorithm maintains a well performance.

6   Conclusion and Future Work

In this paper, we define the problem of discovering CoPES. Our work is different from previous research as we focus on the event stream data. Besides, we integrate the spatial and temporal property as one factor-STSP to identify the CoPES. Our experimental results provide further evidence of the viability of our approach.

In the future, we plan to examine the behavior of our algorithm in more skewed datasets. Besides, we plan to consider other more efficient method to discover CoPES.

Acknowledgement. This research was partially supported by the grants from the Natural Science Foundation of China(No.61070055,91024032,91124001);The Fundamental Research-Fund s for the Central Universities, and the Research Funds of Renmin University(No. 11XNL010); the National 863 High-tech Program (No.2012AA011001,2013AA01320 4).

References

1. Agrawal, R., Srikant, R.: Fast algorithms for mining association rules in large databases. In: VLDB, pp. 487–499 (1994)
2. Estivil-Castro, V., Murray, A.: Discovering associations in spatial data-an efficient medoid based approach. In: Wu, X., Kotagiri, R., Korb, K.B. (eds.) PAKDD 1998. LNCS, vol. 1394, pp. 110–121. Springer, Heidelberg (1998)

3. Morimoto, Y.: Mining frequent neighboring class sets in spatial databases. In: SIGKDD, pp. 353–358 (2001)

4. Lee, C., Lin, C., Chen, M.: Sliding-Window Filtering: An Efficient Algorithm for Incremental Mining. In: CIKM, pp. 263–270 (2001)

5. Estivil-Castro, V., Lee, I.: Data mining techniques for autonomous exploration of large volumes of geo-referenced crime data. Geocomputation, 24–26 (2001)

6. Huang, Y., Shekhar, S., Xiong, H.: Discovering colocation patterns from spatial data sets: A general approach. In: TKDE, pp. 1472–1485 (2004)

7. Yoo, J.S., Shekhar, S., Smith, J., Kumquat, J.P.: A partial join approach for mining co-location patterns. In: GIS, pp. 241–249 (2004)

8. Goldin, D., Millstein, T., Kutlu, A.: Bounded similarity querying for time-series data. In: IANDC, pp. 203–241 (2004)

9. Kalnis, P., Mamoulis, N., Bakiras, S.: On Discovering Moving Clusters in Spatio-temporal Data. In: SSD, pp. 364–381 (2005)

10. Yoo, J.S., Shekhar, S.: A joinless approach for mining spatial colocation patterns. In: TKDE, pp. 1323–1337 (2006)

11. Huang, Y., Zhang, P.: On the Relationships Between Clustering and Spatial Co-location Pattern Mining. In: ICTAI, pp. 513–522 (2006)

12. Gudmundsson, J., Kreveld, M.J.: Computing Longest Duration Flocks in Trajectory Data. In: GIS, pp. 35–42 (2006)

13. Celik, M., Shekhar, S., Rogers, J.P., Shine, J.A.: Mixed-drove Spatiotemporal co-occurrence Pattern Mining. In: TKDE, pp. 1322–1335 (2008)

14. Huang, Y., Zhang, L., Zhang, P.: A Framework for Mining sequential Patterns from Spatio-temporal Event Data sets. IEEE, 433–448 (2008)

15. Xiao, X.Y., Xie, X., Luo, Q., Ma, W.Y.: Density Based co-location pattern discovery. In: GIS, pp. 29–10 (2008)

The Icecite Research Paper Management System

Hannah Bast and Claudius Korzen

Department of Computer Science, University of Freiburg, Germany
{bast,korzen}@informatik.uni-freiburg.de

Abstract. We present Icecite, a new fully web-based research paper management system (RPMS). Icecite facilitates the following otherwise laborious and time-consuming steps typically involved in literature research: automatic metadata and reference extraction, on-click reference downloading, shared annotations, offline availability, and full-featured search in metadata, full texts, and annotations. None of the many existing RPMSs provides this feature set. For the metadata and reference extraction, we use a rule-based approach combined with an index-based approximate search on a given reference database. An extensive quality evaluation, using DBLP and PubMed as reference databases, shows extraction accuracies of above 95%. We also provide a small user study, comparing Icecite to the state-of-the-art RPMS Mendeley as well as to an RPMS-free baseline.

1 Introduction

This paper is about *Icecite*, a new research paper management system (RPMS) that provides the following unique set of features:

(1) Automatic Metadata AND Reference Extraction: Icecite automatically extracts, with accuracies over 95%, bibliographic metadata (title, authors, year, conference, etc.) as well as references from academic research papers uploaded to the system.

(2) On-Click Download of New Papers: When reading a paper, other papers cited or listed in the reference section can be downloaded with a single click. Using the metadata from the reference extraction from (1), Icecite automatically searches the web for the correct PDF and uploads it to the system.

(3) Collaborative Annotation: Research papers can be annotated in the browser using the PDF standard. This ensures, that annotations remain modifiable in all standard (annotation-enabled) PDF viewers. Internally, annotations are kept separately from the PDF files. This enables collaborative annotation with other users in both online and offline mode (when annotating offline, annotations will be synchronized the next time the user goes online).

(4) Offline Availability: Icecite is web-based (no software download required), but papers can be read and annotated also when offline.

(5) Full-Featured Search: With Icecite, all the metadata, references, annotations, full texts as well as the underlying reference databases can be searched interactively (search as you type).

X. Lin et al. (Eds.): WISE 2013, Part II, LNCS 8181, pp. 396–409, 2013.
© Springer-Verlag Berlin Heidelberg 2013

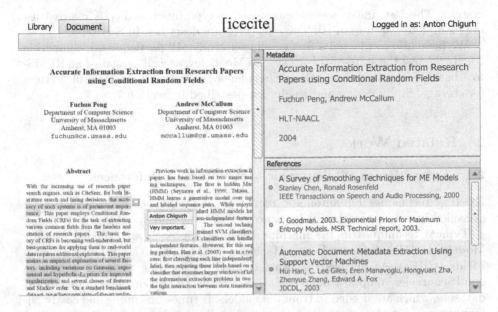

Fig. 1. A screenshot of the *Document View*. The left panel displays the PDF file, the right panels display the metadata (upper right) and the extracted references (lower right). The PDF file can be annotated in the browser using standard PDF annotations. The metadata and references panel can be arbitrarily resized, or hidden to display the PDF file in full screen mode. The references are listed with their full metadata. If no metadata record was found in the reference database, only the extract is displayed (as for the 2nd reference). The documents of the user are organized in a personal library (accessible by clicking the tab "Library" in the header). The colored bullet besides each reference indicates its availability in the user's library. A green bullet means: The document is already stored in the library and can be called by clicking it. A gray bullet means: The reference is not available in the library and can be clicked to import it.

The feature set described above looks quite natural and straightforward for a RPMS. However, *none* of the many existing RPMSs provides this combination of features. In fact, not one of these systems is able to provide even automatic metadata AND reference extraction (with acceptable accuracy). We provide an overview and comparison of fifteen RPMSs in Section 2.

Technically, Icecite combines known techniques in a (more or less) clever way to do what it does. The main idea behind the high-accuracy metadata and reference extraction is a combination of a rule-based recognition (of the passages in the text referring to metadata) with a fast index-based approximate search on a reference database. This is described in more detail in Sections 3 (metadata) and 4 (references). The results of our experimental evaluation, as well as a description of our reference databases are provided in Section 6.

The annotation and offline features are realized using the capabilities of the new HTML5 standard, namely its *Filesystem API* and the *Application Cache*. Annotations are merged using a standard text-based concurrent versioning system. The fast and powerful search-as-you-type functionality is realized using CompleteSearch from [1]. These features are described in Section 5.

We have also conducted a first small user study (12 participants), comparing Icecite against the state-of-the-art system Mendeley, as well as against an RPMS-free baseline approach using Google Scholar for search and the local file system for storage. Study design and results are described in Section 7.

2   Related Work

2.1   Extraction of Bibliographic Metadata and References

Existing techniques for automatic metadata and reference extraction can be classified in two approaches: using *machine learning* and *rule-based*.

Typical techniques in the machine learning approach are: Hidden Markov Models (HMMs), Support Vector Machines (SVMs), and Conditional Random Fields (CRFs). As outlined in Table 1, machine learning approaches achieve good accuracies, of up to around 90%. However the generation of accurate labeled datasets, which are needed to train the models, is time-consuming and costly [9]. Further, machine learning approaches are usually expensive in terms of runtime of the extraction processes [2].

Table 1. Overview of the accuracies of selected machine learning doing metadata extraction (M. Ex.) and/or reference extraction (R. Ex.). The percentage marked * denotes the accuracy of only the title extraction.

Paper	Model	M. Ex.	R. Ex.	Accuracy
Seymore et al. (1999) [16]	HMM	✓		90.1%
Borkar et al. (2001) [3]	HMM		✓	87.3%
Han et al. (2003) [10]	SVM	✓		92.9%
Granitzer et al. (2012) [8]	SVM	(✓)		85.5%*
Peng et al. (2004) [15]	CRF	✓	✓	95.4%
Councill et al. (2008) [5]	CRF		✓	91.6%

Rule-based approaches consist of a set of rules, which are usually derived from human observations (e.g. regarding style information) to identify metadata fields in the headers or reference strings in the bibliography sections of research papers. Pure rule-based approaches are usually faster, but less accurate than machine learning approaches. Beel et al. [2] achieve an accuracy of 77.9% on extracting the titles of research papers by analyzing their font sizes. Guo and Jin [9] combine a rule-based approach with a metadata knowledge base to guide the extraction process of reference metadata. This approach yields a higher average extraction accuracy of 89.1% over all reference metadata fields.

2.2 Record Matching Techniques

Record matching (also called *record linkage*) is the problem of matching a given string (which could be an extracted title or reference string) to the "correct" record in a given database (of titles or references or whatever the application is). This process is usually affected by noisy factors like typing errors, alternative spellings or extraction errors. Most approaches therefore use fuzzy string comparisons to compute the similarity of the given string to selected fields of the database records. Typical similarity measures are: character-based (e.g. *Levenshtein distance* [13], *Smith-Waterman similarity* [17], etc.), token-based, and phonetic-based. See the surveys given in [6] and [11] for more details.

A brute-force comparison of the given string to all database records is usually too slow. Instead, a common approach is to use, in a first step, simplified criteria to quickly obtain a set of candidates of possibly matching candidates. This is often called *blocking* [11]. In a second step, the exact similarities are then computed only for the candidate records. Reasonable blocking strategies are mainly index-based, see the survey [4] for more details. Our approach taken for Icecite also falls in this category. A machine learning approach to blocking is presented in [14].

2.3 Related Applications

Table 2 summarizes and compares the feature sets of fifteen recent RPMSs as well as Icecite. In the following, we discuss a selection of the most powerful of these systems. We distinguish between desktop-based and web-based systems.

The **desktop-based applications** (upper part of Table 2) usually allow to organize research papers in a personal library, listed with extracted metadata. ReadCube is the only system that also provides metadata for bibliographic references of selected papers. However, the references are not actually extracted from the PDF files but fetched from special websites (like *digital libraries*, see below) if available. Automatic search and download of PDF files is supported only by ReadCube, EndNote and Citavi.

Annotating PDF files in a built-in PDF viewer is provided by EndNote, Mendeley, Qiqqa and ReadCube. However, annotations are either not displayed in the PDF when exported (EndNote, Qiqqa and ReadCube) or are "drawn" into the PDF (Mendeley) and thus can not be fully edited anymore in an external PDF viewer. Further, Mendeley does not support a built-in search in external sources to import new research papers easily.

The **web-based applications** (lower part of Table 2) can be further distinguished into (1) reference managers, (2) PDF annotation tools, and (3) digital libraries of academic publishers.

The main purpose of reference managers like *BibSonomy*, *CiteULike*, *EndNote Web*, and *RefWorks* is to manage collections of bibliographic metadata. Besides, BibSonomy and CiteULike also allow to attach PDF files to each record, but the automatic extraction of metadata and references from these PDFs is not supported. Furthermore, in CiteULike, annotating PDF files is only supported for (paying) premium users.

Table 2. Comparison of the feature sets of sixteen RPMSs, eight desktop-based (upper part) and eight web-based (lower part). If a feature is fully provided, it is denoted with "✓". If a feature is partially provided, it is denoted with "(✓)". The listed features are: *(EX-M)*: automatic extraction of metadata; *(EX-R)*: automatic extraction of bibliographic references; *(AUTO-DL)*: automatic search and download of PDF files; *(ANNOT)*: native and colored annotations; *(SHARED)*: data can be shared to collaborate with other users; *(OFFLINE)*: (parts of) the features can be used in offline mode; *(SEARCH)*: search in metadata, full texts, annotations and external sources. *(CLOUD)*: the data can be stored in the cloud to access them from multiple devices.

System/URL	(EX-M)	(EX-R)	(AUTO-DL)	(ANNOT)	(SHARED)	(OFFLINE)	(SEARCH)	(CLOUD)
Citavi www.citavi.com	(✓)	-	(✓)	-	(✓)	✓	(✓)	-
EndNote www.endnote.com	-	-	✓	(✓)	(✓)	✓	✓	✓
EverNote www.evernote.com	-	-	-	-	✓	✓	(✓)	✓
Mendeley www.mendeley.com	✓	-	-	(✓)	✓	✓	(✓)	✓
Papers www.mekentosj.com/papers	(✓)	-	(✓)	(✓)	✓	✓	(✓)	✓
Qiqqa www.qiqqa.com	✓	-	✓	✓	✓	✓	✓	✓
ReadCube www.readcube.com	✓	(✓)	✓	✓	-	✓	✓	-
Zotero www.zotero.org	✓	-	-	-	✓	✓	(✓)	✓
BibSonomy www.bibsonomy.org	-	-	-	-	✓	-	-	✓
CiteULike www.citeulike.org	-	-	-	(✓)	✓	-	✓	✓
EndNoteWeb www.myendnoteweb.com	-	-	-	-	✓	-	(✓)	✓
RefWorks www.refworks.com	-	-	-	-	✓	-	(✓)	✓
A.nnotate www.a.nnotate.com	-	-	-	✓	✓	-	(✓)	✓
Crocodoc personal.crocodoc.com	-	-	-	✓	✓	-	-	✓
WebNotes www.webnotes.net	-	-	-	✓	✓	-	(✓)	✓
Icecite **www.icecite.org**	✓	✓	✓	✓	✓	✓	✓	✓

PDF annotation tools like *A.nnotate*, *Crocodoc* or *WebNotes* focus on annotating and commenting various file types (e.g. PDF files) in collaboration with other users. They are usually implemented in HTML5 or Flash and do not follow the PDF annotation standard, but again, "draw" annotations into the PDF while exporting.

Digital libraries of academic publishers like *ACM Digital Library*[1], *IEEE Xplore*[2], *SpringerLink*[3], etc. provide archives of scientific research papers, including extracted metadata and references. There are two caveats, however.

---

[1] http://dl.acm.org/
[2] http://ieeexplore.ieee.org/
[3] http://link.springer.com/

First, the techniques behind these services are neither published nor publicly accessible, and the extraction accuracy can only be guessed. Second, articles from the same publisher exhibit a homogeneous structure (and sometimes even include explicit meta information) which greatly facilitates the extraction task.

We did not include services like CiteSeerX[4] or Google Scholar[5] in our Table 2 above, because these are global archives and not really RPMSs. However, they also employ techniques for automatic metadata and reference extraction. Their task is harder though, because they lack a reference database. Accuracies reported in [7] are much lower than what we achieve for Icecite.

3   The Extraction of Full Metadata

We proceed in two steps. In the first step, we identify candidates for the title from the given PDF. In the second step, we approximately match these candidates against the titles of the records from our reference database, which is described in more detail in Section 6.

3.1   Title Identification

We use the open source Java tool *PDFBox*[6] to extract text along with characteristic properties like the position, the height, the width and the font of each character, word and text line from PDF files. These properties are then used to identify the title of a research paper.

Definition 1 (Emphasis Score). *Let l_i denote the i-th text line and $fs(l_i)$ denote the font size of l_i. The emphasis score $es(l_i)$ is defined by*

$$es(l_i) = fs(l_i) + \alpha(l_i) + \beta(l_i) \tag{1}$$

where $\alpha(l_i) = \begin{cases} 0.2, l_i \text{ is printed in bold} \\ 0, \text{otherwise} \end{cases}$ *and* $\beta(l_i) = \begin{cases} 0.1, l_i \text{ is printed in italic} \\ 0, \text{otherwise} \end{cases}$

Let $ES_j = \{l_i : es(l_i) = j\}$ denote the set of all text lines with emphasis score j. The most common emphasis score $es_\varnothing$ is then defined by

$$es_\varnothing = \arg\max_k \{|ES_k|\}$$

Let $T = \{l_i : l_i \text{ is member of the title}\}$ be the set of all lines belonging to the title. To identify T, the following assumptions are made:

(A1) All lines $\in T$ are placed in the header (the upper half) of the first page.

(A2) The emphasis score of all lines $\in T$ is $> es_\varnothing$.

Consequently, $es_\varnothing$ is computed for the first page and each line l_i with $es(l_i) \leq es_\varnothing$ is filtered. From all remaining lines in the upper half of the first page, stop words

---

[4] http://citeseerx.ist.psu.edu/
[5] http://scholar.google.com/
[6] http://pdfbox.apache.org/

(like "the", "and", etc.) are removed. Subsequently, the reference database is searched for candidate records, whose title contains (parts of) the remaining words. In the result, the candidate records are sorted by the number of words that they have in common with the extracted title words. Because there may exist candidates with similar titles, the related record is not necessarily the first candidate. To find the related record anyway, each candidate record is evaluated more precisely in the matching process.

3.2   The Matching of Titles

For each record of the top-100 candidates from the title identification, the following scores are computed.

First, the title score $s_t(r) = sim_{SW}(t_r, ex)/sim_{max}(t_r)$ where ex = the lines of the first page's upper half is computed. Here, $sim_{SW}(t_r, ex)$ is the Smith-Waterman similarity between the title t_r of the record r and ex, and $sim_{max}(t_r)$ denotes the maximum achievable similarity for t_r. Note that $s_t(r) \in [0,1]$ and $s_t(r) = 1$ if and only if ex contains t_r completely.

Second, the author score $s_a(r) = \sum_{a_i \in A(r)} s_{a_i}(r)/|A(r)|$ is computed, where $s_{a_i}(r) = sim_{SW}(a_i, ex)/sim_{max}(a_i)$. $A(r)$ is the set of authors of record r. Note that $s_a(r) \in [0,1]$ and $s_a(r) = 1$ if and only if ex contains all authors $\in A(r)$ completely.

Third, the year score $s_y(r) = 0.1$ (if the first page contains the year of r, otherwise $s_y(r) = 0$) and the venue score $s_v(r) = 0.1$ (if the first page contains the venue of r, otherwise $s_v(r) = 0$) are computed.

The total score $s(r)$ is given by $s(r) = s_t(r) + s_a(r) + s_y(r) + s_v(r)$. Finally, the research paper is matched to the record r with the highest score $s(r)$, as long as $s(r)$ exceeds a threshold of 1.5.

4   The Extraction of Bibliographic References

We again proceed in two steps: identification of the individual references from the PDF (again, using PDFBox), and matching of those references against the titles, authors and years of the records from the reference database. Due to the multitude of possible formats for the References section of a paper, the identification step is much more involved now.

4.1   The Identification of Bibliographic References

First of all, the extracted text lines are searched for a proper bibliography section header (e.g. consisting of the word "References", "Literature", "Bibliography", etc.). All lines following such a header are separated into logical blocks by analyzing the *line pitch* of each line to its previous line.

Definition 2 (Line Pitch). *Let l_i denote the i-th text line ($i \geq 0$) and let $y(l_i)$ be the vertical position of l_i in the page. Consider line pairs (l_{i-1}, l_i) sharing the same page. The line pitch $lp(l_{i-1}, l_i)$ between l_{i-1} and l_i is defined by*

$$lp(l_{i-1}, l_i) = y(l_i) - y(l_{i-1})$$

Let $LP_j = \{l_i : lp(l_{i-1}, l_i) = j\}$ denote the set of all lines, whose line pitch to the previous line is j. The most common line pitch $lp_\varnothing$ is then defined by

$$lp_\varnothing = \arg\max_k \{|LP_k|\}$$

Definition 3 (Type of a Reference Line). *Given a sequence of lines representing a reference. The first line is defined as the reference header, the last line as the reference end, and all other lines as the reference body.*

If $lp(l_{i-1}, l_i) > lp_\varnothing$, the lines l_{i-1} and l_i are separated into distinct blocks. If a block consists mainly of digits or if it is a caption (e.g. it starts with the word "Figure", "Table", etc.) it is not meaningful with respect to the reference extraction and is ignored. The type of each line in the remaining blocks is determined by checking it against the following rules:

(1) The line l_i is a *reference header*, if ...
  (a) l_i starts with a reference anchor (like "[1]", "(2)" or "[Smith95]"); *or*
  (b) l_{i-1} is a reference end; *or*
  (c) l_{i-1} (or l_{i+1}) is indented compared to l_i; *or*
  (d) l_i starts with an author and l_{i-1} does not end with an author.
(2) The line l_i is a *reference* end, if ...
  (a) l_{i-1} and l_{i+1} ends up at the same horizontal position and l_i ends before l_{i-1} and l_{i+1}; *or*
  (b) l_{i+1} is a reference header.
(3) The line l_i denotes the *end of the bibliography*, if ...
  (a) l_i is the last line of the document; *or*
  (b) $\lfloor es(l_{i+1}) \rfloor > es_\varnothing$
  (rounded down to allow bold and italic lines within the bibliography).
(4) The line l_i is a *reference body* otherwise.

We assume that all references of a bibliography section share the same inner structure, so that the positions of the metadata fields within the references are consistent. Further, we assume that the authors are the first metadata field in a reference. Rule (1c) implies that, if indentations exists, reference headers are not indented, but references bodies and reference ends. Rule (1d) targets the fact that a listing of authors may cover multiple lines of the reference (a word is identified as a part of an author name if the reference database contains such an author name).

With rule (2a) we assume that the references are formatted as justified text if l_{i-1} and l_{i+1} share the same horizontal end position and that l_i denotes a reference end if it does not fill the whole line.

Once a reference string was identified, stop words are filtered and the remaining words are scanned for title words, authors and the year. Afterwards the reference database is searched for such records, which hold these metadata. Again, the resulting records are evaluated with a more precise scoring scheme.

4.2 The Matching of Bibliographic References

For each record of the top-100 candidates from the references identification, the scores $s_t(r)$, $s_a(r)$, $s_y(r)$ and $s_v(r)$ are computed as for the title matching described in Section 3.2 (with $ex =$ the extracted reference string). Additionally, the pages score $s_p(r) = 1$ (if r defines page numbers and ex contains them, otherwise $s_p(r) = 0$) is computed. The total score $s(r)$ is given by $s(r) = s_t(r) + s_a(r) + s_y(r) + s_v(r) + s_p(r)$. Finally, the extracted reference is matched to the record r with the highest score $s(r)$, as long as $s(r)$ exceeds a threshold of 1.5.

5 Annotation, Offline Mode and Search

We enable the annotation of PDF files in the browser using the Adobe Acrobat Standard plugin. Javascript code is injected into the PDF files, such that annotations can be modified dynamically on opening the PDF file or when synchronizing with the server. The annotations are kept in text files, separate from the PDFs. This allows merging of annotations from different users (for the same paper) using a standard text-based versioning system (we use SVN). When online, Icecite periodically synchronizes with a server and automatically merges all annotations appropriately.

The offline mode is realized with the *Filesystem API* and the *Application Cache*, two features of the new HTML5 standard. The Filesystem API is used to store all library data (PDF files, metadata and annotations) locally on the file system of the client. The Application Cache is used to cache all specific web resources (like HTML files, CSS files, images etc.). If the resources have changed on the server, the browser downloads them and updates the cache automatically.

The search functionality of Icecite is implemented with *CompleteSearch* [1], which supports efficient search-as-you-type functionality. There is an index per user, which is automatically updated as soon as a paper or annotation is added. There is also one index for the reference databases (DBLP and PubMed). All searches can be directed to either of these, or to both at the same time.

6 Experiments

6.1 Experimental Setup

We evaluate the accuracy of our metadata and reference extraction algorithms on two reference databases: DBLP[7] and PubMed[8]. At the time of this writing, DBLP holds 2.1 million metadata records (with title, authors, year, venue, journal, etc.) of publications from the area of computer science and neighboring disciplines. PubMed is an order of magnitude larger, with 22 million metadata records of publications from the life sciences. We want to stress that nothing in

---

[7] http://dblp.uni-trier.de/
[8] http://www.ncbi.nlm.nih.gov/pubmed/

our approach is specific to these reference databases. We expect Icecite to work just as well with any other reference database.

Our test collection consists of 690 randomly selected research papers from DBLP and of 500 randomly selected research papers from PubMed. For all of these, we have determined the correct titles manually. For 91 papers of DBLP (containing 1,012 references) and 34 papers of PubMed (containing 1,235 references), we have also determined the correct reference strings. For each such title and reference string, we have further determined the key of its related metadata record (if available) in the reference database.

The code for the metadata and reference extraction is entirely written in Java, based on the Java library *PDFBox*. The index to browse the reference databases is written in C++. All the tests were run on a single machine with 4 Intel Xeon 2.8 GHz processors and 35GB of main memory, running Ubuntu 9.10 64-bit.

6.2   Extraction Accuracies

First, we have measured the accuracies (the percentage of correct results from the total numbers of results) of both the extraction and the matching algorithms. We have considered an extract as correct, if its Levenshtein distance to the expected extract ex_{gt} is $\leq 0.2 \cdot |ex_{gt}|$. Further, we have considered a matched metadata record as correct, if its key is equal to the key of the expected record or if no record was returned and there is in fact no expected record.

Table 3.   Overview of the extraction accuracies of the metadata and references extraction on DBLP and PubMed. Column 3 provides the number of entities (document or reference) to process in the test collection. There were PDF files which could not be processed by PDFBox; subtracting these gives the numbers in Column 4. Columns 5 and 6 provide the absolute number of correct extractions as well as the percentage with respect to the value in Column 4.

		num.	max.	corr. extracts	corr. matches
Meta.	DBLP	690	679	672 (98.9%)	665 (97.9%)
	PubMed	497	490	474 (96.7%)	468 (95.5%)
Ref.	DBLP	1012	997	974 (97.7%)	951 (95.4%)
	PubMed	1235	1235	1179 (95.5%)	1166 (94.4%)

As shown in Table 3, we achieve very good extraction accuracies for both datasets. There are only few documents for which the extraction or matching process failed. We manually investigated the individual reasons for these few failures. For example, the title extraction failed, if (1) the title in the document was not emphasized compared to other text passages on the first page, (2) the title was not placed in the upper half of the first page, or (3) two titles were placed on the first page and the other title was extracted. The reference extraction failed, if (4) there was no bibliography header, (5) existing reference anchors were not extracted or (6) title words were misleadingly identified as author words. (5) and (6) have led to the wrong identification of the reference line's type. Matching a title failed, if (7) there were multiple variants of a paper in the reference database (but published in an alternative journal) and there was no criterion to distinguish the variants, (8) more words than the title word were extracted (because of their emphasis score) such that records other than the related one were found in the reference database, (9) words were misspelled (mostly due to extraction errors), (10) the title in the document and the title of its related database record did not match exactly. Matching a reference string has failed, if (11) author names were misspelled, (12) the year in the reference did not correspond to the year of its related record, (13) the reference did not have a related record, but there was a record with a related title by the same authors.

6.3   Running Times

We have also evaluated the running times of both of our extraction algorithms.

Table 4. Overview of the runtimes of the metadata and reference extraction from research papers of DBLP and PubMed. The runtimes are broken down into the following subtasks: (1) *loading* a PDF file and analyzing the text lines, (2) *querying* the reference database, (3) *matching* an extract to its related record. The stated runtimes are per document (metadata extraction) respectively per reference (reference extraction; on average).

		total	loading	querying	matching
Meta.	DBLP	137.7ms	31.1ms (23%)	73.1ms (53%)	33.5ms (24%)
	PubMed	479.6ms	44.9ms (9%)	341.3ms (71%)	93.4ms (20%)
Ref.	DBLP	54.2ms	14.7ms (27%)	19.7ms (36%)	19.8ms (37%)
	PubMed	91.4ms	10.2ms (11%)	47.4ms (52%)	33.8ms (37%)

Table 4 shows that our algorithms are fast enough for an interactive experience, even on the very large PubMed reference database. Note that the given times for the metadata extraction are per document, while the times for the reference extraction are per reference. The time for *loading* is needed only once per PDF. The typical time for full metadata and reference extraction from a single PDF is therefore generally below 1 second for DBLP, and 1-2 seconds for PubMed.

7   User Study

We have implemented a fully-functional prototype for Icecite. To assess the user experience with our system, we have conducted a small user study with 12 participants: 1 female, 11 males, all aged between 22 and 30 years. All of them were familiar with web browsing and have not used Icecite before. One half of the participants were asked to compare Icecite against a plain RPMS-free baseline, namely using *Google Scholar* to search for papers and managing the research papers manually on the local file system. The other half of the participants were asked to compare Icecite against a state-of-the-art RPMS, namely *Mendeley*. We have chosen Mendeley, because it considers itself to be "the world's largest social reference management system" [12]. Besides, we believe that Mendeley's feature set comes closest to the state of the art.

Each participant was asked to solve the following set of tasks twice. A participant from the first half would solve it once with Icecite and once with Google Scholar. A participant from the second half would solve it once with Icecite and once with Mendeley.

(T1) Download the paper X and store it in your system.

(T2) Find the paper Y in DBLP and store it in your system.

(T3) Open the first paper and add at least three annotations.

(T4) Log in on a second machine and annotate the paper from both machines.

(T5) As *(T4)*, but with one machine disconnected from the internet.

(T6) *(Only to solve with Icecite and Mendeley)* Export the PDF file and open it in an external viewer. Edit some annotations.

(T7) Choose the first ten references from the paper of *(T1)* and store the respective PDF files into your system.

(T8) Use the available search functions to search for the terms Z.

(T9) Identify the paper of *(T2)* and open it.

In total, there were three variants of this task set, each of them with different entities for X, Y, and Z. Each participant had to solve exactly two variants (one with the one system, one with the other system). The variants were assigned in a permuted form such that each variant was assigned equally often to a system and to the participants. For each task, the participants were asked to estimate the required time and to assign a score from $1 - 5$ indicating the (subjective) satisfaction on completing the task ($1 =$ absolutely dissatisfied, $5 =$ absolutely satisfied). Table 5 summarizes the results of this quantitative feedback. Also, each participant had the opportunity and was encouraged to give (anonymized) general feedback in a free-text field.

Most Liked Features of Icecite. Using Icecite, all participants have enjoyed the automatic extraction of references, and the ability to download a citation or reference with a single click. The possibility to annotate research papers collaboratively in the browser was also positively mentioned by 10 participants.

Most Disliked Features of Google Scholar and Mendeley. On using Google Scholar and Mendeley, the references could not be extracted automatically and referenced papers could not be downloaded on click. Instead, the participants had to download all of them manually. That's why solving task *(T7)* with Google Scholar or Mendeley took much longer than with Icecite. Further, 3 participants disliked that Mendeley does not support multi-colored annotations, and 5 participants disliked that the annotations are not fully modifiable after

Table 5. Breakdown of the results of our user study. For each task *(T1)-(T9)*, the participant's subjective satisfaction and the required time for solving the task with Google Scholar, Mendeley and Icecite is stated. For each task, the best results are emphasized in **bold**.

	Google Scholar		Mendeley		Icecite	
(T1)	4.0	2.4 min	3.8	2 min	**4.1**	**1.3 min**
(T2)	**4.2**	1.4 min	3.7	1.8 min	**4.2**	**1.3 min**
(T3)	3.7	3.5 min	2.7	4.5 min	**4.4**	**2 min**
(T4)	1.2	-	3.0	7.2 min	**4.0**	**2.5 min**
(T5)	1.0	-	3.5	3.4 min	**4.5**	**2 min**
(T6)	-	-	2.2	5 min	**4.6**	**1.5 min**
(T7)	2.2	11.8 min	2.2	15.6 min	**4.1**	**8.3 min**
(T8)	2.0	4.1 min	**4.7**	**1.9 min**	4.0	2.1 min
(T9)	4.0	0.8 min	4.6	1 min	**4.8**	**0.7 min**

exporting a PDF file. It turned out, that the tasks *(T4)* and *(T5)* could not be solved with a reasonable effort using Google Scholar. That's why 5 participants have missed these tasks, if they were asked to solve them using Google Scholar (and that's why the required time for *(T4)* and *(T5)* is denoted by "-" in the respective column in Table 5).

Most Liked Features of Google Scholar and Mendeley. The search function of Mendeley was generally enjoyed and outperformed those of Icecite and Google Scholar *(T8)*. Every participant liked the possibility to jump directly to the position of a query-relevant text-passage in a PDF file. Further, Google Scholar was praised by 3 participants for its simplicity and its quality of search results.

Most Disliked Features of Icecite. 9 participants have complained about the minimal feedback the system is giving to users about what it is currently doing. They asked for more messages of the sort: "logging in", "saving documents/annotations", "synchronizing", etc. 6 participants expressed that existing messages could be more precise, e.g. by specifying the reason, why an import of a referenced research paper has failed. All participants but one were annoyed by small bugs of the search box: after sending a search query, the focus of the search box was lost such that it must be clicked again to modify the query. However, all of these revealed weaknesses are easy to address.

8   Conclusion and Future Work

We have presented Icecite, a fully web-based research paper management system (RPMS) with a unique feature set that has not yet been achieved by any other RPMS. This in particular applies to the automatic metadata and reference extraction, provided by Icecite with accuracies over 95%. We have also verified the benefits of Icecite in a small user study.

We have provided an error analysis of the missing few percents in accuracy. It appears that about half of these errors can be addressed by a further improved identification step (see Sections 3 and 4). The total extraction time per PDF is good (below 1 second for DBLP, 1-2 seconds for PubMed) but could be improved further. However, the current bottleneck here is not our algorithms, but the PDFBox library. Our small user study confirmed that the unique feature set of Icecite, in particular the automatic metadata and reference extraction and the one-click reference downloading, is of great practical value to users.

References

1. Bast, H., Weber, I.: The CompleteSearch Engine: Interactive, Efficient, and Towards IR&DB Integration. In: CIDR, pp. 88–95 (2007)
2. Beel, J., Gipp, B., Shaker, A., Friedrich, N.: SciPlore Xtract: Extracting Titles from Scientific PDF Documents by Analyzing Style Information (Font Size). In: Lalmas, M., Jose, J., Rauber, A., Sebastiani, F., Frommholz, I. (eds.) ECDL 2010. LNCS, vol. 6273, pp. 413–416. Springer, Heidelberg (2010)
3. Borkar, V.R., Deshmukh, K., Sarawagi, S.: Automatic Segmentation of Text into Structured Records. In: SIGMOD Conference, pp. 175–186 (2001)
4. Christen, P.: A Survey of Indexing Techniques for Scalable Record Linkage and Deduplication. IEEE Trans. Knowl. Data Eng. 24(9), 1537–1555 (2012)
5. Councill, I.G., Giles, C.L., Kan, M.-Y.: ParsCit: An Open-source CRF Reference String Parsing Package. In: LREC (2008)
6. Elmagarmid, A.K., Ipeirotis, P.G., Verykios, V.S.: Duplicate Record Detection: A Survey. IEEE Trans. Knowl. Data Eng. 19(1), 1–16 (2007)
7. Giles, C.L., Bollacker, K.D., Lawrence, S.: CiteSeer: An Automatic Citation Indexing System. In: ACM DL, pp. 89–98 (1998)
8. Granitzer, M., Hristakeva, M., Jack, K., Knight, R.: A Comparison of Metadata Extraction Techniques for Crowdsourced Bibliographic Metadata Management. In: SAC, pp. 962–964 (2012)
9. Guo, Z., Jin, H.: Reference Metadata Extraction from Scientific Papers. In: PDCAT, pp. 45–49 (2011)
10. Han, H., Giles, C.L., Manavoglu, E., Zha, H., Zhang, Z., Fox, E.A.: Automatic Document Metadata Extraction Using Support Vector Machines. In: JCDL, pp. 37–48 (2003)
11. Kan, M.-Y., Tan, Y.F.: Record Matching in Digital Library Metadata. Commun. ACM 51(2), 91–94 (2008)
12. Kraker, P., Körner, C., Jack, K., Granitzer, M.: Harnessing User Library Statistics for Research Evaluation and Knowledge Domain Visualization. In: WWW (Companion Volume), pp. 1017–1024 (2012)
13. Levenshtein, V.I.: Binary Codes Capable of Correcting Deletions, Insertions, and Reversals. Soviet Physics Doklady 10, 707–710 (1966)
14. Michelson, M., Knoblock, C.A.: Learning Blocking Schemes for Record Linkage. In: AAAI, pp. 440–445 (2006)
15. Peng, F., McCallum, A.: Accurate Information Extraction from Research Papers using Conditional Random Fields. In: HLT-NAACL, pp. 329–336 (2004)
16. Seymore, K., McCallum, A., Rosenfeld, R.: Learning Hidden Markov Model Structure for Information Extraction. In: AAAI 1999 Workshop on Machine Learning for Information Extraction, pp. 37–42 (1999)
17. Smith, T., Waterman, M.: Identification of Common Molecular Subsequences. Journal of Molecular Biology 147, 195–197 (1981)

A Middleware for Managing Big-Data Flows

Rajeev Gupta[1], Himanshu Gupta[1], Sanjeev Gupta[2],
and Sriram Padmanabhan[2]

[1] IBM Research, India
{grajeev,higupta8}@in.ibm.com
[2] IBM Software Group, SanJose, California
{guptasa,srp}@us.ibm.com

Abstract. Hadoop is being used for various diverse kinds of applications over diverse kinds of data. This makes developing and managing data flows over Hadoop MapReduce a complex task. Various scripting languages such as Hive, Pig, Jaql, etc., have been developed to hide the complexity of MapReduce applications from the user. But, even these high level query languages can get complex over-time and it is a non-trivial task even for a user proficient in these languages to develop, debug, and maintain these scripts. This paper presents a middleware for developing and maintaining MapReduce data flows. This middleware can be used to *E*xtract data from diverse data sources, *L*oad it into distributed file system, and *T*ransform in a format which can be easily analyzed by the subsequent systems in a user friendly manner. MetaOperators are the backbone of our middleware. Using MetaOperators one can express a data-flow only by specifying the relevant inputs rather than worrying about data schema and the query syntax. A data-flow written using such MetaOperators localizes schema specific parts of the query to the MetaOperator parameters making the flow easier to develop, debug, and maintain. Using these MetaOperators we show how one can express operations over hierarchical as well as flat data in a similar manner, *track* data schema as it flows through the operators, and add a *drag-and-drop* GUI layer on top of this framework. This brings MapReduce application development in the realm of middle management.

1 Introduction

Hadoop, an open source implementation of Google's MapReduce [11], has become extremely popular for carrying out analytics over large scale data, e.g., telecom call detail records (CDRs), web-logs, social media data, credit card data, etc. These applications can easily get multiple millions of new records everyday. It may not be suitable to use traditional SQL data warehouses in such scenarios as the cost of storing such large amount of data may be very high. Secondly warehouses only process relational data while many scenarios will contain unstructured and/or semi-structured data. One can process structured, semi-structured, and unstructured data using Hadoop. Hadoop does not enforce any data model, data format, or operation semantic. For example, using Hadoop one

X. Lin et al. (Eds.): WISE 2013, Part II, LNCS 8181, pp. 410–424, 2013.

can get all the users making calls above a certain duration, calculate percentage of twits having positive opinion about a particular product, etc. Such diversity of data model, formats, schema types, etc., makes writing applications over Hadoop very difficult. One can use various procedural languages like Java, Python, etc. to write *map* and *reduce* functions. Various scripting languages like Jaql [7] and Pig [5]; and descriptive SQL-like interfaces like Hive [13] are proposed to query data stored in Hadoop's Distributed File System (HDFS). But there are various problems with using procedural and scripting languages for writing MapReduce applications:

- For writing MapReduce jobs using procedural languages, one has to create key-value pairs from input data and write custom code for each application. This makes the MapReduce code development and maintenance difficult and time consuming requiring highly skilled developers.
- For each MapReduce job to run optimally, one needs to configure a number of parameters, such as number of mappers, number of reducers, size of data block each mapper will process, etc. Finding suitable values of these parameters is not easy.
- An application may require a series of MapReduce jobs. Hence, one needs to write these jobs and schedule them properly.
- One can use descriptive language like Hive to write simple MapReduce applications, but Hive only supports structured data and relational operations.
- Scripting languages like Pig and Jaql can be used to write complex flows of statements. But in these scripting languages, script syntax is function of data model and schema. Hence, the learning curve needed to get an expertise in these scripting languages is steeper vis-a-vis SQL.
- For applications written using procedural languages as well as scripting languages, it is very difficult to visually extract information about data model and application semantics as the whole process of reading data, providing structure to it (or creating entities), processing the data, and writing output is monolithic without any descriptive meta data.
- In absence of meta data, it is very difficult to *describe* output data, debug a program, and reuse data and applications (or parts of applications).

This paper presents a middleware that has been designed for developing, debugging, and maintaining data flows over Hadoop with increased usability. Figure 1 shows various parts of the middleware we are presenting in this paper.

1.1 Contributions and Organization

Contributions of this paper are

- We present a middleware using which data flows involving multiple types of data can be written in a unified manner without worrying about the data model and syntax.

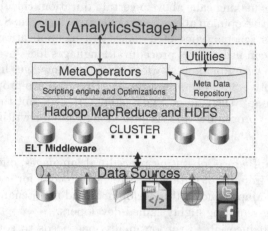

Fig. 1. Middleware Block-diagram

- Our middleware includes a set of MetaOperators which are generic enough to allow a user to develop complex MapReduce data flows. A MetaOperator is designed in a manner so that a user can express her requirements only by specifying the set of relevant inputs (operation to be performed with operands). As a result, **the logic of the data flow is separated from the schemas that are specified for a given data flow instance**.
- A data-flow can be visualized as a sequence of MetaOperator statements. Thus, writing a data-flow reduces to figuring out what operations need to be carried out and what should be the input parameters for these operations irrespective of data schema and format. Hence using our middleware one can develop data flows with *write-as-you-think* philosophy.
- Although one can use any generic programming language like Java to implement these MetaOperators, in this paper, we use Jaql for developing a proof-of-concept, as Jaql provides an easier abstraction of MapReduce programming. The issues and challenges remain the same irrespective of which scripting language is used.
- As an application is a well structured sequence of MetaOperators, the middleware allows one to easily track the data flow. The middleware provides utilities for deriving the data schema and lineage, checking inconsistencies in the input for MetaOperators calls, etc. The middleware provides a catalog to store and query meta data. This catalog can be used to query, for example, how the schema will change after a specified MetaOperator call. This allows for a faster debugging of data-flows.
- One can easily extend the middleware using a GUI tool to develop data flows using a visual drag-and-drop approach. We illustrate this, by integrating IBM InfoSphere DataStage [3] with Hadoop using this middleware.

Section 2 discusses the related work in the areas of user friendly ways of developing data-flows. Jaql data model and operations are introduced in Section

3. This section also outlines difficulty in processing hierarchical data. Section 4 outlines the motivation and important characteristics of MetaOperators. Middleware utilities are explained in Section 4.2. Extensions for IBM InfoSphere DataStage to support Hadoop based data-flows are explained in Section 5. By using real-world data, Section 6 shows that MetaOperator based design is scalable with low overheads. The paper concludes in Section 7.

1.2   Running Example

Our middleware makes unstructured, semi-structured, and hierarchical nested data processing easier. Hence the running example has *books* data which is array of book records. Each book record has details about book, its authors, book reviews, and textual discussions of those reviews. Including reviews makes the data hierarchical with each review having its meta data besides the review text and ratings. Figure 2 shows the example data.

```
books = [{
    publisher : 'Scholastic',
    author : 'J. K. Rowling',
    title : 'Deathly Hallows'
    year : 2007
},
{
    publisher : 'Scholastic',
    author : 'J. K. Rowling',
    title : 'Chamber of Secrets'
    year : 1999,
    reviews : [
        {rating: 8, user: 'rob', review: 'High on my list ...'},
        {rating: 2, user: 'mike', review: 'Not worth the paper ...',
            discussion:
            [{user: 'ben', text:'This is too harsh ... '},
            {user: 'jill', text:'I agree ...'}]}]},
}]
```

Fig. 2. Example hierarchical data

Besides using this data for explaining various concepts, we use real data from diverse domains such as credit card transactions, web-logs at IBM web servers, warehouse data for a retail chain, etc., to show that our middleware is broad-based and gives good performance for these diverse data types.

2   Related Work

A number of studies have looked at various methods for assisting users in formulating complex SQL queries. Various visual query building tools have been developed [2,4,8] which are targeted to users who struggle with *select-from-where* paradigm. QueRIE [10] system recommends queries based on patterns developed using queries authored by other users. SnipSuggest [12] is a SQL auto-complete system which considers a partial query that a user has typed so far and generates recommendations for completing the query. This is done by maintaining queries issued so far and drawing an inference from them. Our middleware does not assist users in completing complex MapReduce queries, rather, it provides a new framework using which users can easily express her data processing tasks. Techniques developed in this middleware can be used to design the corresponding system for SQL as well. We are not aware of any study which help users in formulating Jaql/Pig queries.

A number of systems have been developed for large scale data processing using MapReduce. Hive [13] provides HiveQL, a subset of SQL, with features like *from* clause sub-queries, various types of *joins*, *group bys*, *aggregations*, and *create table as select*, which all make HiveQL very SQL like and hence requires minimum effort to migrate to Hadoop infrastructure. Pig [5] provides a script-like syntax for developing data flows that are evaluated using MapReduce. Pig provides a sweet spot between SQL and MapReduce by providing high-level data manipulation constructs, which can be assembled in an explicit data-flow, and interleaved with custom MapReduce functions. Jaql [7] is a functional data query language whose data model is based on Java Script Object Notation (JSON). Jaql is a general purpose data-flow language that manipulates semi-structured data. We do not propose a new language, rather, a framework which may use these scripting languages to easily develop and debug MapReduce applications.

While Hive, Pig, and Jaql are data processing languages, a number of other studies have looked at aiding the development of MapReduce tasks, e.g., Google's FlumeJava [9], the PACTs programming model [6], Cascading project [1], etc. FlumeJava handles pipelines of *map* and *reduce* functions and makes them easy to test, and run efficiently. PACT programming model is a generalization of MapReduce programming model and aims to carry out data processing tasks cannot be represented as *map* and *reduce* tasks or which are hard to represent. Cascading is a library of Java APIs used for defining and executing MapReduce workflows on Hadoop without having to *think* in MapReduce. We also define a middleware which can be implemented in any of the scripting languages so that a data flow can be specified in a user friendly manner. Compared to Cascading, our ELT middleware works on scripting languages providing higher level abstraction and meta data utilities.

3   Data Processing Tasks and Jaql

Considering that Jaql is a Turing complete language (hence one can implement any functionality), we take Jaql as the reference scripting language for this paper. This section summarizes the data model and operations supported by Jaql. While designing the middleware we ensure that these operations are well supported without the complexities of Jaql syntax.

3.1   Jaql Data Model (JDM)

Like JSON, Jaql data model consists of atomic types (integer, double, string, boolean, null, etc.) and complex types (arrays and records (structures)). Unlike JSON, JDM's atomic types are extended with binary, date, and a few other useful data types. A record consists of a set of keys and their associated values. An array value is an ordered list of atomic values. Thus, JDM can model data that ranges from simple and flat to complex and nested hierarchical. Figure 2 shows the example JDM records for *books* data. In the *books* records attributes *publisher*, *author*, *title*, *year*, and *reviews* can be seen as depth 1 attributes. Child nodes of *reviews* are depth 2 attributes. The attributes *reviews* and *discussion* are array of records. Unlike relational format, all the JSON records need not have the same schema.

3.2   Data Processing Using Jaql

Jaql provides users with a declarative syntax to carry out SQL like operations on data. Jaql has six core functionalities: filter, transform, group, join, sort, and expand. These functionalities have the same meaning as their SQL counterparts. Operation *transform* is similar to SQL's select clause allowing the users to carry out a projection or apply a function to all items of an input. The *expand* operation takes an array of arrays as input, and unnests one level creating an array of input elements. For example, consider the following code snippet **C1:**

```
books2 = books → filter $.year ≤ 2000;
AllReviews = books2 → expand $.reviews;
AllReviews → group by rating = $.rating into
    {Rating : rating, count : count($[*])};
```

In this snippet, all the books published before the year 2000 are filtered out; all reviews are collected in variable *AllReviews*; and and the number of reviews for each rating are counted.

Jaql *functions* enable re-usability of a Jaql script. A set of expressions can be encapsulated as a function, this function can be parametrized, and invoked from anywhere in a Jaql script. Body of a Jaql function *TestFunc* and an example function call is shown in **C2:**

```
TestFunc = fn(books, year)
    (books → filter $.year ≤ year);
TestFunc(books, 2000);
```

One can think of various operations to be applied to a nested data (e.g., book reviews). For example consider the following code snippet:

C3: For each discussion, count the number of comments made by every user and add this information as the new attribute *DiscussionStats* with the parent being *reviews*.

$books \rightarrow transform \ \{\${* - .reviews\},$
$reviews : \$.reviews \rightarrow transform \ \{\${*\},$
$DiscussionStats : \$.discussion \rightarrow group \ by \ user = \$.user$
$\quad\quad into\{user, commentCount : count(\$[*])\}\}\};$

In C3, the group-by operation is needed to be applied on the attribute *discussion* which is a depth 2 attribute. This nested operations makes the script complex and difficult to understand. Our middleware abstractions aim to lift such references to the MetaOperator parameters so that one can easily specify the operation to performed along with required inputs. Further, field name changes, or more complex structural changes, can be easily dealt with. Thus, using MetaOperators make it easier to manipulate nested data and accommodate the structural variations that arise from related but distinct data sources (e.g., *books* data from some another source), as explained next.

4   MetaOperators

As seen from the code snippet C3, a statement for processing hierarchical data may become pretty complex and such complex statements frequently occur in work-flows processing hierarchical data. A hierarchical operation may often change the data schema, for example, by adding or dropping a certain attribute, or by changing the level of an attribute etc. It hence becomes difficult to parse, understand or debug the logic of the scripts of such work-flows. One needs to carefully read every statement and infer its intent to make out the overall objective of the script. This process is even more difficult for a person who is not the author of the script, especially more so if the user does not have enough experience with the writing MapReduce applications using scripting languages.

The aim of our middleware is to provide a new approach using which a user can express a data-flow easily. A user need not write complex statements and hence need not have a detailed knowledge of the scripting languages. Secondly, it is easy to understand and parse a workflow script generated using our middleware. The backbone of our middleware is the concept of MetaOperators. We identify the common functionalities and define a MetaOperator for each of these functionalities. A user, instead of writing a scripting statement for using a functionality, makes a call to the corresponding MetaOperator defined in our middleware. The intuition behind a MetaOperator is that a user identifies the necessary inputs for carrying out an operation and provides these inputs to the MetaOperator. The MetaOperator in turn carries out the corresponding operation. For example, instead of writing a statement for aggregating a data in C3, a user can make a call to a MetaOperator *MetaAggregation*:

C4: For each discussion, count the number of comments made by every user. Add this information as the new attribute *DiscussionStats* with the parent being *reviews*.

$path = [\text{``reviews''}, \text{``discussion''}]; keys = [\text{``user''}];$
$functions = [fn(r)count(r)];$
$newAttributes : [\text{``commentCount''}]; aggOutputAttr : \text{``DiscussionStats''};$
$params = \{path : path, keys : keys, functions : functions,$
$\qquad aggOutputAttr : aggOutputAttr, newAttributes : newAttributes\};$
$MetaAggregation([books], params);$

This is the equivalent middleware statement for the code-snippet C3. Note that instead of writing a complex statement like C3, a user identifies the relevant inputs. The inputs required for an aggregation operation on a hierarchical data are as follows:

- **path:** Which attribute in the hierarchical schema is being aggregated? The variable *path* specifies the path from the root node in the schema tree to the attribute being aggregated. In C3, we are carrying aggregation on the attribute *discussion*. This attribute is a depth 2 attribute as shown in Figure 2 with its ancestors being *reviews*.
- **keys:** What are the keys on which the input data is grouped by? In C3, the aggregation is being done on attribute *user*. As a group-by operation may need more than one key, the variable *keys* is an array.
- **functions:** What is the aggregation function being used? In C3, we perform a *count* for each distinct value of user. As a group-by statement may involve more than one aggregation function (e.g, max, min, avg etc), the variable *functions* is an array.
- **newAttributes:** These attributes hold the output of individual aggregation operation for each distinct value of aggregation-key. In C3, the count of comments by each user is held by attribute *commentCount*. The size of the array *newAttributes* is hence same as the size of the array *functions*.
- **aggOutputAttr:** This attribute is only required in case of hierarchical data. If the group-by operation is being done on a nested attribute, we need to insert a new attribute in the data schema which will hold the output of aggregation. This attribute will be a sibling of the node on which aggregation is being carried out. In C3, we introduce a new attribute *DiscussionStats* which holds the output of aggregation on the attribute *discussion*.

Different MetaOperators may need different sets of input parameters. The details of these input-parameters is made available to the users a-priori. Note that the task of specifying a set of input parameters correctly is much easier vis-a-vis writing a complex Jaql statement correctly. A MetaOperator statement hence is much simpler vis-a-vis the corresponding operation in Jaql. As a result, the users not proficient in Jaql, can also easily specify an analytic task on Hadoop.

MetaAggregation is a MetaOperator our middleware implements which carries out the aggregation on the input data. All MetaOperators have a common interface **MetaOperator(input_data, parameters)** irrespective of the operation being performed and schema of the input data. By providing a common interface MetaOperators hide the complex implementation from the user. Each MetaOperator has two parameters:

- **input_data:** The attribute *input_data* specifies the input stream(s) or data sequence for the MetaOperator. The *input_data* may represent one or more input streams, e.g., MetaJoin may need two input streams.
- **parameters:** The attribute *parameters* is a JSON formatted string and specifies the parameters needed to carry out an operation. The structure of *parameters* (i.e., the set of the key-value pairs) is different for different MetaOperators.

We next outline the various functionalities for which we developed the Meta-Operators in our middleware.

4.1   Types of MetaOperators

1. **Core Operators:** These operators correspond to various SQL operations, performed over flat data, extended to more complex data types. *MetaFilter*, *MetaAggregation*, *MetaSort*, *MetaTransform*, and *MetaJoin* have the semantics similar to their SQL counter part. Besides these, *MetaExpand* is specifically useful for array field types.
2. **IO operators:** These operators are used for reading data from various data sources and writing data to various destination types. *MetaRead* and *MetaWrite* have parameters using which one can read/write files from local disk, remote disk, or HDFS; and files with different formats– text input, CSV, etc.; from different data sources (e.g., a database, a web source). One can use a custom input format to read HDFS files with different formats. Similarly, one can provide an output format for *MetaWrite* to write on a file, database, etc. For example, MetaRead for reading custom format has four extra attributes in the *params* structure: *adapter* (to identify record boundaries), *format*(class to read custom input records), *converter* (class to convert custom record into JSON), and *configurator*(class to read format specific configurations).
3. **Miscellaneous Operators:** We have defined some other MetaOperators to perform certain functions over unstructured and streaming data types. *MetaAnnotation* can be used to provide a structure to unstructured natural language text. We have encoded annotators to extract out commonly used annotations such as phone_numbers, addresses, IBM_product names, person names, etc., from a given text or fields of the input data. User can also specify regular expressions to extract other types of annotations. Another MetaOperator, *MetaWindow*, can be used to process input data with window semantics. Width of the window can be temporal, number of records, or one can also specify start and stop functions for creating a customized window. Other miscellaneous MetaOperators include *MetaMove* and *MetaCopy* for hierarchical data restructuring.
4. **Vertical Specific Operators:** These operators are designed for a specific vertical, e.g., operators for telecom data, scientific data, etc. These operators can be *enabled* for the vertical specific applications. Scientific data arises in a number of scenarios like weather modeling, earth modeling, etc. Scientific

data contains the values of attributes (e.g., temperature, pressure) at various latitude, longitude, elevation, time, etc. We implemented MetaOperators for carrying out a number of analytic tasks like pattern search, histogram and anomaly computation, spatial and interval joins, etc. For example, MetaP-atternSearch can be used to find data points which are within a threshold distance of the pattern provided. The input for this MetaOperator is spatio-temporal data in the form of a 2-dimensional array of attribute-values, defined over a pre-defined range of latitude and longitudes.

5. **Custom Operators:** Considering that Hadoop is being used for diverse kinds of applications and input data, we have made the middleware expandable, where users can define their own MetaOperators. These MetaOperators should have the same signature so that the operators can be used for schema inferencing, debugging support, meta data management, etc., as explained Section 4.2. For help in defining custom MetaOperators, we defined two extensible MetaOperators– one can be used to expose a Java function as data flow operator and another to encapsulate any Jaql function. Aim of the MetaOperator based middleware is to design most data flows with minimum number of MetaOperators. Still one can use these extensibility features for domain specific operations.

As proof-of-concept, we implemented these MetaOperators using Jaql because of its abstraction capabilities, described in Section 3, and its Turing completeness. For example, implementing these operators using Hive (no support for hierarchical processing), and Pig (inefficient support for hierarchical processing with parameter passing) will be very difficult. The MetaOperators implementation using Jaql can use the build-in compiler of Jaql for compiling and efficient running of the code. In contrast, if we use a generic language like Java for implementing MetaOperators, we had to write our own compiler and optimization algorithms. Implementation using Jaql was challenging specifically for hierarchical data processing as input data schema and operation *parameters* are not known a-priori. The principal challenge proved to be how to execute an operation on a nested attribute. This required us to traverse the data, reach the given attribute (i.e, the attribute provided in the parameter *path*) and perform the given operation on the attribute. For each ancestor of that attribute, we perform different functionalities depending on whether that attribute is an *array* or a nested *record*. We used the recursion capability of Jaql to do the same. We optimized this recursive implementation for each MetaOperator for performance so that the overhead of our middleware (with respect to native Jaql code) is minimum.

4.2   Advantages of MetaOperators

Main advantages of our MetaOperators based middleware can be summarized as *ease of developing and maintaining big data applications* and *code and data reuse*. A MetaOperator based script will be a series of MetaOperator calls with more transparent code and data schema at the operator boundaries. There is support for processing hierarchical data in a manner similar to the flat data. A user can simply look at the details of *parameters* to infer the semantics of the operation. This results in easier debugging and management of data-flows. Further, by virtue of operation semantics becoming transparent, MetaOperator based scripts can be reused easily by different users for different applications. One can easily derive data lineage (how data was generated?) in our middleware based data-flows. This make Hadoop data being reused across applications and users, which, otherwise not very easy to do.

As MetaOperators bring out functionality and properties of every operation in open, we can extract meta data about each MetaOperator which in turn can be used to extract meta data about a data-flow. Such meta data can be used for schema propagation in a data-flow, searching flows for some specified functionality (which can enable reusing parts of flow), checking input properties, etc.

For each MetaOperator we implemented a utility which checks for the validity of input parameters. This utility performs the checks such as: each *parameters* should contain all the required attributes with their required types, the *path* attribute on which the intended operation is being carried out should be an array, etc. For example, each *MetaAggregation* operator should contain *keys* whereas *functions* is optional. One can just group the records using the *keys* without applying any aggregation function.

For every MetaOperator we implemented a schema propagation utility that computes the schema of the output data that will be obtained after a call to the MetaOperator for a given input data/schema and the corresponding MetaOperator *parameters*. Each of these functions takes two parameters: (a) input data or schema, and (b) the *parameters* passed to the MetaOperator call. One can use this utility to get schema after a series of MetaOperator calls in a data flow.

5   Adding GUI to the Middleware

We have discussed how our MetaOperators based design paradigm makes it easy for a user to design MapReduce data-flows. But still one has to write MetaOperators based script remembering all the required and optional parameters. To further improve usability, we next show how other software systems can easily be plugged into the middleware to extend their capabilities. We illustrate this by discussing extensions of IBM InfoSphere DataStage [3], AnalyticsStage, using the middleware. IBM InfoSphere DataStage is a platform for integrating relational data across multiple data sources and loading (after ETL) it to data warehouses. It has a visual front-end (GUI)– canvass and set of *stages* and *links*. Users can drag and drop stages and links over canvass to define ETL flows. Each stage, requiring a certain number of input properties, describes a specific operation. Two stages are joined by a directed edge implying that the output of the source stage is fed as an input to the destination stage. We extend relational operations for their hierarchical counter-parts and create new stages for non-relational operations. For each such stage, we have the corresponding MetaOperators calls. For example, *TextAnnotation* stage can be used in text analysis scenarios for parsing natural language text and extracting structured/semi-structured entities. Figure 3 shows an example of a data flow over DataStage. In this example,

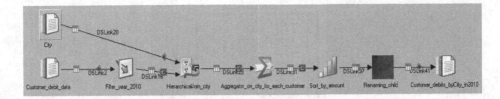

Fig. 3. Data flow over AnalyticsStage

one reads *Customer_debit_data* and filters out the data corresponding to year 2010. The resultant data is hierarchical data with each *debit* record containing *city_id*. This data is joined with *City* data. The join output is aggregated on *city* for each customer and output is presented in the user specified format. A compiler reads the data-flow, parse each stage with its input and output links, generates a MetaOperators script, and runs the script over Hadoop MapReduce.

6   Performance Results

We performed various experiments to show that (1) All the MetaOperators of the middleware are scalable; and (2) Compared to a script (or MapReduce code) written by an expert (having written more than 10000 lines of MapReduce code and/or more than 2000 lines of scripts), middleware (with AnalyticsStage) provide query response time with very low overhead.

Experiment setup: The experiments were run over a 16 node Hadoop cluster built using Blade Servers with four 3 GHz Xeon processors having 8GB memory and 200 GB SATA drives. These machines run Red Hat Linux 5.2. The software stack comprises of Hadoop-0.20.2 with HDFS.

For the first experiment we used books data (Figure 2) as input where each *book* record had between 0 and 5 reviews. The number of discussion for every review were distributed between 0 and 15. Total number of authors and users were 5000. Ten million such records occupied approximately 20 GB space on disk. We bench-marked three hierarchical queries.

SQ1: Insert a new attribute *textLength* which measures the count of characters of the discussion text. This involves a call to *MetaTransform* operator.
SQ2: Carry out a group-by operation on attribute *discussion* and count the number of comments for every user.
SQ3: Join *discussion* attribute with a set of *user* records describing user details such as user name, educational details, geography, etc.

Number of book records were varied from 2 million to 10 million and the time taken for the MetaOperator execution was noted. Results from Figures 4 show that the Jaql implementation of our middleware is scalable. It can be seen that for processing up to 5 times data, MetaOperators take around 4 times the response time compared to that for processing of 2 million records. This

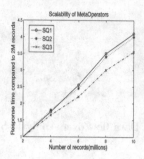

Fig. 4. Scalability Analysis

shows that our implementation has some fixed cost component besides the cost increasing sub-linearly with increasing data sizes. This is an important result for big data processing.

In the next experiment, we compared the response time for data flows written using AnalyticsStage with the same flows written by a user proficient in Jaql. The proficient user is expected to write script which leads to better response times. We considered three use cases for this experiment:

Credit Card Fraud Detection: In this job, we get credit card transactions data, and create tuples of consecutive transaction over the same card. These tuples were used to create indices over amounts of transactions, their geographical area, and time difference between them. These indices can further be used to identify potentially fraud transactions.

Distributed Store Chain: In this typical ETL job, we collect transactional data from distributed store chain and create various summaries of them.

Web-log Processing: The third job involves semi-structured/unstructured data processing. In this job, web logs are collected from various sources, fields like date, client IP address, geography, url, referrer, etc., are extracted and data is cleaned for further analytics.

Figure 5 compares the execution time taken for the three jobs along with the time taken for the expert written queries. AnalyticsStage has to perform extra processing because of two reasons: firstly, MetaOperators have to determine the hierarchical level where the intended operation is performed; and secondly, the MetaOperator code which is called for a particular query depends on whether any of the ancestors (represented using *path*) is an array or a record. Still, the response times for AnanlyticsStage job and expert written scripts were found to be roughly same. This shows that our middleware can be used to express job-semantics succinctly, and the generated query does not result into additional or redundant map-reduce tasks. Thus, we can say that the middleware can indeed be used to build and execute real-life job flows.

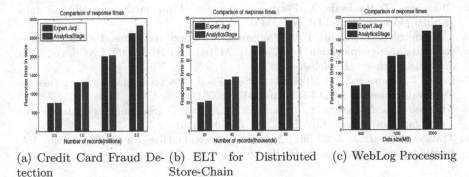

(a) Credit Card Fraud Detection

(b) ELT for Distributed Store-Chain

(c) WebLog Processing

Fig. 5. Comparison of execution time for expert and AnalyticsStage query

7 Conclusions and Future Work

In this paper we presented a middleware and associated utilities aiming at making the process of designing MapReduce data flows easier and more convenient; especially for a user not proficient in MapReduce query languages. We discussed various features the middleware provides, the support for hierarchical and textual data, schema handling and debugging, meta data management, etc. We also discussed a proof-of-concept implementation of MetaOperators and showed that the implementation provides scalable results. Currently we are exploring and incorporating capabilities for storing and handling various data-flow meta-data statistics.

References

1. http://www.cascading.org
2. IBM Cognos Software, http://cognos.com
3. IBM Infosphere DataStage,
   http://www-01.ibm.com/software/data/infosphere/datastage/
4. Microstreategy, http://www.microstrategy.com
5. Pig, http://pig.apache.org
6. Battre, D., Ewen, S., Hueske, F., Kao, O.: Nephele/pacts: A programming model and execution framework for web-scale analytical processing. In: SoCC, pp. 119–130 (2010)
7. Beyer, K., Ercegovac, V., Gemulla, R., Balmin, A., Eltabakh, M., Kanne, C., Ozcan, F., Shekita, E.J.: Jaql: A scripting language for large scale semistructured data analysis. In: VLDB, pp. 1272–1283 (2011)
8. Catarci, T., Costabile, M.F., Levialdi, S., Batini, C.: Visual query systems for databases: A survey. Journal of Visual Languages and Computing 8(2), 215–260 (1997)

9. Chambers, C., Raniwala, A., Perry, F., Adams, S.: Flumejava - easy, efficient data-parallel pipelines. In: PLDI, pp. 363–375 (2010)
10. Chatzopoulou, G., Eirinaki, M., Polyzotis, N.: Query recommendations for inter-active database exploration. In: Winslett, M. (ed.) SSDBM 2009. LNCS, vol. 5566, pp. 3–18. Springer, Heidelberg (2009)
11. Dean, J., Ghemawat, S.: MapReduce: Simplified data processing on large clusters. Communications of ACM 51(1), 107–113 (2008)
12. Khoussainova, N., Kwon, Y., Balazinska, M., Suciu, D.: Snipsuggest:context-aware autocompletion for sql. In: VLDB, pp. 22–33 (2010)
13. Thusoo, A., Sarma, J.S., Jain, N., Shao, Z., et al.: Hive - a petabyte scale data warehouse using hadoop. In: ICDE, pp. 996–1005 (2010)

WYSIWYM Authoring of Structured Content Based on Schema.org

Ali Khalili and Sören Auer

University of Leipzig, Institute of Computer Science, AKSW Group,
Augustusplatz 10, D-04009 Leipzig, Germany
lastname@informatik.uni-leipzig.de
http://aksw.org

Abstract. Structured data is picking up on the Web, particularly in the search world. Schema.org, jointly initiated by Google, Microsoft, and Yahoo! provides a hierarchical set of vocabularies to embed metadata in HTML pages for an enhanced search and browsing experience. RDFa-Lite, Microdata and JSON-LD as lower semantic techniques have gained more attention by Web users to markup Web pages and even emails based on Schema.org. However, from the user interface point of view, we still lack user-friendly tools that facilitate the process of structured content authoring. The majority of information still is contained in and exchanged using unstructured documents, such as Web pages, text documents, images and videos. This can also not be expected to change, since text, images and videos are the natural way how humans interact with information. In this paper we present RDFaCE as an implementation of WYSIWYM (What-You-See-Is-What-You-Mean) concept for direct manipulation of semantically structured content in conventional modalities. RDFaCE utilizes on-the-fly form generation based on Schema.org vocabulary for embedding metadata within Web documents. Furthermore, it employs external NLP services to enable automatic annotation of entities and to suggest URIs for entities. RDFaCE is written as a plugin for TinyMCE WYSIWYG editor thereby can be easily integrated into existing content management systems.

Keywords: WYSIWYG, WYSIWYM, Semantic Content Authoring.

1 Introduction

Structured data is picking up on the Web, particularly in the search world. Semantic structuring of content provides a wide range of advantages compared to unstructured information. It facilitates a number of important aspects of information management:

- For *search and retrieval* enriching documents with semantic representations helps to create more efficient and effective search interfaces, such as faceted search [18] or question answering [9].

X. Lin et al. (Eds.): WISE 2013, Part II, LNCS 8181, pp. 425–438, 2013.
© Springer-Verlag Berlin Heidelberg 2013

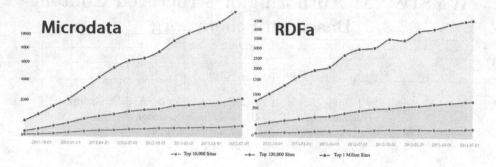

Fig. 1. Microdata and RDFa web usage trends over a historical time period on a large selection of Websites queried by BuiltWith (http://trends.builtwith.com)

- In *information presentation* semantically enriched documents can be used to create more sophisticated ways of flexibly visualizing information, such as by means of semantic overlays as described in [3].
- For *information integration* semantically enriched documents can be used to provide unified views on heterogeneous data stored in different applications by creating composite applications such as semantic mashups [2].
- To realize *personalization*, semantic documents provide customized and context-specific information which better fits user needs and will result in delivering customized applications such as personalized semantic portals [15].
- For *reusability* and *interoperability* enriching documents with semantic representations facilitates exchanging content between disparate systems and enables building applications such as executable papers [11].

Schema.org, jointly initiated by Google, Microsoft, and Yahoo! provides a collection of shared schemas that webmasters can use to markup their pages in order to enable enhanced search and browsing experiences recognized by major search providers. RDFa, Microdata and JSON-LD as lower semantic techniques have gained more attention by Web users to markup Web pages and even emails based on Schema.org (cf. [10] and Figure 1). However, in order for users to truly benefit from this semantic techniques, we need ways to author, visualize and explore unstructured *and* semantic information in a user-friendly manner. The majority of information still is contained in and exchanged using unstructured documents, such as Web pages, text documents, images and videos. This can also not be expected to change, since text, images and videos are the natural way how humans interact with information.

In this paper we present WYSIWYM (What-You-See-Is-What-You-Mean) concept for direct manipulation of semantically structured content in conventional modalities. Our WYSIWYM concept formalizes the binding between semantic representation models and UI elements for authoring, visualizing and exploration. As an implementation of WYSIWYM model suitable for lower semantic techniques, we showcase RDFaCE interface. RDFaCE utilizes on-the-fly form generation based on Schema.org vocabulary for embedding metadata within Web documents. Furthermore, it employs external NLP services to enable automatic annotation of entities and to suggest URIs for entities. RDFaCE is written as a plugin for TinyMCE WYSIWYG editor thereby can be easily integrated into existing content management systems.

The remainder of this article is structured as follows: In Section 2, we describe the background of our work and discuss the related work. Section 3 describes the fundamental WYSIWYM concept proposed in the paper. In Section 4, we introduce RDFaCE as an implemented WYSIWYM interface. Section 5 explains some use cases of WYSIWYM authoring of structured content based on Schema.org. Finally, Section 6 concludes with an outlook on future work.

2   Related Work

Binding data to UI elements. There are already many approaches and tools which address the binding between data and UI elements for visualizing and exploring semantically structured data. Dadzie and Rowe [4] present the most exhaustive and comprehensive survey to date of these approaches. For example, *Fresnel* [12] is a display vocabulary for core RDF concepts. Fresnel's two foundational concepts are lenses and formats. Lenses define which properties of an RDF resource, or group of related resources, are displayed and how those properties are ordered. Formats determine how resources and properties are rendered and provide hooks to existing styling languages such as CSS.

Parallax, Tabulator, Explorator, Rhizomer, Sgvizler, Fenfire, RDF-Gravity, IsaViz and *i-Disc for Topic Maps* are examples of tools available for visualizing and exploring semantically structured data. In these tools the binding between semantics and UI elements is mostly performed implicitly, which limits their versatility. However, an explicit binding as advocated by our WYSIWYM model can be potentially added to some of these tools.

In contrast to the structured content, there are many approaches and tools which allow binding semantic data to UI elements within unstructured content (cf. our comprehensive literature study [6]). As an example, *Dido* [5] is a data-interactive document which lets end users author semantic content mixed with unstructured content in a web-page. Dido inherits data exploration capabilities from the underlying *Exhibit*[1] framework. *Loomp* as a prove-of-concept for the *One Click Annotation* [1] strategy is another example in this context. Loomp is a WYSIWYG web editor for enriching content with RDFa annotations. It employs a partial mapping between UI elements and data to hide the complexity of creating semantic data.

WYSIWYG and WYSIWYM. The term *WYSIWYG* as an acronym for What-You-See-Is-What-You-Get is used in computing to describe a system in which content (text and graphics) displayed on-screen during editing appears in a form closely corresponding to its appearance when printed or displayed as a finished

---

[1] http://simile-widgets.org/exhibit/

product. WYSIWYG text authoring is meanwhile ubiquitous on the Web and part of most content creation and management workflows (e.g. content management systems , weblogs, wikis). However, the WYSIWYG model has been criticized, primarily for the verbosity, poor support of semantics and low quality of the generated code and there have been voices advocating a change towards a *WYSIWYM* (What-You-See-Is-What-You-Mean) model [17,16].

The first use of the WYSIWYM term occurred in 1995 aiming to capture the separation of presentation and content when writing a document. The *LyX editor*[2] was the first WYSIWYM word processor for structure-based content authoring. Instead of focusing on the format or presentation of the document, a WYSIWYM editor preserves the intended meaning of each element. For example, page headers, sections, paragraphs, etc. are labeled as such in the editing program, and displayed appropriately in the browser. Another usage of the WYSIWYM term was by Power et al. [13] in 1998 as a solution for *Symbolic Authoring*. In symbolic authoring the author generates language-neutral "symbolic" representations of the content of a document, from which documents in each target language are generated automatically, using *Natural Language Generation* technology. In this What-You-See-Is-What-You-Meant approach, the language generator was used to drive the user interface (UI) with support of localization and multilinguality. Using the WYSIWYM natural language generation approach, the system generates a feed-back text for the user that is based on a semantic representation. This representation can be edited directly by the user by manipulating the feedback text.

The WYSIWYM term as defined and used in this paper targets the novel aspect of integrated visualization, exploration and authoring of unstructured and semantic content. The rationale of our WYSIWYM concept is to enrich the existing WYSIWYG presentational view of the content with UI components revealing the *semantics* embedded in the content and enable the exploration and authoring of semantic content. Instead of separating presentation, content and meaning, our WYSIWYM approach aims to integrate these aspects to facilitate the process of *Semantic Content Authoring*.

Schema.org Tools. Schema.org is an effort initiated by the popular search engines Bing, Google and Yahoo! on June 2011 to define a broad, Web-scale and shared vocabulary focusing on popular concepts. It stakes a position as a *middle* ontology that does not attempt to have the scope of an *ontology of everything* or go into depth in any one area. A central goal of having such a broad schema all in one place is to simplify things for mass adoption and cover the most common use cases [14]. Much of the vocabulary on schema.org is inspired by existing vocabularies such as *FOAF*[3], *GoodRelations*[4], *OpenCyc*[5], *rNews*[6], etc.

---

[2] http://www.lyx.org/
[3] http://www.foaf-project.org
[4] http://www.heppnetz.de/projects/goodrelations
[5] http://www.cyc.com/platform/opencyc
[6] http://dev.iptc.org/rNews

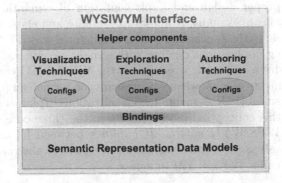

Fig. 2. Schematic view of the WYSIWYM model

Schema.org annotations can be written using markup attributes in HTML such as *RDFa*[7] and *Microdata*[8] or in pure JSON as *JOSN-LD*[9]. Google recently published two Webmaster tools namely *Data Highlighter* and *Structured Data Markup Helper* (SDMH)[10] for content annotation based on Schema.org. SDMH is very similar to RDFaCE described in this paper but with the following differences: RDFaCE unifies the authoring and publishing of the structured content on the Web. It can be integrated into the existing content management systems to deal with the structured content from the scratch. On the contrary, SDMH as a standalone tool offers a separate additional step to publish metadata on websites. The RDFaCE approach is very flexible and can be configured to cover all the schemas on Schema.org as well as other existing vocabularies. SDMH at the moment only supports a limited set of schemas (i.e. Articles, Events, Local Businesses, Movies, Products, Restaurants, Software Applications, TV Episodes) from Schema.org. RDFaCE also offers automatic content annotation feature which is not supported by SDMH.

3   WYSIWYM Concept

In this section we introduce the fundamental WYSIWYM (What-You-See-Is-What-You-Mean) concept and formalize key elements of the concept. Formalizing the WYSIWYM concept has a number of advantages: First, the formalization can be used as a basis for design and implementation of novel applications for authoring, visualization, and exploration of semantic content. The formalization serves the purpose of providing a terminology for software engineers, user interface and domain experts to communicate efficiently and effectively. It provides insights into and an understanding of the requirements as well as corresponding

---

[7] http://www.w3.org/TR/rdfa-syntax/

[8] http://www.w3.org/TR/microdata/

[9] http://json-ld.org/

[10] https://www.google.com/webmasters/markup-helper/

UI solutions for proper design and implementation of semantic content management applications.p Secondly, it allows to evaluate and classify existing user interfaces according to the conceptual model in a defined way. This will highlight the gaps in existing applications dealing with semantic content.

Figure 2 provides a schematic overview of the WYSIWYM concept. The rationale is that elements of a knowledge representation formalism (or data model) are connected to suitable UI elements for visualization, exploration and authoring. Formalizing this conceptual model results in the three core definitions (1) for the abstract *WYSIWYM model*, (2) *bindings* between UI and representation elements as well as (3) a concrete instantiation of the abstract WYSIWYM model, which we call a *WYSIWYM interface*.

Definition 1 (WYSIWYM model). *The WYSIWYM model is a quintuple* (D, V, X, T, H) *where:*

- *D is a set of semantic representation data models, where each $D_i \in D$ has an associated set of data model elements E_{D_i};*
- *V is a set of tuples (v, C_v), where v is a visualization technique and C_v a set of possible configurations for v;*
- *X is a set of tuples (x, C_x), where x is an exploration technique and C_x a set of possible configurations for x;*
- *T is a set of tuples (t, C_t), where t is an authoring technique and C_t a set of possible configurations for t;*
- *H is a set of helper components.*

Semantic representation models are conceptual data models to express the meaning of information thereby enabling representation and interchange of knowledge. Based on their expressiveness, we can roughly divide popular semantic representation models into the three categories *tree-based*, *graph-based* and *hypergraph-based*. Each semantic representation model comprises a number of representation elements, such as various types of entities and relationships. For example, elements of a typical graph-based data model are:

- Instances – e.g. `Warfarin` as a drug.
- Classes – e.g. `anticoagulants drug` for Warfarin.
- Relationships between entities (instances or classes) – e.g. the `interaction` between `Aspirin` as an antiplatelet drug and `Warfarin` which will increase the risk of bleeding.
- Literal property values – e.g. the halflife for the Amoxicillin.
  - Value – e.g. `61.3` minutes.
  - Language tag – e.g. `en`.
  - Datatype – e.g. `xsd:float`.

Visualization. The primary objectives of visualization are to present, transform, and convert semantic data into a visual representation, so that, humans can read, query and edit them efficiently. We divide existing techniques for visualization of knowledge encoded in text, images and videos into the three categories *Highlighting*, *Associating* and *Detail-view*. Highlighting includes UI techniques which are used to distinguish or highlight a part of an object (i.e. text, image or video) from the whole object. For example, *Framing and Segmentation* (i.e. different borders, overlays and backgrounds), *Text formatting* (i.e. different colors, fonts, text size, etc.) and *Marking* (i.e. icons appended to text or image) are some highlighting techniques. Associating deals with techniques that visualize the relation between some parts of an object. *Line connectors* and *Arrow connectors* are two examples of associating techniques. Detail-view includes techniques which reveal detailed information about a part of an object. *Callouts* as strings of text connected to a part of text giving information about that part are an example of detail-view techniques.

Exploration. To increase the effectiveness of visualizations, users need to be capable to dynamically explore the visual representation of the semantic data. The dynamic exploration of semantic data will result in faster and easier comprehension of the targeted content. *Zooming* (i.e. changing the scale of the viewed area in order to see more or less detail), *Faceting* (i.e. accessing information organized according to a faceted classification system), *Bar layouts* (i.e. using vertical nested bars to indicate the hierarchy of entities) and *Expandable callouts* (i.e. interactive and dynamic callouts to explore the properties and relations of an entity) are some examples of available exploration techniques.

Authoring. Semantic authoring aims to add more meaning to digitally published documents. If users do not only publish the content, but at the same time describe what it is they are publishing, then they have to adopt a structured approach to authoring. A semantic authoring UI is a human accessible interface with capabilities for writing and modifying semantic documents. *Form editing*, *Inline editing*, *Drawing*, *Drag and drop*, *Context menu* and *(Floating) Ribbon editing* are some examples of authoring techniques.

Helper components. In order to facilitate, enhance and customize the WYSIWYM model, we utilize a set of helper components, which implement crosscutting aspects. A helper component acts as an extension on top of the core functionality of the WYSIWYM model. For example, *Automation* is a helper component which means the provision of facilities for automatic annotation of text, images and videos to reduce the need for human work and thereby facilitating the efficient annotation of large item collections. *Recommendation* means providing users with pre-filled form fields, suggestions (e.g. for URIs, namespaces, properties), default values etc. These facilities simplify the authoring process, as they reduce the number of required user interactions. Moreover, they help preventing incomplete or empty metadata.

The WYSIWYM model represents an abstract concept from which concrete interfaces can be derived by means of bindings between semantic representation model elements and configurations of particular UI elements.

			Graph-based (e.g. RDF)					
			Instance	Class	Relationships between entities	Literal property values		
						Value	Language tag	Datatype
UI categories		UI techniques						
Visualization	Highlighting	Framing and segmentation (borders, overlays, backgrounds)	■					
		Text formatting (color, font, size etc.)	▓					
		Marking (appended icons)	■					
	Associating	Line connectors			■	*		
		Arrow connectors			■	*		
	Detail view	Callouts (infotips, tooltips, popups)	■			■		
Exploration		Zooming	■	■				
		Faceting	■	■				
		Bar layout	■			■		
		Expandable callouts		■	■			
Authoring		Form editing	■			■		
		Inline edit	■					
		Drawing	■	▓	■			
		Drag and drop	■		▓			
		Context menu	■		▓	■		
		(Floating) Ribbon editing	■					

* If value is available in the text.

No binding □   Partial binding ▓   Full binding ■

Fig. 3. Possible bindings between user interface & RDF semantic representation model

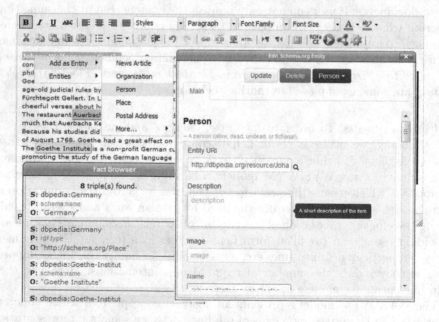

Fig. 4. An screenshot of RDFaCE authoring environment

Definition 2 (Binding). *A binding b is a function which maps each element of a semantic representation model e (e $\in E_{D_i}$) to a set of tuples (ui, c), where ui is a UI technique ui (ui $\in V \cup X \cup T$) and c is a configuration c $\in C_{ui}$.*

Figure 3 gives an example of possible bindings between the user interface (rows) and RDF semantic representation model (columns)[11]. These possible bindings were obtained from a comprehensive survey of existing tools and an analysis of possible connections between a specific UI element and a semantic model element. The shades of gray in a certain cell indicate the suitability of a certain binding between a particular UI and data model element. Each binding can have a configuration to customize its implementation. For example, special borders or background styles, bar styles or a related icons can be defined for each semantic entity type.

Once a selection of data models and UI elements was made and both are bound to each other in a binding, we attain a concrete instantiation of our WYSIWYM model called WYSIWYM interface.

Definition 3 (WYSIWYM interface). *An instantiation of the WYSIWYM model I called WYSIWYM interface is a hextuple $(D_I, V_I, X_I, T_I, H_I, b_I)$, where:*

- *D_I is a selection of semantic representation data models ($D_I \subset D$);*
- *V_I is a selection of visualization techniques ($V_I \subset V$);*
- *X_I is a selection of exploration techniques ($X_I \subset X$);*
- *T_I is a selection of authoring techniques ($T_I \subset T$);*
- *H_I is a selection of helper components ($H_I \subset H$);*
- *b_I is a binding between a particular occurrence of a data model element and a visualization, exploration and/or authoring technique[12].*

4   RDFaCE WYSIWYM Interface for Structured Content Authoring Based on Schema.org

RDFaCE (RDFa Content Editor)[8] is a WYSIWYM interface for semantic authoring of textual content. It is implemented on top of the *TinyMCE*[13] rich text editor. RDFaCE extends the existing WYSIWYG user interfaces to facilitate semantic authoring within popular CMSs, such as blogs, wikis and discussion forums. The RDFaCE implementation is open-source and available for download together with an explanatory video and online demo at http://rdface.aksw.org. Since releasing the RDFaCE, the tool has been downloaded over 2000 times and the online demo

---

[11] A more complete list of bindings are available at [7].

[12] Note, that we limit the definition to one binding, which means that only one semantic representation model is supported in a particular WYSIWYM interface at a time. It could be also possible to support several semantic representation models (e.g. RDFa and Microdata) at the same time. However, this can be confusing to the user, which is why we deliberately excluded this case in our definition.

[13] http://www.tinymce.com

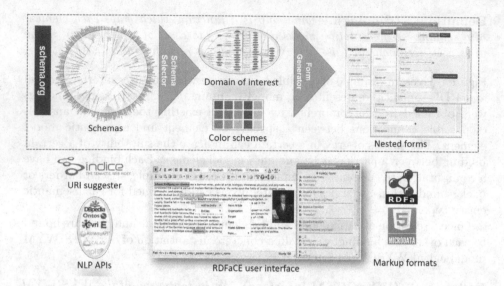

Fig. 5. Configuration steps in RDFaCE

page has received more than 4000 unique visits. RDFaCE as a WYSIWYM instantiation supports Microdata and RDFa serializations of RDF data model based on the bindings defined in Figure 3. For visualization, it uses framing using backgrounds to highlight the semantic entities within text content. It also supports callouts in form of dynamic tooltips to reveal the type and property values of semantic entities. For exploration, it supports faceting based on the type of entities, properties and values. For authoring, it employs different methods such as form editing, context menu and ribbon editing(cf. Figure 4). As helper component, RDFaCE utilizes automation and recommendation based on external Web services.

4.1   Configuration

Figure 5 presents the configuration steps in RDFaCE. The first step is to model the user's domain of interest by selecting a subset of Schema.org schemas. For example user might select NewsArticle, Person, Organization and Place schemas as his desirable schemas. For each schema, the range of properties will be checked in order to include derivative schemas as well (e.g. PostalAddress and Country for the Place schema). The results of this step is an input JSON file which describes the selected schemas together with their corresponding properties. For visualization of schemas, we need to assign unique colors to the selected schemas. We use an algorithm to automatically generate a light color scheme for the schemas. The color scheme is available as CSS styles and is easily configurable by users.

The next step is to generate appropriate forms based on the selected schemas. Form inputs are created based on the corresponding data type defined as range

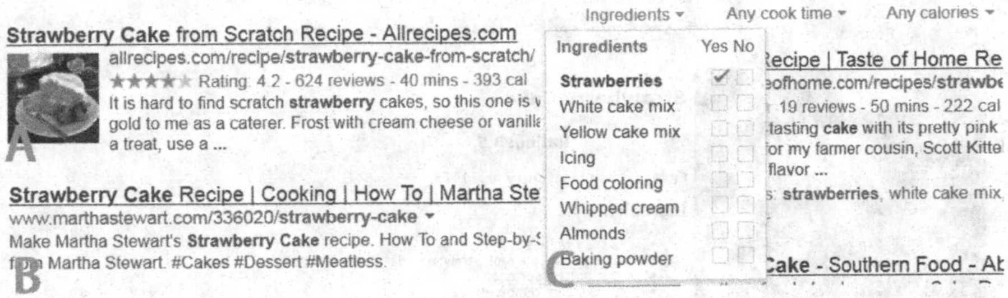

Fig. 6. Search results improved by rich snippets. A: enhanced recipe, B: normal recipe, C: browsing recipes by ingredients, cook time and calories.

of the schema properties. For example we add Datepicker UI for properties with Date as their range. These forms are then used to add metadata into the text.

The final step in configuration is to select the desired markup format (e.g. RDFa or Microdata) as well as desired NLP APIs (e.g. DBpedia Spotlight) for automatic annotation of content. Users can select multiple NLP APIs and determine how they want to combine the results. The combination can be performed based on the agreement between two or more of the involved APIs. Users are also able to set a confident level for automatic annotation and can limit the type of recognized entities to only annotate specific entities like Persons and Places.

4.2 One-Click Annotation

The annotation steps in RDFaCE are very straightforward. RDFaCE follows the *One Click Annotation* [1] strategy. For automatic annotation, user only needs to press the corresponding RDFaCE button. For manual annotation, user should select some parts of the text and using the context menu, he can choose the appropriate annotations for the selected entities. Context menu is created dynamically based on the user selection. For example if user's selection is inside a NewsArticle schema, user will only see the properties related to the NewsArticle.

RDFaCE includes a fact browser which will reveal the RDF annotation embedded in the text. User can also use tools like Google structured data testing tools[14] to check the validity of embedded annotations.

5 Use Cases: Wordpress and Rich Text Snippets

On-page markup based on Schema.org enables search engines to increasingly understand the information on Web pages and provide richer search results. *Rich Text Snippets* as an example of on-page markup provides an immediate advantage and motivation for Web users to embed structured content into their documents. Rich snippets comprise a wide range of schemas such as *Breadcrumbs*,

---

[14] http://www.google.com/webmasters/tools/richsnippets

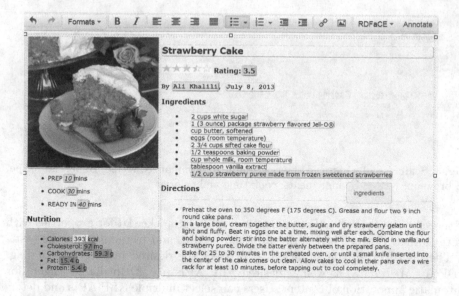

Fig. 7. Using RDFaCE to annotate recipes based on `Schema.org` recipe schema

Events, Music, Organizations, People, Products, Recipes, Review ratings, Reviews, Software Applications and *Videos*. Web documents which are annotated based on these schemas will attract more attention by people searching the Web due to the richness of presented information. Figure 6 shows as example of enhanced search results for recipes powered by rich snippets.

RDFaCE comes with a *Wordpress* plugin[15] to enable Weblog authors to create rich snippets based on Schema.org. Wordpress is an open source Weblog tool and publishing platform which is often customized into a Content Management System (CMS). Wordpress is used by 58.5% of websites with known CMS (i.e. 18.8% of all websites[16]). Wordpress uses TinyMCE as its content editor. That makes it easy to add the RDFaCE plugin for WYSIWYM authoring of structured content within this CMS. With the integration of RDFaCE into the Wordpress, the availability of semantically annotated content on the Web can be substantially increased.

As an example scenario, Figure 7 presents the RDFaCE WYSIWYM interface employed to annotate a sample recipe rich snippet. The user simply selects the parts of the text and annotates them using the corresponding schema from Schema.org. On the background, RDFaCE generates the corresponding RDFa or Microdata markup based on the following rules:

– if the selected text already has an HTML tag, metadata will be added as new attributes for the current tag (e.g. Code 1.1 line 2 or 6).

---

[15] http://wordpress.org/plugins/rdface/
[16] W3Techs.com, 17 July 2013.

- if the selected text does not have an HTML tag, a new `<SPAN>` or `<DIV>` tag with corresponding attributes will be created (e.g. Code 1.1 line 1 or 3).
- if no text is selected, a new `<META>` tag with corresponding attributes will be created (e.g. Code 1.1 line 17 or 18).

Code 1.1. Example of Microdata annotations generated by RDFaCE

```
1   <div itemscopeitemtype="http://schema.org/Recipe">
2       <h2 itemprop="name">Strawberry Cake</h2>
3       <p>By <span itemprop="author">Ali Khalili</span>, July 8, 2013 </p>
4       <h4>Ingredients</h4>
5       <ul>
6           <li itemprop="ingredients">2 cups white sugar</li>
7           <li>...</li>
8       </ul>
9       <p>Preparation time: 10 mins</p>
10      <p>Cooking time: 30 min</p>
11      <p>Ready in 40 min</p>
12      <p>
13          <span itemscope itemtype="http://schema.org/NutritionInformation"
                  itemprop="nutrition">
14              Calories: <span itemprop="calories">393 kcal</span>
15          </span>
16      </p>
17          <meta itemprop="dateCreated" content="2013-07-08">
18          <meta itemprop="prepTime" content="PT15M">
19  </div>
```

6   Conclusion

Bridging the gap between unstructured and semantic content is a crucial aspect for the ultimate success of semantic technologies. With RDFaCE we presented in this article an approach and its implementation of a WYSIWYM (What You See Is What You Mean) concept for direct manipulation of structured content in conventional modalities. RDFaCE employs Schema.org schemas to promote structured content authoring among a wide range of Web users.

We see the work presented in this article as an initial step in a larger research agenda aiming at simplifying the authoring and annotation of semantically enriched textual content. Regarding future work we envision to extend the RDFaCE implementation to support other modalities such as images and multimedia objects. We also envision to perform an extensive usability evaluation of RDFaCE in order to improve its usability as well as functionality.

Acknowledgments. We would like to thank our colleagues from AKSW research group for their helpful comments and inspiring discussions during the development of RDFaCE. This work was supported by a grant from the European Union's 7th Framework Programme provided for the project LOD2 (GA no. 257943).

References

1. One Click Annotation. CEUR Workshop Proceedings, vol. 699 (February 2010)
2. Ankolekar, A., Krötzsch, M., Tran, T., Vrandecic, D.: The two cultures: mashing up web 2.0 and the semantic web. In: WWW 2007, pp. 825–834 (2007)
3. Burel, G., Cano1, A.E., Lanfranchi, V.: Ozone browser: Augmenting the web with semantic overlays. . CEUR WS Proceedings, vol. 449 (June 2009)
4. Dadzie, A.-S., Rowe, M.: Approaches to visualising linked data: A survey. Semantic Web 2(2), 89–124 (2011)
5. Karger, D.R., Ostler, S., Lee, R.: The web page as a wysiwyg end-user customizable database-backed information management application. In: UIST 2009, pp. 257–260. ACM (2009)
6. Khalili, A., Auer, S.: User interfaces for semantic authoring of textual content: A systematic literature review. To appear in the Journal of Web Semantics (2013)
7. Khalili, A., Auer, S.: Wysiwym – integrated visualization, exploration and authoring of unstructured and semantic content. Submitted to the Semantic Web Journal (2013)
8. Khalili, A., Auer, S., Hladky, D.: The rdfa content editor. In: IEEE COMPSAC 2012 (2012)
9. Lopez, V., Uren, V., Sabou, M., Motta, E.: Is question answering fit for the semantic web? a survey. Semantic Web? Interoperability, Usability, Applicability 2(2), 125–155 (2011)
10. Mika, P., Potter, T.: Metadata statistics for a large web corpus. In: Bizer, C., Heath, T., Berners-Lee, T., Hausenblas, M. (eds.) LDOW. CEUR Workshop Proceedings, vol. 937, CEUR-WS.org (2012)
11. Muller, W., Rojas, I., Eberhart, A., Haase, P., Schmidt, M.: A-r-e: The author-review-execute environment. Procedia Computer Science 4, 627–636 (2011); ICCS 2011
12. Pietriga, E., Bizer, C., Karger, D.R., Lee, R.: Fresnel: A browser-independent presentation vocabulary for RDF. In: Cruz, I., Decker, S., Allemang, D., Preist, C., Schwabe, D., Mika, P., Uschold, M., Aroyo, L.M. (eds.) ISWC 2006. LNCS, vol. 4273, pp. 158–171. Springer, Heidelberg (2006)
13. Power, R., Power, R., Scott, D., Scott, D., Evans, R., Evans, R.: What you see is what you meant: direct knowledge editing with natural language feedback (1998)
14. Ronallo, J.: HTML5 Microdata and Schema.org. The Code4Lib Journal 16 (February 2012)
15. Sah, M., Hall, W., Gibbins, N.M., Roure, D.C.D.: Semport - a personalized semantic portal. In: 18th ACM Conf. on Hypertext and Hypermedia, pp. 31–32 (2007)
16. Sauer, C.: What you see is wiki – questioning WYSIWYG in the Internet age. In: Proceedings of Wikimania 2006 (2006)
17. Spiesser, J., Kitchen, L.: Optimization of html automatically generated by wysiwyg programs. In: WWW 2004, pp. 355–364 (2004)
18. Tunkelang, D.: Faceted Search (Synthesis Lectures on Information Concepts, Retrieval, and Services). Morgan and Claypool Publishers (June 2009)

Probabilistic n-of-N Skyline Computation over Uncertain Data Streams[*]

Wenjie Zhang[1], Aiping Li[2], Muhammad Aamir Cheema[1], Ying Zhang[1],
and Lijun Chang[1]

[1] University of New South Wales, Sydney, NSW, Australia
[2] National University of Defense Technology, China
{zhangw,macheema}@cse.unsw.edu.au, apli1974@gmail.com,
{yingz,ljchang}@cse.unsw.edu.au

Abstract. Skyline operator is a useful tool in multi-criteria decision making in various applications. Uncertainty is inherent in real applications due to various reasons. In this paper, we consider the problem of efficiently computing probabilistic skylines against the most recent N uncertain elements in a data stream seen so far. Specifically, we study the problem in the n-of-N model; that is, computing the probabilistic skyline for the most recent n ($\forall n \leq N$) elements, where an element is a probabilistic skyline element if its skyline probability is not below a given probability threshold q. Firstly, an effective pruning technique to minimize the number of uncertain elements to be kept is developed. It can be shown that on average storing only $O(\log^d N)$ uncertain elements from the most recent N elements is sufficient to support the precise computation of all probabilistic n-of-N skyline queries in a d-dimension space if the data distribution on each dimension is independent. A novel encoding scheme is then proposed together with efficient update techniques so that computing a probabilistic n-of-N skyline query in a d-dimension space is reduced to $O(d \log \log N + s)$ if the data distribution is independent, where s is the number of skyline points. Extensive experiments demonstrate that the new techniques on uncertain data streams can support on-line probabilistic skyline query computation over rapid data streams.

1 Introduction

Skyline analysis has been shown as a useful tool in multi-criterion decision making. Given a certain data set D, an element $s_1 \in D$ dominates another element $s_2 \in D$ if s_1 is better than s_2 in at least one aspect and not worse than s_2 in all other aspects. The skyline on D comprises of elements in D that are not dominated by any other element from D.

Uncertain data analysis is an important issue in many emerging important applications, such as sensor networks, trend prediction, moving object management, data

---

[*] Wenjie Zhang was partially supported by ARC DE120102144 and DP120104168. Aiping Li (Corresponding Author) was partially supported by State Key Development Program of Basic Research of China (No. 2013CB329601) and National Key Technology R&D Program (No. 2012BAH38B-04). Muhammad Aamir Cheema was partially supported by ARC DE130101002 and DP130103405. Ying Zhang was partially supported by ARC DP110104880, DP130103245 and UNSW ECR grant PSE1799.

X. Lin et al. (Eds.): WISE 2013, Part II, LNCS 8181, pp. 439–457, 2013.
© Springer-Verlag Berlin Heidelberg 2013

cleaning and integration, economic decision making, and market surveillance. Uncertainty is inherent in such applications due to various factors such as data randomness and incompleteness, limitation of equipment, and delay or loss in data transfer. In many scenarios, uncertain data is collected in a streaming fashion. Uncertain streaming data computation has attracted significant research attention and the existing work mainly focuses on aggregations, top-k queries [1–3], etc.

Skyline computation over uncertain streaming data has many applications and has been studied in [4, 5]. For instance, in an on-line shopping system products are evaluated in various aspects such as *price*, *condition* (e.g., brand new, excellent, good, average, etc), and *brand*. A customer may want to select a product, say laptops, based on the multiple criteria such as low price, good condition, and brand preference. In the application, each seller is also associated with a "trustability" value which is derived from customers' feedback on the seller's product quality, delivery handling, etc; the trustability value may be regarded as the "occurrence" probability of the product since it represents the probability that the product occurs exactly as described in the advertisement in terms of delivery and quality. For simplicity, we assume that a customer only prefers a particular brand and remove the brand dimension from ranking. Table 1 lists four qualified results. Both L_1 and L_4 are skyline points regarding (price, condition), L_1 is better than (dominates) L_2, and L_4 is better than L_3. Nevertheless, L_1 is posted long time ago, and the trustability of L_4 is quite low. In such applications, customers may want to continuously monitor on-line advertisements by selecting the candidates for the best deal - skyline points. Clearly, we need to "discount" the dominating ability from offers with too low trustability. Moreover, too old offers may not be quite relevant. We model such an on-line selection problem as probabilistic skyline against sliding windows by treating on-line advertisements as an uncertain data stream (see Section 2 for details) such that each data element (advertisement) has an occurrence probability. Moreover, different users may have different favorite thresholds of the number N of most recent elements to monitor. Therefore, it is important for an information provider (system) to organize the most recent N elements in an effective way, so that any "n-of-N skyline" queries (the computation of the skyline of the most recent n ($\forall n \leq N$) elements) can be processed efficiently.

Table 1. Laptop Advertisements

Product ID	Time	Price	Condition	Trustability
L_1	107 days ago	$ 550	excellent	0.80
L_2	5 days ago	$ 680	excellent	0.90
L_3	2 days ago	$ 530	good	1.00
L_4	today	$ 200	good	0.48

[6, 7] are the first attempts to investigate skyline computation on certain sliding windows while [6] is the first paper to tackle such a problem in the n-of-N model. To the best of our knowledge, this paper is the first work to study skyline queries in context of n-of-N model over uncertain data streams. Our contribution can be summarized as follows.

1. We formally define the problem of probabilistic skyline computation over uncertain data streams regarding the n-of-N model.
2. An efficient pruning technique has been developed to minimize the number $\mathcal{N}$ ($\mathcal{N} \leq N$) of uncertain elements to be kept in the most recent N elements for processing all probabilistic n-of-N queries. We showed that in a d-dimensional space $\mathcal{N} = O(\log^d N)$ if the data distribution on each dimension is independent.
3. A novel encoding scheme with linear size $O(\mathcal{N})$ on the stored elements is developed, together with the efficient update algorithms based R-tree and interval tree techniques. This encoding scheme effectively reduces the time complexity for processing a probabilistic n-of-N skyline query to $O(\log \mathcal{N} + s)$ from $O(n \log n)$ for $d = 2, 3$ and $O(n \log^{d-2} n)$ for $d \geq 4$, where s is the number of skyline points.
4. Extensive experiments indicated that the new techniques can accommodate on-line computation against very rapid data streams.

The rest of the paper is organized as follows. In section 2, we formally define probabilistic skyline queries over uncertain data streams regarding the n-of-N model and provide necessary preliminaries. Section 3 presents the minimum candidate set to process probabilistic n-of-N skyline queries and the encoding scheme. Section 4 provides the techniques for continuously maintaining the indexing structures for the candidate set. Results of comprehensive performance studies are discussed in section 5. Related works are summarized in Section 6 and section 7 concludes the paper.

2 Background

We present problem definition and necessary preliminaries in this section.

2.1 Problem Definition

For two exact d-dimensional elements u and v, u dominates v, denoted by $u \prec v$, if $u.i \leq v.i$ for every $1 \leq i \leq d$, and there exists a dimension j with $u.j < v.j$. Given a set of elements, the *skyline* consists of all elements which are not dominated by any other elements. In many applications, a data stream is *append-only*; that is, there is no deletions of data elements involved. In this paper, we study the skyline computation problem restricted to the append-only data stream model. In a data stream, elements are positioned according to their relative arrival ordering and labeled by integers. Note that the position/label $\kappa(a)$ means that the element a arrives $\kappa(a)$-th in the data stream.

In an uncertain data stream DS, each element $a \in DS$ has a probability $P(a)$ ($0 < P(a) \leq 1$) to occur where $a.i$ (for $1 \leq i \leq d$) denotes the i-th dimension value. Given a sequence DS of uncertain data elements, a *possible world* W is a subsequence of DS. The probability of W to appear is $P(W) = \Pi_{a \in W} P(a) \times \Pi_{a \notin W}(1 - P(a))$. Let ω be the set of all possible worlds, then $\sum_{W \in \omega} P(W) = 1$. We use $SKY(W)$ to denote the set of elements in W that form the skyline of W. The probability that an element a appears in the skylines of the possible worlds is $P_{sky}(u) = \sum_{a \in SKY(W), W \in \omega} P(W)$. $P_{sky}(a)$ is called the *skyline probability* of a. Equation 2 below can be immediately verified.

$$P_{sky}(a) = P(a) \times \Pi_{a' \in DS, a' \prec a}(1 - P(a')) \tag{1}$$

Denote N as the size of the sliding window, namely we only keep the recent N elements in the data stream. In the rest of the paper, we abuse $P_{sky}(a)$ to denote the skyline probability for a to be a skyline element within the recent N elements. Note that a is also within the most recent N elements P_N. Namely,

$$P_{sky}(a) = P(a) \times \Pi_{a' \in P_N, a' \prec a}(1 - P(a')) \qquad (2)$$

In n-of-N model, skyline computation is supported for any window length n with $n \leq N$. Suppose an element a is within the most recent n ($n \leq N$) elements, we denote $P_{sky,n}(a)$ as the skyline probability of a computed regarding the most recent n elements only. Namely,

$$P_{sky,n}(a) = P(a) \times \Pi_{\kappa(a), \kappa(a') \geq M-n+1, a' \prec a}(1 - P(a')) \qquad (3)$$

where M is the totaly number of elements in the data stream so far and $\kappa(a), \kappa(a') \geq M - n + 1$ implicates that a and a' are within the most recent n elements. Clearly we have $P_{sky,n}(a) \geq P_{sky}(a)$, which implies that even a is not a skyline element in the sliding window with size N, it may still be a skyline element in the most recent n elements for some $n \leq N$. We denote $\mathcal{P}n$-of-N query as the query to retrieve probabilistic skyline elements against any most recent n ($n \leq N$) elements in the data stream regarding a given probability threshold.

Problem Statement. Given a data stream DS in which each uncertain element $a \in DS$ is associated with an occurrence probability $P(a)$ ($0 < P(a) \leq 1$) indicating the likelihood that a exists in DS. We say an element a is a probabilistic skyline element within the most recent n elements if $P_{sky,n}(a) \geq q$. A $\mathcal{P}n$-of-N query retrieves the probabilistic skyline elements within the most recent n ($\forall n \leq N$) data elements in the data stream DS.

2.2 Preliminaries

n-**of-**N **Model.** As an important method to support query processing over different thresholds of window size, n-of-N model is firstly proposed in [8] to efficiently maintain quantile summaries. We will investigate the problem of effectively organising the most recent N elements in an uncertain data stream seen so far, so that the computation of probabilistic skyline against any most recent n ($n \leq N$) elements can be processed efficiently. Note that a *sliding window* model [9] is a special case of the n-of-N model where $n = N$.

Stabbing Queries. Given a set of m intervals and a *stabbing* point p in the 1-dimensional space, the *stabbing query* is to find all intervals which contain p. By the interval tree techniques in [10], a stabbing query can be processed in $O(\log m + l)$ where l is the number of intervals in the result. By storing an interval only in the tree node that is the lowest common ancestor (LCA) of the two end points of the interval, the space complexity of the interval tree is $O(m)$. It has been also shown that the time complexity of an update (insertion or deletion) to an interval tree is amortized to $O(\log m)$ per deletion or insertion. Note that the intervals here can be closed, half closed, or open at both ends.

n-of-N Skyline Query over Exact Data Streams. [6] studies skyline computation over exact sliding windows following the n-of-N model. It is observed that over exact data streams, if an element a is dominated by a newer element a', then a will never be a skyline for any recent n ($n \in N$) elements since a' expires later than a. It is also proved that in such a case removing a from the data stream will not affect computation of n-of-N skyline queries processing. Thus, minimum candidate set R_N comprises of elements in the data stream which are not dominated by newer elements.

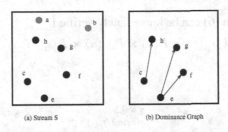

(a) Stream S   (b) Dominance Graph

Fig. 1..Dominance Graph

Example 1. As shown Figure 1 (a), assume the elements a, b, ..., h arrive at time 1, 2, ..., 7, respectively. Since a and b are both dominated by elements newer than them, the candidate set is $\{c, e, f, g, h\}$.

In Figure 1 (a), g is dominated by c and e. It is noticed that if the dominance relation $e \rightarrow g$ is released due to the expiration of e then the dominance $c \rightarrow g$ has already been released since c expires earlier than e. Therefore, it is only necessary to keep $e \rightarrow g$ to hold a "lock" on g. In R_N, a dominance relation $e' \rightarrow e$ is *critical* if and only if e' is the youngest one (but older than e) in R_N, which dominates e; that is, $\kappa(e')$ is maximized among all the elements (other than e), in R_N, dominating e. A dominance graph G_{R_N} is constructed where the edge set consists of all critical dominance relations. Figure 1 (b) depicts the dominance graph of Figure 1(a). The encoding scheme is as follows: 1) every edge $e' \rightarrow e$ in G_{R_N} is represented by the interval $(\kappa(e'), \kappa(e)]$, and 2) each root e in G_{R_N} is represented by the interval $(0, \kappa(e)]$. Thus, an element $e \in R_N$ is in the answer of an n-of-N query ($n \leq N$) if and only if $\kappa(e)$ is the right end of an interval $(a, \kappa(e)]$ that contains $M - n + 1$. Based on such a scheme, the problem of computing an n-of-N query is converted to the *stabbing query* problem with the stabbing point $M - n + 1$. Namely, stab the intervals by $M - n + 1$, and then return the data elements e such that each $\kappa(e)$ is the right end of a stabbed interval.

Example 2. Regarding the example in Figure 1, the dominant graph can be encoded by the following intervals: $(0, 3], (0, 4], (3, 7], (4, 5]$, and $(4, 6]$. When $n = 6, M - n + 1 = 2$ as $M = 7$. Clearly, the intervals $(0, 3]$ and $(0, 4]$ are the results of stabbing query; consequently, c and e are the skyline elements for the most recent 6 elements among the 7 already arrived elements.

Various Dominating Probabilities in Uncertain Data Streams. For each element a in the sliding window DS, we use $P_{new}(a)$ to denote the probability that none of the elements in the sliding window which are newer than a (i.e., arrives later than a) dominates a; that is,

$$P_{new}(a) = \Pi_{a' \in DS, a' \prec a, \kappa(a') > \kappa(a)} (1 - P(a')) \tag{4}$$

Note that $\kappa(a') > \kappa(a)$ means that a' arrives after a. We use $P_{old}(a)$ to denote the probability that none of the elements which are older than a (i.e., arrives earlier than a) dominates a; that is,

$$P_{old}(a) = \Pi_{a' \in P_N, a' \prec a, \kappa(a') < \kappa(a)} (1 - P(a')) \tag{5}$$

The following equation (6) can be immediately verified.

$$P_{sky}(a) = P(a) \times P_{old}(a) \times P_{new}(a). \tag{6}$$

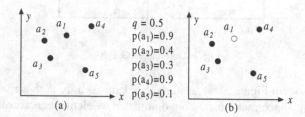

Fig. 2. A Sequence of Data Elements

Example 3. Regarding the example in Figure 2(a) where the occurrence probability of each element is as depicted, assume that $N = 5$, and elements arrive according the element subindex order; that is, a_1 arrives first, a_2 arrives second, ..., and a_5 arrives last. $P_{new}(a_4) = 1 - P(a_5) = 0.9$ and $P_{old}(a_4) = (1 - P(a_2))(1 - P(a_3))(1 - P(a_1)) = 0.042$, and $P_{sky}(a_4) = P(a_4)P_{new}(a_4)P_{old}(a_4) = 0.034$. If $N = 4$, a_1 expires once a_5 arrives as shown in Figure 2 (b). Then $P_{old}(a_4) = (1 - P(a_2))(1 - P(a_3)) = 0.42$ and $P_{sky}(a_4) = 0.34$.

3 Minimizing the Number of Uncertain Elements and the Encoding Scheme

In this section, we first minimize the number of uncertain elements to be kept for processing all $\mathcal{P}n$-of-N queries. Then, we present an effective encoding scheme on the stored elements to support efficient $\mathcal{P}n$-of-N query processing.

3.1 Minimizing the Number of Elements

As introduced in Section 2.2, in an exact data stream, an element e is "redundant" if it is dominated by a *newer* element e'. In an uncertain data stream DS, if an uncertain element e is dominated by a *newer* uncertain element e', e could still be a probabilistic skyline point regarding a given probability threshold q.

Example 4. In Figure 3, there are 7 uncertain elements a, b, c, d, f, g, and h. The order of the elements in the stream is the alphabetic order. The occurrence probability of each element and the probability threshold q are as illustrated in the figure. As shown, element c is dominated by a newer element h, however, as $P_{new}(c) = 1 - P(h) = 0.9$, $P_{old}(c) = 1$, $P_{sky}(c) = 0.63 > q$. Thus, g is a probabilistic skyline in the sliding window within the most recent 7 elements.

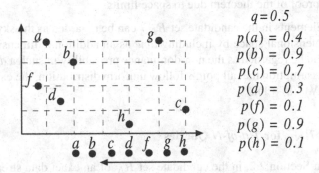

$$q = 0.5$$
$$p(a) = 0.4$$
$$p(b) = 0.9$$
$$p(c) = 0.7$$
$$p(d) = 0.3$$
$$p(f) = 0.1$$
$$p(g) = 0.9$$
$$p(h) = 0.1$$

Fig. 3. Uncertain Data Stream

Example 4 shows that modeling redundant elements in an uncertain data stream requires further analysis on the probabilities associated with each element. Remind that $P_{new}(e)$ refers to the probability that uncertain element e is not dominated by any elements newer than it. Let R_N denote the most recent N elements, and $R_{N,q}$ denote the set of elements in the most recent N elements with P_{new} values not smaller than q; that is,

$$R_{N,q} = \{e | e \in P_N, P_{new}(e) \geq q\} \tag{7}$$

In [5], $R_{N,q}$ is proved to be the minimum set of elements to be maintained to correctly answer probabilistic skyline queries over the most recent N elements in the data stream.

Theorem 1. *$R_{N,q}$ is the minimum set of elements to be maintained to correctly compute probabilistic skyline queries for sliding window size of N regarding probability threshold q.*

The proofs of Theorem 1 can be found in [5] based on the following properties of $R_{N,q}$.

- Processing skyline query based on $R_{N,q}$ only will not miss any skyline points.
- For a skyline point, its skyline probability computed based on $R_{N,q}$ only is equal to that computed based on all most recent N elements.
- $R_{N,q}$ is the minimum set of points to guarantee correct retrieval of skylines within the most recent N elements.

The following theorem states that $R_{N,q}$ is also the minimum set of elements to be maintained to correctly compute skyline queries for any recent n elements where $n \leq N$, namely, to correctly retrieve results for $\mathcal{P}n\text{-}of\text{-}N$ queries.

Theorem 2. $R_{N,q}$ *is the minimum set of elements to be maintained to correctly compute* $\mathcal{P}n\text{-}of\text{-}N$ *queries regarding probability threshold* q.

We omit the proof of the theorem due to space limits.

Size of $R_{N,q}$. Elements in the candidate set $R_{N,q}$ can be regarded as the skyline points in the $d+1$ dimensional space by including *time* as an additional dimension. This is because P_{new} can be regarded as the non-dominance probability in such a $d+1$ space. Thus, with the assumption that all points follow uniform distribution, the expected size of $R_{N,q}$ is $ln^d(N)/(d+1)!$.

3.2   Encoding $R_{N,q}$ for $\mathcal{P}n\text{-}of\text{-}N$ Queries

As introduced in Section 2.2, in the candidate set R_N of an exact data stream, we say an element e' dominates e is a critical dominance relation if e' is the youngest element (yet older than e) that dominates e. For a value n $(n \leq N)$, if e' is not within the most recent n elements (i.e., $\kappa(e') < M - n + 1$ where M is the total number of element seen so far), e is a skyline element.

Similar to the philosophy in encoding the candidate set R_N for $n\text{-}of\text{-}N$ queries over exact data streams, we aim to identify the most *critical* dominance relationship for elements inside the candidate set $R_{N,q}$ for uncertain sliding windows. Remind that the skyline probability of an element e within the most recent N elements consists of two parts besides its own occurrence probability, $P_{old}(e)$ representing the probability that e is not dominated by any element older than e and $P_{new}(e)$ representing the probability that e is not dominated by any elements newer than it. Furthermore, similar to $P_{sky,n}(a)$ which refers to the skyline probability of an element a computed regarding the most recent n elements in Equation 3, P_{old} value of an element could also be defined for the most recent n $(n \leq N)$ elements as follows, given that $\kappa(a) \geq M - n + 1$, namely a is also within the most recent n elements.

$$P_{old,n}(a) = \Pi_{a' \in DS, \kappa(a) \geq M-n+1, \kappa(a') \geq M-n+1, \kappa(a) > \kappa(a'), a' \prec a}(1 - P(a')) \quad (8)$$

The following equation is immediate since all elements newer than a are within the most recent n elements if a is within the most recent n elements. Hereafter, if discussing $P_{sky,n}$ or $P_{old,n}$ values for an element a it is assumed that a is within the most recent n elements

$$P_{sky,n}(a) = P(a) \times P_{old,n}(a) \times P_{new}(a). \quad (9)$$

Example 5. Continue with the example in Figure 2, assume $N = 5$ and $n = 3$, namely we are interested in only the most recent three elements a_3, a_4 and a_5. $P_{old,3}(a_4) = 1 - P(a_3) = 0.7$, $P_{new}(a_4) = 1 - P(a_5) = 0.9$, so $P_{sky,3} = 0.567$.

Consider an increase in the value n $(n \leq N)$. $P_{old,n}(e)$ is non-increasing with the increase of value n since more elements older than e may be included in the most recent n elements and contribute to the P_{old} value. On the other hand, $P_{new}(e)$ does not change with the value of n because elements which contribute to $P_{new}(e)$ are all newer than e. Thus, to determine the critical dominance relation for e in an uncertain data stream is to locate the element e_c with $\kappa(e_c) = M - n_c + 1$ making $P_{sky,n_c}(e) \geq q$ invalid, where n_c is minimized (or $\kappa(e_c)$ is maximized). We use $e_c \xrightarrow{c}_{q} e$ to denote that e_c probabilistically critically dominates e regarding the probability threshold q. Namely,

$$n_c = \arg\min_n P_{sky,n}(e) < q$$

Clearly, for any value $n < n_c$, e is a probabilistic skyline element within the most recent n elements.

Example 6. In Figure 3, for element g, $P_{new}(g) = 1 - P(h) = 0.9$. When $n = 5$, namely, within the most recent 5 elements, $P_{old,5}(g) = (1 - P(f)) \times (1 - P(d)) = 0.63$, and $P_{sky,5}(f) = 0.5103 > q$; When $n = 6$, $P_{old,6}(g) = (1 - P(f)) \times (1 - P(d)) \times (1 - P(b)) = 0.063 < q$. Thus, $n_c = 6$ is the minimum number of elements in the sliding window to make f unqualified to be a probabilistic skyline element. Equally speaking, b is the youngest element in the sliding window which dominates g and after the expiration of which g is a probabilistic skyline element. Element b probabilistically critically dominates g, namely $b \xrightarrow{c}_{q} g$, as $\kappa(b) = M - n_c + 1 = 2$.

Once the critical dominance relation is determined for an uncertain element e, we can have the dominance graph $G_{R_{N,q}}$ which is an edge set consisting all probabilistic critical dominance relations. Note that for an element e in the candidate set $R_{N,q}$, if $P(e) \times P_{new}(e) < q$, e does not have a critical dominance relation available since e is not a skyline element for any value of n $(n \leq N)$. However, we still need to keep e in $R_{N,q}$ as shown in Theorem 2 because deleting e will affect the skyline probability calculation for other elements in $R_{N,q}$. Based on $G_{R_{N,q}}$, given a value of n $(n \leq N)$, e is a skyline element for n if either of the following two conditions hold.

- e is a root in the dominance graph $G_{R_{N,q}}$, or
- there is an edge $e_c \xrightarrow{c}_{q} e$ in $G_{R_{N,q}}$, such that e_c arrives earlier than the n-th most recent element (i.e., $\kappa(e_c) < M - n + 1 \leq \kappa(e)$).

The encoding scheme for $G_{R_{N,q}}$ is as follows. 1) Every edge $e_c \xrightarrow{c}_{q} e$ in $G_{R_{N,q}}$ is represented by an interval $(\kappa(e_c), \kappa(e)]$. 2) Each root e in $G_{R_{N,q}}$ is represented by the interval $(0, \kappa(e)]$. Let $I_{R_{N,q}}$ denote the interval tree on the intervals obtained by the encoding scheme on $G_{R_{N,q}}$. So, an element e in $G_{R_{N,q}}$ is the answer of a $\mathcal{P}n$-of-N query $(n \leq N)$ if and only if $\kappa(e)$ is the right end of an interval that contains $M - n + 1$. The problem of computing $\mathcal{P}n$-of-N query is thus converted to the *stabbing query* problem with stabbing point $M - n + 1$ as discussed in Section 2.2. Namely, stab the intervals in $I_{R_{N,q}}$ by $M - n + 1$, and then return the data elements e such that $\kappa(e)$ is the right end of a stabbed interval.

Example 7. In Figure 3, $M = 7$ since there are 7 elements in the stream so far. Suppose $N = 6$. The candidate set $R_{N,q}$ consists of all recent 6 elements b, c, d, f, g and h since the P_{new} value of each element is not below the threshold q. Only the elements with occurrence probabilities not smaller than q are considered when computing probabilistic dominance relations, i.e., b, c, g. Element b is dominated by two newer elements d and f, and $P_{sky}(b) = P(b) \times (1 - P(d)) \times (1 - P(f)) = 0.567$, so b is a root in the dominance graph $G_{R_{N,q}}$. c is dominated by newer element h with $P_{sky}(c) = P(c) \times (1 - P(h)) = 0.63$, so c is a root in $G_{R_{N,q}}$. g is dominated by newer element h and older elements b, d, f. From Example 6, b probabilistically critically dominates g ($b \xrightarrow{c}{q} g$). So the interval tree $I_{R_{N,q}}$ consists the following intervals by encoding the dominance relations in the dominance graph $G_{R_{N,q}}$: $(0, 2], (0, 3]$ and $(2, 6]$. If $n = 5$ (to retrieve probabilistic skylines within the recent 6 elements), we stab the interval tree $I_{R_{N,q}}$ with $M - n + 1 = 3$ and c, g will be returned as the final results.

Note that in exact data streams, any non-redundant element in R_N is a skyline point for some $n \leq N$. However, in uncertain streams, this statement no longer holds. For instance, an element e may have an occurrence probability lower than q, disabling it from being a skyline point for any n values. However, we still need to keep e in $R_{N,q}$ if its $P_{new}(e)$ value is above q. This is because as proved in [5], removing such elements may incur incorrect probabilistic skyline results.

Time Complexity. The number of intervals kept in $G_{R_{N,q}}$ is $O(|R_{N,q}|)$ since $G_{R_{N,q}}$ is a forest. Thus, the stabbing query which retrieves the results for $\mathcal{P}n\text{-}of\text{-}N$ queries runs in $O(log|R_{N,q}|) + s$ where s is the number of probabilistic skyline points within the most recent n elements.

4   Maintaining $R_{N,q}$ and the Encoding Scheme

In the sliding window model, when a new element e_{new} arrives, the window slides to accommodate e_{new}, and the oldest element e_{old} moves out of the window range and should be removed. These may also trigger updates in the non-redundant element set $R_{N,q}$ and the dominance interval tree $I_{R_{N,q}}$. Algorithm 1 describes the overall framework to handle the key issues while the window slides for the uncertain stream.

Algorithm 1. Framework

---

1 **while** a new element e_{new} arrives **do**
2   **if** $\kappa(e_{new}) \leq N$ **then**
3   Updates introduced by e_{new};
4   **else**
5   Updates introduced by e_{new};
6   Updates introduced by e_{old};
7   Deletion of e_{old};
8   Insertion of e_{new};

As shown in Algorithm 1, when the sliding window is not yet full (i.e., e_{new} is the i-th element and $i \leq N$), we only need to handle the updates introduced by e_{new} and insertion of e_{new}, where insertion of e_{new} identifies the qualification of e_{new} regarding the candidate set $R_{N,q}$ and $I_{R_{N,q}}$. After the window is full, we need to further address the deletion of the oldest element e_{old} (i.e., the element which arrives $(M - N + 1)$-th in the stream) from the sliding window as well as the updates introduced by the deletion of e_{old}. In the following subsections we discuss the three major steps, *insertion of e_{new}*, *updates introduced by e_{new}*, and *updates introduced by e_{old}*, respectively. Deletion of e_{old} is trivial since we only need to delete e_{old} from the candidate set $R_{N,q}$ and interval tree $I_{R_{N,q}}$ if necessary. Naively processing these steps requires a sequential scan of elements in $R_{N,q}$.

4.1 Insertion of e_{new}

Since the most recent element e_{new} is not dominated by any element newer than it, $P_{new}(e_{new}) = 1$ and we insert e_{new} into the aggregate R-tree indexing the candidate set $R_{N,q}$. Next, if $P(e_{new}) \geq q$ the skyline probability of e_{new} should be explored to determine its probabilistic dominance relation. Otherwise (i.e., $P(e_{new}) < q$), the identification of probabilistic dominance relation is not necessary since e_{new} has no chance to be a probabilistic skyline element for any $n \leq N$.

Remind that to determine the critical dominance relation for e_{new} is to locate the element e_c with $\kappa(e_c) = M - n_c + 1$ making $P_{sky,n_c}(e_{new}) \geq q$ invalid, where n_c is minimized (or $\kappa(e_c)$ is maximized). A naive way to do so is to firstly sort all elements in $R_{N,q}$ decreasingly according to timestamps of elements; then scan the sorted elements and update $P_{sky}(e_{new})$ by multiplying $1 - P(e)$ if an element $e \prec e_{new}$. Each element dominating e_{new} in the process will be kept in a list called critical dominance list of e_{new}, denoted as $L_c(e_{new})$ which is decreasingly sorted based on timestamps. The scan stops when the first element e_c with timestamp $\kappa(e_c) = M - n_c + 1$ making $P_{sky,n_c}(e_{new}) < q$ is encountered.

Considering that elements in $R_{N,q}$ are indexed by an R-tree. We propose to use the best first search paradigm on R-tree to determine the critical dominance relation for e_{new}. For a node v in the R-tree indexing $R_{N,q}$, we record the maximum timestamp $\kappa(v)$ of all descendent elements of v. A max heap H based on $\kappa(v)$ is built to keep the nodes to be expanded. We denote the lower left corner of the minimal bounding box (MBB) of v as v^{lower}. The criteria to expand a node is $v^{lower} \prec e_{new}$. Otherwise (i.e., $v^{lower} \nprec e_{new}$), no elements from v dominates e_{new} and v will not be expanded. We terminate if the heap is empty, or the current element under investigation e_c with timestamp $\kappa(e_c) = M - n_c + 1$ makes $P_{sky,n_c}(e_{new}) < q$. If such an element is not found, e_{new} is a probabilistic skyline element for the time interval $(0, \kappa(e_{new}]$.

Algorithm 2 depicts above steps. Remind that Algorithm 2 is invoked only when $P(e_{new}) \geq q$. Starting from the root node of $R_{N,q}$, child entries of an intermediate entry v are inserted into the max heap if $v^{lower} \prec e_{new}$ (**Line** 6). If v is a data element and the updated skyline probability of e_{new} remains above q after considering the dominance of v, v is inserted into the critical dominance list of e_{new} (**Line** 10); $P_{sky}(e_{new})$ is also updated accordingly (**Line** 11). Otherwise (i.e., $P_{sky}(e_{new})$ below q), the algorithm terminates with the critical dominance relation identified.

Algorithm 2. Critical Dominance of e_{new}

---

1 max Heap $H \leftarrow$ root of R-tree indexing $R_{N,q}$;
2 **while** H is not empty **do**
3 | $v = H.top()$;
4 | $H.pop()$;
5 | **if** v is an intermediate node AND $v^{lower} \prec e_{new}$ **then**
6 | | insert children entries of v into H;
7 | **else**
8 | | **if** $v \prec e_{new}$ **then**
9 | | | **if** $P_{sky}(e_{new}) * (1 - P(v)) \geq q$ **then**
10 | | | | insert v into $L_c(e_{new})$;
11 | | | | $P_{sky}(e_{new})* = 1 - P(v)$;
12 | | **else**
13 | | | insert $(\kappa(v), \kappa(e_{new})]$ into $I_{R_{N,q}}$;
14 | | | found = true; exit;
15 **if** found = false **then**
16 | insert $(0, \kappa(e_{new})]$ into $I_{R_{N,q}}$;

4.2   Updates Introduced by e_{new}

Next we handle the updates introduced by e_{new}. If an element e is dominated by e_{new}, we need to update its P_{new} and P_{sky} probability which may render them invalid as a citizen in the non-redundant set $R_{N,q}$ or in the dominance graph $G_{R_{N,q}}$. First of all we retrieve the set of elements dominated by e_{new}, denoted as $D_{e_{new}}$. The following algorithm describes the key update issues related to e_{new}. We maintain a priority query Q initialized to the root node of R-tree R. Q is prioritized according to the levels of nodes, i.e., nodes of higher levels are accessed first.

Algorithm 3. Dealing with $D_{e_{new}}$

---

1 **for** $\forall e \in D_{e_{new}}$ **do**
2 | **if** $P_{new}(e) \times (1 - P(e_{new})) < q$ **then**
3 | | delete e from $R_{N,q}$;
4 | | **if** e is the end of an interval in $G_{R_{N,q}}$ **then**
5 | | | delete the interval;
6 | **else if** $P_{sky}(e) \times (1 - P(e_{new})) < q$ **then**
7 | | update the probabilistic critical dominance relation of e;
8 | | $P_{new}(e) = P_{new}(e) \times (1 - P(e_{new}))$;

Lines 2-5 in Algorithm 3 handle the first case when P_{new} value of e degrades to be smaller than q with the contribution of $1 - P(e_{new})$. In this case we remove e from $R_{N,q}$ and if it has a critical dominance relation captured in $G_{R_{N,q}}$, it is also deleted. **Lines** 6-8 deal with the case where e survives the citizenship test in $R_{N,q}$ but $P_{sky}(e)$ becomes below q after multiplying $1 - P(e_{new})$. In this case the probabilistic critical dominance relationship will be re-calculated by visiting the critical dominance list $L_c(e)$ sequentially,

Note that if an element e is deleted from $R_{N,q}$ in the first case ($P_{new}(e) \times (1 - P(e_{new})) < q$), we do not need to update the information of elements dominated by e. We formally prove this in the following lemma.

Lemma 1. *If $e_{new} \prec e$ and $P_{new}(e) \times (1 - P(e_{new})) < q$ after the arrival of the new element e_{new}, e could be removed from $R_{N,q}$ without updating the dominating probabilities of elements dominated by e.*

Proof. For an element e' and $e \prec e'$, first suppose $\kappa(e') < \kappa(e)$, namely, e' is older than e. Since $e \prec e'$, all elements dominating e also dominates e', so $P_{new}(e') \times (1 - P(e_{new})) < q$. e' should also be removed from $R_{N,q}$; if $\kappa(e') > \kappa(e)$, since $e \prec e'$, the skyline probability of e' computed within the most recent $M - \kappa(e) + 1$ elements, $P_{sky, M - \kappa(e) + 1}(e')$, after multiplying $1 - P(e_{new})$, must be smaller than q. Thus, the critical dominance relationship of e', $e_c \xrightarrow{c}{q} e'$, will be re-computed in Algorithm 3 and $\kappa(e_c) > \kappa(e)$, which means deleting e does not affect the critical dominance relationship and dominating probabilities of e'.

4.3   Updates Introduced by e_{old}

For the expired element e_{old}, we first remove it from the candidate set $R_{N,q}$ and the dominance graph if necessary. Next, for each element e dominated by e_{old}, the P_{old} and P_{sky} values of e change and the critical dominance relation might also change. Algorithm 4 illustrates the process. **Line** 1 deletes e_{old} from $R_{N,q}$ and **Line** 3 removes the critical dominance relation from $G_{R_{N,q}}$. Note that we do not need to check every element dominated by e_{old} to remove the effect of P_{old}. Instead, only those which are critically dominated by e_{old} will be actually affected by e_{old}. This is because e_{old} is the oldest element in $R_{N,q}$ and may not contribute to the critical dominance relation of all elements.

Algorithm 4. Expiration of Old Element e_{old}

---

1 $R_{N,q} := R_{N,q} - \{e_{old}\}$;
2 **if** $(0, \kappa(e_{old})] \in I_{R_{N,q}}$ **then**
3 $\quad$ remove $(0, \kappa(e_{old})]$ from $I_{R_{N,q}}$;

4 **for** $\forall (\kappa(e_{old}), \kappa(e)] \in I_{R_N}$ **do**
5 $\quad$ update $(\kappa(e_{old}), \kappa(e)]$ to $(0, \kappa(e)]$;
6 $\quad$ $P_{sky}(e)$ is updated by discounting $1 - P_{old}$;

---

5   Performance Evaluation

In this section, we present the results of a comprehensive performance evaluation of our techniques. As mentioned earlier, there is no existing technique specifically designed to support efficient computation of n-of-N skyline queries over uncertain sliding windows. In our performance study, we implement the most efficient main-memory algorithm for skyline queries over uncertain data streams [5] and use it as a benchmark algorithm to evaluate our techniques.

All algorithms proposed in the paper are implemented in standard C++ with STL library support and compiled with GNU GCC. Experiments are conducted on a PC with Intel Xeon 2.4GHz dual CPU and 4G memory under Debian Linux. In our implementation, MBBs of the uncertain objects are indexed by an R-tree with page size 4096 bytes.

Real Dataset. The real dataset is extracted from the stock statistics from NYSE (New York Stock Exchange). We choose 2 million stock transaction records of Dell Inc. from December 1st 2000 to May 22nd 2001. For each transaction, the average price per volume and total volume are recorded. This 2-dimensional dataset is referred to as *stock* in the following. To evaluate the techniques over uncertain sliding windows, we randomly assign a probability value between 0 and 1 to each transaction; that is, probability values follows *uniform* distribution. Elements arrival order is based on their transaction time.

Synthetic Dataset. We evaluate our techniques against the 3 most popular synthetic benchmark data, *correlated*, *independent*, and *anti-correlated* [11]. We evaluate our techniques against the space dimensions from 2 to 5. To evaluate the techniques over uncertain sliding windows, we use two models *uniform* and *normal* to assign occurrence probability to each element. In *uniform* distribution, the occurrence probability of each element takes a random value between 0 and 1, while in the *normal* distribution, the mean value of occurrence probabilities P_μ varies from 0.1 to 0.9 and standard deviation S_d is set to 0.3. The occurrence probability distribution follows *uniform* distribution by default unless otherwise specified. We assign a random order for elements arrival in a data stream.

The following algorithms are evaluated in this subsection for $\mathcal{P}n$-of-N queries.

q-sky: The query processing algorithm for probabilistic skyline queries over uncertain sliding windows in [5].

pnN: Our query processing algorithm (Section 3.2) for probabilistic n-of-N queries; that is, the stabbing query processing algorithm.

pmnN: Our algorithms (Section 4)for continuously maintaining the data structures for supporting probabilistic n-of-N queries.

5.1   Evaluating Query Algorithm: pnN

In this set of experiments, we fix $N = 10^6$ and randomly choose 1000 different n values varying from 1000 to 10^6. Each n is thus mapped to a $\mathcal{P}n$-of-N query with $N = 10^6$ to evaluate the query processing algorithm pnN. The processing time reported in Figure 4 is the average of the bucket of 1000 queries. As there is no existing work supporting skyline query processing over uncertain sliding windows with variable length, we

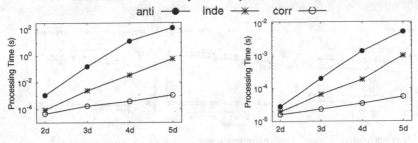

Fig. 4. q-sky vs pnN

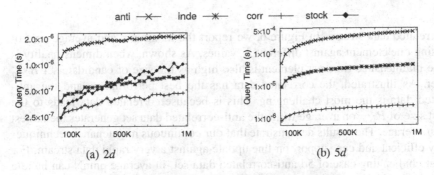

(a) 2d

(b) 5d

Fig. 5. Performance of pnN against Different n

naively search each candidate kept in the q-sky algorithm [5] and test if it is a probabilistic element over the recent n elements; pnN utilizes the stabbing query processing
algorithm over the interval set $I_{R_{N,q}}$. In Figure 4, we vary dimensionality from 2 to
5, and evaluate both q-sky and pnN over the three synthetic datasets anti-correlated,
independent, and correlated. As shown, both q-sky and pnN have a better performance
over corr dataset and pnN is up to 2 orders of magnitude faster than q-sky. In the more
challenging anti dataset, pnN is up to 5 orders of magnitude faster. In the remaining
of this subsection, we no longer evaluate the performance of q-sky in our performance
study since pnN significantly outperforms q-sky.

Figure 5 reports the impact of different n values on query processing time. The space
dimensionality is fixed to 2 and 5 respectively. We also record the average query processing time of 1000 queries. The results in Figure 5 show that the query techniques for
$\mathcal{P}n$-of-N queries are not very sensitive to the value of n. On the other hand, dimensionality and data distribution have a greater impact over the efficiency.

5.2 Efficiency of Maintenance Techniques: pmnN

In this subsection, we report the efficiency of the maintenance techniques for $\mathcal{P}n$-
of-N queries over uncertain sliding windows. The dimensionality is fixed to 2 and
5, and N varies from 100k to 1M. For each of the space dimensions, we generate
three data streams where the spatial distribution follows correlated, independent, and
anti-correlated, respectively. The real data stream *stock* is also studied along with the

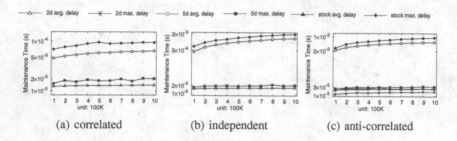

(a) correlated   (b) independent   (c) anti-correlated

Fig. 6. pmnN Performance against Different N

anti-correlated data stream. In Figure 6, we report the maximum and average cost of processing one element against different N values. As shown, when dimensionality is high the maintenance time per element is also high as the size of candidate set $R_{N,q}$ is larger. As illustrated, the correlated data has the best performance and the anti-correlated data is the most challenging. This is because correlated data leads to the smallest size of $R_{N,q}$ on average while the anti-correlated data set generates the largest $R_{N,q}$ on average. The results demonstrate that our continuous maintenance techniques are very efficient and can support on-line update against a very rapid data stream. For the most challenging case of 5d anti-correlated data set, in average pmnN can handle the stream speed of about 500 elements per second.

5.3 Scalability Evaluation

We evaluate the scalability for the proposed techniques to handle a number of $\mathcal{P}n\text{-}of\text{-}N$ queries regarding various parameters. We choose $N = 10^6$, and limit the data set size to 2×10^6. For each space dimension d ($1 \leq d \leq 5$), we generate two streams (independent and anti-correlated) with 2×10^6 data elements. The *stock* data is reported along with anti-correlated data.

The scalability of our algorithms is recorded as follows. We randomly generate 2×10^6 $\mathcal{P}n\text{-}of\text{-}N$ queries and randomly assign them among the most recent 1M elements. Then, we run the pmnN algorithm to continuously maintain the data structures and run pnN for processing $\mathcal{P}n\text{-}of\text{-}N$ queries. We record the processing time between two consecutive data elements which includes both the time of processing the queries and the time to maintain the data structures. Since such time is too short to be captured, we use average time for processing 1000 elements as the processing time. In Figure 7, we vary the number of points from 10^6 to 2×10^6. As illustrated, the proposed techniques could support queries over very rapid data streams with the arrival speed higher than 10K per second when dimensionality is lower (2 and 3). When dimensionality is higher (4 and 5), our techniques could still support data stream with a medium arrival speed at 200 elements per second even in the most challenging scenario of 5d anti correlated data set.

We also report the impact of the expected occurrence probability (P_μ) in normal distribution and the probability threshold (q) on the scalability of $\mathcal{P}n\text{-}of\text{-}N$ query processing. Figure 8 illustrates that the query processing techniques perform better with

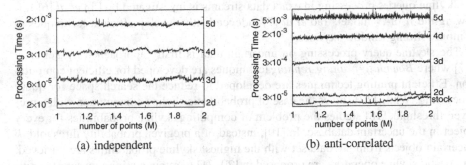

(a) independent

(b) anti-correlated

Fig. 7. Overall Performance

the increase of P_μ. This is because when the occurrence probabilities of uncertain elements are large, it is less likely for an element to be a probabilistic skyline point and thus the size of candidate set $R_{N,q}$ is smaller. Figure 9 shows that the processing time decreases with the increase of probability threshold q also because less elements are in the candidate set $R_{N,q}$.

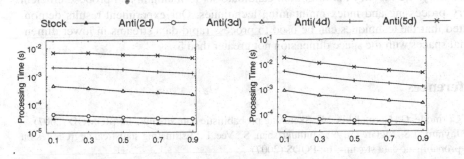

Fig. 8. Processing Time vs P_μ

Fig. 9. Processing Time vs q

6   Related Work

Börzsönyi et al [11] first study the skyline operator in the context of databases and propose an SQL syntax for the skyline query. They also develop two computation techniques based on *block-nested-loop* and *divide-and-conquer* paradigms, respectively. Another *block-nested-loop* based technique SFS (*sort-filter-skyline*) is propose d by Chomicki et al [12], which takes advantage of a pre-sorting step. SFS is then significantly improved by Godfrey et al [13]. The *progressive* paradigm that aims to output skyline points without scanning the whole dataset is firstly proposed by Tan et al [14]. It is supported by two auxiliary data structures, *bitmap* and *search tree*. Kossmann et al [15] present another progressive technique based on the nearest neighbor search technique. Papadias et al [16] develop a *branch-and-bound* algorithm (BBS) to progressively output skyline points based on R-trees with the guarantee of minimal I/O cost.

Skyline queries processing in exact data streams is investigated by Lin *et al* [6] following the n-of-N model. Tao *et al* [7] independently develop efficient techniques to compute sliding window skylines.

The skyline query processing on uncertain data is firstly approached by Pei *et al* [17] where *Bounding-pruning-refining* techniques are developed for efficient computation. Efficient pruning techniques are developed to reduce the search space for query processing. While [17] solves the case of probabilistic skyline computation with a pregiven threshold, [18] studies the problem of computing skyline probabilities for every object in the uncertain database. In [19], instead of a pregiven probability threshold, k uncertain objects from the data set with the highest skyline probabilities are retrieved. Stochastic skyline operators are proposed in [20, 21] to retain a minimum set of candidates for all ranking functions in the light of expected utility principles.

7   Conclusions

In this paper, we presented novel techniques for on-line skyline computation over the most recent n elements (for any $n \leq N$) in an uncertain data stream in a probability threshold fashion. Each element in the data stream is associated with an occurrence probability. We identify the minimum candidate set to maintain and propose efficient query processing and index maintaining techniques. Our experiment results demonstrated that the techniques can be used to process rapid data streams in lower dimensional spaces with the space dimension not greater than 5.

References

1. Cormode, G., Garofalakis, M.: Sketching probabilistic data streams. In: SIGMOD (2007)
2. Jayram, T.S., McGregor, A., Muthukrishan, S., Vee, E.: Estimating statistical aggregrates on probabilistic data streams. In: PODS (2007)
3. Jin, C., Yi, K., Chen, L., Yu, J.X., Lin, X.: Sliding-window top-k queries on uncertain streams. In: VLDB (2008)
4. Ding, X., Lian, X., Chen, L., Jin, H.: Continuous monitoring of skylines over uncertain data streams. In: Information Sciences (2012)
5. Zhang, W., Lin, X., Zhang, Y., Wang, W., Yu, J.X.: Probabilistic skyline operator over sliding windows. In: ICDE (2009)
6. Lin, X., Yuan, Y., Wang, W., Lu, H.: Stabbing the skye: Efficient skyline computation over sliding windows. In: ICDE (2005)
7. Tao, Y., Papadias, D.: Maintaining sliding window skylines on data streams. In: TKDE (2006)
8. Lin, X., Lu, H., Xu, J., Yu, J.X.: Continuously maintaining quantile summaries of the most recent n elements over a data stream. In: ICDE, pp. 362–374 (2004)
9. Babcock, B., Babu, S., Datar, M., Motwani, R., Widom, J.: Models and issues in data stream systems. In: PODS, pp. 1–16 (2002)
10. Mehlhorn, K.: Data Structures and Algorithms: 3. Multidimensional Searching and Computational Geometry. Springer (1984)
11. Börzsönyi, S., Kossmann, D., Stocker, K.: The skyline operator. In: ICDE, pp. 421–430 (2001)

12. Chomicki, J., Godfrey, P., Gryz, J., Liang, D.: Skyline with presorting. In: ICDE, pp. 717–816 (2003)
13. Godfrey, P., Shipley, R., Gryz, J.: Maximal vector computation in large data sets. In: VLDB (2005)
14. Tan, K.-L., Eng, P.-K., Ooi, B.C.: Efficient progressive skyline computation. In: VLDB, pp. 301–310 (2001)
15. Kossmann, D., Ramsak, F., Rost, S.: Shooting stars in the sky: An online algorithm for skyline queries. In: VLDB, pp. 275–286 (2002)
16. Papadias, D., Tao, Y., Fu, G., Seeger, B.: An optimal and progressive algorithm for skyline queries. In: SIGMOD, pp. 467–478 (2003)
17. Pei, J., Jiang, B., Lin, X., Yuan, Y.: Probabilistic skylines on uncertain data. In: VLDB 2007 (2007)
18. Atallah, M.J., Qi, Y.: Computing all skyline probabilities for uncertain data. In: PODS (2009)
19. Zhang, Y., Zhang, W., Lin, X., Jiang, B., Pei, J.: Ranking uncertain sky: the probabilistic top-k skyline operator. In: Information Systems (2011)
20. Lin, X., Zhang, Y., Zhang, W., Cheema, M.A.: Stochastic skyline operator. In: ICDE 2011 (2011)
21. Zhang, W., Lin, X., Zhang, Y., Cheema, M.A., Zhang, Q.: Stochastic skylines. In: TODS (2012)

An Approach for Sponsored Search Auctions Based on the Coalitional Game Theory

Wenlin Xu[1], Kun Yue[1,2,*], Jin Li[2,3], Liang Duan[1], Suiye Liu[4], and Weiyi Liu[1]

[1] Department of Computer Science and Engineering, School of Information Science and Engineering, Yunnan University, Kunming, China
[2] Key Laboratory of Software Engineering of Yunnan Province, Kunming, China
kyue@ynu.edu.cn
[3] School of Software, Yunnan University, Kunming, China
[4] Department of Economics, State University of New York at Binghamton, United States

Abstract. Sponsored search auctions play a crucial role in the Internet advertising. By considering the mutual interactions among advertisers in sponsored search auctions, we propose a game-theory based method for advertisers cooperating with each other in a sponsored search auction. First, we propose a cooperation bid strategy for advertisers' coalition, which could make the utility of the coalition increased and be obtained in linear time. Then, we prove the coalitional game of advertisers has a non-empty core containing the Shapley value. Following, we use an approximate Shapley value to distribute the coalition's utility among advertisers in the coalition. Experiments results verify the efficiency and effectiveness of our method.

Keywords: Sponsored search auction, VCG mechanism, Coalitional game theory, Shapley value.

1   Introduction

Sponsored search auction helps search engine companies generate multibillion dollar revenue by selling navigation service to advertisers. About 6 billion dollars was produced by sponsored search auction in the US alone in the first half of 2010 [1]. Actually, the sponsored search auction is not only crucial to search engine companies, but also important to other business, such as bidding campaign consulting firms, big management software firms and keyword selecting firms, etc.

The function of a sponsored search auction is straightforward. When a user types a query in a search engine, an auction could be run to determine which sponsored links will appear and what price will be paid by an advertiser once the ad is clicked. Various mechanisms are used in sponsored search auctions, among which the most popular are GSP (Generalized Second Price) and VCG (Vickrey-Clarke-Groves). The allocation of the VCG is identical to that of the GSP, but the payment is different. The VCG mechanism is a canonical method that incentivizes advertisers to bid truthfully

---

* Corresponding author.

X. Lin et al. (Eds.): WISE 2013, Part II, LNCS 8181, pp. 458–468, 2013.

[2]. In this paper, we assume that each advertiser's bid reveals their true valuation function in the auctions, and consequently we consider the domain of sponsored search auctions under the VCG mechanism.

In recent years, sponsored search auctions have been studied extensively. Pin et al. [3] tried to predict the click-through-rates (CTRs) in sponsored search auctions and Hafalir et al. [4] designed a mechanism which is beneficial to both the search engine and advertisers. With huge competition in this field, how to increase the utility has become an imminent problem for both advertisers and search engines.

It is known that each advertiser's utility in sponsored search auctions not only depends on advertiser himself but also on other advertisers, while the conflict among advertisers co-exists with their coordination. This exactly forms a game when we consider how to increase the utility in sponsored search auctions. Game theory, including coalitional game theory and non-cooperate game theory, is the study of mathematical model for conflict and cooperation between decision makers and has become an effective tool to describe strategic interactions in real world application [5]. Coalitional game theory differs from its non-cooperate counterpart, and in the former there is a strong incentive for players to work together to receive the largest total utility [6]. Game theory has been used to increase the utility of advertisers or search engines recently.

Some researches applied non-cooperate game theory to improve the search engine's utility by finding out the Nash equilibrium in sponsored search auctions [7]. Some other researches enhanced the utility of search engines by cooperation, where collaborative behaviors were modeled at multiple levels of abstraction, from an organization down to a single user's strategies [8]. Ceppi et al. [9] presented that multiple domain-specific search engines could provide a complete and precise result to the user by cooperating with each other. By this way, search engines could attract more users, but only the cooperation among search engines was considered without advertisers. Somanchi et al. [10] demonstrated that the cooperation between search engine and advertisers could help a search engine retain more advertisers, and increase the utility of the search engine and relevant advertisers. By this method, the probability of cooperation among advertisers was considered, but the concrete mechanism that the advertisers cooperate with each other was still not addressed.

Therefore, in this paper we propose a cooperate approach to increase the utility of advertisers in a sponsored search auction based on the coalitional game theory. To establish a coalitional game model for advertisers' cooperation, two problems have to be addressed:

(1) How to increase the utility of the advertisers' coalition?
(2) How to find an optimal utility distribution method for advertisers in a coalition?

For the first problem, it is necessary to design a utility and efficient cooperation bid strategy that each advertiser in the coalition must satisfy. Moreover, it is expected that the utility cooperation bid strategy should make the whole coalition's utility increased and could be executed in short time.

For the second problem, it is necessary to design a utility distribution method, and two objectives must be pursued: fairness and stability. For instance, the Shapley value divides the value fairly in a certain sense while the core divides the value in a stable way. It is pointed out [2] that when a coalitional game is convex, it always has a non-empty core containing the Shapley value. By this property of convex coalitional game, we could use the Shapley value to distribute the utility in a stable and fair way.

Taking the above interpretations as motivation, the main contributions of this paper can be summarized as follows:

- We provide a cooperation bid strategy for the advertisers' coalition and demonstrate that the cooperation bid strategy can make the whole coalition's utility increased and be executed in linear time. Furthermore, we prove the advertisers' coalition is a convex one.
- We propose an approximate Shapley value to compute each advertiser's utility in the coalition since the complexity of Shapley value computation is #P-hard [11]. Then, we prove that the approximate Shapley value can be obtained in polynomial time and the result of distribution obtained by the approximate Shapley value is quite close to that obtained by the exact Shapley value.
- We implement the proposed algorithms and make preliminary experiments to test the feasibility of our method.

The rest of the paper is organized as follows: In Section 2, we present preliminaries and state the problem of this paper. In Section 3, we give the cooperation bid strategy for advertisers in the coalition. In Section 4, we give an algorithm to distribute the utility among advertises in the coalition. In Section 5, we show experimental results and performance studies. In Section 6, we conclude and discuss our future work.

2   Preliminaries and Problem Statement

2.1   Preliminaries

First, we introduce the VCG mechanism and give the definition of coalitional game and Shapley value as the basis for later discussions.

There is a set of n advertisers $A=\{a_1, a_2, ..., a_n\}$ competing for k slots. Each advertiser a_i has a nonnegative valuation $v_i \geq 0$ per click and $v_1 > v_2 > ... > v_n$. The set of CTR (click-through-rate) of each slot is $R=\{r_i \mid 1 \leq i \leq k\}$ and $r_1 > r_2 > ... > r_k$. The VCG mechanism accepts a bid b_i ($b_1 > b_2 > ... > b_n \geq 0$) from each advertiser a_i, and the per-click price q_i of a_i to the slot i is defined as [12]:

$$q_i = \frac{\sum_{j=i+1}^{k+1} b_j * r_{j-1} - b_j * r_j}{r_i}$$

(1)

where b_j denotes the j-th highest bid (advertiser a_i pays nothing when $i > k$).

The utility of advertiser a_i for a click is defined as [12]:

$$u_i = v_i - q_i$$

(2)

Definition 1 [13]. A coalitional game with transferable utility in normal characteristic form is (N, v), where $N=\{1, 2, ..., n\}$ is the set of agents, and $v: 2^N \to R$. For each coalition which is a subset of agents $S \subseteq N$, $v(S)$ is the value of the coalition S, which is the total utility that the member of S can achieve by coordinating and acting together.

Theorem 1 [14]. A coalitional game is monotonically increasing if for all coalitions $S_1 \subseteq S$, $v(S_1) \le v(S)$ always holds.

Definition 2 [15]. There exists a unique function $Sh(S, i)$ which satisfies the following form $Sh(S,i) = \sum_{S_1} \dfrac{(|S_1|-1)!(n-|S_1|)!}{n!}[v(S_1)-v(S_1-\{i\})]$, where $i \in S_1$, S, $S_1 \subseteq S$, and $|S_1|$ and n is the number of players in S_1 and that in S.

2.2 Problem Statement

In order to illustrate the VCG mechanism in sponsored search auctions, we first give the following example.

Example1. Let $A=\{a_1, a_2, a_3, a_4, a_5, a_6\}$ be the set of advertisers competing for 4 (i.e., $k=4$) slots, and $C=\{40, 30, 20, 10\}$ be the set of click numbers in a unit time for each slot. For ease of exposition, we use click numbers to replace CRTs, and use p_i as the payment of advertiser a_i in a unit time in this paper. Let $B=\{4.7, 4.4, 4.3, 4.2, 4.0, 3.8\}$ be the set of bids of advertisers in the auction. According to the VCG mechanism, advertiser a_1 gets the first slot and the utility of a_1 in a unit time is $u_1=v_1*c_1-p_1=40*4.7-169=19$. Similarly, we can obtain all the advertisers' allocation, payments, and utility, given in Table1.

Table 1. VCG mechanism

Name	Slot	Value function	Payment	Utility
a_1	1	188.0	169.0	19.0
a_2	2	132.0	125.0	17.0
a_3	3	86.0	82.0	4.0
a_4	4	42.0	40.0	2.0
a_5	0	0	0	0
a_6	0	0	0	0

Note that it is possible for some advertisers to form a coalition since advertisers can cooperate with each other during the process of auction. For the advertisers that form a coalition, it is necessary to solve the following two problems: what cooperation bid strategy should be designed and how to distribute the utility of the coalition?

3   Designing a Cooperation Bid Strategy for Advertisers in a Coalition

According to Equation (2), the utility of an advertiser is equal to the advertiser's valuation function minus its payment. Under the VCG mechanism, the payment of an advertiser depends on other advertisers whose bids are lower than that of this advertiser. We divide the advertisers in coalition into two parts: the first part is the advertiser who has the highest original bid and the second part is the rest advertisers in the coalition. The basic idea of cooperation bid strategy is as follows: the first part of the coalition does not change its original bid, but the bids of the second part are set to the smallest under the condition that all of the advertisers in the coalition can obtain the same slots as they are not in the coalition. The cooperation bid strategy is described by the following algorithm.

Algorithm 1. Cooperation Bid Strategy

```
Input:
A = {a₁,a₂,…,aₙ}, the set of advertisers
S = {a₁ˢ,a₂ˢ,…,aᵣˢ} the coalition and S⊂A
k, the number of ad slots
Output:
S', the coalition with the bids reported by the
advertisers in coalition
Steps:
sort A and S by the decreasing order of b
for each advertiser
  if aᵢˢ.b<aₖ₊₁.b then
     remove aᵢˢ from S
end for
for i=|S| to 2 do
     aᵢ.b← aᵢ₊₁.b+0.01
     aᵢˢ.b← aᵢ.b
end for
return S
```

In Algorithm 1, the two for-loops are the most time-consuming operation, which could be done linearly in $O(n)$ time, where n is the number of advertisers in coalition. The obtained bid strategy can make the coalition's utility increased actually, stated in Theorem 2.

Theorem 2. The utility of the whole coalition S is increased by using Algorithm 1.

Proof. For n advertisers competing for k slots, we denote b_i as the original bid of advertiser a_i, and $b_x > b_y > b_z$. We denote $S=\{a_x, a_y, a_z\}$ as a coalition, b_i^* as the reported bid of advertiser a_i. We can obtain $b_x^* = b_x$, $b_y^* = b_{y+1}+0.01$ and $b_z^* = b_{z+1}+0.01$ from Algorithm 1. Thus, when forming the coalition S, the payment of advertiser a_x is:

$$p_x^* = b_{x+1}c_x + b_{x+2}c_{x+1} + \cdots + b_{k+1}c_k - b_{x+1}c_{x+1} - b_{x+2}c_{x+2} - \cdots - b_k c_k$$

$$= p_x - [(c_{y-1} - c_y)(b_y - b_y^*) + (c_{z-1} - c_z)(b_z - b_z^*)] \tag{3}$$

where p_x is the payment of advertiser a_x in unit time when not form the coalition S.

From Equation (3), we can prove $p_x \geq p_x^*$, $p_y \geq p_y^*$ and $p_z \geq p_z^*$. Thus, the utility of coalition S is increased since the payment of coalition S is decreased while the allocation has no change.

4   Distributing the Utility of the Coalition

Suppose the auction has k slots, and a set of n advertisers, denoted $A=\{a_1, a_2, \ldots, a_n\}$, takes part in the auction. We use $C=\{c_1, c_2, \ldots, c_k\}$ to present different slots since the difference among slots is just the click numbers. We examine a certain coalition $S \subseteq A$, whose utility is as follows:

$$u(S) = \sum_{a_i \in S} u_i = \sum_{a_i \in S} v_i * c_i - \sum_{a_i \in S} p_i \qquad (4)$$

If the advertisers in S decide to cooperate with each other, they can form a coalition and use the cooperation bid strategy of Algorithm 1, changing the payments to $P^* = \{p_1^*, \ldots, p_n^*\}$. Thus, we have $p_i^* \leq p_i$ and the following utility is generated for the advertisers in the coalition in a unit time:

$$u^*(s) = \sum_{a_i \in S} u_i^* = \sum_{a_i \in S} v_i * c_i - \sum_{a_i \in S} p_i^* \qquad (5)$$

Definition 3 [14]. Given a VCG sponsored search auction, the utility $u(S)$ $(S \subseteq A)$ of a coalition is $u(S) = u^*(S)$.

In this definition, u maps advertisers' coalition to the utility that they can achieve, and S is restricted to the subset of the advertisers whose original bids are the $k+1$ highest.

Definition 4 [14]. A game is convex if for any A, $B \subseteq I$ we have $u(A \cup B) \geq u(A) + u(B) - u(A \cap B)$.

It is known that the Shapley value can be used to distribute the utility of a coalition in a stable and fair way, only if the coalitional game in sponsored search auctions is convex [2], which is guaranteed by Theorem 3.

Theorem 3. The coalitional game is a convex coalitional game.

Proof. By Definition 3, the coalitional game in sponsored search auctions is increasing (monotone). For the increasing situations, the game is convex (by Definition 4) if for any $S' \subseteq S$ and advertiser a_r (a_r is not in S), we have $u(S' \cup \{a_r\}) - u(S') \leq u(S \cup \{a_r\}) - u(S)$. We define the marginal contribution of advertiser a_r to the coalition S as $\Delta_{a_r}^S = u(S \cup \{a_r\}) - u(S)$. Thus, the coalitional game in sponsored search auctions is a convex game when $\Delta_{a_r}^{S'} \leq \Delta_{a_r}^S$. By Definition 3, we have

$$\Delta_{a_r}^S = u^*(S \cup \{a_r\}) - u^*(S) = v_r * c_r + \left(\sum_{a_i \in S} p_i^* - \sum_{a_i \in S \cup a_r} p_i^* \right) \qquad (6)$$

Then by Equation (6), we have

$$\Delta_{a_r}^{S} - \Delta_{a_r}^{S'} = \sum_{a_i \in S \cup \{a_r\}} p_i^* - \sum_{a_i \in S} p_i^* - \left(\sum_{a_i \in S \cup \{a_r\}} p_i^* - \sum_{a_i \in S} p_i^* \right).$$

We denote $S = \{a_1, a_2, \ldots, a_t\}$ $(t \le n)$ and their bids satisfy $b_1 > b_2 > \ldots > b_t$. We denote $S' = \{a_1, a_2, \ldots, a_m\} \subseteq S$ $(m \le t)$ and advertiser a_r is not in S. By Equation (4), we have

$$\Delta_{a_r}^{S} - \Delta_{a_r}^{S'} = \sum_{i < r, a_i \in S-S'} (c_{r-1} - c_r)(b_r - b_r^*) + \sum_{j > r, a_j \in S-S'} (c_{j-1} - c_j)(b_j - b_j^*) \tag{7}$$

By Equation (7), we have $\Delta_{a_r}^{S} \le \Delta_{a_r}^{S}$. Therefore, the coalitional game in sponsored search auctions is a convex game.

Following, we give an approximate algorithm for calculating the Shapley value.

Algorithm 2. Distributing Utility

Input:
$A = \{a_1, a_2, \ldots, a_n\}$, the set of advertisers
S, the given coalition
$u(S)$, the utility of coalition S
a_i, the given advertiser
Output:
$Sh(S, a_i)$, the utility of a_i obtained in S
Variables:
S_1, S_2, the subsets of S and $S_1 \cup S_2 = S$
S^*, a new coalition that replaces S
Steps:
$S^* \leftarrow S$
Decompose S^* into S_1 and S_2 where $a_i \in S_1$
while $S^* \ne a_i$ **do**
    $u(S_1) \leftarrow Sh(S^*, S_1)$
    $S^* \leftarrow S_1$, $u(S^*) \leftarrow u(S_1)$
    Decompose S^* into S_1 and S_2 where $a_i \in S_1$
end while
$Sh(S, a_i) \leftarrow u(S^*)$
return $Sh(S, a_i)$

For a coalition with n advertisers, the Shapley value $Sh(S^* = (S_1, S_2), S_1)$ will be calculated in $O(n \log_2 n)$ time.

5   Experimental Results

5.1   Experiment Setup

The data used for the experiments was obtained from Yahoo's sponsored search logs [16], consisting of bid and click logs sampled over a few months. The data contains roughly 18 million records, and each record is formed from keywordID, advertiserID, bid and the number of click in one month. All the test data were stored in MS SQL Server 2008 and all the codes were written in Java. The machine configurations are as follows: AMD Athlon64 X2 3800+ CPU, 2GB of main memory, running Windows 7 Ultimate 32-bit operating system.

5.2 Utility and Efficiency of the Cooperation Strategy

We tested Algorithm 1 to verify the utility and efficiency of the cooperation strategy. In the current experimental environment, this test was conducted on the advertisers' coalition with various numbers of advertisers: 20, 40, 60, 80, 100 and 120. It can be seen from Fig. 1 that the utility of the coalition is increased by the cooperation bid strategy. In Fig. 2, n represents the number of advertisers that take part in the auction, and k represents the number of the advertisement slots. It can be seen that the cooperation bid strategy can be computed in linear time with the increase of advertisers, which verifies the efficiency of Algorithm 1.

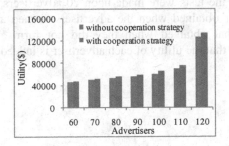

Fig. 1. Utility of advertisers' coalition

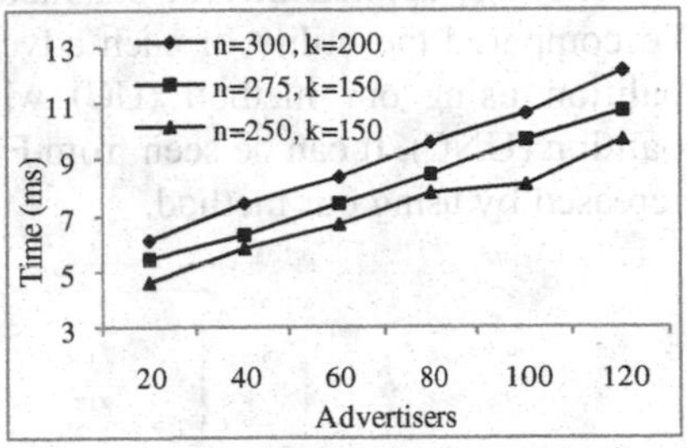

Fig. 2. Execution time of cooperation strategy with the increase of advertisers

5.3 Performance of the Approximation Method for Distributing Utility

By Algorithm 2, we can obtain the approximation solution to distributing the utility in the coalition. The performance of an approximation method was evaluated in terms of the following two criteria: approximation error and time complexity. We tested the performance of Algorithm 2 by recording the results upon the approximation errors and execution time of Algorithm 2, shown in Fig. 3 and Fig. 4 respectively. It can be seen from Fig. 3 that the approximation error of Algorithm 2 is basically lies in the interval of [0, 0.1]. This means the result of Algorithm 2 is quite close to that obtained by exact Shapley value algorithm [6]. In Fig. 4, n represents the number of advertisers who take part in the auction, and k presents the number of the slots. It can be seen that the execution time is increased linearly with the increase of the advertisers. In particular, the execution time is 0.045s and is 5.516s when the number of advertisers is 5 and 80 respectively, which verifies the efficiency of Algorithm 2.

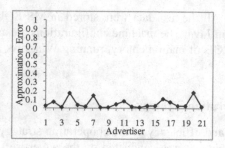
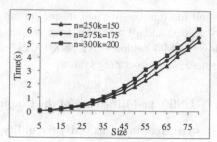

Fig. 3. Approximation error of Algorithm 2

Fig. 4. Execution time of Algorithm 2 with the increase of advertisers in coalition

5.4   Applicability Test

To verify the applicability of our method, the tests were made upon 20 advertisers. We compared the utility of each advertiser obtained when the advertisers formed a coalition using our method (UC) with that obtained when they did not form a coalition (UNC). It can be seen from Fig. 5 that the utility of each advertiser is indeed increased by using our method.

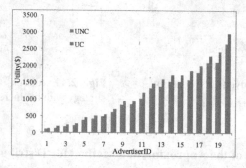

Fig. 5. The utility of advertiser in and out of coalition

To sum up, the experimental results and performance studies given in this section show that our method for increasing the advertisers' utility in sponsored search auctions is efficient and effective.

6   Conclusion and Future Work

To make the advertisers' utility increased, we focused on a cooperation bid strategy under the VCG mechanism, and proved that the cooperation bid strategy could easily be obtained in linear time and increase the coalition's utility. We gave an approximate Shapley value to distribute coalition's utility.

To improve the efficiency of Algorithm 2, propose the methods for combining coalitions, and exam GSP mechanism in sponsored search auctions are our future work. Meanwhile, we have assumed that advertisers in coalition completely trust each other, but in fact some advertisers in coalition may not abide by the coalitional game approach we designed, which means that it is worthwhile to design a punishment mechanism.

Acknowledgement. This work was supported by the National Natural Science Foundation of China (61063009, 61163003, 61232002), the Ph.D Programs Foundation of Ministry of Education of China (20105301120001), the Yunnan Provincial Foundation for Leaders of Disciplines in Science and Technology (2012HB004), the Natural Science Foundation of Yunnan Province (2011FB020), the Foundation of Key Program of Department of Education of Yunnan Province (2011Z015, 2013Z049), and the Foundation of the Key Laboratory of Software Engineering of Yunnan Province (2012SE205).

References

1. Gatti, N., Lazaric, A., Trovo, F.: A Truthful Learning Mechanism for Contextual Multi-Slot Sponsored Search Auctions with Externalities. In: Faltings, B., Brown, K.L., Ipeirotis, P. (eds.) Proc. of ACM-EC 2012, Valencia, Spain, pp. 605–622. ACM (2012)
2. Bachrach, Y., Zadimoghaddam, M., Key, P.: A Cooperative Approach to Collusion in Auctions. Journal of SIGecom Exchanges 10(1), 17–22 (2011)
3. Pin, F., Key, P.: Stochastic Variability in Sponsored Search Auction: Observations and Models. In: Shoham, Y., Chen, Y., Roughgarden, T. (eds.) Proc. of ACM-EC 2011, San Jose, CA, USA, pp. 61–77. ACM (2011)
4. Hafalir, I.E., Ravi, R., Sayedi, A.: A near Pareto optimal auction with budget constraints. Journal of Games and Economic Behavior 74(1), 699–708 (2012)
5. Nash, J.F.: Non-cooperative games. Annals of Mathematics (54), 286–295 (1951)
6. Fatima, S.S., Wooldridge, M., Jennings, N.R.: A linear approximation method for the Shapley value. Journal of Artificial Intelligence 172(1), 1673–1699 (2008)
7. Yao, S., Mela, C.F.: A Dynamic Model of Sponsored Search Advertising. Journal of Marketing Science 30(1), 447–468 (2011)
8. Dorn, C., Taylor, R.N.: Architecture-Driven Modeling of Adaptive Collaboration Structures in Large-Scale Social Web Applications. In: Wang, X.S., Cruz, I., Delis, A., Huang, G. (eds.) WISE 2012. LNCS, vol. 7651, pp. 143–156. Springer, Heidelberg (2012)
9. Ceppi, S., Gatti, N., Gerding, E.H.: Mechanism Design for Federated Sponsored Search Auctions. In: Burgard, W., Roth, D. (eds.) Proc. of AAAI 2011, San Francisco, California, USA, pp. 1323–1324. AAAI Press (August 2011)
10. Somanchi, S., Nittala, C., Narahari, Y.: A Novel Bid Optimizer for Sponsored Search Auctions Using Cooperative Game Theory. In: Proc. of IAT 2009, Milan, Italy, pp. 435–438. IEEE (2009)
11. Shapley, L.S.: Cores of convex games. International Journal of Game Theory (1), 12–26 (1971)
12. Babaioff, M., Roughgarden, T.: Equilibrium Efficiency and Price Complexity in Sponsored Search Auctions. In: Proc. of Workshop on Ad Auctions 2010, Cambridge, MA, USA (2010)

13. Zlotkin, G., Rosenschein, J.: Coalitioin Cryptography and Stability Mechanism for Coalition Formation in Task Oriented Domains. In: Roth, B.H., Korf, R.E. (eds.) Proc. of AAAI 1994, Seattle, WA, USA, pp. 432–437. AAAI Press (1994)

14. Bachrach, Y.: Honor Among Thieves-Collusion in Multi-Unit Auctions. In: Hoek, W., Kaminka, G.A., Lesperance, Y., Luck, M., Sen, S. (eds.) Proc. of AAMAS 2010, Toronto, Canada, pp. 617–624. IFAAMAS (2010)

15. Shapley, L.S.: Contributions to be Theory of Games. Princeton University Press (1953)

16. Yahoo! (2013), http://webscope.sandbox.yahoo.com/catalog.php?datatype=a

Current Attitude Prediction Model
Based on Game Theory

Zhan Bu[1,2], Chengcui Zhang[2], Zhengyou Xia[1], and Jiandong Wang[1]

[1] College of Computer Science and Technology,
Nanjing University of Aeronautics and Astronautics, China
{buzhan,zhengyou_xia,aics}@nuaa.edu.cn
[2] Computer and Information Sciences, The University of Alabama at Birmingham, USA
{zhanb,zhang}@cis.uab.edu

Abstract. Social interactions on online communities involve both positive and negative relationships: people give feedbacks to indicate friendship, support, or approval; but they also express disagreement or distrust of the opinions of others. One's current attitude to the other user in online communities will be affected by many factors, such as the pre-existing viewpoints towards given topics, his/her recent interactions with others and his/her prevailing mood. In this paper, we develop a game theory based method to analyze the interactive patterns in online communities, which is the first in its kind. The performance of this prediction model has been evaluated by a real-world large-scale comment dataset, and the accuracy reaches 82%.

Keywords: online community, relationship, current attitude, game theory.

1 Introduction

Traditional online community researches have mainly considered positive relationships only. Most of those researches focus on the friend-networking sites, such as MySpace, Facebook, and Google+ [2-4]. Negative relationships are seldom found on those sites, because the interactive users might be friends in real world. Recently a number of researchers have begun to investigate negative as well as positive relationships in online communities. For example, users on Wikipedia can vote for or against the nomination of others [5]; users on Epinions can express trust or distrust of others [6]; and participants on Slashdot can declare others to be either "friends" or "foes" [7]. However, all of the above works exhibit a common problem -the lack of explicit labeling, making it difficult to reliably determine the sentiment of a given interaction.

A few recent studies examined the interactive patterns using data that can be gathered from online communities, for instance, writing messages to other users [1, 11-12, 15-17]. In our most recent study [1, 10], we roughly identify several terms or phrases from the public discussions as either supportive or opposing. Every term/phrase is assigned with a value between 0 and 1 according to their tone manually. A higher value corresponds to a greater degree of support; if the phrase is neutral,

X. Lin et al. (Eds.): WISE 2013, Part II, LNCS 8181, pp. 469–478, 2013.

we assigned it a value of 0.5. Thus, every phrase has an associated numerical "trust". For a given comment from one ID to the other, we can determine the implicit orientation by counting the number of positive or negative words in it (if there are several emotional words in one comment, we take the average).

Accordingly, the **semantic strength** from user i to user j under a given topic p can be calculated as:

$$s^p(i,j) = (\sum_{k=1}^{n_{i,j}^p} s^p(i,j,k)) / (n_{i,j}^p) \tag{1}$$

Where $s^p(i,j,k)$ is the implicit orientation of one comment from i to j under the topic p, and $n_{i,j}^p$ is reply number from i to j under the topic p.

In reality, one's current attitude to the other user in online communities will be affected by many factors, such as the pre-existing viewpoints towards given topics, his/her recent interactions with others and his/her prevailing mood. In this paper, we develop a game theory based method to analyze the interactive patterns in online communities, which is the first in its kind. Our contributions are as follow:

1) In online communities, if both users A and B send a positive comment, then their happiness will get a positive promotion. If just user A sends a positive comment (user B sends a negative comment), user A's happiness will decrease, and the happiness of user B will increase. Finally, if both users A and B send a negative comment to each other, their happiness will get a negative promotion. Therefore, we can model the user interaction as a game, in which, the two players are solely concerned with maximizing their own payoffs. In the end of this game, user A will choose the optimal strategy to maximize his/her own payoff. The optimal strategy taken by user A can be used to predict his/her current attitude to user B.

2) To evaluate the performance of our current attitude prediction model, we crawled the comment data from Tianya website, and built a test set according to existing user discussion lists. The accuracy of our model in predicting one's current attitude to the other is 82%.

2   Game Theory Based User Interaction Model

Suppose that user A is an online community user, he can interact with other users by writing comments to each other. If user A sends a positive comment to user B, user B might give user A a feedback, which user A hopes is also positive. In this case, the happiness user A received will be mainly decided by the implicit orientation of the comment from user B to user A, which is $s(B,A)$. While if user A sends a negative comment to user B, user A might not look forward a positive feedback from user B. The happiness user A gets will be mainly decided by the attitude to user B of himself. For instance, A is strongly against user B, the semantic strength $s(A,B)$ from user A to user B will be very low. In this case, user A's happiness will be measured as $1 - s(A,B)$. Thus, user A has two possible options for how to behave: to send a positive comment, or to send a negative one. The challenge in reasoning about this is that the happiness of user A in online communities with the outcome depends not just on his own decisions but on the decisions made by every participant. Therefore, the interactive users constitute a complete description of each player's happiness with each of the possible outcomes of a game. In the end of this game, user A will choose his optimal strategy to maximize his own happiness. The optimal strategy taken by user A will be used to measure the current attitude from him to user B.

2.1   Interactive Game Model

Game theory is a study of strategic decision making. More formally, it is the study of mathematical models of conflict and cooperation between intelligent rational decision-makers [13]. A game consists of three components: a set of players, the strategy set for each player and a utility function for each player measuring the degree of "happiness" of the player. Then, we can use a tuple $G = [\kappa, \{A_k\}, R_k]$ to represent a game, where $\kappa = \{1, ... K\}$ is the set of players, A_k is the set of actions (strategies) available to user k, and R_k is the utility function for user k. Taking the above user interaction as an example, there are two players which are A and B. And their sets of actions are the same which are to send a positive comment, or to send a negative one. The basic utility functions for user A are $s(B, A)$ or $1 - s(A, B)$ which depends on the decisions made by everyone. The situation is symmetric when we consider the basic utility of user B. The underlying principles of user interactions in online communities are as follow: if both users A and B send a positive comment, then their payoffs will get a positive promotion. If just user A sends a positive comment (user B sends a negative comment), user A's payoff will decrease, and the payoff of user B will increase. Finally, if both users A and B send a negative comment to each other, their payoffs will get a negative promotion. As one's pre-existing viewpoints towards given topics and emotional changing are hard to capture, in this paper, both positive and negative promotions of payoff will be randomly assigned in a given range.

There is a simple tabular way to summarize all these outcomes, as follows. We represent user A two choices-to send a positive comment, or to send a negative comment-as the rows of a 2×2 table. We represent user A's partner's user B two choices as the columns. So each box in this table represents a decision by each of the two users. In each box, we record the happiness/payoff they each receive: first user A's, then user B's. Writing all this down, we have the table shown in Fig. 1. Now user A needs to figure out what to do: sends a positive comment, or sends a negative one. Clearly, user A's happiness/payoff depends not just on which of these two options he choose, but also on what user B decides. Therefore, as part of user A's decision, A has to reason about what user B is likely to do. Thinking about the strategic consequences of his own actions, where user A needs to consider the effect of decisions by user B. Our interest is in reasoning about how users in online communities will behave in this interactive game.

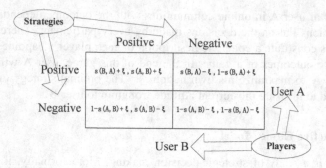

Fig. 1. User interaction model based on game theory

2.2 Optimal Strategy

Once the payoffs have been defined, interactive users will select their optimal strategies according to the above three assumptions. If a_A^i, $a_A^i \in A_A$ is a strategy chosen by user A, and a_B^j, $a_B^j \in A_B$ is a strategy chosen by user B, then there is an entry in the payoff matrix corresponding to the pair of chosen strategies (a_A^i, a_B^j). We will write $R_A(a_A^i, a_B^j)$ to denote the payoff to user A as a result of this pair of strategies, and $R_B(a_A^i, a_B^j)$ denote the payoff to user B as a result of this pair of strategies. Then, we used two fundamental concepts that will be central to our discussion of the interactive game.

The first concept is the idea of **a (strict) best response**: it is the best choice of one user, given a belief about what the other player will do. In our interactive game, we say a strategy a_A^* for user A is **a (strict) best response** to a strategy a_B^j for user B is a_A^* produces at least as good a payoff as any other strategy paired with a_B^j:

$$R_A(a_A^*, a_B^j) \ge R_A(a_A', a_B^j) \tag{2}$$

$$R_A(a_A^*, a_B^j) > R_A(a_A', a_B^j) \tag{3}$$

for all other strategies a_A' of user A. When A has a strict response to a_B^j for user B, this is clearly the strategy s/he should select when faced with a_B^j.

The second concept is the idea of **a (strictly) dominant strategy**. We say that a dominant strategy of user A is a strategy that is a best response to every strategy of user B:

$$R_A(a_A^*, a_B') \ge R_A(a_A', a_B') \tag{4}$$

$$R_A(a_A^*, a_B') > R_A(a_A', a_B') \tag{5}$$

Thus, we made the observation that if a user has a strictly dominant strategy, then we can expect him/her to use it. As shown in Fig. 2, suppose user A posted a topic, in which he called on China government to implement quality education. User B met this post and discussed with him about this topic. In their recent interaction, user B sends a comment, the semantic strength of which is $s^p(B,A,t)$, then user A gives a feedback to user B, the semantic strength of this feedback is $s^p(A,B,t)$. How should user B do next? In this case, user B may meet three different situations: 1) a game in which user B has a strictly dominant strategy; 2) a game in which only user A has a strictly dominant strategy; 3) a game in which neither user has a strictly dominant strategy. We will respectively discuss these three situations next.

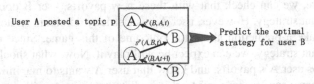

Fig. 2. An example of the recent interaction between user A and user B

2.2.1   A Game in Which User B Has a Strictly Dominant Strategy

Suppose user B sends a positive comment to user A to express her agreement. Taking the semantic analysis of her comment as in [1, 12], we learn that the semantic strength from B to A is 0.8. Then, user A also sends a positive comment to user B, in which he gave B a good compliment, the semantic strength from A to B is 0.7. According to the interactive model in Section 2.1, we can capture this situation with numerical payoffs as shown in Fig. 3(a). ξ represents the positive and negative promotions of payoff decided by their pre-existing viewpoints towards given topics and prevailing moods. As one's pre-existing viewpoints towards given topics and emotional changing are hard to capture, in this paper, both positive and negative promotions of payoff will be randomly assigned in a given range. In this example, we unified set ξ as 0.1, and the final payoff table will be acquired as shown in Fig. 3(b).

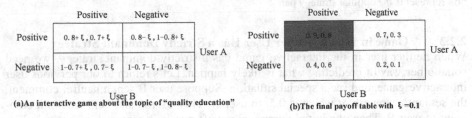

(a)An interactive game about the topic of "quality education"

(b)The final payoff table with ξ =0.1

Fig. 3. A game in which user B has a strictly dominant strategy, the strategy pair in red box represents the optimal strategy pair

As $R_B(Positive, Positive) > R_B(Postive, Negative)$ and $R_B(Negative, Positive) > R_B(Negative, Negative)$, to send a positive comment is a strictly dominant strategy for user B. So it is easy to reason about what user B is likely to do.

2.2.2   A Game in Which only User a Has a Strictly Dominant Strategy

There are many situations where the structure of the game and the resulting behavior looks very different. Indeed, even simple changes to a game can change it from the above interactive game to something more benign. For example, suppose that user B also sends a positive comment to user A, and additionally gave a few supplemental points. The semantic strength from user B to user A is 0.7. However, user A sends a negative feedback to user B, because he thinks that the supplemental points are not reasonable. In this case, the semantic strength from user A to user B is 0.3. The numerical payoffs in this situation are captured as shown in Fig. 4(a). This time, we randomly assign a value between 0 and 0.2 to ξ. Then we can check that the payoff matrix has changed as shown in Fig. 4(b).

Furthermore, we can check that with these new payoffs, user B does not have a strictly dominant strategy. However, user A has a strictly dominant strategy. Still, it is not hard to make a prediction about the outcome of this game. Since user A has a strictly dominant strategy, we can expect he will play it. Now, what should user B do? If user B knows user A's payoffs, and know that user A wants to maximize her happiness/payoff, then user B can confidently predict that user A will send a negative comment to her. Then, since to send a negative comment is the strict best response by user B when user A sends a negative comment, we can predict that user B will send a negative comment. So our overall prediction of interaction in this game is to send a negative comment by user B.

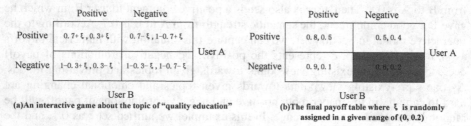

(a)An interactive game about the topic of "quality education"

(b)The final payoff table where ξ is randomly assigned in a given range of (0, 0.2)

Fig. 4. A game in which only user A has a strictly dominant strategy, the strategy pair in red box represents the optimal strategy pair

2.2.3   A Game in Which Neither User Has a Strictly Dominant Strategy

When neither user in the interactive game has a strictly dominant strategy, we need some other way of predicting what is likely happen. Let's return to our pervious user interactive game, and see a special situation. Suppose user B sent a neutral comment, the semantic strength of which is 0.5, to user A. Then user A also sent a neutral comment to user B. Then, the initial numerical payoff matrix is captured as shown in Fig. 5(a). We still randomly assign a value between 0 and 0.2 to ξ and the final numerical payoffs as shown in Fig. 5(b).

If we study how the payoffs in this example, we see that neither user has a dominant strategy. So how should we reason about the outcome of play in this game?

A complete definition of how a player will play a game will be called a **pure strategy**. For a two-player game, if there is a strictly dominant strategy, we can find the pure strategies of both players. However, if neither user has a strictly dominant strategy, we should consider the **mixed strategies**. In this model, the possible strategies of user A are numbers p between 0 and 1, and p means that user A is committing to send a positive comment with the probability p, and to send a negatvie one with the probability 1-p. Similarly, the possible strategies for user B are numbers q between 0 and 1. Once we have done this, we can then rank outcomes accoding to their associated number. As in this example, if user A chooses to send a positive comment while user B chooses a probability of q, then the expected payoff to user A is

$$E_A(Positive) = (0.7)(q) + 0.3(1-q) = 0.3 + 0.4q \qquad (6)$$

Similarly, if user A chooses to send a negative comment while user B chooses a probability of q, then the expected payoff to user A is

$$E_A(Negative) = (0.6)(q) + 0.4(1-q) = 0.4 + 0.2q \qquad (7)$$

Then we can get the unique Nash equilibrium [14] for the miexed-strategy (the **Mixed- Strategy Equilibrium**) for this example, just have

$$E_A(Positive) = E_A(Negative) \qquad (8)$$

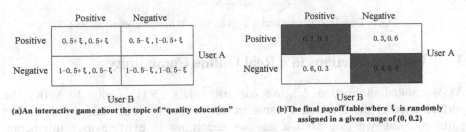

(a)An interactive game about the topic of "quality education"

(b)The final payoff table where ξ is randomly assigned in a given range of (0, 0.2)

Fig. 5. A game in which neither user has a strictly dominant strategy, the strategy pairs in red box are Nash equilibriums

or in other words q=0.5. The situation is symmetric when we consider things from user B's point of view, and evaluate the payoffs from a play of probability p by user A. We also have p=0.5. Thus, the pair of strategies p=0.5 and q=0.5 is the only possibility for a Nash equilibrium. In his famous paper [14], John Forbes Nash proved that there is an equilibrium for every finite game.

Above all, we analyze all the three suitations in the two-player interavtive game. We can get a pair of pure strategies in the first two suitations, and get a pair of mixed strategies in the last suitations. The optimal strategy taken by user B can be used to predict her current attitude to user A.

2.3   Relationship Predication

The interactive game in online environment has a feature that the two interactive users will choose their actions independently accoding to their recent interacions. That means, their actions are not simultaneously given. As shown in Fig. 6, user A posts a topic p, then a reply article from user B commnets on this topic so as to continue the discussion. The semantic strengths between these two users are sited on every directed link. At the end of their discussion about this topic, we can predict the optimal strategy (to send a postive comment, or to send a negative comment) for user

A by considering his latest interactions with user B, which are $s^p(A,B,n^p_{A,B})$ and $s^p(B,A,n^p_{B,A})$. We can also predict the optimal strategy for user B by considering interctions $s^p(B,A,n^p_{B,A}-1)$ and $s^p(A,B,n^p_{A,B})$. Actually, we have already known the strategy for user B as is sited on the last link of this discussion. We can use $s^p(B,A,n^p_{B,A})$ to verify the performance of our interactive game model. Similarly, we can use one's $t+1$ th action to verify the performance of the interactive game model in the t th interaction with his/her partner. The overall performance of our interactive game model will be discussed in Section 3.

Fig. 6. User interactions under a given topic p in online environment

3   User Interactions in a Real Online Community

As we stated in Section 2.3, we can use one's $t+1$ th action to verify the performance of the interactive game model in the t th interaction with his/her partner. Then, we can build a test set according to existing user interaction history. Let's go back to the user interaction example in Section 2.3, we assume that user A will send only one comment to user B if s/he receives the recent feedback from user B. Thus, we can predict user A's $t^p_{A,B}+1$ th action using the interaction game model in Section 2.2 based his recent interactions with user B, namely $s^p(A,B,t^p_{A,B})$ and $s^p(B,A,t^p_{B,A})$. In this case, user A will have an optimal strategy, $optimal\_strategy^p(A,B,t^p_{A,B}+1)$.

We can compare $optimal\_strategy^p(A,B,t^p_{A,B}+1)$ with $s^p(A,B,t^p_{A,B}+1)$ using a simple judgment function, which is:

$$\delta(x,y) = \begin{cases} 1 & x \geq 0.5, y \geq 0.5 \ or \ x \leq 0.5, y \leq 0.5 \\ 0 & otherwise \end{cases} \qquad (9)$$

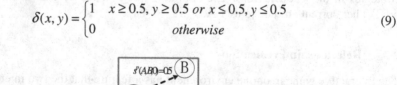

Fig. 7. The simplest discussion list which can be used to evaluate the performance of our model

We need to consider a special case in the above interaction example. As shown in Fig. 7, there is only two comments in the discussion list between users A and B. In this case, to predict the optimal strategy for user A, we assume that the pre-exsiteing attitude from user A to user B is neutral, namely, $s^p(A,B,0) = 0.5$. The the two-comment discussion list is the simplest one which can be used to evaluate the performance of our interactive game model.

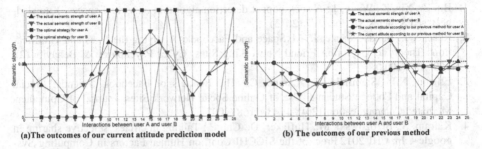

(a)The outcomes of our current attitude prediction model   (b) The outcomes of our previous method

Fig. 8. Comparisons of the current attitude prediction model and our previous method [12]

We study a real discussion list in our Tianya data. User A first post a topic, and then user B add in to have a discussion with user A. Fig. 8 shows a comparsion of the current attitide attitude prediction model based on the recent interaction and our previous method by averaging the semantic strengths in one's previous t times interaction. As shown in Fig. 8(a), we predict one's current attitude by considering the recent interaction with his/her parter. And we use one's $t+1$ th action to verify the performance of the interactive game model in his/her t th interaction. For the total twenty-four times prdictions, we have successed twenty times, the accuracy of the current attitude prediction model is 83%. In our previous work [12], we measure one's attitude to the other by simply averaging the semantic strengths in one's pre-vious t times interactions. We also compare the outcomes of this method with the actual semantic strengths of the two users. As shown in Fig. 8(b), we have successed 13 times. The accuracy of this method is only 54%.

When we apply our current attitude prediction method to the entire Tianya data, we get an accuracy of 82%, which is very high. We also implemented the same evaluation work to our previous method. The accuracy of this method in the entire Tianya data is only 63 %; obviously, the current attitude prediction model performs much better.

4   Conclusion

Recently, online communities have become a supplemental form of communication between people. Relations on online communities often reflect a mixture of positive (friendly) and negative (antagonistic) interactions. In this paper, we develop a game theory based method to analyze the interactive patterns in online communities, which is the first in its kind. The performance of this prediction model has been evaluated by a real comment data, and the accuracy reaches 82%. Our study uncovered a number of interesting findings, some of which are related to the specific nature of online community environments.

Acknowledgments. This work was supported by Jiangsu Innovation Program for Graduate Education (Project NO: CXZZ12_0162) and the Fundamental Research Funds for the Central Universities.

References

1. Bu, Z., Xia, Z., Wang, J.: A sock puppet detection algorithm on virtual spaces. KnowledgeBased Systems 37, 366–377 (2013)
2. Raacke, J., Bonds-Raacke, J.: MySpace and Facebook: Applying the Uses and Gratifications Theory to Exploring Friend-Networking Sites. Cyber Psychology & Behavior (2), 169–174 (2008)
3. Ellison, N.B., Steinfield, C., Lampe, C.: The Benefits of Facebook 'Friends:' Social Capital and College Students' Use of Online Social Network Sites. Journal of Computer-Mediated Communication 12(4), 1143–1168 (2007)
4. Kairam, S., Brzozowski, M., Huffaker, D., Chi, E.: Talking in circles: selective sharing in google+. In: CHI 2012 Proc. of the SIGCHI Conf. on Human Factors in Computing Systems, pp. 1065–1074 (2012)
5. Burke, M., Kraut, R.: Mopping up: modeling wikipedia promotion decisions. In: Proc. of the 2008 ACM Conf. on Computer Supported Cooperative Work, pp. 27–36 (2008)
6. Brzozowski, M.J., Hogg, T., Szabo, G.: Friends and foes: ideological social networking. In: Proc. of the SIGCHI Conf. on Human Factors in Computing Systems, pp. 817–820 (2008)
7. Kunegis, J., Lommatzsch, A., Bauckhage, C.: The slashdot zoo: mining a social network with negative edges. In: Proc. of the 18th Intl.Conf. on WWW, pp. 741–750 (2009)
8. Heider, F.: Attitudes and cognitive organization. Journalof Psychology 21, 107–112 (1946)
9. Guha, R.V., Kumar, R., Raghavan, P., Tomkins, A.: Propagation of trust and distrust. In: Proc. of the 13th Intl.Conf. on World Wide Web, pp. 403–412 (2004)
10. Bu, Z., Zhang, C., Xia, Z., Wang, J.: A fast parallel modularity optimization algorithm (FPMQA) for community detection in online social network. Knowledge Based Systems (June 28, 2013)
11. Pak, A., Paroubek, P.: Twitter as a Corpus for Sentiment Analysis and Opinion Mining. In: Proc. of the Seventh Conf. on International Language Resources and Evaluation LREC 2010, Valletta, Malta, European Language Resources Association ELRA (May 2010)
12. Xia, Z., Bu, Z.: Community detection based on a semantic network. Knowledge Based Systems (26), 30–39 (2012)
13. R. B. Myerson, Game Theory: Analysis of Conflict, p. 1. Chapter-preview links, pp. vii–xi. Harvard University Press (1991)
14. Nash, J.: Non-Cooperative Games. The Annals of Mathematics 54(2), 286–295 (1951)
15. Pang, B., Lee, L.: Opinion Mining and Sentiment Analysis. Foundations and Trends in Information Retrieval 2, 1–135 (2008)
16. Go, A., Huang, L., Bhayani, R.: Twitter sentiment analysis. Final Projects from CS224N for Spring 2008/2009 at the Stanford Natural Language Processing Group (2009)
17. Read, J.: Using emoticons to reduce dependency in machine learning techniques for sentiment classification. In: ACL. The Association for Computer Linguistics (2005)

Towards Automatic Client-Side Feature Reuse

Josip Maras[1], Jan Carlson[2], and Ivica Crnković[2]

[1] University of Split, Croatia
josip.maras@fesb.hr
[2] Mälardalen University, Sweden
{jan.carlson,ivica.crnkovic}@mdh.se

Abstract. Client-side applications often contain similar features and facilitating reuse could offer considerable benefits in terms of faster development. Unfortunately, due to the specifics of prevailing technologies, the techniques and tools used to support reuse are not as advanced as in other software engineering disciplines and the main method of reuse is still copy-pasting code. Copy-paste reuse can introduce a number of different types of errors that are time-consuming to detect and fix. In this paper we present an automatic method for feature reuse in client-side web applications. We identify problems that occur when introducing code from one application into another, present a set of algorithms that detect and fix those problems and perform the actual code merging. We have evaluated the approach on four case study applications, and the results show that the method is capable of performing feature reuse.

Keywords: Web applications, Reuse, Client-side Analysis.

1 Introduction

From the user's perspective, the behavior of a client-side application is composed of distinguishable parts, i.e. features, that manifest at runtime. Similar features are often used in a large number of applications, and facilitating their reuse offers significant benefits in terms of easier development. However, the client-side domain does not offer any widely used feature-reuse method, and code is usually copy-pasted to the new application. Copy-paste reuse can be complex and error-prone. Usually it is hard to identify code for reuse and introduce it into the new application without errors, and there is need for systematic reuse.

A feature is an abstract notion representing a distinguishable part of the system behavior that manifests at runtime triggered by the user [2]. Since client-side web applications are highly dynamic event-driven applications where the majority of code is executed as a response to user-generated events, identifying the exact implementation details of a certain feature is difficult and time-consuming. In our previous work [4] we have introduced a client-side dependency graph that captures dependencies that exist in a client-side web application. The dependency graph is built during the feature identification process [4] by analyzing an execution of a particular scenario demonstrated by the user. By using the

X. Lin et al. (Eds.): WISE 2013, Part II, LNCS 8181, pp. 479–488, 2013.

feature identification process, we are able to identify the implementation details
of individual features.

In this paper we present a method for code-level feature reuse in client-side
web development. We have specified how to reuse features, and when can that
reuse be considered successful. Naturally, when merging two code bases a number of problems can arise. We have identified those problems and have developed
algorithms capable of detecting and fixing them. The approach has been evaluated by performing the reuse process on four case study web applications, and
the evaluation shows that the method is capable of performing feature reuse.

2   Reuse Process Overview

The goal of the method is to enable code-level feature reuse from one client-side
application into an already existing application. Let A and B be two client-side
applications, each defined with HTML, CSS, and JavaScript code and resources:
$\langle H_A, C_A, J_A, R_A \rangle$ for application A, and $\langle H_B, C_B, J_B, R_B \rangle$ for application B.

An application offers a set of features F, and when a user performs a certain scenario s_i, a feature f manifests. A feature is implemented by a subset of
the application's code and resources. However, identifying the exact subset is
a challenging task: code responsible for the desired feature is often intermixed
with irrelevant code, and there is no trivial mapping between the code and the
application running in the browser. In our previous work [4], we have developed
a method that can, by analyzing the execution of a scenario causing the manifes-
tation of a feature f_a, identify the subset of the application $\langle h_a, c_a, j_a, r_a \rangle$ that
implements the feature. The goal of the reuse method is to enable the inclusion
of code and resources $\langle h_a, c_a, j_a, r_a \rangle$ of f_a from application A into the application
B, thereby obtaining a new application B' that offers both the feature f_a from
A, and the features F_B from B. We consider **reuse successful** if, in the final B'
application, the scenario s_a causing the manifestation of f_a can be repeated with
the same presentational and behavioral characteristics on h_a, and all scenarios
S_B can be repeated with the same presentational and behavioral characteristics
on H_B. This means that, in order for the reuse to be correct, there should not be
any feature "spilling" – the feature f_a, in the context of B', should not operate
on parts of application originating from application B (nor should features from
B operate on parts of the application originating from A). With regard to presentational characteristics this means that CSS rules c_a, when included in $C_{B'}$,
should only be applied to HTML nodes h_a included in $H_{B'}$ (similarly, C_B should
only be applied to H_B). For the preservation of behavioral characteristics, code
j_a should only interact with h_a, c_a, r_a, and J_B with H_B, C_B, R_B.

2.1   Process Description

Due to the fact that client-side web applications are highly dynamic event-driven
UI applications, we have based the process on the dynamic analysis of application
execution while performing scenarios that capture the behavior of individual

applications. As input, the process (Figure 1) receives the whole code of the application A from which a feature f_a will be extracted, a scenario s_a that invokes the feature f_a, a selector that specifies the part of the web page where f_a manifests; the whole code of the web application B where the feature will be reused into, a set of scenarios S_B that capture the behavior of the application, and a reuse position that specifies where the feature will be reused.

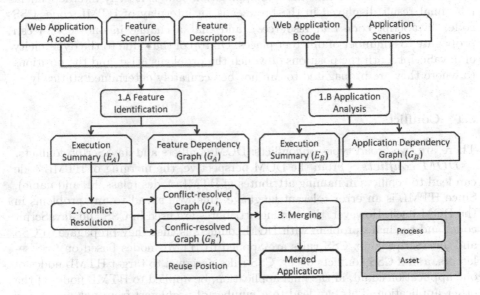

Fig. 1. The process of extracting and reusing features

The reuse process starts by invoking the *feature identification* process [4] (1.A, Figure 1) for application A, and by analyzing the execution of application B (1.B, Figure 1). The feature identification process executes the application with the scenario event trace as a guideline, logs an execution summary, builds a dependency graph, and automatically identifies the subset of the application A that implements the feature f_a. Similarly, the application analysis process analyzes the execution of application B, logs an execution summary and builds a dependency graph. When merging two code bases a number of conflicts can occur. For this reason, all artifacts generated in the previous phases are used as input to the *conflict resolution* phase which automatically fixes the conflicts (or notifies that the fix can not be applied). Next, the modified, conflict-resolved dependency graphs, along with the reuse position are used as inputs to the merging phase where the code of the two applications is merged, and reuse is achieved.

3   Conflict Resolution

Including code and resources of f_a into B changes the situation in both the feature code and the application code, primarily because merging code creates a new page whose DOM is different from what is expected by the code of each individual application. This difference can create a number of problems that are complicated by the fact that the web application code is heavily interdependent (the final result displayed in the browser is an interplay of HTML code, CSS code, JavaScript code and resources), and that a change in one section can propagate to a number of different places. On top of that, due to the dynamicity of JavaScript, both the positions on which the problems arise, and the positions to where they are propagated to can not be accurately determined statically.

3.1   Conflicts

There are two broad types of conflicts: DOM conflicts and JavaScript conflicts.

DOM conflicts – From the DOM perspective, the merging of HTML code can lead to conflicts in naming attributes of HTML nodes (class, id, and name). Since HTML is an error tolerant language, this won't lead to any problems in the DOM itself. However, the naming attributes are used in CSS and JavaScript code, and the main problem with DOM conflicts is that they propagate to CSS and JavaScript code. CSS rules are applied to HTML nodes based on CSS selectors, and if CSS conflicts occur, CSS rules designed to target HTML nodes of one application could, in the final application, be applied to HTML nodes of the other application. This can lead to a number of problems: from not preserving the visual properties, to changing the values of code expressions that access the element's visual properties. JavaScript code interacts with the DOM and accesses HTML nodes by using queries similar to CSS selectors. This means that if there are conflicts in HTML, JavaScript expressions that query the DOM of the page can return different results in the context of the final application than in the original contexts. These differences can lead to a number of errors.

JavaScript conflicts – Apart from conflicts that propagate from HTML and CSS, JavaScript code can introduce a number of errors, mostly because of the use of global variables. JavaScript has different types of global variables, and from the perspective of conflict-handling they can be divided into three groups: *i) standard global variables* created by declaring variables in the global scope, or by extending the global window object (writing to a non-registered identifier, or adding a new property to the window object); *ii) Built-in object extensions* – extending built-in objects (e.g. the Math object, String, Array prototypes); and *iii) Event-handling variables* used by the browser to register event handlers (e.g. onload, onmousemove properties of the global window object). *Standard global variables* can cause naming conflicts; *Built-in object extensions* can cause naming conflicts within the extended objects, and errors can be introduced when iterating over object properties; and *Event-handling variables* can cause problems with property overriding. Similar to CSS type selectors, JavaScript type DOM queries can come into conflict and a different set of elements, compared to the originating applications, can be returned in the context of the final application.

Conflicts can also occur between resources (images, fonts, videos, files, etc.) if there exist resources with the same identifiers in both applications. These types of conflicts can propagate to HTML, CSS, and JavaScript code, and have to be tracked and handled.

3.2 Fixing Conflicts

The following sections describe algorithms for fixing conflicts in different parts of the application. The process is composed of two steps: *i)* fixing conflicts that arise due to changes in the DOM structures of both applications, and *ii)* fixing problems that happen when merging two JavaScript code bases. Since conflicts can occur both statically and dynamically, all possible conflicts can not be accurately detected with static analysis, and fixes performed with simple string renamings, without taking into consideration the semantics of the changed expressions, can only handle a subset of possible problems (and even then, we can not be sure if they are applied to correct expressions). This is why we heavily rely on client-side dependency graphs and execution summaries to identify the exact conflict positions, and the positions to where these conflicts propagate.

Fixing DOM conflicts – Algorithm 1 describes the process of detecting and fixing conflicts that arise when merging HTML code of two different applications, but that also propagate to CSS and JavaScript. The main idea of the algorithm is to identify all static or dynamic code positions that can cause conflicts due to the fact that a new page will be created by merging the DOMs of both applications. This means replacing conflicted HTML attributes, expanding HTML nodes, and modifying both the CSS rules and JavaScript DOM queries in order to localize them in a way that they only interact with nodes from their respective applications. As input the algorithm receives the dependency graph *fGraph* and execution summary *fExe* of A for scenario u_a, and the dependency graph *bGraph* and the execution summary *bExe* of B for U_B.

Algorithm 1. handleDOM(*fGraph, fExe, bGraph, bExe*)

1: *attrConflicts* ← getHtmlAttrsConflicts(*fExe, bExe*)
2: *resources* ← getResources(*fGraph*)
3: **for all** *item* : merge(*attrConflicts, resources*) **do**
4:     *new* ← getName(*item, fExe, bExe*)
5:     **for all** *pos* : getUsagePos(*item, fExe*) **do**
6:         **if** isInHtml(*pos*) OR isInCss(*pos*) **then**
7:             replaceVal(*pos, item, new*)
8:         **else if** isInJs(*pos*) **then**
9:             replaceDomStrLit(*pos, item, new, fGraph*)
10: expandNds(getName('f', *fExe, bExe*), *fGraph, fExe*)
11: expandNds(getName('b', *fExe, bExe*), *bGraph, bExe*)

The algorithm finds all conflicts in HTML node named attributes and all used resources, and generates a new, unconflicting name for each item. An item can be used in a number of different positions: in HTML code as node attributes, in CSS code as selectors or key values, and in JavaScript code as assignment or call expressions. (e.g. assignment expressions that modify node attributes, or DOM querying call expressions). If the usage position is in HTML or CSS code then the old value in the feature code is simply replaced with the new, unconflicting value. If the usage position is in JavaScript code, the process traverses the dependency graph and attempts to find the string literal that matches the old value and replace it with the new value. If the string literal can not be found (e.g. is constructed by concatenating strings), then a comment notifying that a conflict was not handled is added to the access position.

The process of handling DOM conflicts continues in line 10, Algorithm 1, by handling conflicts with selectors in CSS and JavaScript. The main idea is to make the selectors more specific by limiting them only to parts of the DOM that match the originating application. Unconflicting names are generated with calls to the *getName* function and are added as attributes to enable differentiation between nodes originating from different applications. Type selectors are expanded so they target only nodes they have targeted in the originating applications (both for CSS selectors, and JavaScript DOM queries).

Algorithm 2. handleJs(*fGraph, fExe, bGraph, bExe*)

1: **for all** *cnflct* : getStandardGlobalConflicts(*fExe, bExe*) **do**
2:     renameIdDeps(getDecl(*cnflct, fGraph*), getNewName(*cnflct, fExe, bExe*))

3: **for all** *objExt* : getBuiltInObjExtns(*fExe*) **do**
4:     addSkipIterationToAllPropertyIters(getPropertyIters(*objExt, bExe*))
5:     **if** hasConflicts(*objExt, fExe, bExe*) **then**
6:         renameIdDeps(getDecl(*objExt, fGraph*), getNewName(*objExt, fExe, bExe*))

7: **for all** *objExt* : getBuiltInObjExtns(*bExe*) **do**
8:     addSkipIterationToAllPropertyIters(getPropertyIters(*objExt, fExe*))

9: *conflicts* ← getConflictedHandlers(*fExe, bExe*)
10: **if** sizeOf(*conflicts*) != 0 **then**
11:     addInitConflictHandlerObjectAsTopNode(*bGraph*)
12:     expandWithConflictHandlerCode(conflicts)
13:     addHandlerInvokerCodeAsLastBodyNode(*bGraph*)

Fixing JavaScript conflicts – The main goal of Algorithm 2 is to detect and fix JavaScript conflicts that arise due to global variable naming conflicts, and due to the modifications of the globally accessible objects that can change the behavior of additionally included code. Since conflicts can occur both dynamically and statically, as input the algorithm receives the dependency graphs and execution summaries from both applications. The algorithm starts by finding conflicts regarding standard global variables: for each conflicted variable, a new unconflicting identifier is generated, and all usage positions of that identifier in the feature code are replaced (if it is not possible, a warning is added). Next, the algorithm is dealing with conflicts that arise by extending built-in objects. For the object extensions in feature code, the algorithm traverses all code positions in application B that iterate over the extended objects and adds a statement that will skip the iteration over the properties extended by the other application. The algorithm proceeds by checking if there are any conflicts with the object extensions done in application B, and if there are, the property names in the feature code are replaced with unconflicting names. The process goes similarly for the application B with the exception that there is no need for handling naming conflicts. Finally, event handler conflicts are handled by inserting code that creates an event-handler-tracker object that keeps track of all registered handlers, replacing conflicting code expressions in both applications with code that reroutes the handler registration and deregistration to the event-handler-tracker, and inserting code that invokes the necessary handlers.

4   Merging Code

Once all conflicts have been detected and fixed or reported, the process is finished by merging the code of the feature and the application (Algorithm 3). The main idea is to perform the merge of both the header and body nodes of each application, and then to move the feature nodes to the designated position, without introducing errors. Algorithm 3 works by taking the head children and the body children of the feature graph from application A and appending them to the head and the body node of the application B. Next the HTML nodes that define the feature are selected from the graph with the goal of moving them to a new position defined with the $rSlctr$ selector. Since not only feature HTML nodes are selected in the feature identification process (others might be included due to dependencies) it is not always possible to perform the moving of nodes without introducing errors. Some CSS selectors that apply styles to feature nodes, or JavaScript DOM queries, could be structurally dependent on the position of the feature nodes in the page hierarchy, and by moving the feature nodes errors are introduced. In this case we detect and report the error positions (lines 7, 8). Also, due to DOM queries, when moving feature nodes it is necessary to maintain the relative position of the feature script nodes towards the feature nodes (line 11).

5   Case Studies

The evaluation of the approach is based on four case study applications divided into two groups: $i)$ Group 1, applications 1 and 2 that use the most wide-spread third-party JavaScript library – jQuery; and $ii)$ Group 2, applications 3 and 4 developed with the second most-wide spread JavaScript library – MooTools[1]. With these four applications we have created a set of four case studies with a goal to test whether our method is capable of performing automatic feature

---

[1] http://w3techs.com/technologies/history_overview/javascript_library/all

---

Algorithm 3. merge(*fGraph, bGraph, fSlctr, rSlctr*)

1: **for all** *hChild* : getHeadChildren(*fGraph*) **do**
2:     **if** isLinkScriptStyle(*hChild*) **then**
3:         appendToHeadNode(*hChild, bGraph*)
4: **for all** *bodyChild* : getBodyChildren(*fGraph*) **do**
5:     appendToBodyNode(*bodyChild, bGraph*)
6: *featrNds* ← getFeatureNodes(*fSlctr, fGraph*)
7: **for all** *pos* : getStructSlctrsPos(*featrNds, fGraph*) **do**
8:     addComment(pos, "Error - struct query")
9: moveNodes(*rSlctr, featrNds*)
10: *scripts* ← getFeatureScriptsAffctdPos(*fSlctr, rSlctr, bGraph*)
11: updatePosition(*rSlctr, scripts, featrNds*)

---

reuse in different situations (e.g. is the process able to include a feature developed with the jQuery library into the application developed with the MooTools library, and vice versa). Based on the features from each application we have specified Selenium tests[2] that test the correctness of the features in the final application. The case study applications, their tests, and the results are available at: *www.fesb.hr/~jomaras/download/reuseCaseStudies.zip*.

Table 1. Case studies: Lines of code (LOC); Feature (F); HTML modifications (H), CSS modifications (C), JavaScript modifications (J) ; Time – process execution time in seconds

#	App A	App B	A-LOC	B-LOC	F-LOC	H (A;B)	C (A;B)	J (A;B;B')	Time	Success
1	App 1	App 2	10554	12031	1083	30;0	12;201	9;0;0	179	Y
2	App 1	App 3	10554	5112	1083	29;0	12;11	9;0;0	159	Y
3	App 4	App 2	9083	12031	1258	46;0	23;201	6;0;0	195	Y
4	App 4	App 3	9083	5112	1258	44;0	23;22	N/A	168	N

Table 1 shows the results of 4 reuse case studies. For each reuse we present the total number of lines of code in the application from which a feature was extracted (A-LOC), lines of code of the application where the feature will be reused into (B-LOC), total LOC of the feature extracted with the feature identification process (F-LOC), number of changes done to the HTML code (H), CSS code (C), and JavaScript code that were performed by the process to resolve conflicts; and the total running time of the reuse process[3]. All experiments have been performed with a plugin to the Firefox browser – Firecrow[4], which implements all algorithms described in the paper.

In all cases but one, the method was able to introduce a feature from one application into another. However, in order to achieve this, some modifications

---

[2] http://docs.seleniumhq.org/
[3] Firefox 20, Core i7, 1.73GHz, 6 GB RAM.
[4] https://github.com/jomaras/Firecrow

of the application source code were necessary. As can be seen from the Table 1 the majority of these modifications was concerned with resolving HTML naming attributes conflicts, and conflicts that arise due to overriding CSS styles. As far as JavaScript conflicts go, conflicts in the library methods were found and fixed.

The failing Case 4, related to reusing a feature from an application with the MooTools library into another application that also uses the MooTools library, has a number of problems that have occurred in the JavaScript code. The problems are related to extensions of built-in prototypes – the method was able to identify the conflicting positions, but the tool was unable to perform the necessary corrections, due to the particularities of how these extensions were implemented.

6   Related Work

There are a number of approaches that support reuse: HunterGatherer [7], Internet Scrapbook [8], HTMLviewPad [9], and Web Mashups [6] in the web application domain; and G&P [3] and Jigsaw [1] in the Java domain.

HunterGatherer [7], Internet Scrapbook [8], and HTMLviewPad [9] are similar approaches mostly related to clipping and reusing fragments of Web pages. Since these approaches were developed in 1990's and early 2000, when web page development was not so dynamic on the client side, their usability in the current web development is quite limited. These approaches are mostly limited to reusing HTML element such as text fragments and forms, and make no attempts to also include CSS and JavaScript. Web mashups [6] are web applications that combine information and services from multiple sources on the web. The main advantage of mashups is that they enable the creation of new applications by integrating services offered by third-party providers. The main difference between mashups and our approach is that mashups foster reuse on a service-level, while we specifically target reuse on the code level.

In the more general domain of Java applications, G&P [3] is a reuse environment composed of two tools: Gilligan and Procrustes, that facilitates pragmatic reuse tasks. Gilligan allows the developer to investigate dependencies from a desired functionality and to construct a plan about their reuse, while Procrustes automatically extracts the relevant code from the originating system, transforms it to minimize the compilation errors and inserts it into the developer's system. In this domain there is also a tool called Jigsaw [1] which facilitates small-scale reuse of source code. The main difference between our approaches is that G&P and Jigsaw are approaches that statically analyze Java applications – while the ideas and end goals are similar, their methods can not be used in the highly dynamic, multi-paradigm environment of client-side web applications.

This work is a continuation of our previous work [5] where we have described an approach for extracting and reusing web application UI controls by combining simple analysis and profiling information. Due to not having a way of precisely capturing dependencies between different parts of the application, the reuse process was primitive, and was more focused on extraction than on handling reuse problems. This is why in this work, by providing a detail description of possible conflicts, and by basing the process on the client-side dependency graph [4], we are able to handle conflicts that arise both statically and dynamically.

7   Conclusion

In this work we have shown how to achieve code-level feature reuse in client-side web application development. We have defined what exactly feature reuse is, and when can it be considered successful. Naturally, when reusing code from one application into another application a number of problems can occur – we have identified those problems and have developed algorithms both for their handling and for merging code bases. Finally, by testing the method on four non-trivial case-study applications, we have shown that the method is capable of identifying and handling conflicts, and performing actual reuse.

For future work, we recognize that providing user specified scenarios is time-consuming, and we plan to include a method for automatic generation of usage scenarios. We also plan to expand the reuse process to include the server-side – the client-side and the server-side are parts of the same whole, and should be treated as such. Also, for this research we have considered a particular kind of client-side feature reuse – the reuse of behavior on a structure. However, one might argue that support for global application behavior reuse should also be provided. This would require more advanced behavior analysis and code modification because a wider range of problems can occur when allowing code originating from one application to operate on the DOM of the other application.

References

1. Cottrell, R., Walker, R.J., Denzinger, J.: Jigsaw: a tool for the small-scale reuse of source code. In: ICSE, pp. 933–934. ACM (2008)
2. Eisenbarth, T., Koschke, R., Simon, D.: Locating features in source code. IEEE Transactions on Software Engineering 29(3), 210–224 (2003)
3. Holmes, R., Walker, R.J.: Semi-Automating Pragmatic Reuse Tasks. In: ASE, pp. 481–482. IEEE Computer Society Press (2008)
4. Maras, J., Carlson, J., Crnkovic, I.: Extracting client-side web application code. In: WWW 2012, pp. 819–828. ACM, New York (2012)
5. Maras, J., Štula, M., Carlson, J.: Reusing web application user-interface controls. Web Engineering, 228–242 (2011)
6. Murugesan, S.: Understanding web 2.0. IT Professional 9(4), 34–41 (2007)
7. Schraefel, M.C., Zhu, Y., Modjeska, D., Wigdor, D., Zhao, S.: Hunter Gatherer: Interaction Support for the Creation and Management of Within-Web-Page Collections. In: WWW, pp. 172–181 (2002)
8. Sugiura, A., Koseki, Y.: Internet scrapbook: creating personalized world wide web pages. In: HCI, pp. 343–344. ACM (1997)
9. Tanaka, Y., Ito, K., Fujima, J.: Meme Media for Clipping and Combining Web Resources. World Wide Web 9, 117–142 (2006)

Solving Complex Decision-Making Problems through Agent-Matrices Cooperation

Hao Lan Zhang[1], Jiming Liu[2], Yong Tang[3], and Chaoyi Pang[1]

[1] NIT, Zhejiang University, Ningbo, Zhejiang Province, China
[2] Hong Kong Baptist University, Kowloon Tong, Hong Kong
[3] South China Normal University, 55, West Zhong Shan Avenue, China
haolan.zhang@nit.zju.edu.cn, jiming@comp.hkbu.edu.hk,
ytang@scnu.edu.cn, chaoyi.pang@csiro.au

Abstract. Making decisions in a dynamic environment is complicated and indefinite because of various unpredictable factors and large volumes of information. It has become a key issue for modern organizations to have efficient systems and tools to support rational decision-making in problem-solving processes, particularly, for problems with a lack of pre-defined structure. This paper provides a novel agent-based framework for Decision Support Systems (DSS) based on the analysis of the major problems in existing DSS. The proposed multi-agent-based framework, namely the Agent-Matrices framework, provides an open, efficient and flexible architecture for DSSs. The Agent-Matrices framework overcomes the compatibility and connectivity problems in most traditional DSS applications. This framework utilises Matrices to allow agents to acquire information, self-upgrade, perform tasks, travel to other Matrices, and be reused. This paper primarily introduces the methods used for coordinating the Matrices and agents to solve a complex problem.

Keywords: Agent-based Systems, DSS, Agent-Matrices.

1 Introduction

Agent architectural design has become a primary issue in the intelligent agent area as more and more concrete developments has emerged in the intelligent agent area such as JACK (agent development environment) [1], KAoS [2], etc. Existing agent architectural design methodologies brought forth several basic questions including: How would agent architectural design impact on the efficiency of an agent-based application? What are the major factors affecting the agent architectural design methodologies? Researchers in the field are pursuing answers to these questions.

In this paper we suggest several new methods for agent architectural design based on Agent-Matrices framework, which includes the concept of Matrices, the core components of the framework, etc. In this framework, agents are assembled through a mediator, i.e. the Matrix, which can coordinate agents to solve a complex problem.

X. Lin et al. (Eds.): WISE 2013, Part II, LNCS 8181, pp. 489–498, 2013.

The reminder of this paper is organized as follows: the next section describes the core components in the Agent-Matrices framework. Section 3 introduces the Matrices concept. Section 4 illustrates the structural design of Matrices. Section 5 introduces the unified Matrices structure and the last section concludes the research findings.

2 The Matrix Concept

The Matrix concept in the Agent-Matrices framework suggests a new method for agent cooperation and coordination, which incorporates and extends the agent society concept into the Matrix design. Unlike the middle-agent-based DSS, the Matrix not only acts as a middle-agent or an agent facilitator but also as a living platform for agents. The Matrix provides a small community environment for agents, where agents can self-upgrade and cooperate with other agents.

The unified Matrices structure provides a larger and more comprehensive structure that allows the cooperation among various Matrices. This unified Matrices structure extends the agent society concept that heterogeneous agents are distributed in various agent groups (communities) and each group is dominated by a Matrix. These groups, i.e. Matrices, form a large scale of interconnected Matrix network as shown in Figure 1.

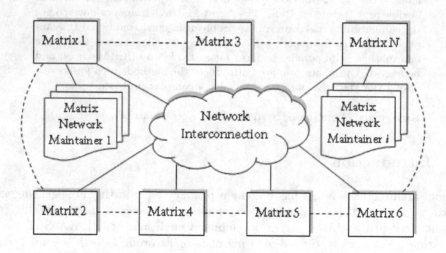

Fig. 1. Matrices Interconnection

2.1 Components in the Agent-Matrices Framework

An Agent-Matrices system basically consists of four major components, which include a number of agents, a number of Matrices, several databases (or knowledge-bases) that can be accessed by the Matrices, and an Agent-Matrices network maintainer for managing the connections between the Matrices and agent transportations. Table 1 lists the core components deployed in the Agent-Matrices framework and the major sub-components used in each of these core components. An Agent-based Matrix normally assembles a group of agents to solve the problems for a specific organization. Beyond the advantages of utilizing a flexible and open structure in the Agent-Matrices framework, the unified Matrices structure can support the cooperation between local agents and external (remote) agents.

Table 1. Core Components of Agent-Matrices

Core Components	Major Sub-Components
Matrix	Matrix register, Matrix control panel, Matrix learning centre
Agents	BDI Model, Capability register, Travel control centre, DSC container
Databases	Data, rules, information, etc
Matrix Maintainer	Matrix distribution information centre

This unified Matrices structure allows the heterogeneous agents in different domains (Matrices) to work cooperatively to solve complex problems without redundantly developing new agents for organizations. For instance, an agriculture department's Agent-Matrices-based DSS temporarily needs the population statistics of a specific region for a specific task analysis. It is costly to develop a new agent or several new agents to perform this task. Fortunately, a similar Agent-Matrices-based agent has been developed in a statistical company and this agent is also plugged into the company's Agent-Matrices-based Matrix. Therefore, the agriculture department could just send their requests to the statistical company's Agent-Matrices-based DSS for the results. This framework is cost-effective for the organizations that temporarily need outsourcings.

2.2 The Structural Design of Matrices

Based on the concepts of the Matrix, the four fundamental components in the Matrix are listed, which include:

(1) the agent society that provides a virtual space to agents;
(2) the Matrix learning centre that acquires information, i.e. new Domain-Specific-Component (DSC) items from the external environment;
(3) the Matrix control panel which is the core part that mainly handles agent matching, requests processing, and resource allocation;
(4) the DSC Usage Centre that allows the functional components to be used/reused by agents.

Each Agent-Matrices-based agent carries a DSC container, which holds a number of DSC items. Each DSC item can perform one (or several) specific task(s) and produce results; and each item can be plugged/ unplugged into/from the DSC container, which offers a flexible structure for upgrading agents' capabilities. These DSC items in an agent represent the agent's problem-solving capabilities. The Matrix control panel manages the agent society through establishing relationships with agents. An agent society can be regarded as the aggregation of a set of agent-relationships.

The Matrix-Agent connection design is based on a hybrid network topology. The connections between agents and Matrix are centralised, whereas the connections between agents are decentralised. The major role of the Matrix is to organise the agents to accomplish the tasks sent by users.

3   A Unified Matrices Structure for Matrices Cooperation

Each Matrix in the Agent-Matrices framework is connected by a number of agents that can deal with various problems in different domains. However, not every request can be solves by the internal agents in a Matrix. Once a request cannot be solved in a Matrix, the Matrix will forward the request to its nearest or most familiar Matrix for further processing. In this paper, a unified Matrices structure is employed based on the previous work [9, 10]. This is the fundamental motivation of establishing the unified Matrices structure in the Agent-Matrices framework.

3.1   Matrix Searching Algorithms for Service-Provider Matrices

Two primary guidelines are deployed for searching a service-provider Matrix based on a requesting Matrix in a unified Matrices structure [3], including the Most familiar partner method and the Supplemental partner method. The following sections incorporate the Agent-Matrices framework with the previous work conducted for unified Matrix design structure [10].

There are two situations when dealing with a request for cooperation (searching for external Matrices): one situation is to search for a partner Matrix that has similar capabilities to the requesting Matrix; the other situation is to find a partner Matrix that can provide some services that the requesting Matrix does not have. For instance, when an accounting agent in a company's Matrix system requires last year's regional taxation statistics, then it would look for another accounting agent that could provide regional statistical data, and the cooperation between these two agents might happen regularly. In this case, we use the most familiar partner method.

The following example explains the matching process. There is a huge natural disaster in the region where this company is located and the company's accounting agent requires a climate forecast report to analyse: (1) whether it is necessary to inject more funds to strengthen the company's buildings; (2) the amount of funds. In this situation, the requesting Matrix is looking for a very unfamiliar service provider Matrix, which has very different functionalities. Therefore, the supplemental partner method is employed.

3.2 Most Familiar Partner Method

We use a Matrix capability description table to describe the functionalities and capabilities of a Matrix as shown in Table 2.

Table 2. Matrix Capability Descriptions

Agent Functionality Keywords	Agent ID	Domain
Salary statistics, superannuation expenses, facility maintenance, etc.	Agent 1	Accountant
Warehouse stocks, sales condition, transportation expenses, etc.	Agent 2	Accountant
....	....	....
Market share history, market trends, major rivals' products, etc.	Agent i	Marketing

This table contains the information on all the connected agents' capabilities and the domain information, which indicates agents' major tasks or responsibility. Each agent's capability is described by several keywords and these keywords also represent the capabilities of the Matrices as Figure 2 shows [10].

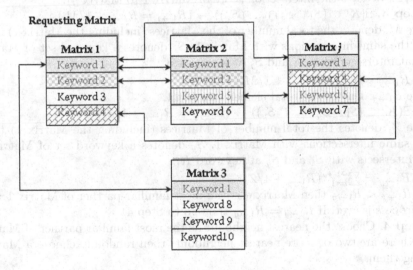

Fig. 2. Intersections of Matrices Capabilities [3]

In Figure 2, we aggregate all the keyword descriptions of the other Matrices, which intersect with Matrix 1's keyword descriptions. If there is one keyword intersection between the requesting Matrix and a corresponding Matrix, we add a score, namely the intersection score, for these two joined Matrices, expressed as $R_{i \to j}$, which means Matrix i and j intersect and their intersection score is $R_{i \to j}$.

If a corresponding Matrix has the most keyword intersections with a requesting Matrix (compared to other Matrices), then this Matrix is regarded as the most familiar partner of the requesting Matrix.

When there are two or more corresponding Matrices having the same number of keyword intersections with a requesting Matrix, then the Matrix that has keyword intersections least shared by other Matrices is selected as the most familiar partner. Thus, the calculation of the most familiar partner method can be expressed as follows. In the following example, we search a most familiar partner of Matrix 1.

Step 1. $\forall S_i \in \Omega, \forall K^j \in S_i$
where $i = 1, 2, n$. S_i denotes a set of keyword aggregation of Matrix i; Ω denotes the total aggregation of all keywords in all Matrices. $j = 1, 2, m$; j denotes the keyword number in a set; K_j denotes the keyword j in a keywords set; Φ denotes empty set.

$$\exists (S_1 \cap S_j \neq \Phi) \to (R_{1 \to j} = |S_1 \cap S_j|)$$
$$\exists (S_1 \cap S_j \neq \Phi) \to (R_{1 \to j} = 0)$$

where S_1, S_j denote the keywords set 1, j of Matrix 1, j, respectively. $R_{i \to j}$ denotes the intersection score of Matrix 1 to Matrix j.

Step 2. If $R_{i \to j}$ is the maximum value among all the other intersection scores with Matrix 1, then Matrix j is the most familiar partner of Matrix 1 and the most familiar partner process is completed. If $\exists (R_{1 \to j} = R_{1 \to g})$ then go to Step 3 ($R_{i \to g}$ denotes the intersection score of Matrix 1 to Matrix g).

Step 3. $\exists (K^i \subset (S_1 \cap S_j \cap ... \cap S_x)) \to (R_{1 \to j} = R_{1 \to i} + 1/M_i)$
where M_i denotes the total number of the Matrices (including the Matrix 1) that have the same intersections with Matrix 1. S_x denotes a keyword set of Matrix X that intersects with S_1 and S_j at keyword K_i.

$$R_{1 \to j} = \sum_{i=1}^{S_{1 \to j}} (R_{1 \to i} + 1/M_i)$$

where $S_{1 \to j}$ denotes the total elements in Set 1.

$$\exists (K^i \subset (S_1 \cap S_g \cap ... \cap S_y)) \to (R_{1 \to g} = R_{1 \to i} + 1/Z_i)$$

where Z_i denotes the total number of Matrices (including the Matrix 1) that have same intersections with Matrix 1. S_y denotes a keyword set of Matrix y that intersects with S_1 and S_g at keyword K_i.

$$R_{1 \to g} = \sum_{i=1}^{S_{1 \to g}} (R_{1 \to i} + 1/Z_i)$$

If $R_{1 \to j} < R_{1 \to g}$ then Matrix g is the most familiar partner of Matrix 1 and the process is over. If $R_{1 \to j} = R_{1 \to g}$ then go to Step 4.

Step 4. Choose the nearest neighbour as the most familiar partner of Matrix 1. If there are two or more nearest neighbours then randomly choose a Matrix among them.

3.3 Supplemental Partner Method

In previous work [10], the supplemental partner method has been introduced. In such a method, a service-provider Matrix of a requesting Matrix is choosen that has minimum keywords intersections with the requesting Matrix. The supplemental partner method seeks a service provider Matrix that has a minimum intersection score. We still use the same example to search a supplemental partner for a specific Matrix. The detailed steps are described as follows, which mainly based on the previous work [10].

Step 1. It is the same procedure as in the most-familiar-partner method .
$$\forall S_i \in \Omega, \forall K^j \in S_i$$
$$\exists (S_1 \cap S_j \neq \Phi) \rightarrow (R_{1 \rightarrow j} = |S_1 \cap S_j|)$$
$$\exists (S_1 \cap S_j \neq \Phi) \rightarrow (R_{1 \rightarrow j} = 0)$$

Step 2. If $R_{1 \rightarrow j}$ is the minimum value among all the other intersection scores with Matrix 1, then Matrix j is the supplemental partner of Matrix 1 and the supplemental partner process is completed.

If $\exists (R_{1 \rightarrow j} = R_{1 \rightarrow j} = 0)$ then choose a nearest partner among these same intersection score Matrices. If there are two or more nearest neighbours then randomly choose a Matrix among them.

If $\exists (R_{1 \rightarrow j} = R_{1 \rightarrow j} > 0)$ then go to Step 3 ($R_{1 \rightarrow g}$ denotes the intersection score of Matrix1 to Matrix g).

Step 3. In this step, the supplemental partner method seeks a minimum intersection score.
$$R_{1 \rightarrow j} = \sum_{i=1}^{S_{1 \rightarrow j}} (R_{1 \rightarrow i} + 1/M_i)$$
$$R_{1 \rightarrow g} = \sum_{i=1}^{S_{1 \rightarrow g}} (R_{1 \rightarrow i} + 1/Z_i)$$
If $R_{1 \rightarrow j} > R_{1 \rightarrow g}$ then Matrix g is the supplemental partner of Matrix 1 and the process is over. If $R_{1 \rightarrow j} = R_{1 \rightarrow g}$ then go to next step.

Step 4. Choose a nearest partner among these same intersection score Matrices. If there are two or more nearest neighbours, then randomly choose a Matrix among them.

The most familiar partner and supplemental partner methods are basically for matching capabilities and functionalities among Matrices. For more specific agent capability search and agent relationship management, we use a novel Agent-rank algorithm. The agent-rank algorithm is the most efficient means to search a service provider for a request. The most familiar partner and supplemental partner methods help to narrow the searching range for the agent-rank algorithm. These two methods based on Matrix matching are complementary; the combination of these two methods could enhance the matching efficiency and accuracy.

4   The Optimised Model and Performance Evaluation

The unified Matrices structure introduced in the previous sections has its limitations on central control and fault-tolerance. These limitations lead to the circumstances of low efficiency in the matching process, massive traffic chaos in the cooperation process and vulnerability to partial system breakdown. In the Agent-Matrices framework, an optimised model, namely Super-node model, is introduced.

4.1   Super-Node Model in the Matrices Cooperation

Based on the early work [10], a super-node model is suggested to tackle the above problems. The super-node model [4] has been extensively applied to many online systems, such as KaZaA [5], SkypeNet [6], etc. In a super-node model, a number of nodes are selected as super nodes, which manage a limited number of other nodes. The major advantages of the super-node model include reducing time and bandwidth for search, enhancing manageability, etc.[4]. This model avoids mesh peer-to-peer connections, which minimizes the probability of the occurrence of concurrency. The concurrency problem often causes repetitions of communications, which increases network traffic volume.

The criteria of determining whether a computing terminal is suitable to be a super-node are quite simple, which include: (1) it must have a high bandwidth connection; (2) it needs to have a reasonable computing capability; (3) it should be flexible and efficient in entering and exiting a network. Generally, each super-node Matrix is directly connected to a Matrix network maintainer.

4.2   Performance Evaluation

The Agent-Matrices framework employs the optimised methods including the most familiar partner method, supplemental partner method, and super-node model. Figure 3 and 4 show the performance comparison between optimised and non-optimised Agent-Matrices-based framework. The following figures illustrate the performance comparison.

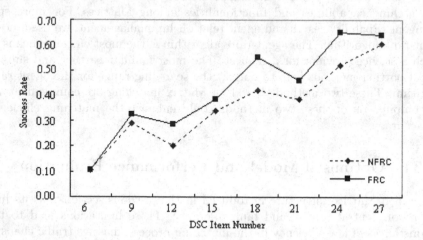

Fig. 3. Success rate comparison based on NFRC and FRC (200 requests)

The performance comparison clearly indicate that the most familiar partner and supplemental partner methods adopted in the Agent-Matrices framework can successfully improve the matching success rate and reduce the matching

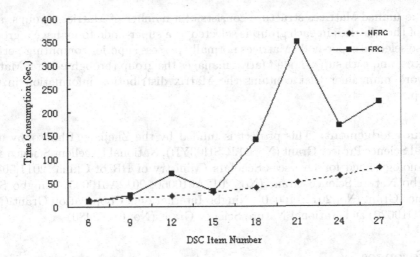

Fig. 4. Time consumption based on NFRC and FRC (200 requests)

time consumption. The application of the super-node model in the Unified Matrices Framework enhances the fault-tolerance capability and the system manageability. In the unified Matrices structure, the super-nodes are those selected Matrices that play a coordinator role in processing the other Matrices' requests. These super-nodes receive the requests from the other Matrices and search the corresponding Matrices.

5   Conclusion

Autonomy oriented computing (AOC) unifies the methods for effective analysis, modelling, and problem-solving in complex systems [7, 8]. This paper addresses AOC-by-prototyping and AOC-by-self-discovery issues through introducing the Matrix concept [7].

The Matrix provides a virtual platform for intelligent agents through managing Matrix-agent relationships in an agent society. The Matrix maintains a superior transmission performance in its virtual platform through combining centralized and decentralized topologies. The Matrix also presents seemly mobility as the peer-to-peer connections between agents only exist temporarily.

The Matrix employs a Domain-Specific-Component usage centre to assure that every agent is upgradeable. Different from agent mediators and facilitators in other middle-agent-based systems, a Matrix in the Agent-Matrices [9, 10, 11] framework not only coordinates its internal agents but also provides its internal agents with various information resources, such as the DSC items, data resources, etc. An agent can perform self-upgrade through obtaining new DSC items from the Matrix and replacing dated items. This Matrix-agent connection design simplifies the self-upgrade processes of an agent, which are rather complicated in many agent-based systems.

The unified Matrices structure consists of a number of Matrices groups; and one of the Matrices in each group is selected as a super-node to manage a group. These selected super-node Matrices normally possess superior computing performances; and each super-node Matrix manages the group through using a Matrix network maintainer that contains the Matrix distribution information in the group.

Acknowledgement. This project is funded by the Zhejiang Philosophy and Social Science Project Grant (No. 11JCSH03YB), National Excellent Science and Technology Fund for Overseas Scholars (Ministry of HR of China, 2011[508]), Ningbo Nature Science Grant (No. 2012A610060, 2012A610025), Ningbo Soft Science Grant (No. 2012A10050), Ningbo International Cooperation Grant (No. 2012D10020) and National Nature Science Grant (No. 61272480).

References

1. Evertsz, R., Fletcher, M., Frongillo, R., Jarvis, J., Brusey, J., Dance, S.: Implementing Industrial Multi-agent Systems Using JACK. In: Dastani, M., Dix, J., El Fallah-Seghrouchni, A. (eds.) PROMAS 2003. LNCS (LNAI), vol. 3067, pp. 18–48. Springer, Heidelberg (2004)
2. Bradshaw, J.M., Dutfield, S., Benoit, P., Woolley, J.D.: KAoS: Toward an Industrial-Strength Generic Agent Architecture. In: Software Agents, pp. 375–418. AAAI/MIT Press, Cambridge (1997)
3. Padgham, L., Winikoff, M.: Prometheus: A Methodology for Developing Intelligent Agents. In: Proc. of AAMAS, Bologna, Italy, pp. 37–38 (2002)
4. Fiorano, Software Inc.: Super-Peer Architectures for Distributed Computing. Technical report, Fiorano Copyrights (2007)
5. KazaA, Inc.: How Peer-to-Peer (P2P) and Kazaa Software Works. Technical report, KazaA Copyrights (2006)
6. Skype, Inc.: Skype Guide for Network Administrators. Technical report, Skype Copyrights (2005)
7. Liu, J., Jin, X., Tsui, K.C.: Autonomy Oriented Computing: From Problem Solving to Complex Systems Modeling. pp. 8–10. Springer (2004)
8. Liu, J., Jin, X., Tsui, K.C.: Autonomy oriented computing (AOC): Formulating computational systems with autonomous components. IEEE Transaction on Systems, Man and Cybernetics, Part A: Systems and Humans 35(6), 879–902 (2005)
9. Zhang, H.L., Leung, C.H.C., Raikundalia, G.K.: Topological analysis of AOCD-based agent networks and experimental results. Journal of Computer and System Sciences 74(2), 255–278 (2008)
10. Zhang, H.L., Leung, C.H.C., Raikundalia, G.K.: Matrix-Agent Framework: A Virtual Platform for Multi-agents. Journal of Systems Science and Systems Engineering 15(4), 436–456 (2006)
11. Zhang, H.L., Pang, C., Li, X., Shen, B., Jiang, Y.: A Topological Description Language for Agent Networks. In: Sheng, Q.Z., Wang, G., Jensen, C.S., Xu, G. (eds.) APWeb 2012. LNCS, vol. 7235, pp. 759–766. Springer, Heidelberg (2012)

A Cloud System for Community Extraction from Super-Large Scale Social Networks

Zhiang Wu, Haicheng Tao, Youquan Wang, Changjian Fang, and Jie Cao*

Jiangsu Provincial Key Laboratory of E-Business,
Nanjing University of Finance and Economics, Nanjing, China
{zawuster,haicheng.tao,youq.wang}@gmail.com,
{Changjian.Fang,Jie.Cao}@njue.edu.cn

Abstract. This demo showcase the Community Extraction Cloud (CEC) system. The key idea is to drop weak-tie nodes by efficiently extracting core nodes based on the novel concept of asymptotically equivalent structures (AES) and parallel AES mining algorithm. Meanwhile, to facilitate storing and processing of massive networks, several cloud computing technologies including HDFS, Katta, and Hama are seamlessly integrated into CEC system.

1 Introduction

As real-life social networks evolving into super-large scales, community detection becomes increasingly challenging, even though a sea of research efforts have been devoted. Since there exist a large number of weak-tie or overlapping nodes in super-large scale network, the network often cannot be partitioned into several crisp communities. To remedy this, *community extraction* has been proposed [4] to extract tight and meaningful communities with only core nodes from massive social networks. Along this line, this paper showcase a cloud system designed for extracting communities from super-large scale networks (a.k.a. CEC, Community Extraction Cloud). As shown in Fig. 1, the CEC system consists of three layers, i.e., data layer, processing layer, and presentation layer. We proceed to introduce these three layers as follows.

Data layer is responsible for storing and managing network data. Each network is represented as a *market basket transaction* file, where each line corresponds to a node and items in this line are neighbor nodes of that node. Since our target is super-large scale networks, Hadoop Distributed File System (HDFS) is employed, and then a big file is split into many small files. To speed up retrieving a specific node and its neighbors, we utilize Katta to create distributed indexes on each small file.

Processing layer executes the core task of CEC, that is, to extract closely-knit communities from super-large social networks. Our COSCOM (COSine-pattern based COMmunity extraction) framework is equipped on this layer. Generally, COSCOM consists of three main phases: (1) to extract asymptotically equivalent structures (AES) using CoPaMi, which will be elaborated in

X. Lin et al. (Eds.): WISE 2013, Part II, LNCS 8181, pp. 499–502, 2013.

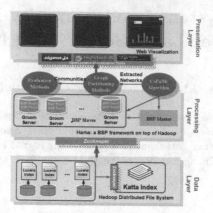

Fig. 1. The system architecture of CEC

Sect. 2; (2) to assemble all the nodes from the extracted AESs and partition them into communities using existing community detection methods such as Kmeans, METIS, Fast Newman, etc; (3) to evaluate the quality of the detected communities using various validation measures. To facilitate the parallel mining, Bulk Synchronous Parallel (BSP) model and one of its implementation framework called Hama are adopted in CEC.

Presentation layer provides user interaction and visualization results according to Internet explorer. The user can upload or select networks to be processed. Two kinds of settings are provided for extracting the core network derived from AESs: the first one for professional users is to set two thresholds τ_G, τ_F, and another one for common users is simply to set the percentage of extracted nodes. Then, before starting partitioning, CEC also provides two kinds of settings: the first one is to set the number of communities C, and another one is to select automatically determine C in which CEC will return several communities with maximal *modularity*. During this process, original network, extracted sub-networks, and communities will be visualized to provide intuitive view for users. Meanwhile, the quality of extracted communities in terms of modularity, and some topological statistics both in original and extracted network such as degree, clustering coefficient, eigenvector centrality, betweenness centrality, etc will be exhibited.

2   Method Details

In this section, we briefly introduce the key idea of CEC, i.e., AES and the parallel AES mining algorithm.

2.1   Asymptotically Equivalent Structures

The nodes in *structural equivalence* must have exactly the same friends, and thus tend to form a very tight community. This idea, however, confront one problem. That is, structural equivalence is often too restrictive for real-life social networks. To meet this challenge, we first reformulate the concept of structural equivalence. As we know, a node set $S = \{i_1, \cdots, i_{|S|}\}$ is structurally equivalent if $N_1 = \cdots = N_{|S|}$, or equivalently, $r_p = 1$. Let $r_p = |\bigcap_{q=1}^{|S|} N_q|/|N_p|$ characterize the ratio of common friends for node i_p, $1 \leq p \leq |S|$. To relax the concept of structural equivalence, we can set r_p as a measure but lower its threshold from one to a proper level. Moreover, to filter out weak-tie nodes, we should further demand that a node set should have a certain number of common friends. Therefore, we can derive the following two measures:

$$G(S) = \sqrt[|S|]{\prod_{p=1}^{|S|} r_p}, F(S) = \frac{|N_1 \cap \cdots \cap N_{|S|}|}{n}. \tag{1}$$

Based on G and F, we have the following definition:

Definition 1. *A node set S is an asymptotically equivalent structure if $G(S) \geq \tau_G$ and $F(S) \geq \tau_F$, where $\tau_G, \tau_F \in [0, 1]$ are given thresholds.*

So our task here is to extract all the AES from a large-scale network. The most beautiful part of an AES, however, is that it is a *cosine pattern* in essence. To understand this, let us consider the adjacency matrix of an undirected network (denoted as A), where $A_{pq} = 1$ ($p \neq q$) if there is an edge between node i_p and node i_q and 0 otherwise. Accordingly, $\sum_{q=1}^{n} A_{pq} = |N_p| = d_p$, $1 \leq p \leq n$. Let $\mathcal{T}_A$ be the transaction data set transformed from A, we now reformulate G and F from a pattern mining perspective as follows. Given a node set S, $|N_1 \cap \cdots \cap N_{|S|}| = |\{t_p | S \subseteq t_p, 1 \leq p \leq n\}| = \sigma(S)$, where $t_p = \{i_q | A_{pq} = 1, 1 \leq q \leq n\}$ is the pth transaction in $\mathcal{T}_A$, and $\sigma(S)$ is the support count of S in $\mathcal{T}_A$. As a result, we have $F(S) = s(S)$, i.e., the support of S. Moreover, it is easy to show $G(S) = \sigma(S)/\sqrt[|S|]{\prod_{p=1}^{|S|} \sigma(\{i_p\})} = s(S)/\sqrt[|S|]{\prod_{p=1}^{|S|} s(\{i_p\})} = \cos(S)$, i.e., the cosine value of S. To sum up, we have the following proposition:

Proposition 1. *Given the thresholds τ_G and τ_F, to extract all the asymptotically equivalent structures from a network $\mathcal{G}$ is equivalent to mine all the cosine patterns from the corresponding adjacency matrix A, with $\tau_c = \tau_G$ and $\tau_s = \tau_F$.*

2.2 Parallel Cosine Pattern Mining Algorithm

Since mining AESs is equivalent to mining cosine patterns, we present a novel Cosine PAttern MIning (CoPaMi) algorithm which employs a FP-growth-like [2] procedure. However, the key difference between CoPaMi and FP-growth lies in that CoPaMi employs cosine similarity to prune sub-trees, but retains the pruning effect of support. Cosine similarity can be used for pruning, since it holds the conditional anti-monotone property (CAMP), an extension of the anti-monotone property. More details about CAMP can be found in [3].

To cope the challenge led by the drastically increase of the data scale, CoPaMi should be extended to support parallelized mining from disk. Fortunately, FP-tree is apt to be decomposed into several smaller sub-trees. In light of the aggressive decomposition of FP-tree [1] and BSP model, we present parallelized implementation of CoPaMï, which works as follows:

1. A big file is split into several small files of which the size is 64M according to the setting of HDFS. Then, the first round of BSP peers are started, and each peer works on a small file. Each peer creates Lucene index for each line and obtains tuples $(i_k, \sigma_l(i_k))$ indicating the node i_k and its support in the local file.
2. After BSP master collects all outputs of peers, it aggregates all tuples and sort nodes by support, so that F_1 is obtained. Note that since the range of support is between 0 and n, counting sort can be applied to obtain F_1 with the time complexity $O(n)$.
3. Before starting the second round of BSP, the master should partition F_1 into K groups, where K is the number of BSP peers. Each group corresponds to aggregated frequent items in [1], and then projection can be done to obtain sub-FP-trees. Note that the index on each node is essential to this step, since it needs to repeatedly retrieve a specific line for constructing sub-trees.
4. Every BSP peer invokes CoPaMï on its sub-FP-tree for mining AES. BSP master finally collect the outputs of all BSP peers to get AESs of the original large-scale network.

3   Conclusions

This demo showcase a cloud system called CEC for community extraction from super-large scale social networks. CEC has two notable features: (1) the novel concept of AES and its parallel mining algorithm are employed to isolate weakly connected nodes; (2) multiple cloud computing technologies are integrated for storing and processing massive social networks.

Acknowledgments. We thank all of teachers and students in JSELAB who have put their efforts to CEC. This research was partially supported by NSFC (Nos. 71072172, 61103229), National Key Technologies R&D Program of China (Nos. 2013BAH16F01, 2013BAH16F04), Industry Projects in Jiangsu S&T Pillar Program (No. BE2012185), and the Priority Academic Program Development of Jiangsu Higher Education Institutions (PAPD).

References

1. Grahne, G., Zhu, J.: Fast algorithms for frequent itemset mining using fp-trees. IEEE Trans. Knowl. Data Eng. 17(10), 1347–1362 (2005)
2. Han, J., Pei, J., Yin, Y.: Mining frequent patterns without candidate generation. In: Proceedings ACM SIGMOD, pp. 1–12. ACM Press, Dallas (2000)
3. Wu, J., Zhu, S., Liu, H., Xia, G.: Cosine interesting pattern discovery. Information Sciences 184(1), 176–195 (2012)
4. Zhao, Y., Levina, E., Zhu, J.: Community extraction for social networks. Proceedings of the National Academy of Sciences of the USA 108(18), 7321–7326 (2011)

An Educational Tool for an Interactive Faceted Exploration of DBpedia Life Sciences Data

Stefan Negru and Sabin C. Buraga

Faculty of Computer Science, A.I. Cuza University of Iasi, Romania
{stefan.negru,busaco}@info.uaic.ro

Abstract. As the volume of information on the Web and the number of Linked Open Data sources increase considerably, users need easy to use Web-based tools and applications to help them navigate, interpret and make use of the available information/data. In this context, we introduce RDFSpecies, as an educational tool for interactive exploration of Semantic Web data. Our approach is targeted at users that have no experience or little knowledge of RDF model and Semantic Web technologies. We make use of different interaction metaphors to aid them in navigating, exploring and visualizing DBpedia data.

Keywords: Semantic Web, RDF, Faceted Navigation, DBpedia.

1 Introduction

As more and more companies and governmental agencies make their data available on the Web, in the spirit of the Linked Open Data [4] movement, end-users find it difficult to grasp and navigate and make use of all the available information.

As Dadzie et al. [3] points out "making sense of such data presents a huge challenge to the research community", mainly because the data publicly available was designed for end-user consumption. Hence tasks like information seeking, visualization and interactive exploration are key in "making sense of data".

Despite the fact that RDF[1] became a standard model for data interchange on the Web, one major of its major step-back is that RDF tools are counterintuitive and not user friendly. As a further matter querying languages like SPARQL[2], although "offer premises of exploring and navigating through this kind of data, they are difficult to use and require prior knowledge of the datasets and data model.

As a step towards providing solutions to these new challenges, we developed a client-side Web application – which we will refer to as **RDFSpecies** – focused on interactive exploration of *DBpedia*(http://dbpedia.org/About) data. The RDFSpecies project is based on metaphors such as direct manipulation of information, visual exploration, and faceted navigation, and at the same time providing immediate feedback to the user's actions.

---

[1] http://www.w3.org/RDF/

[2] http://www.w3.org/TR/rdf-sparql-query/

X. Lin et al. (Eds.): WISE 2013, Part II, LNCS 8181, pp. 503–506, 2013.

Related Work: The majority of current approaches [1,2,3,5] provide solutions regarding means of visualizing, manipulation and searching for semantic Web data. Furthermore are focusing on facilities regarding interactive navigation and visualization of complex data sets without a clear understanding of the needs of non-expert end-user. Additionally, [6] points out "faceted browsing is a popular choice to support user navigation over the Semantic Web RDF data".

2   RDFSpecies

The **RDFSpecies** project in centered around an RDF dataset extracted from the dinosaurs list, depicting the connections between prehistoric animals based on certain anatomic features, living environment, food diet, time-line etc. From the beginning, **RDFSpecies** was designed as an educational tool for Semantic Web classes, in order to ease the process of presenting Semantic Web concepts and technologies to students.

RDF-Based Data Modeling: To provide and interactive method of querying the data, we adopted HTML5[3] and a responsive design approach along with several JavaScript libraries such as JQuery[4], Raphael[5] and Processing[6] for the user interface. The RDFQuery[7] library was utilized for creating a local triple-store (using a simplified version of the Turtle syntax – http://www.w3.org/TR/turtle/) of the data and a the same time querying the dataset.

Although our initial aim was to retrieve all the information (in real-time) from *DBpedia* linked data, we found that making all the information was confusing for the users. Users could not understand the majority of the (abstract) concepts and properties that are presented in the *DBpedia* knowledge base. Thus on the client side we performed certain pre-processing tasks and stored various data, in order to eliminate the noisy data (which end-user would find confusing). At the same time, we added some properties (e.g. *dino:onContinent, dino:hasHabitat, dino:diet* etc.) and concepts (e.g. *North America, Plains* etc.) to facilitate the presentation of the Semantic Web data, and to assist the users in their information seeking process. Other vocabularies utilized include FOAF (http://xmlns.com/foaf/spec/) and PROV-O (http://www.w3.org/TR/prov-o/).

User Interface: The *RDFSpecies*[8] user interface follows the center stage plus toolbox design pattern [7] and divided into two main parts (Figure 1): **Menu** and **Canvas**.

---

[3] http://www.w3.org/TR/html5/
[4] http://jquery.com/
[5] http://raphaeljs.com
[6] http://processingjs.org/
[7] http://code.google.com/p/rdfquery/
[8] Video demonstration: http://youtu.be/trqS_-EtsY4

The **Menu** provides functionalities such as: search by *Name*, select *Habitat*, select *Continent*, select *Diet*, select *Size*, select *Weight*, select *Timeline* or *Reset* filter options. The **Canvas** is the main user interaction area, where we can explore the individual elements grouped by certain criteria such as: *Habitat*, *Continent*, *Diet* and *Timeline*, but also obtain additional information on each element.

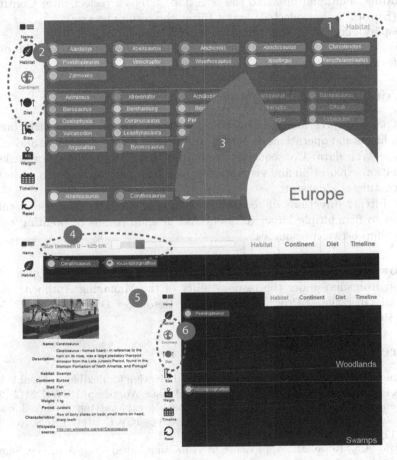

Fig. 1. A Navigation Scenario step by step

An individual information element – in fact, the subject of a RDF triple – is represented via a different color from all the other elements along with a label which represents the RDF data from the dataset. Each element is sorted alphabetically and grouped by the selected display criteria in the center stage.

A Navigation Scenario: Using *RDFSpecies*, the user would like to view additional information about view all the dinosaurs which lived on the European continent, have a size between 0 and 625 meters and list them by Habitat. Even it is a simple query, a regular user must know both the SPARQL syntax and semantics and, additionally, how data is modeled and stored. A visual representation of such a scenario proposed above is found in Figure 1.

1. Selecting the filter which gives the order for the displayed elements – Habitat selected;
2. Adding additional filters to the selection – steps 2 (select filter Continent), 3 (Europe option selected) and 4 (filter the results by size):
3. Display information regarding found elements – step 5;
4. The user can further refine the search until (s)he can discover the desired information – step 6.

3   Conclusions and Future Work

In this paper we presented the RDFSPecies – an Web application which combines search, facets and operations on sets of resources in order to navigate and explore semantic Web data. Consequently, our approach is well suited for tasks like information exploration and visualization making it easy for users to understand the structure of a dataset.

Our further directions of research are focused on extending the developed solution to fit multiple larger datasets and develop means of pointing out the relationship between resources.

Acknowledgments. This work was partially supported by the European Social Fund in Romania, under the responsibility of the Managing Authority for the Sectorial Operational Program for Human Resources Development 2007-2013 [grant POSDRU/107/1.5/S/78342].

References

1. Brunetti, J.M., Gil, R., Garcia, R.: Facets and pivoting for flexible and usable linked data exploration. In: Interacting with Linked Data Workshop, ECWE 2012 (2012)
2. Cohen, M., Schwabe, D.: Support for reusable explorations of linked data in the semantic web. In: Harth, A., Koch, N. (eds.) ICWE 2011. LNCS, vol. 7059, pp. 119–126. Springer, Heidelberg (2012)
3. Dadzie, A.S., Rowe, M.: Approaches to visualising linked data: A survey. Semantic Web 2(2) (2011)
4. Heath, T., Bizer, C., Berners-Lee, T.: Linked data – the story so far. International Journal on Semantic Web and Information Systems 5(3), 1–22 (2009)
5. Heim, P., Lohmann, S., Stegemann, T.: Interactive relationship discovery via the semantic web. In: Aroyo, L., Antoniou, G., Hyvönen, E., ten Teije, A., Stuckenschmidt, H., Cabral, L., Tudorache, T. (eds.) ESWC 2010, Part I. LNCS, vol. 6088, pp. 303–317. Springer, Heidelberg (2010)
6. Maali, F., Loutas, N.: SPARQL 1.1 and RDF faceted browsing. Tech. rep., DERI (August 2012)
7. Tidwell, J.: Designing Interfaces: Patterns for Effective Interaction Design. O'Reilly (2005)

DataEx: Interactive Relationship Explorer of Freebase Knowledge Repository

Stefan Negru and Sabin C. Buraga

Faculty of Computer Science, A.I. Cuza University of Iasi, Romania
{stefan.negru,busaco}@info.uaic.ro

Abstract. In this paper we present DataEx, a HTML5 client-side Web application, which aims to make Web data more accessible by utilizing visualization as an useful approach to explore and navigate through large amounts of data and ultimately emphasize the relations between them. In order to provide several examples, we choose Freebase as a knowledge repository and its API in order to process data in real-time. In our case, the user interaction with such a large knowledge space is facilitated by different navigation and visualization techniques. Furthermore the main focus of the application lies on highlighting and analyzing the existing relationships between the data from the Freebase knowledge repository.

Keywords: Information Visualization, Big Data, Semantic Web, Freebase.

1 Introduction

While the data available on the Web is constantly increasing due to the large number of repositories (for both structured and unstructured data), end-users find it difficult to handle the available information.

In this context, Dadzie et al. [2] points out the need of "making sense of such data presents a huge challenge to the research community, a challenge which is compounded further by the drive to produce data... for end user consumption". Thus issues like information and relationships search, visualization and interactive exploration are essential for this process of sensemaking[8].

As a step towards providing solutions to these new challenges, we designed and implemented a client-side Web application – **DataEx** – focused on interactive exploration of large knowledge repository. The **DataEx** application is built on top of *Freebase*[1]. *Freebase* as described by Bollacker et al. [1] is a structured "database system designed to be a public repository of the world's knowledge" and it consists of user harvested metadata. It was built in the spirit of Linked Open Data [3] movement, and consists of over 40 million topics (entities), more than 30,000 properties and over 1.9 billion of facts about those topics – as of July 2013.

Given such a large repository, **DataEx** provides the user with techniques to visualize and explore direct and indirect relations between data, but also

---

[1] http://www.freebase.com/

X. Lin et al. (Eds.): WISE 2013, Part II, LNCS 8181, pp. 507–510, 2013.
© Springer-Verlag Berlin Heidelberg 2013

view specific instances. This in-depth exploration of such a large number of relationships is relevant to the humans in the information-seeking process, but also in decision-making activities.

2   DataEx: From Design to Deployment

Related Work: While related applications such as [4,5,7] and others are available, our solution focuses on a client-side Web application in order to highlight the data relations from a knowledge repository. Our approach is well suited for tasks like data navigation and visual exploration, thus making it easier for users to explore such a repository.

DataEx is a HTML5[2] client-side Web application which allows users to finds and view specific relations between data. Main advantages and contributions are the big data processing and an interactive graph visualization that displays relationships between multiple instances. Moreover the application is based on open technologies and libraries and can be configured applied on other repositories.

Data Modeling and Computing Relations: The largest data structure is called *Domain*[3], which specifies the general meaning of a data instance (e.g. "Music" or "Computers"). It contains collections of data, called *Types* (e.g. for "Music" some types are "Musical Artist" or "Music Genre"). Inside a *Type*, there are collections of data, where each individual pieces of data (which we labeled) *Instances* reside (e.g. for "Musical Artist", an instance is "Bob Dylan").

For each *Instance* there may exist a set of *Properties*, which specify a relationship with another *Instance* or any kind of data (e.g. the "Bob Dylan" instance has the "instruments played" property with values such as: "Acoustic guitar", "Electric guitar" etc.). These set of *Properties* specific to a certain *Type* are utilized to establish relationships between different *Instances* – thus we labeled them as *Connections* (or *Connection* groups).

DataEx provides two options of exploring data: one is by user input of a search term (suggestions are provided via an auto-complete functionality) and the other is by browsing through knowledge base repository in order to identify a desired domain of interest.

In order to visualize certain relations between items of interest, the search terms are transformed into MQL (Metadata Query Language) queries, and then sent to *Freebase. Freebase* API sends a response in JSON (JavaScript Object Notation) format. This response is processed and stored as temporary objects, then for each *Property* of any *Instance* a list of *Connections* will be created (e.g. for "Musical Artist", some *Connections* identified are "Album", "Track", "Genre"). These *Connections* type relationships can be browsed and visualized as a graph (using the sigma.js[4]. JavaScript library, which draws graphs using the HTML5 canvas element) from the **DataEx** application.

---

[2] http://www.w3.org/TR/html5/
[3] Freebase Basic Concepts available at: http://goo.gl/bjcWf
[4] http://sigmajs.org/

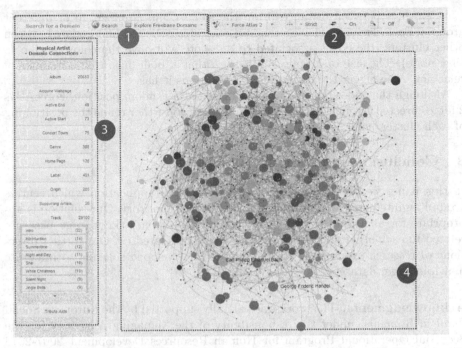

Fig. 1. DataEx application User Interface

User Interface: DataEx[5] is divided into four sections (Figure 1):

1. **Section 1** - Provides the basic functionalities of the **DataEx** application: searching and browsing the Freebase Domains;
2. **Section 2** - Offers interactive functions for graph visualization of the *relations* (**Section 4**) and includes options for changing the graph layout, displaying different types of node connections (strict, related or bidirectional), controlling the number of label shown and also Fisheye zooming;
3. **Section 3** - After the relations have been made between the instances, all of them will be displayed in this section. The user can toggle between visualizing only relations and searching viewing details on demand for specific instances;
4. **Section 4** - When the user selects one or more *relation* group(s) to be displayed, in this section a graph will be drawn and filled with the corresponding instances of the selected *relation* group(s).

Section 3 and **Section 4** are mainly used to provide the user with several techniques for exploring and interacting with the data. While **Section 3** focuses on the navigation through *Types* and *Instances*, the other section adds means for a visual exploration of the relations between data.

In graph visualization from **Section 4**, the nodes and edges will be grouped in clusters, each cluster representing one *Instance* and its associated relationships

---

[5] Video demonstration: http://youtu.be/MrI-A2EDXUY

(relationships between instances from the same *Connection* group or different groups). Each cluster will have a random generated color, to differentiate between other clusters (the edges associated to one note will have the color specific to that node). The size of a node emphasizes that a node has a high number of relations – the bigger the node the more relations it has.

Although the nodes in the graph are drawn at a random position, ForceAtlas 2 force-directed layout algorithm [6] can be used to further refine the positioning of each cluster/node.

3   Conclusions and Future Work

In this paper we presented the **DataEx** – an Web application which centers around an interactive exploration of a knowledge repository. Our solution is appropriate for analyzing how pieces of information from a knowledge base are related to each other and the properties they are related by. Our further directions of research are focused on extending the developed solution to integrate multiple larger datasets.

Acknowledgments. This work was partially supported by the European Social Fund in Romania, under the responsibility of the Managing Authority for the Sectorial Operational Program for Human Resources Development 2007-2013 [grant POSDRU/107/1.5/S/78342].

References

1. Bollacker, K., Evans, C., Paritosh, P., Sturge, T., Taylor, J.: Freebase: a collaboratively created graph database for structuring human knowledge. In: International Conference on Management of Data, SIGMOD 2008. ACM (2008)
2. Dadzie, A.S., Rowe, M.: Approaches to visualising linked data: A survey. Semantic Web 2(2) (2011)
3. Heath, T., Bizer, C., Berners-Lee, T.: Linked data - the story so far. International Journal on Semantic Web and Information Systems 5(3) (2009)
4. Heim, P., Lohmann, S., Stegemann, T.: Interactive relationship discovery via the semantic web. In: Aroyo, L., Antoniou, G., Hyvönen, E., ten Teije, A., Stuckenschmidt, H., Cabral, L., Tudorache, T. (eds.) ESWC 2010, Part I. LNCS, vol. 6088, pp. 303–317. Springer, Heidelberg (2010)
5. Hirsch, C., Hosking, J., Grundy, J.: Interactive visualization tools for exploring the semantic graph of large knowledge spaces. In: Workshop on Visual Interfaces to the Social and the Semantic Web, VISSW 2009. CEUR-WS.org (2009)
6. Jacomy, M., Heymann, S., Venturini, T., Bastian, M.: Forceatlas2, a graph layout algorithm for handy network visualization, Paris (2011),
   http://www.medialab.sciences-po.fr/fr/publications-fr
7. Lehmann, J., Schüppel, J., Auer, S.: Discovering unknown connections - the dbpedia relationship finder. In: Proceedings of the 1st Conference on Social Semantic Web, CSSW 2007 (2007)
8. Weick, K.E., Sutcliffe, K.M., Obstfeld, D.: Organizing and the process of sensemaking. Organization Science 16(4) (2005)

Disambiguating Authors
in Academic Search Engines

Long Zhang[1], Guohua Chen[1,2,*], Yong Tang[1], and Zurui Cai[1]

[1] South China Normal University, Guangzhou, 510631, Guangdong, China
[2] Shenzhen Engineering Laboratory for Mobile Internet Application Middleware
Technology, Shenzhen, 518060, Guangdong, China
{amberlife,JoJo_Tsoi}@qq.com, chguohua@gmail.com, ytang@scnu.edu.cn

Abstract. Author name ambiguity is a common problem in current
academic search engines. It presents a great challenge: first, it's not convenient for researchers to effectively access the academic publications;
second, the author publication profile cannot be induced, thus further
meaningful semantic analysis for the author cannot be conducted. In
this paper, we present a coauthorship based model to disambiguate the
authors, and apply this model into a large academic search engine –
Scholat Search, which proves to be effective and efficient.

Keywords: author name ambiguity, coauthorship, search engine.

1 Introduction

Researchers cannot conduct academic research without the help of various kinds
of digital libraries or academic search engines, such as Google Scholar, Microsoft
Academic Search. We often meet the author name ambiguity problem when using
digital libraries or academic search engines. When we key an author's name, these
systems often returns a complete list of the publications with that name. And it
takes a large part of queries for this kind of searching[1]. It has following negative
impacts: 1. For normal researchers, it's not convenient to focus on the exact
author he is actually interested; 2. For research managers, it's hard to determine
author achievements, thus it may causes confusions when make a decision in the
situations such as position promotion or research grant; 3. It cannot establish
the publication list for the author, and thus cannot conduct a further semantic
analysis, such as determining the author's academic performance, concluding
research interests, or extracting his research teams, which is extremely useful in
current academic social network sites. Therefore, it's very important to solve the
author ambiguity problem.

2 Related Works

Many research works were conducted to attack this problem in recent years.
They can be divided into 3 categories:

---

* Corresponding author.

X. Lin et al. (Eds.): WISE 2013, Part II, LNCS 8181, pp. 511–514, 2013.
© Springer-Verlag Berlin Heidelberg 2013

1. Classification Methods[2]. It is viewed as a classification problem in these methods. Each actual author is treated as a distinct class. Then a classification model is obtained using the labeled training set. This model is the disambiguation function we wanted. The advantage of this method is that it can often get a more accurate result, but it needs to label many publications for each author, which is impractical for large digital libraries.
2. Clustering Methods[3]. A similarity function is introduced to calculate the distance between any pair of authors to be disambiguated. Then a clustering algorithm is selected to group the similar authors together. Each cluster corresponds to an actual author.
3. Probabilistic Models Method. It is treated as an probabilistic problem. The dependencies between actual authors and the author references are depicted in a probabilistic model, such as HMRF[4] or Naive Bayes[5], and then iteratively calculate the paramenters in the model using EM or Gibbs Sampling methods.

According to recent studies, the coauthorship plays an very valuable role in disambiguating authors[6,7]. And it can be easily extracted from the co-author list in the publication. In this paper, we present an coauthorship based model to solve the problem. The details are given in the following section.

3 System Architecture

3.1 Overview

Six-degree world is a well known theory in social networks. It illustrates the latent power of "friends". Coauthorship is also a kind of friendship and we expect to connect authors into a whole network. Then we can check each others' distance in this author network, and this distance can be used to disambiguate authors.

We propose a new idea to construct the author network. We split it into two phases. The authors who co-write a paper can be viewed a tight academic circle. First, we cluster these circles into atom clusters, and then we connect the atom clusters together.

3.2 Building of the Atom Clusters

An atom cluster means we have a high confidence that the name references within it refer to the same actual author[8]. We alternatively use the word "name reference" and "paper" according to the context. The basic idea is that we first cluster the authors with high confidence, and hope these atom clusters can give a guide in the following clustering.

To keep the atom clusters with high accuracy, we adopt the coauthorship feature. We calculate the sum of common coauthor name's differentiation and check it with a THRESHOLD. We add name references into the atom cluster only when the total name differentiation is above the THRESHOLD. After evaluation, we set THRESHOLD to be 0.1 in the system.

The initial atomic cluster may exists many fragments due to the initial setting. We construct an atom cluster for each paper at first, and then we iteratively compare each pair of them. If the similarity is above the THRESHOLD, then we merge this pair. If none is merged, then the algorithm finishes. We use the union-find set to keep it efficient. The Pseudocode is illustrated below:

Input: A list of papers P which all share same author name
Output: A list of atomic cluster list C
1 Create an empty atomic cluster list c=$\{\Phi\}$;
2 **for** *each paper p_i in P* **do**
3 | create an atom cluster c_i and add it to C;
4 **end**
5 **while** *true* **do**
6 | **for** *each pair of atom clusters c_i and c_j* **do**
7 | | **if** *similarity between c_i and c_j > THRESHOLD* **then**
8 | | | merge c_i and c_j;
9 | | **end**
10 | **end**
11 | **if** *atom cluster list doesn't change* **then**
12 | | break;
13 | **end**
14 **end**
15 return C;

Algorithm 1: Atom cluster Construction

3.3 Extend Atom Clusters

As stated above, each name has a series of atom clusters. We use the direct coauthorship only, it may keep high precision but the drawback is that it tends to split one actual author into several atom clusters. Thus it's necessary to connect these atom clusters together.

Several authors co-write a paper, and each author corresponds to one specific atomic cluster for that author name. Then we can conclude that these atom clusters have an association with each other via that paper. This association information can be utilized to further improve the atomic cluster aggregation. For example,for the author name a, it has several atomic clusters $A_a^1, A_a^2, ..., A_a^k$. If A_a^i and A_b^j are associated with another author name's atomic cluster A_b^k, then we can infer that A_a^i and A_a^j correspond to the same community and we can merge the two atom clusters.

4 Demonstration

This model is applied to a famous Chinese scholar social network, Scholat[1]. When a user registers in Scholat, the system will recommend papers which may belongs to him. A screenshot of our system is shown in Fig.2. In this figure, clusters of papers are pushed to the user with the name 'Long Zhang', and the

---
[1] http://www.scholat.com

user can choose which cluster belongs to him. We also provide an open API via http://www.scholat.com/nameDis/ to view any author's disambiguation result. In the future work,we will use other features or extend the whole network in the coauthorship to further improve the recall rate in the system.

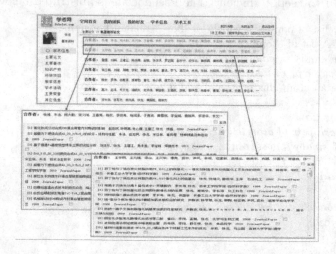

Fig. 1. Author Disambiguation Result

Acknowledgments. This paper is supported by National Nature Science Foundation of China (No. 61272067 and 60970044), Guangdong Province Science and Technology Foundation(No.S2012030006242, 2011B080100031, 2011168005 and 2011A091000036), Project on the Integration of Industry, Education and Research of Guangdong Province(No.2011B090400145)

References

1. Torvik, V.I., Smalheiser, N.R.: Author name disambiguation in medline. ACM Transactions on Knowledge Discovery From Data 3, 1–29 (2009)
2. Han, H., Zha, H., Li, C., Tsioutsiouliklis, K., Giles, C.L.: Two supervised learning approaches for name disambiguation in author citations (2004)
3. Han, H., Zha, H., Giles, C.L.: Name disambiguation in author citations using a k-way spectral clustering method (2005)
4. Zhang, D., Tang, J., Zi Li, J., Wang, K.: A constraint-based probabilistic framework for name disambiguation (2007)
5. Han, H., Xu, W., Zha, H., Giles, C.L.: A hierarchical naive bayes mixture model for name disambiguation in author citations (2005)
6. Fan, X., Wang, J., Pu, X., Zhou, L., Lv, B.: On graph-based name disambiguation. J. Data and Information Quality 2, 10 (2011)
7. Su Kang, I., Hoon Na, S., Lee, S., Jung, H., Kim, P., Kyung Sung, W., Hyeok Lee, J.: On co-authorship for author disambiguation. Information Processing and Management 45, 84–97 (2009)
8. Wang, F., Li, J., Tang, J., Zhang, J., Wang, K.: Name disambiguation using atomic clusters. In: WAIM 2008, pp. 357–364 (2008)

A Toolkit for Simplified Web-Services Programming

Moshe Chai Barukh and Boualem Benatallah

School of Computer Science & Engineering,
University of New South Wales, Sydney – Australia
{mosheb,boualem}@cse.unsw.edu.au

Abstract. The Internet has truly transformed into a global deployment and development platform. For example, Web 2.0 inspires large-scale collaboration; Social-computing empowers increased awareness; as well as Cloud-computing for virtualization of resources. As a result, developers have thus been presented with ubiquitous access to countless web-services. However, while this enables tremendous automation and re-use opportunities, new productivity challenges have also emerged: The same repetitive, error-prone and time consuming integration work needs to get done each time a developer integrates a new API. In order to address these challenges, we designed and developed *ServiceBase*, a "programming" knowledge-base to abstract, organize, incrementally curate and thereby re-use service-related programming-knowledge. Empowered by this knowledge we then provide: (a) A set of APIs that expose a common and high-level interface to developers for integrating services in a simplified manner; (b) An extended version of the GIT repository, creating a *plug-n-play* environment for services; (c) A mind-map based visualization tool to help explore the base.

1 Introduction

It is no doubt that the inception of the Service Oriented Architecture (SOA) paradigm has significantly empowered the capabilities for modern software engineering. Moreover, coupled with other recent advances in web-technology, such as Web 2.0, Social and Cloud computing, the Internet has delivered developers with ubiquitous access to a plethora of rich and logical services along with computing resources, data sources, and tools – and the potential to build powerful systems. For instance, *ProgrammableWeb* now lists more than 6,700 APIs in its directory. Moreover, in order to increase the dispersion of APIs, organizations such as *Mashery*[1] and *Apigee*[2] are building on these trends to provide platforms for the management of APIs. However, while advances in Web-service and services composition have enabled tremendous automation and reuse, new productivity challenges have also emerged. The same repetitive, error-prone and time consuming integration work needs to get done each time a developer integrates a new API. Moreover, the heterogeneity associated with services also means service programming has remained a technical and complex task. For example, the developer require sound understanding of the different service types and access-methods, as well as being able to format input data,

---

[1] http://www.mashery.com
[2] http://apigee.com

X. Lin et al. (Eds.): WISE 2013, Part II, LNCS 8181, pp. 515–518, 2013.
© Springer-Verlag Berlin Heidelberg 2013

or parse and interpret output data in the various formats, (e.g. XML, JSON, SOAP, Multimedia, HTTP, etc.). In addition, programmers may also need to develop additional functionality such as: user management, authentication-signing, access control and third-party authorization; tracing as well as version management, etc.

In order to address these challenges, we have designed and developed *ServiceBase*, a "programming" knowledge-base, where common service-related low-level logic can be abstracted, organized, incrementally curated and thereby re-used by other application-developers. Architecturally, we have drawn inspiration from *Freebase* and *Wikipedia*, where just as encyclopediatic information is distributed in the form of user-contributed content, similarly, technical knowledge about services could be both populated and shared amongst other developers. Empowered by that knowledge, we then provide a set of APIs that expose a common and high-level interface to developers for integrating services in a simplified manner, despite their underlying heterogeneity. This is facilitated by the implemented service-bus middleware that translates high-level methods to more concrete service calls, by looking-up and then building the necessary information from the knowledge-base. While this itself significantly simplifies service-programming, we have been motivated to provide further support in the form of a supporting programming toolkit for service-based programming. In particular we implement two extensions: (a) To ease access, navigation and exploration of the service-base from within a typical programming environment, we implement an extended version of the GIT repository, creating a plug and play environment of services. This means new services can be added to the knowledge-base (i.e. "plugged"), and registered services can be utilized in application development (i.e. "played"); and (b) In order to help visualize service meta-data, (for example determining what are the message-parameters and formats for some service), we have implemented a mind-map visualization of the service knowledge-base.

Inevitably, our work provides the potential for a vast increase in application-development productivity and greater code-maintainability. We have endeavored to fill the gap amongst non-highly skilled programmers who often resort to homebrewed systems; or even skilled programmers who often end up re-implementing systems (or part of systems) due to being unable to locate a suitable API, or if it being too hard to understand how to use. In order to substantiate our results, we demonstrate in this paper how a service-oriented application can be built using various services that cover the 3 main different service-types, (WSDL, REST & ATOM/RSS), and implemented faster and in much fewer lines of code than if using a traditional coding approach.

2   *ServiceBase* System Architecture

As illustrated in Figure 1, *ServiceBase* has been realized in three layers of abstraction: (i) Firstly, the *Programming Knowledge-Base* acts as the central repository for storing all service-related programming information. (ii) The *ServiceBus* then exposes a set of Java (client-library) and RESTful APIs that enable simplified programmatic access to registered services. Finally, (iii) we provide a *toolkit* in order to initiate a supportive programming environment. Consisting of: an extended GIT-Repository that enables *plug-n-play* capabilities via a command-line interface; as well as a web-based mind-map visualization GUI for exploring and traversing the graph service-base.

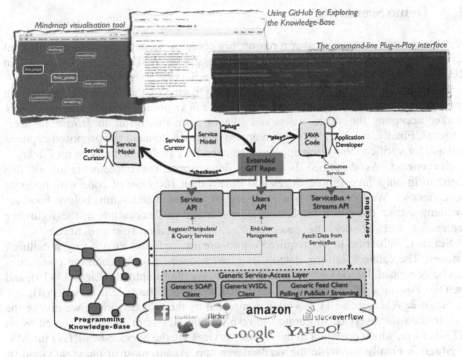

Fig. 1. ServiceBase System Architecture

The programmatic interface to *ServiceBase* offers the following APIs:

The **UsersAPI** provides a means for end-users and apps to register and identify. Identity management is achieved via *Facebook* login credentials as well as native login. 3rd-party application interaction is achieved via OAuth 2.0 Provider interface.

The **ServicesAPI** is used to *register* new services into the knowledge base; as well as *search*, *explore*, *update* and *delete* service-models that are already registered.

The **ServiceBusAPI** is the main gateway for application-developers to interact with typical request-reply ops via the `invoke(..)` methods. The main innovation is the common programming interface provided to interact with services, irrespective of their underlying heterogeneity (e.g. JSON/REST vs. XML/REST vs. SOAP/WSDL).

The **StreamsAPI** is used to enable interaction with real-time feeds-services. It provides a `subscribe(..)` method to any registered feed. The API then exposes three modes for working with subscriptions: (a) It provides *querying* (i.e. pull) of feed events; (b) It enables setting up *listeners* (i.e. asynchronous callback push) of events; and (c) It enables *navigating* streams using `pause()`, `stop()` and `rewind(date)`.

The **Plug-n-Play API** is used to even further simplify service-programming. We have done this by extending the standard GIT commands with `"plug"` (helps to automate adding new services); and `"play"` (generates a code-stub to use a service). Usage shown below:

```
git plug <service_model_file>

git play [-operation | -feed [-get | -callback]]
         <service_name><operation_name>|<feedtype_name>
         <destination_location>
```

3   Demo Scenario

We demonstrate the toolkit over a reasonably complex scenario involving 3 different service-types: Namely, Amazon *MTurk* expressed in WSDL, *Flickr* expressed in REST/ATOM; and Apache *HBase* Database-as-a-Service, expressed in REST.

Description. We want to build an application to employ the crowd for determining appropriate captions for photos added to some *Flickr* photo-stream. Upon an *MTurk* worker accepting the work, we would like to store this result in *HBase* for later retrieval. Finally, we want to write an app to query-on-demand the proposed captions and ask for verification – upon approval the captions can then be updated on *Flickr*.

Screencast. As described below, we demonstrate the implementation of this scenario in only three simple stages and in less than *100 lines* of code with no extra dependencies. We have only shown snippets of generated code-stubs below, however we demonstrate the entire implementation scenario and execution in the following screencast: http://www.cse.unsw.edu.au/~mosheb/servicebase.html

Stage 1. Subscribe to the required photo-stream feed, and then set-up a callback listener. The callback is formulated using two services, and for each the code stubs can be generated. I.e. (i) Listening to the stream for photo-uploads (Lines 11–19); and then (ii) Posting a HIT on *Mturk* (Lines 1–10). Therefore, the code in (i) calls (ii).

Stage 2. When each HIT has been completed by the crowd-worker, we invoke the *HBase* data-store service to curate the results. The HIT is in fact represented as an HTML form, which post data to the *HBase* services via the Service-bus RESTful API.

Stage 3. Finally, we write the verifier web-app. Again, most of the code could be auto-generated via the play function: i.e. getting the photos from *HBase* and displaying to the verifier. Then if verified, updating the photo by posting to *Flickr*.

```
1.   public class MTurk_CreateHit {
2.       public static Message invoke(String title, String desc, File qn,
             String duration, String lifetime, String reward) {

3.           WSDLService mturk = ServiceAPI.getService("AmazonMTurk");
4.           TupleField hit = mturk.getOperationType("createHit")
                               .getInputMessage().getField();

5.           hit.getField("title").setValue(title);
6.           hit.getField("description").setValue(desc);
             ...
7.           hit.getField("question").setValue(qn);

8.           return ServiceBus.invoke(acs_key, mturk, "createHit", hit);
9.       }
10. }
                                    ◆ ◆ ◆
11. public class Flickr_GroupPoolFeed {
12.     public static void callback(Field field, String sub_id){

13.         TupleField photo = ((TupleField)field);
14.         String id = photo.get("photo_id").getValueAsString();
15.         String url = photo.get("photo_url").getValueAsString();

16.         File question_html = getHtmlQuestion(id, url);
17.         MTurk_CreateHit
               .invoke("Describe an Image",      //title
                   "Tagging of an Image",        //description
                   question_html,                //question HTML
                   "86400",                      //duration = 1 day
                   "7",                          //lifetime = 7 days
                   "1"                           //reward = 1 dollar
18.     }
19. }
```

KPMCF: A Learning Model
for Measuring Social Relationship Strength

Youliang Zhong, Xiaoming Zheng, Jian Yang,
Mehmet A. Orgun, and Yan Wang

Department of Computing, Macquarie University, NSW 2109, Australia
{youliang.zhong,xiaoming.zheng,jian.yang,mehmet.orgun,yan.wang}@mq.edu.au

Abstract. We present the demonstration of the *KPMCF* model for measuring relationship strength, highlighting the concepts of the model, two experimental applications, and a comparison with other methods.

1 Introduction

Social relationship strength is one of the most important research topics in social network analysis, measuring how strong or weak the relationships are among the users in a social network. In particular, a theory named "The Strength of Weak Ties" was initially introduced in [1], whereas a quantitative measurement using multiple indicator techniques was explored in [3].

While social networking services on the web are getting increasingly popular in recent years, social relationships have significantly changed from those in the old days. Firstly, it becomes normal for a user to have hundreds or even thousands of "friends" in social networks. The so-called "friends" in the virtual space range from actual close friends to casual acquaintances on the web [6]. Furthermore, it is even possible to obtain not only users' interaction data but also users' profile information through social networks. These changes bring greater opportunities of, yet challenges to, the research on social relationship strength.

The principle of homophily in social networks suggests that, people tend to build connections with other people having similar profile characteristics. On the other hand, the stronger the connections, the higher the likelihood that more interactions occur between people. Based on these assumptions, the research on social relationship strength expands in several directions. Some efforts refined the granularity of the relationship strength in online social networks, while others utilized interaction data to predict relationship strengths. In the meantime, several studies exploited profile or interaction information to estimate social relationship strengths among users. In those studies, either only one type of a profile or interaction data was used, or a single method was employed to handle all different forms of data sources. Therefore, restricted users' overall relationship strength learnt from comprehensive social information.

Several recent studies indicated that users' social interactions affected their behaviors on the web, and such information could improve the quality of social network analysis, for instance, in recommendation tasks [2,8]. Motivated by these studies, we have developed a *Kernelized Probabilistic Matrix Co-Factorization*

X. Lin et al. (Eds.): WISE 2013, Part II, LNCS 8181, pp. 519–522, 2013.

(KPMCF) model to learn users' overall relationship strength in a social network [7]. The model simultaneously captures users' profile and social interaction information by integrating *Matrix Co-Factorization* with *Multiple Kernels* methods. The main contributions of our work are summarized as follows:

– We present a new probabilistic matrix co-factorization model integrated with kernels of social interactions. The model incorporates users' interactions into the factorization process, thus resulting in latent matrices depending on all the input resources.
– We formulate users' relationship strength in the learnt latent feature space. While most matrix factorization methods focus on minimizing the Root Mean Square Error or RMSE against a target resource, in *KPMCF*, the users' relationship strength is learnt from various profile data, maintaining a balanced RMSE over multiple resources.
– We conduct experiments to investigate the concepts of our model, and to compare with other state-of-the-art methods. Our experimental applications show that the learnt social strength helps achieve higher performance than other methods in terms of *PRC* and *ROC*, owing to the balanced RMSE.

This paper describes the demonstration of the *KPMCF* evaluation system. In section 2, we briefly review the model and the evaluation system. The dataset and scenarios of our demonstration are described in section 3.

2   KPMCF Model and Evaluation System

2.1   Overview of KPMCF Model

Fig. 1 shows the graphical model of *KPMCF*, following the plate notation of graph models. In the figure, p_{ij}^c, p_{ik}^b and p_{is}^t denote the observed profile data (i.e., countries, labels and tags), whereas c_j, b_k, t_s and u_i denote the latent vectors of *countries*, *labels*, *tags* and *users* respectively. On the corners of the plates, n_c, n_b, n_t and n_u are the sizes of the latent vectors. Out of the plates, K_C, K_B, K_T and K_U stand for the kernels assigned to the corresponding latent vectors, whereas σ_{p^c}, σ_{p^b} and σ_{p^t} for the parameters of the distributions (i.e., Gaussian distribution) set over the profile data.

The algorithm of *KPMCF* firstly constructs users' kernel K_U through social interaction information, and then co-factorizes multiple profile matrices to obtain the users' latent matrix U. Based on U, the *KPMCF* calculates pair-wise relationship strengths among the users [7].

2.2   Evaluation System

We have developed an evaluation system of the *KPMCF* model by *MATLAB*. The core part of the system is the implementation of our kernelized matrix co-factorization algorithm. To facilitate our evaluation under various conditions, we have built an interactive interface called "KPMCF Evaluation Panel", having two Tabs: "Relationship Strength Measurement" (Fig. 2) and "Application Evaluation" (Fig. 3).

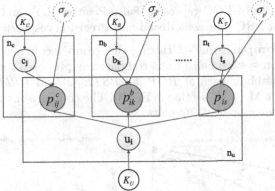

Fig. 1. Graphical model of *KPMCF*

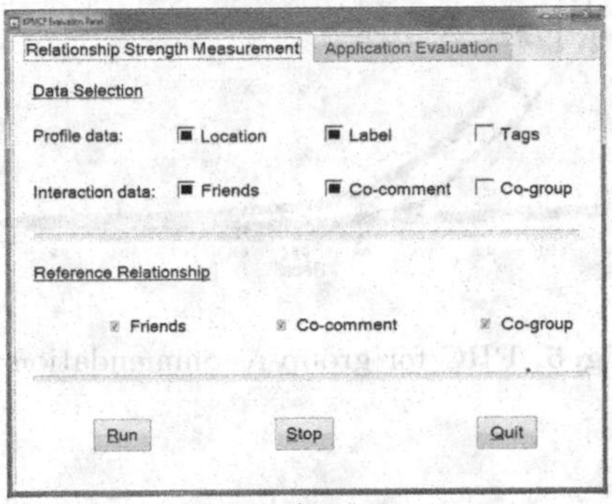

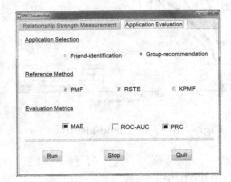

Fig. 2. Relationship Strength Measurement

Fig. 3. Application Evaluation

3   Demonstration

In the demonstration we make use of a real dataset of the *Flickr* web site [4]. From the original dataset, we further extract five matrices for learning users' social relationship strength and evaluating application performance.

The evaluation system provides demo users with three scenarios: (1) concepts of the *KPMCF* model, (2) experimental applications, and (3) performance comparisons.

Concepts of *KPMCF* Model - Through the "Data Selection" section in *KPMCF* Evaluation Panel (Fig. 2), users are allowed to choose various profiles or interaction matrices to learn social relationship strength. The derived strengths can be used in experimental applications and performance comparisons.

Experimental Applications - The evaluation system includes two experimental applications: friend-identification and group-recommendation (selected through "Application Selection" section in Fig. 3). The experiment results can be examined by *ROC* or *PRC* curves, such as those in Fig. 4 and Fig. 5 (where the x- and y-coordinates of ROC represent False-Positive-Rate and True-Positive-Rate, and those of PRC represent Recall and Precision respectively).

Performance Comparisons - Utilizing the group recommendation application, the evaluation system helps users compare the proposed model with other state-of-the-art methods, such as *PMF* [5], *RSTE* [2] and *KPMF* [8] (selected through "Reference Method" section in Fig. 3). The comparisons can be observed in *Command Window* or by *PRC* plots.

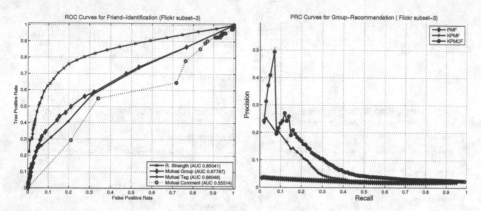

Fig. 4. ROC for friend-identification   **Fig. 5.** PRC for group-recommendation

References

1. Granovetter, M.: The strength of weak ties. American Journal of Sociology, 1360–1380 (1973)
2. Ma, H., Zhou, D., Liu, C., Lyu, M.R., King, I.: Recommender systems with social regularization. In: Proceedings of the Fourth ACM International Conference on Web Search and Data Mining, pp. 287–296. ACM (2011)
3. Marsden, P., Campbell, K.: Measuring tie strength. Social Forces 63(2), 482–501 (1984)
4. McAuley, J., Leskovec, J.: Image labeling on a network: Using social-network metadata for image classification. In: Fitzgibbon, A., Lazebnik, S., Perona, P., Sato, Y., Schmid, C. (eds.) ECCV 2012, Part IV. LNCS, vol. 7575, pp. 828–841. Springer, Heidelberg (2012)
5. Salakhutdinov, R., Mnih, A.: Probabilistic matrix factorization. In: Advances in Neural Information Processing Systems 20, pp. 1257–1264 (2008)
6. Van House, N.A.: Flickr and public image-sharing: distant closeness and photo exhibition. In: Proceedings of CHI 2007, pp. 2717–2722. ACM (2007)
7. Zhong, Y., Du, L., Yang, J.: Learning social relationship strength via matrix cofactorization with multiple kernels. In: Huang, G. (ed.) WISE 2013, Part I. LNCS, vol. 8180, pp. 15–28. Springer, Heidelberg (2013)
8. Zhou, T., Shan, H., Banerjee, A., Sapiro, G.: Kernelized probabilistic matrix factorization: Exploiting graphs and side information. In: Proceedings of SDM 2012, pp. 403–414. SIAM (2012)

PEACE-Ful Web Event Extraction and Processing*

Tim Furche[1], Giovanni Grasso[1], Michael Huemer[2],
Christian Schallhart[1], and Michael Schrefl[2]

[1] Department of Computer Science, Oxford University,
Wolfson Building, Parks Road, Oxford OX1 3QD
`firstname.lastname@cs.ox.ac.uk`
[2] Department of Business Informatics – Data & Knowledge Engineering,
Johannes Kepler University, Altenberger Str. 69, Linz, Austria
`lastname@dke.uni-linz.ac.at`

Abstract. PEACE, our proposed tool, integrates complex event processing and web extraction into a unified framework to handle web event advertisements and to run a notification service atop. Its bitemporal schemata distinguish occurrence and detection time, enabling PEACE to deal with updates and delayed announcements, as often occurring on the web. To consolidate the arising event streams, PEACE combines simple events into complex ones. Depending on their occurrence and detection time, these complex events trigger actions to be executed.

We demonstrate PEACE's capabilities with a business trip scenario, involving as raw events business trips, flight bookings, scheduled flights, and flight arrivals and departures. These events are scrapped from the web and combined into complex events, triggering actions to be executed, such as updating facebook status messages. Our demonstrator records and reruns event sequences at different speeds to show the system dealing with complex scenarios spanning several days.

1 Introduction

If an event is published today, it is published on the web. And if so, the announcement is likely to change over time, as more precise and up-to-date information becomes available. Checking such continuously changing event streams is tedious and stressful, especially, if more than one event source is involved, requiring coordination among individual events. Thus, one would assume that event notification has been addressed already: On the one hand, there is a clear need for web event processing, as our life style is more dynamic and interactive than ever before. On the other hand, all necessary information is readily available, most people wear an Internet enabled mobile device at all times, and hence, one would suppose that events need only to be extracted, checked for some application specific criteria, and delivered to the user.

But until now, no integrated solution exists for this task, except for isolated solutions dedicated to specific application domains. We argue that a good solution for web

---

* The research leading to these results has received funding from the European Research Council under the European Community's Seventh Framework Programme (FP7/2007–2013) / ERC grant agreement DIADEM, no. 246858. Michael Huemer has been supported by a Marietta Blau Scholarship granted by the Austrian Federal Ministry of Science and Research (BMWF) for a research stay at Oxford University's Department of Computer Science.

X. Lin et al. (Eds.): WISE 2013, Part II, LNCS 8181, pp. 523–526, 2013.

```
  doc("http://www.flightarrivals.com")
2 //a#panel0/{click /}//form#qbaForm/descendant::field()[1]/{$airport }
  /following::field()[3]//option{select }/following::field()[1]/{click /}
4 /(/descendant::a[string(.)='Next >'][1]/{click /})*
    //table#flifo//tr[position()>1]/self():<FlightArrival>
6     [./td[1]:<fromLoc=string(.)>]         [./td[2]:<flightNo=string(.)>]
      [./td[3]/div:<flightDay=string(.)>]   [.:<toLoc=$airport>]
8     [./td[3]/text()[1]:<occTime=toUnixTime(.)>]
```

Fig. 1. OXPATH Wrapper

event processing needs to integrate complex event processing and web extraction, dealing with frequent changes and retractions. We demonstrate PEACE *(Processing Event Ads into Complex Events)* as a possible solution. PEACE takes an event model to determine available event types and attributes, and based thereupon, **(1)** wrappers to extract events from multiple sources, **(2)** complex event queries to aggregate the raw events into complex events, and **(3)** action executors to perform the resulting actions.

Contrasting previous work, PEACE is fully *bitemporal,* distinguishing detection and occurrence time, i.e., the time when events are extracted and supposedly take place. PEACE allows for different reactions, depending on the detection and occurrence time in relation to the current time. While complex events [3] have been studied extensively, existing systems like [1,6] deal only with some aspects of the bitemporality. Most systems, as [2,4], drop this distinction in identifying occurrence and detection time.

2 Extracting and Processing Events from the Web with PEACE

Once an event model with its event classes and attributes is fixed, PEACE takes wrappers, complex event specifications and action executors to extract, process and react upon events. **(1)** The wrappers for event extraction are given in OXPATH, an XPATH-based language for highly scalable web data extraction [5]. Aside extracting web events, we can also integrate other sources for events, such as a local database. **(2)** The complex events are specified in BICEPL, our SQL-based language to define complex events as **SELECT** statements, extended with constructs for expressing constraints over occurrence and detection time. BICEPL specifications also include publication events, issued when complex events are introduced, updated, or revoked. **(3)** Finally, these publication events trigger OXPATH action executors, e.g., notifying users or changing a booking.

In the following, we give an example for each those three component types. These examples are directly taken from the demonstration. **Event Extractors.** The wrapper in Figure 1 fetches its target page (Line 1), selects options to obtain all current flight arrivals at a parameterized airport, and submits the form (Lines 2-3). The wrapper deals with paginated results by repeatedly clicking on "next" (Line 4). On each result page, FlightArrival objects are extracted altogether with their attributes (Lines 5–7) for every entry on the page. Matching the event model for the scenario, each FlightArrival results in an event tuple. **Complex Event Specifications.** In Figure 2 we show the complex event specification for ArrivedAtDestination indicating when a business person arrives at the final trip destination. ArrivedAtDestination has explicit attributes flightNo,

```
1  CREATE COMPLEX EVENT CLASS ArrivedAtDestination (flightNo TEXT, flightDay
      TEXT, location TEXT) ID (flightNo, flightDay)
2    AS SELECT fa.flightNo, fa.flightDay, fa.toLoc
3      FROM BusinessTrip bt, FlightBooking fb, Flight f, FlightArrival fa
4      WHERE bt.tripTitle = fb.tripTitle AND fb.bookingId = f.bookingId AND
5            f.connFlNo = 'NULL' AND f.connFlDay = 'NULL' AND
6            f.flightNo = fa.flightNo AND f.flightDay = fa.flightDay
7      OCCURRING AT fa
8   PUBLISH ArrivedAtDestinationOnTime
9      CASE LATE (0s, 1h) ArrivedAtDestinationLate
10     CASE RETROACTIVECHANGE ArrivedAtDestinationChanged
11     CASE REVOKE ArrivedAtDestinationRevoked;
```

Fig. 2. Event Definition for Complex Event ArrivedAtDestination

flightDay and location, along with implicit timing information, using flightNo and flightDay as key (Line 1). In BICEPL a complex event is defined with an extended SQL SELECT statement involving constituent events. In our example these constituent events are BusinessTrip, FlightBooking, Flight and FlightArrival (Line 3). These events are joined and filtered (Lines 4-6) and defined to occur when the constituent FlightArrival event occurs (Line 7). Further, complex event descriptions contain event publication statements to control the event published in case the complex event occurs on time, at maximum one hour late, has retroactively changed, or was revoked (Lines 8-11). **Action Executors.** For each publication event produced, PEACE runs action executors, such as a parameterized OXPATH expressions filling a form to send an SMS.

3   PEACE-Ful Business Trip Management

Our demonstration scenario deals with business travelers and their arrangements. To keep up with tight schedules, business travelers observe and quickly react upon many events, e.g., upon learning about flight delays, they need to check for connecting flights, or change hotel reservations. This can be time consuming and stressful, but having PEACE, all this hassle can be dealt with automatically! Aside managing the daily disasters of business travel, PEACE can be even configured to keep friends and family posted about an ongoing trip by updating facebook status messages.

Event Extractors. We implement our scenario with 5 event types from 3 different sources, partly extracted from the web via OXPATH wrappers. **(1)** BusinessTrip events are extracted from Google Calendar, providing a unique reference for each trip. **(2)** FlightBooking and Flight events are stored locally and describe a booking with all connecting flights of a one-way trip. **(3)** FlightArrival and FlightDeparture events for all relevant flights are extracted from flightarrivals.com. We show the wrapper for FlightArrivals in Figure 1. **Complex Event Specifications.** Atop these subscribed raw events, we specify 5 complex event types in BICEPL. **(A)** E1DaysToBusinessTrip indicates that a trip begins in 1 day; **(B)** E3DaysToFlightDeparture that a departure is going to take place in 3 days; **(C)** MissedConnectingFlights that a connection is unreachable due to a canceled or delayed flight. **(D)** UncatchableConnection captures connection

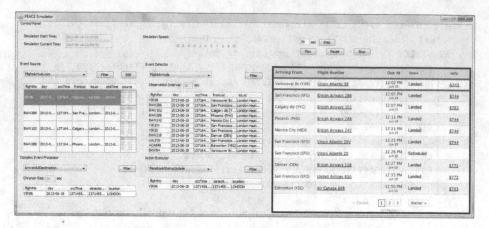

Fig. 3. PEACE Simulator

flights which are missed. (**E**) `ArrivedAtDestination` indicates that a business traveler supposedly arrived at the destination of the trip. This specification is depicted in Figure 2. **Action Executors.** Complex events trigger various actions, for example upon an `E1DaysToBusinessTrip` event, the system notifies the business traveler via email, and upon an `ArrivedAtDestination` event, the system updates the traveler's facebook status. In case of `E3DaysToFlightDeparture`, the referred flight data is checked to be correct and still valid. Moreover, action executors *control* the event extractors currently deployed: For example, a `E3DaysToFlightDeparture` event triggers the setup of event extractors for `FlightArrivals` and `FlightDepartures` for the flight under scrutiny.

Simulation and Visualization. We showcase our scenario with the PEACE simulator and visualizer, illustrated in Figure 3. Presenting interesting cases, we not only run the system live but also on recorded event streams over a couple of days. The simulator runs the entire system at different speeds, skimming over uneventful phases and carefully analyzing more turbulent phases. With the simulator, we trace events, occurring on event sources, then being extracted, leading to complex events, and thus taken actions.

References

1. Adi, A., Etzion, O.: Amit - the situation manager. VLDB J. 13 (2004)
2. Cugola, G., Margara, A.: TESLA: a formally defined event specification language. In: DEBS (2010)
3. Cugola, G., Margara, A.: Processing flows of information: From data stream to complex event processing. ACM Comput. Surv. 44 (2012)
4. Eckert, M., Bry, F.: Rule-based composite event queries: the language XChange^{EQ} and its semantics. Knowl. Inf. Syst. 25 (2010)
5. Furche, T., Gottlob, G., Grasso, G., Schallhart, C., Sellers, A.J.: OXPath: A language for scalable data extraction, automation, and crawling on the deep web. VLDB J. 22 (2013)
6. Luckham, D.: Event Processing for Business. John Wiley & Sons, Inc. (2012)

KNOWLE: Searching News in the Search Pattern of Knowledge Flow

Xiao Wei[1,2], Xiangfeng Luo[1], Qing Li[2], and Jun Zhang[1,2]

[1] School of Computer Engineering and Science, Shanghai University, China
[2] Computer Department, City University of Hong Kong, Hong Kong
{xwei,luoxf,junzhang_shu}@shu.edu.cn, qing.li@cityu.edu.hk

Abstract. Search Pattern is one of the important factors that influence the efficiency of web searching. In this paper, we demonstrate a news search service named KNOWLE which provides efficient search for users by taking advantage of the search pattern of knowledge flow. KNOWLE clusters web news into news topics, organizes the news topics into news association link networks, serves users in a form of news association knowledge flow, which can help users understand a news event by providing its causes and effects.

Keywords: Search Pattern, Search Engine, Association Link Network, News Knowledge Flow.

1 Introduction

Search Pattern is one of the important factors that influence the efficiency of web searching [1]. Traditional search engines provide the search pattern of keyword matching, which often returns thousands of webpages for a user query and makes the low efficiency of searching in some scenarios, such as the user cannot provide appropriate keywords, the user wants to get the associated contents, and so on. To overcome the above shortcomings, recent years, new types of search patterns have emerged to improve the efficiency of searching, including the narrow search [2], which helps the user refine his/her search step by step; the associated search [3], which helps the user broaden his/her search by recommending associated contents to him/her; and the topic-based search [4], which serves the user at coarse grained semantic granularity. To increase the efficiency of news searching, we have developed a search engine called KNOWLE, which works at a new search pattern of knowledge flow search. When a news event takes place, for example, users may want to know more information about the event, such as the cause(s), the effect(s), etc. Knowledge flow search can satisfy this kind of requirements well. KNOWLE gathers news from news websites, clusters them into news topics, mines the association relations among news topics, builds the association link network (ALN [6]) of news topics, and serves users in the form of news association knowledge flow. When a user searches news, KNOWLE first locates the most similar topic in the news topics ALN, and then generates several news knowledge flows from the ALN based on the users' command.

X. Lin et al. (Eds.): WISE 2013, Part II, LNCS 8181, pp. 527–530, 2013.

With the help of these news knowledge flows, the user can find the causes and effects of a news event and get its whole picture.

To the best of our knowledge, the search pattern of knowledge flow is novel. Previous researches mainly focus on organizing news into hierarchies or graphs, and few of them use the knowledge flow as the search pattern to interact with users. In [5], news articles are connected by the concurrent words/phrases, which is essentially similar relation. When a user inputs two news articles, a series of news articles will be selected to connect the two inputs. KNOWLE is quite different from it in the kind of semantic relation, the knowledge flow selection algorithm, the interface, and the interaction process of searching. Distinctive features of KNOWLE include the following:

- The new search pattern named knowledge flow search, which provides a new in-teractive paradigm between the user and the search engine and increases the effi-ciency of news searching.
- The multi-layered association link network index of news documents, which has rich semantics and is the foundation of news knowledge flow generating.
- The news knowledge flow generating algorithm based on news association link network, which can dynamically generate personalized knowledge flow according to the user's search.

2   System Overview

The system architecture of KNOWLE is shown in Fig.1, which combines an index of multi-layered association link network (MALN), a news crawler module, a webpages pretreatment module, a news topics clustering module, an association link network (ALN) building mechanism, a knowledge flow generating mechanism, and a knowledge flow search interface. We briefly introduce some components of KNOWLE as following.

- **Multi-layered Association Link Network Index (MALN)**: The index structure of KNOWLE is designed as multi-layered association link networks, as shown in the right part of Fig.1. Layer 1 is the ALN of all keywords in the system, which holds the semantic relations among keywords and is the basis of keywords semantic computing. Layer 2 is the ALN of webpages. In layer 2, webpages belonging to a news topic are organized as an ALN respectively. Layer 3 is the ALN of news topics, which holds the association relations among news topics and is the foundation of news knowledge flow generating.
- **Association Link Network (ALN) Building:** ALN is a semantic link network whose nodes can denote all kinds of web resources (news event, news topic, webpage and so on) and the arc between a pair of nodes denotes their association relation and the weight of the arc denotes the strong of the association relation. ALN is optimized to have good characters, such as small-world and scale-free property. The details of ALN can be found in our previous work [6].

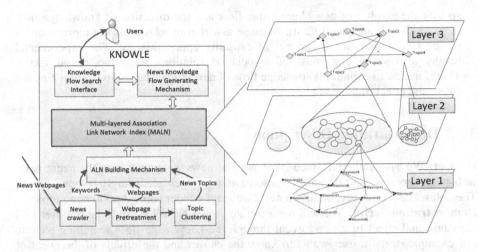

Fig. 1. The architecture of KNOWLE (shown in the left of this figure) and its multi-layered association link network index structure (shown in the right of this figure)

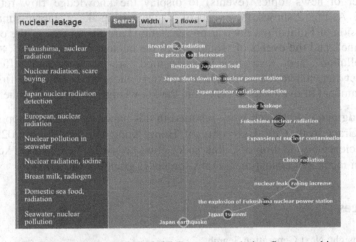

Fig. 2. The interface of KNOWLE news knowledge flow searching

- **News Knowledge Flow Generating:** News knowledge flow is a series of news topics with strong association relation and time-order relation. To select a knowledge flow (as shown in Fig.2) from thousands of paths in the ALN of news topics, KNOWLE first selects several news topics according to the input keywords, among which the most similar one is selected as the core node of the knowledge flow and the others act as the initial backbone nodes of the knowledge flow. Based on the backbone, some candidate paths can be selected from the ALN. The following step is to optimize and reshape the selected paths according to time-order of news and contents of topics, including replacing some nodes and reordering the nodes on each path. Finally, the selected paths are ranked and form the final knowledge flows to fit the user's special requirement. For example, a user can specify the number of news knowledge flow and the direction of knowledge flow (e.g. width-first and depth-first). Because association relation is the representation of the causal relation, it can act as causality approximately. The experimental results also show that most of association relations have the causal factor. Therefore, the association knowledge flow of news topics can help the user find the cause and effect of an event.

3   Demonstration Description

Dataset: We have gathered, since 2009, popular news webpages from the major news websites every day, and organized these news webpages accordingly to news topics. The dataset has about seven million news webpages under some 30,000 news topics .

Demonstration Script: We will demonstrate how KNOWLE could help a user find the cause and effect of a news event through the knowledge flow searching pattern. For example, when a user wants to know the causes and the effects of the event of nuclear leakage in Japan, he/she can first input the keyword 'nuclear leakage' and select the target news event. Then the user can set a parameter for the system to limit the number of news topics (events) to display the knowledge flow returned by KNOWLE. As Fig. 2 shows, the events of "Japan earthquake", "Japan tsunami", and "the explosion of Fukushima nuclear power station" are the causes of the event "nuclear leakage", and the events of "shutdown of the nuclear power station", "spread of nuclear pollution", and "restricting Japanese food" are the consequent effects. Note that the user can change the search directions (e.g. width-first and depth-first) and/or adjust the amount of events to present from the knowledge flows at any time.

Acknowledgments. Research work reported in this paper was jointly supported by a Strategic Grant from City University of Hong Kong (Project No. 7002912), the National Science Foundation of China under grant no. 61071110, and the key project of shanghai municipal education commission under grant no. 13ZZ064.

References

1. ICSTI Insight: Next Generation Search,
   http://www.icsti.org/img/pdf/insight_2010_july.pdf
2. Kaczmarek, A.L.: Clustering by Directions algorithm to narrow search queries. In: IEEE Conference on Human System Interactions, pp. 689–694 (2008)
3. Sohn, S.Y., Kim, Y.: Searching customer patterns of mobile service using clustering and quantitative association rule. Expert Systems with Applications 34(2), 1070–1077 (2008)
4. Buntine, W., Lofstrom, J., Perkio, J., et al.: A scalable topic-based open source search engine. In: Proc. of IEEE/WIC/ACM International Conference on Web Intelligence, pp. 228–234 (2004)
5. Shahaf, D., Guestrin, C.: Connecting the dots between news articles. In: Prof. of ACM SIGKDD International Conference on Knowledge Discovery and Data Mining, pp. 623–632. ACM (2010)
6. Luo, X., Xu, Z., Yu, J., Chen, X.: Building Association Link Network for Semantic Link on Web Resources. IEEE T. Automation Science and Engineering. 8(3), 482–494 (2011)

Document Analytics through Entity Resolution[*]

João Santos, Bruno Martins, and David S. Batista

Instituto Superior Técnico and INESC-ID, Lisboa, Portugal
{joao.d.santos,bruno.g.martins,dsbatista}@ist.utl.pt

Abstract. We present a prototype system for resolving named entities, mentioned in textual documents, into the corresponding Wikipedia entities. This prototype can aid in document analysis, by using the disambiguated references to provide useful information in context.

Keywords: Text Mining, Information Extraction, Entity Resolution.

1 Introduction

Even as structured databases and semantic knowledge bases become prevalent, a substantial amount of human knowledge is still available as free-form text. News articles can, for instance, contain information about entities (i.e., people, organizations, and locations) and their relationships. For humans, reading and understanding large volumes of text is a time consuming task, and so automatically extracting and organizing this information is in high demand.

We present a prototype system that enhances textual documents by identifying references to people, places, or organizations, afterwards disambiguating these references by linking them to identifiers in a knowledge base such as Wikipedia. This paper briefly introduces a web-based demonstration of this prototype system, also outlining its main components.

2 Entity Resolution

Named entity resolution is an important text analytics problem that has been getting an increasing attention [2]. The problem involves two separate sub-tasks, namely (i) entity identification, and (ii) entity disambiguation. The first sub-task is deeply related to Named Entity Recognition (NER), a problem that has been thoroughly studied in the Natural Language Processing community. In our system, entity identification is performed through the Stanford NER system, with models trained for the English (i.e., the standard model distributed with

[*] This research draws on work from two projects supported by Fundação para a Ciência e Técnologia (FCT), namely the projects with references PTDC/EIA-EIA/109840/2009 (SInteliGIS) and UTA-Est/MAI/0006/2009 (REACTION). The work reported on this paper was also supported through INESC-ID's multianual funding provided by the FCT (PEst-OE/EEI/LA0021/2013).

X. Lin et al. (Eds.): WISE 2013, Part II, LNCS 8181, pp. 531–534, 2013.
© Springer-Verlag Berlin Heidelberg 2013

the tool[1]), Spanish (trained with the CoNLL-02 data[2]) and Portuguese (trained with data from the CINTIL corpus[3]) languages.

The second sub-task involves re-expressing the identified entity references into a standard unambiguous format (i.e., mapping each entity reference, previously recognized by the NER system, to an identifier specific to the real-world concept that is being referred to in the text). In our system, the mappings from entity references to real-world concepts are made through a knowledge base built from Wikipedia[4] and DBpedia[5]. Wikipedia is a collaborative wiki-based encyclopedia that covers almost all areas of human knowledge, with articles written in standard prose that are mostly intended for human consumption. On the other hand, DBpedia is a project concerning with the extraction of structured information from Wikipedia articles, representing this information in a machine-readable semantic graph using Resource Description Framework triples. For building the knowledge base, we essentially used the entities from the English, Portuguese, or Spanish versions of Wikipedia, categorized in the DBpedia structured ontology as entities corresponding to either people, organizations, or locations.

The method for performing entity disambiguation follows the general methodology from systems participating in the TAC-KBP yearly-challenge on named entity disambiguation [2], and it involves the following main tasks:

1. **Query Expansion**: Entities may be referenced by several alternative names, some of which more ambiguous than others. Given a reference, we apply expansion techniques that try to identify other names, in the source document, that reference the same entity. We specifically consider two simple mechanisms, namely one that finds alternative names by looking for a textual pattern that corresponds to a set of capital words followed by the alternative name inside parentheses (i.e., finding expressions like *United States (US)*), or vice-versa, and another that looks for longer entity mentions in the source text (i.e., *Union of Soviet Socialist Republics* as an expansion for *USSR*).
2. **Candidate Generation**: This step filters the Knowledge Base (KB) entries that might correspond to the query, based on string similarity. Some of Wikipedia's link structure (e.g., disambiguation pages, redirects, anchors, etc.) is also used to obtain alternative names. We specifically return the top 50 most likely entries in the KB (i.e., those whose name(s) are more similar to the entity reference, and whose textual descriptions are similar to the support text), according to a retrieval model supported by a Lucene[6] index.
3. **Candidate Ranking**: This step sorts the retrieved candidates according to the likelihood of being the correct referent, using the LambdaMART learning to rank algorithm as implemented in the RankLib[7] software library. Three

---

[1] http://nlp.stanford.edu/software/CRF-NER.shtml
[2] http://www.cnts.ua.ac.be/conll2002/ner/
[3] http://cintil.ul.pt/cintilfeatures.html#corpus
[4] http://www.wikipedia.org/
[5] http://dbpedia.org/
[6] http://lucene.apache.org/
[7] http://people.cs.umass.edu/~vdang/ranklib.html

ranking models (i.e., for disambiguating entities in English, Spanish and Portuguese texts) were trained to optimize accuracy over sets of disambiguation examples that were automatically gathered from Wikipedia (i.e., we used hypertext anchors from links towards entities in the knowledge base, occurring in Wikipedia documents different from those in the knowledge base). These models leverages on a rich set of features for representing each candidate, including (i) candidate authority features, (ii) textual similarity features, (iii) topical similarity features, (iv) name similarity features, (v) entity-based features, (vi) geospatial features, and (vii) document coherence features. Space restrictions prevent us from showing the complete set of features here, but the reader can refer to a previous paper describing our participation in the 2011 edition of TAC-KBP, where an extensive set of experiments is reported with an early version of the system [1].

4. **Candidate Validation**: This step decides whether the top ranked referent is an error, resulting from the fact that the correct referent is not given in the knowledge base, through a Random Forest classifier that reuses the features from the ranking model, and that also considers some additional features for representing the top ranked result (e.g., the candidate ranking score, or results from outlier tests over the ranked lists of candidates).

The most innovative aspects of our named entity resolution system relate to the usage of the LambdaMART state-of-the-art learning to rank method, and to the extensive set of features that was considered.

Our prototype presents the entity resolution functionality through a web-based interface, through which users can input the text where entities are to be resolved. The system replies with an XML document encoding the entities occurring in the text, together with the results for their disambiguation. A stylesheet builds a web page from this XML document, where the named entities in the original text appear linked to the corresponding Wikipedia page. Through tooltips, the user can quickly access overviews on the entities that were referenced in the text (e.g., we show photos representing the referenced entities, elementary metadata attributes about the entities, and maps with pins for the latitude and longitude coordinates of the referenced locations).

Figures 1 and 2 present two screenshots of the web-based interface for our entity recognition and disambiguation system. Figure 1 shows the main screen of the service. There are two options to introduce the desired input text to be disambiguated, namely an option to paste it into a text box, and another to upload a file containing the target text. Before submitting this text to the entity linking system, the language needs to be chosen, and the system supports English, Spanish, and Portuguese. Figure 2 presents a screenshot for the output of the system, i.e. the result that is produced from the input textual document. All entities recognized by the NER model apperar highlighted, with each color representing a different entity type. These entities are also linked to the corresponding Wikipedia page, according to the disambiguations made by the system. A tooltip with relevant additional information about each entity pops up when the user moves the mouse over the entity reference.

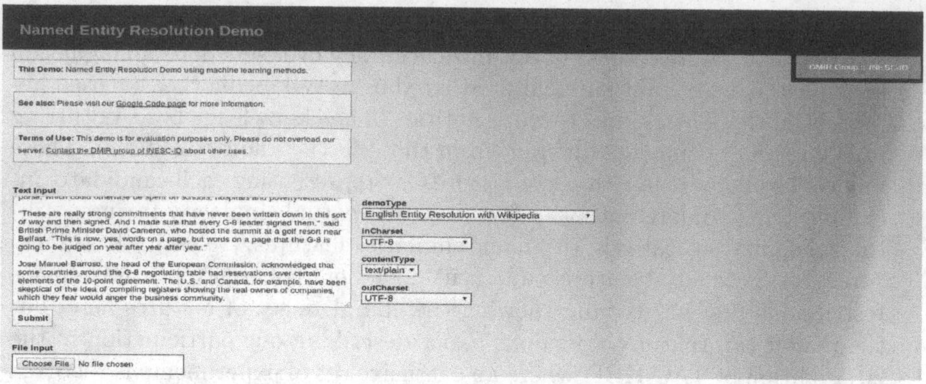

534   J. Santos, B. Martins, and D.S. Batista

Fig. 1. Data entry form for the named entity resolution system

Fig. 2. Output generated by the named entity resolution system

3   Conclusions

We demonstrate an end-to-end system for extracting and disambiguating named entities in textual documents, which combines recent developments in information extraction and natural language engineering. This system integrates many freely available open-source components (e.g., Stanford NER, Lucene, etc.) and offers scalability for processing large datasets.

Many opportunities exist for extending the system. For future work, we intend to incorporate relationship extraction into our information extraction pipeline. We also plan to integrate graph visualization methods into our system, in order to support the visual analysis of co-occurrences or of other types of relations between entities, as extracted from large document collections.

References

1. Anastácio, I., Calado, P., Martins, B.: Supervised learning for linking named entities to wikipedia pages. In: Proceedings of the Text Analysis Conference (2011)
2. Rao, D., McNamee, P., Dredze, M.: Entity linking: Finding extracted entities in a knowledge base. In: Poibeau, T., et al. (eds.) Multi-source, Multi-lingual Information Extraction and Summarization. Springer (2011)

Imagilar: A Real-Time Image Similarity Search System on Mobile Platform

Bicheng Luo[1], Zi Huang[2], Hongyun Cai[2], and Yang Yang[2]

[1] Software Institute, Nanjing University, China
[2] School of Information Technology and Electrical Engineering,
The University of Queensland, Australia
lbc10@software.nju.edu.cn, {uqzhuang,h.cai2,uqyyan10}@uq.edu.au

Abstract. With the rapid development of mobile intelligent devices and wireless communications, users are gradually changing the way of consuming interesting content from the traditional personal computers to smart phones. Hence, we introduce a brand-new content-based image similarity search system which runs on mobile platform in real time. This paper outlines the system which has several novel components, including multi-feature composition, multi-feature indexing, and customized user interface with auxiliary Web data display.

1 Introduction

Integrated with digital camera's functionality, mobile phones have been widely used as photographing tools in order to record some wonderful moments and share them with others. Recently, many image search applications have shown up in the mobile software ecosystem. However, most of such applications search images based on textual information (e.g., tag, image title and caption), which makes it difficult for users to search for images without any a priori knowledge (e.g., finding similar images to an image of an unknown landmark). Although a few content-based image search systems have emerged, they focus more on reducing wireless communication cost or formulating user intent [1].

In this paper, we demonstrate a mobile system for real-time content-based image similarity search, called *Imagilar*, which effectively supports the scenario of capturing a photo to search on mobile devices. This system has several novel features. Firstly, motivated by the observations that different image categories are often better perceived with different types of visual features and exhibit different visual/spatial patterns, we generate multiple features and adaptively compose different visual features for queries to be searched, according to the customized image categories and importance of different features. For instance, color feature could be satisfactory to be used to find similar natural scenic photos, while local interest points could be more helpful to find landmark or product photos than global features. Secondly, we apply a recent multi-feature hashing method [2] to index all the visual features to achieve real-time response. In this method, all feature vectors are encoded into binary codes and image similarities

X. Lin et al. (Eds.): WISE 2013, Part II, LNCS 8181, pp. 535–538, 2013.
© Springer-Verlag Berlin Heidelberg 2013

are effectively approximated by aggregating the Hamming distances measured from different feature spaces, based on the highly efficient XOR operation on binary codes. Thirdly, from the returned image list based on the content similarity, users have the flexility to further discover relevant information of the selected images by searching their associated text from Web, including social websites like Wikipedia and professional webistes like products' homepages. Our dataset contains about one million images crawled from Flickr, most of which carry textual information such as tags and titles. Note that in our system, queries are not associated with any textual data.

2   System Overview

The system contains three major components: image multi-feature composition, multi-feature indexing, and customized user interface.

2.1   Multi-feature Composition

In the system, we consider two types of features, including color/texture, and local interest point. For color/texture feature, we use 168-d joint composite descriptor (JCD) [4] which describes an image by integrating both color and texture information into a single descriptor. This is a global feature for image representation. To consider more discriminative features, we use TOP-SURF descriptor [3] which integrates SURF descriptor with visual word representation to describe an image. In our feature database, we generate 1000 visual words as the dictionary. A frequency histogram of occurring visual words in an image can be generated. We then apply the tf-idf weighting to evaluate the importance of the visual words in the histograms. Therefore, an image can be finally represented by a 1000-d tf-idf histogram derived from its local TOP-SURF descriptors.

With regard to similarity measures, the classical Euclidean distance is used to compute the distance on JCD. For the tf-idf histogram, we define the normalized cosine similarity d_{cos} between two tf-idf histograms T_a and T_b as $d_{cos}(T_a, T_b) = 1 - \frac{T_a \cdot T_b}{|T_a||T_b|}$. A distance of 0 means the histograms are identical. The (dis)similarities of different features are linearly combined with tunable weights. Top-k most similar images with smallest distances to the query image are returned as the results.

It has been ascertained that in some of the image categories, better retrieval results are achieved by using one feature, while in others by using another. Our database images are collected from Flickr mainly in three categories, including landmark, scenery, and commodity. Our experimental results show that search based on color/texture feature can return quite satisfactory results for scenery images, while tf-idf histogram derived from local interest points performs much better than color/texture feature for landmark and commodity images. Our system has default settings on the weight combinations of different features for different image categories, tuned by extensively experimental results. Given a query image and a search category, the corresponding default weight combination is adaptively applied to compose the weighted search space for comparing the images in the category. For searching the whole database covering all the categories, a default weight composition on all the features is also tuned for the best retrieval results. Meanwhile, users also have the flexibility to dynamically update the weights of different features to generate desirable results and view the differences of the results.

2.2   Multi-feature Indexing

To enable real-time response from large-scale image databases, efficient matching of images described by multiple features has to be achieved. Hashing has shown its high efficiency for fast similarity search. Given that the search is performed on multiple features in our system, we extend existing Multiple Feature Hashing (MFH) [2] to index the multiple features simultaneously. To generate effective hash functions, MFH preserves the local structure of each individual feature and also globally considers all the local structures in the optimization. However, it assumes that all features have equal importance. Here we extend it by allowing different features to have different importance in learning the hash functions, resulting in better results. Furthermore, we also generate hash functions on each individual feature so that the importance of different features can be dynamically changed for an arbitrary query image. With the obtained hash functions, all the database images can be encoded into binary codes offline. Given an query image, its multiple features are firstly extracted. The hash functions then map the query image into binary codes, followed by performing binary code matching with database binary codes to get the most similar images in the Hamming space. Typically, the binary code length for a single feature is small, e.g. 32 bits, mainly depending on the data size and feature dimensionality. Binary code matching in memory is extremely fast due to the highly efficient XOR operation in computing the Hamming distance, which contributes to the achievement of real-time search. Our image dataset consist of about 1 million images downloaded from Flickr. Each image is described by two types of features, as discussed in last section. The average query response time for the system is about tens of milliseconds. Due to the space limit, detailed experiment results are not presented in this paper.

2.3   User Interface

The server side of the system is implemented with Java EE, while the client is implemented with iOS SDK under XCode. The main functionalities of the system are shown in Figure 1(a). Users can either choose to select a photo from their album or take a new photo as a query image. Meanwhile, users have the options to choose a category (i.e., landmark, scenery, or commodity) or the whole database (i.e., general) to search.

Users can also customize the weights of different features before the search starts, as shown in Figure 1(b). For simplicity, we use a horizontal scroll bar to indicate the relative importance of local interest point, also called point of interest (POI), and color/texture. By moving the scroll bar, different results can be displayed to users.

From the returned results, users can also pick any image to perform further search based on their associated textual data to have more information on the image. Existing search engines can be used for this purpose. Our system will select a few top ranked results and display them to users. Figure 2 shows an example, where a user selects an image about "Ajax Floor Cleaner" and other related information can also be shown up on the mobile screen by extracting the results from existing search engines, such as Google Product Search.

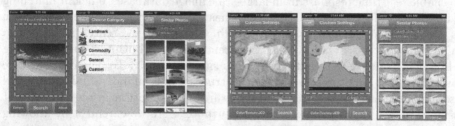

(a) System functionalities   (b) Customize panel and results

Fig. 1. User interface

Fig. 2. Auxiliary data display

3   Summary

In this demonstration, we describe a mobile system which achieves real-time image search based on multiple content similarities. Different features have different importance in the retrieval task for the best results and users also have the flexibility to dynamically tune the importance of different features. The system has been developed on iOS platform and tested extensively for its search effectiveness and efficiency.

References

1. Li, H., Wang, Y., Mei, T., Wang, J., Li, S.: Interactive multimodal visual search on mobile device. IEEE Transactions on Multimedia 15(3), 594–607 (2013)
2. Song, J., Yang, Y., Huang, Z., Shen, H.T., Hong, R.: Multiple feature hashing for real-time large scale near-duplicate video retrieval. ACM Multimedia (2011)
3. Thomee, B., Bakker, E.M., Lew, M.S., Lew, M.S.: Top-surf: a visual words toolkit. ACM Multimedia (2010)
4. Zagoris, K., Chatzichristofis, S., Papamarkos, N., Boutalis, Y.: Automatic image annotation and retrieval using the joint composite descriptor. In: PCI (2010)

Author Index